Heat and Mass Transfer in Microchannels

Heat and Mass Transfer in Microchannels

Guest Editors

Xin Xiao
Xuan Zhang
Long Zhang

MDPI

Basel • Beijing • Wuhan • Barcelona • Belgrade • Novi Sad • Cluj • Manchester

Guest Editors

Xin Xiao
College of Environmental Science and Engineering
Donghua University
Shanghai
China

Xuan Zhang
Department of Energy and Power Engineering
Beijing Institute of Technology
Beijing
China

Long Zhang
Department of Energy and Power Engineering
Beijing Institute of Technology
Beijing
China

Editorial Office
MDPI AG
Grosspeteranlage 5
4052 Basel, Switzerland

This is a reprint of the Special Issue, published open access by the journal *Micromachines* (ISSN 2072-666X), freely accessible at: https://www.mdpi.com/journal/micromachines/special_issues/heat_transfer_in_microchannels.

For citation purposes, cite each article independently as indicated on the article page online and as indicated below:

Lastname, A.A.; Lastname, B.B. Article Title. *Journal Name* **Year**, *Volume Number*, Page Range.

ISBN 978-3-7258-7669-3 (Hbk)
ISBN 978-3-7258-7670-9 (PDF)
https://doi.org/10.3390/books978-3-7258-7670-9

Contents

About the Editors

Xin Xiao

Xin Xiao completed his PhD at Shanghai Jiao Tong University and is currently an associate professor at Donghua University, China. His research mainly includes latent thermal energy storage and thermal management and carbon dioxide capture via hydrate, thermal, and moisture management with fabric. He has led or participated in more than 20 research projects funded by the EU, NSFC, etc., with an accumulated funding value in excess of ¥60 million, and claimed 6 patents. He has published more than 100 scientific research papers, including more than 80 papers with SCI/EI indexes, with a total of more than 4000 citations. He has been awarded the JSPS fellowship, the Marie Curie Fellowship, the Shanghai "Pujiang" Talent award, the Shanghai "Overseas High-level" Talent award, the IChemE&CCST Young Investigator Award, and Distinguished Young Scholar in 2025 for Energy and Built Environment Journal award; Xiao has also won the first prize of the Natural Science Award of Yunnan Province and the third prize of the China Business Federation and the Young Excellent Award of Chinese Society of Engineering Thermophysics. He was also included in the Top 2% Global Scientists Ranking from 2023 to 2025. Xiao is also an editorial board member for several journals, including *Biomass Future*, *Exploration*, *Carbon Neutrality*, *Ecoenergy*, and *Engineered Science*.

Xuan Zhang

Xuan Zhang is an Associate Professor at the Beijing Institute of Technology. He received his Ph.D. degree from Tsinghua University in 2019 and then worked at Nanyang Technological University as a Postdoctoral Research Fellow until 2022. He focuses on the mechanisms and applications of heat/mass transfer and fluid flow coupling processes, especially phase change problems (e.g., icing/frosting/freezing, condensation/evaporation/boiling, solidification/melting). He has published/co-authored 110+ SCI journal articles, with 10 being recognized as ESI Highly Cited/Hot, Featured, Editor's Pick, Front Cover, and CNKI Highly Cited papers, and 5 books/chapters. He has also been invited to deliver 10 reports at conferences and applied for/authorized 20+ patents and 10+ software systems. His works have been cited over 3200+ times in Google Scholar, with an H-index of 32, and have been reported 50+ times by famous science and technology media, such as *People's Daily*, *China Science Daily*, *The New York Times*, and *EurekAlert*. He has been listed among the World's Top 2% Scientists by Stanford University/Elsevier (since 2023) and the 0.5% Top Scholars by ScholarGPS Rankings (since 2025). He has received 10+ grants from the National Natural Science Foundation of China, Ministry of Science and Technology of China, etc. Zhang has served as an Associate Editor for *Particulate Science and Technology*; an Editor Board Member for *Scientific Reports*; an Early-Career/Youth Editor Member for *Coatings*, *Exploration*, etc.; a Guest Editor for *Applied Thermal Engineering*, *Micromachines*, etc.; a Conference Committee Member for Frosting-Icing 2025, MSTP-2024, etc.; and a Reviewer for 70+ academic journals. He has been recognized with the following awards: the First Prize of Science and Technology Progress Award of Shandong Province, Wu Zhonghua Outstanding Graduate Student (Chinese Society of Engineering Thermophysics), Outstanding Doctoral Thesis (Tsinghua University), Highly Cited Doctoral Thesis (CNKI), and Best Presentation Awards.

Long Zhang

Long Zhang is an Assistant Professor at the Beijing Institute of Technology. His research interests include the coupled mechanisms of heat and mass transfer and fluid flow, specifically phase change problems such as frosting, icing, and solidification, alongside key technologies in heat pump and refrigeration systems. He has led over 10 projects funded by the National Natural Science Foundation of China, the Ministry of Industry and Information Technology, the Beijing Municipal Science and Technology Commission, and key state laboratories. Zhang has published over 90 SCI/EI journal papers, including more than 40 as the first or corresponding author with two ESI highly cited papers, and authored three English books. As of March 2026, he has over 2000 Google Scholar citations and an H-index of 25. He holds over 20 patents and software copyrights, has co-authored three industry standards, and has delivered eight invited academic presentations. His honors include the First Prize of Shandong Provincial Science and Technology Progress, the First Prize from the Architectural Society of China, and the Outstanding Doctoral Dissertation Nomination Award from The Hong Kong Polytechnic University, and he has been listed among the Elsevier/Stanford World's Top 2% Scientists for three consecutive years.

Preface

The miniaturization of components in mechanical and electronic equipment and systems has been becoming increasingly popular in recent decades. Heat and mass transfer processes, usually accompanied by multiphase flow, are crucial for the normal operation of such micro/nano-equipment and systems and thus attract a great deal of attention. Significant effort has been devoted to deepening our understanding of complex heat and mass transfer processes and advancing the development of related applications in microchannels. For example, phase change (e.g., boiling and condensation), nanofluids, and micro-/nano-structured surfaces are applied to enhance heat transfer and achieve a potentially higher critical heat flux. The aim of this Special Issue of *Micromachines*, "Heat and Mass Transfer in Microchannels", is to present recent advances in heat and mass transfer in microchannels, including but not limited to associated theoretical analyses, experimental measurements, numerical simulations, and practical applications.

Xin Xiao, Xuan Zhang, and Long Zhang
Guest Editors

micromachines

Editorial

Editorial for the Special Issue on Heat and Mass Transfer in Microchannels

Xin Xiao [1], Xuan Zhang [2,*] and Long Zhang [2,*]

[1] College of Environmental Science and Engineering, Donghua University, Shanghai 201620, China; xin.xiao@dhu.edu.cn
[2] Department of Energy and Power Engineering, School of Mechanical Engineering, Beijing Institute of Technology, Beijing 100081, China
* Correspondence: xuan.zhang@bit.edu.cn (X.Z.); long.zhang@bit.edu.cn (L.Z.)

1. Introduction to the Special Issue on Heat and Mass Transfer in Microchannels

Research on heat and mass transfer in microchannels represents the cutting-edge frontier in modern high-density electronic cooling, advanced energy systems, and precision microfluidic manipulation [1–3]. By deeply integrating heat and mass transfer science with micro-/nano-fabrication technologies, it achieves highly efficient energy and material transport within extreme spatial constraints, underpinning a series of disruptive technologies that range from chip-level thermal management [4] to biosensing [5].

The strategic significance of this field lies not only in its irreplaceable compactness and efficiency, but also in its ability to reveal and harness unique physical mechanisms that dominate at the micro-scale, such as pronounced surface-tension effects [6], Marangoni convection [7], and capillary forces [8]. This lays the foundation for fundamentally innovating thermal management strategies and serves as a key driver for the continued miniaturization and performance leap of devices.

However, to fully unlock the potential of microchannel systems, a profound understanding and mastery of their inherent complexities is essential. This involves uncovering the dynamic transport laws at multiphase interfaces [9], analyzing the fine-tuning effect of complex geometries on flow and heat transfer [10], mastering the coupled multi-physics behavior during phase-change processes [11] and under extreme conditions [12], and promoting the innovative application of novel functional working fluids [13] and high-performance structures.

Therefore, the pursuit of efficient [14], stable, and controllable heat and mass transfer through microchannel technology has become a mission of both great scientific value and urgent engineering need. This requires that the design principles of microchannels be deeply coordinated with mesoscopic physical mechanisms, advanced materials [15], and fabrication processes. It is precisely this interdisciplinary integration that continuously pushes the field to break through bottlenecks and explore new technological frontiers.

The 17 original research articles included in this Special Issue can be clearly categorized into four main directions, based on their research subjects and orientations: fundamental studies on single-phase flow [Contributions 2, 5, 8, 15, 17], fundamental studies on multiphase flow [Contributions 1, 3, 4, 13, 14, 16], and applied research on both single-phase and multiphase flow [Contributions 6, 7, 9–12], as illustrated in Figure 1. This classification framework systematically presents the complete research spectrum from fundamental flow principles to engineering solutions.

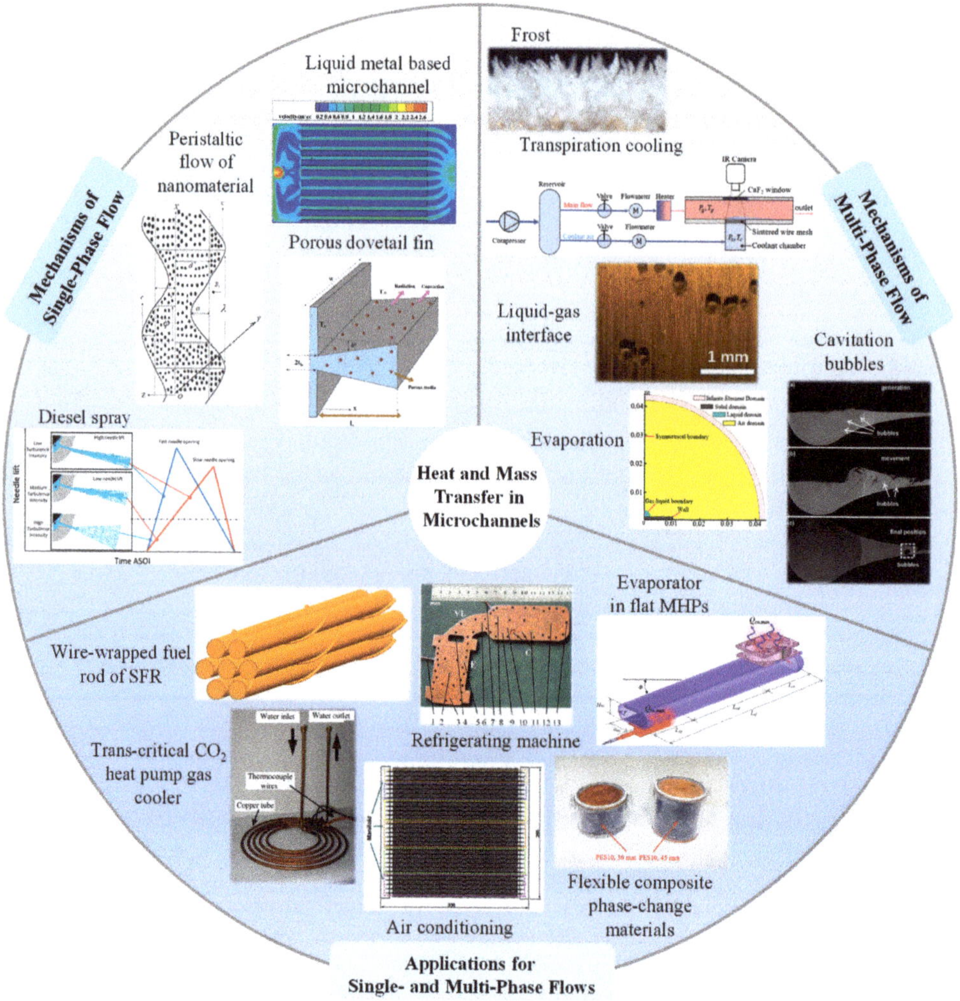

Figure 1. Topics covered in the Special Issue titled "Special Issue on Heat and Mass Transfer in Microchannels".

In summary, the research presented in this Special Issue deepens the understanding of fundamental mechanisms in microscale flow and heat transfer and provides key insights and solutions for the design of next-generation microchannel heat sinks, micro-heat exchangers, and high-performance microfluidic systems. Future research will continue to advance toward intelligent design, cross-scale integration, and reliable applications in even more extreme environments.

2. Mechanisms of Single-Phase Flow

This group investigates the fundamental transport behavior in single-phase flows under complex conditions. Studies explore the influence of needle valve dynamics on near-nozzle spray characteristics [Contribution 2], analyze magnetohydrodynamic nanofluid flow with nonlinear radiation and viscous dissipation [Contribution 5], and examine heat transfer in porous fins with internal heat generation [Contribution 8] and entropy generation in rotating nanofluid flows [Contribution 15]. The performance limits of liquid

https://doi.org/10.3390/mi17010126

metal microchannel heat sinks under high temperatures are also evaluated numerically [Contribution 17]. Together, these works provide deeper insights into convective heat transfer, flow control, and the impact of multi-physics coupling in single-phase systems.

Gao et al. [Contribution 2] reports that a variety of needle-motion profiles are used in diesel injectors. However, it is unclear as to what the underlying mechanism that determines the needle-motion profiles is, and how the needle-motion profiles affect the spray dynamics. Examining how the spray dynamics will change if only altering the needle valve opening speed or closing speed while all other parameters are kept the same has been of significant interest. The different needle-motion profiles are obtained using a piezo nozzle (Nozzle #P) and a solenoid nozzle (Nozzle #S), which have identical nozzle geometry. By utilizing the X-ray imaging technique, it is observed that the average needle valve speed of Nozzle #P is 51% higher at the opening stage but 17% lower at the closing stage than Nozzle #S. When the needle valve lift is low (approximately 200 μm), the needle valve opening speed has a crucial effect on spray dynamics. The faster needle valve opening of Nozzle #P results in a 42% larger spray spreading angle and 34% lower spray velocity at the downstream field. The spray dynamics may be controllable by properly designing the needle-motion profiles in the scenarios of the low needle lifts. However, when the needle valve is sufficiently open (approximately over 200 μm), almost identical spray characteristics are observed, regardless of the needle-motion profiles.

Bafakeeh et al. [Contribution 5] numerically studies the radiated flow of magnetized viscous fluid that is subject to the viscous dissipation phenomenon. The radiative phenomenon is addressed with nonlinear relations. Further, analysis is performed by using the slip effects and convective thermal flow constraints. The transformed problem is numerically evaluated using the Keller Box method. The physical parameter effects—such as the magnetic parameter for the velocity profile, the Prandtl number, the Brownian motion parameter, and the Biot number for the energy profile and Lewis number—and the thermophoresis parameter for the concentration profile are discussed. The obtained results suggest applications in enhancing the heat transfer phenomenon, thermal system, energy generation, heat transmission devices, power generation, chemical reactions, etc.

Sowmya et al. [Contribution 8] uses a variety of methodologies to explore heat transport enhancement; using the fin approach to inspect heat transfer characteristics is one such effective method. In a broad range of industrial applications, including heat exchangers and microchannel heat sinks, fins are often employed to improve heat transfer. Encouraged by this feature, the present research is concerned with the temperature distribution caused by convective and radiative mechanisms in an internal heat-generating porous longitudinal dovetail fin (DF). The Darcy formulation is considered for analyzing the velocity of the fluid passing through the fin, and the Rosseland approximation determines the radiation heat flux. The heat transfer problem of an inverted trapezoidal (dovetail) fin is governed by a second-order ordinary differential equation (ODE); to simplify it to a dimensionless form, nondimensional terms are utilized. The generated ODE is numerically solved by using the spectral collocation method (SCM) via a local linearization approach. The effect of different physical attributes on the dimensionless thermal field and heat flux is graphically illustrated. As a result, the temperature in the dovetail fin transmits in a decreasing manner for growing values of the porosity parameter. For elevated values of heat generation and the radiation-conduction parameter, the thermal profile of the fin displays an increase, whereas an increment in the convection–conduction parameter downsizes the thermal dispersal. It is found that the SCM technique is very effective and more conveniently handles the nonlinear heat transfer equation. Furthermore, the temperature field results from the SCM-based solution are in very close accordance with the outcomes published in the literature.

 https://doi.org/10.3390/mi17010126

Ali et al. [Contribution 15] discuss entropy generation analysis for a peristaltic flow in a rotating medium with generalized complaint walls. The goal of the current analysis is to understand the fluid flow phenomena that are particular to micro devices. Nano materials with a size of less than 100 nm have applications in micro heat exchangers to cool electronic circuits, blood analyzers, biological cell separations, etc. For this study, they consider the effects of radiation, viscous dissipation, and heat flux on the flow of nanomaterial inside a cylindrical micro-channel. To investigate the slip effects on the flow, they use the second order slip condition for axial velocity, the first order slip condition for secondary velocity, and the thermal slip conditions. The flow is governed by partial differential equations (PDEs), which are turned into a system of coupled ordinary differential equations (ODEs) that are highly non-linear and numerically solved using the NDSolve command in Mathematica. The impacts of different involved parameters on the flow field are investigated with the aid of graphical illustrations. Entropy generation and the Bejan number are given special attention, and it is found that they decrease as the Hartman number, rotation, and radiation parameters increase.

Developments in applications such as rocket nozzles, miniature nuclear reactors, and solar thermal generation pose high-density heat dissipation challenges. In these applications, a large amount of heat must be removed in a limited space under a high temperature. Wu et al. [Contribution 17] propose liquid metal-based microchannel heat sinks to handle this kind of cooling problem. Using a numerical method, the flow and heat transfer performances of liquid metal-based heat sinks with different working fluid types, diverse microchannel cross-section shapes and various inlet velocities are studied. By solving the 3D steady and conjugate heat transfer model, they find that among all the investigated cases, lithium and circle are the most appropriate choices for the working fluid and microchannel cross-section shape, respectively. Moreover, the inlet velocity had a great influence on the flow and heat-transfer performances. From 1 m/s to 9 m/s, the pressure drop increases as much as 65 times, and the heat-transfer coefficient is enhanced by about 74.35%.

3. Mechanisms of Multi-Phase Flow

This category delves into the core interfacial and phase-change physics that are dominant at micro-scales. Research topics include the significant enhancement of evaporation via the Marangoni effect [Contribution 1], unique condensation frosting characteristics at plate edges [Contribution 3], mass transfer mechanisms for gas replenishment on hydrophobic surfaces [Contribution 4], particle deposition interactions near film cooling holes [Contribution 13], the non-uniform cooling efficiency of sintered wire mesh [Contribution 14], and the dynamic behavior of cavitation bubbles under actuation [Contribution 16]. These studies reveal the critical role of interfacial forces and phase interactions in multiphase systems.

Liu et al. [Contribution 1] built a coupled thermal mass model of droplet evaporation and test the accuracy of the numerical model through theoretical results to study the droplet heat and mass transfer law in the droplet evaporation process. In order to study the influence of the Marangoni effect on the droplet evaporation process and the effects of different initial droplet radii and an ambient temperature on the temperature and flow, fields within the droplet are compared. From this result, it can be seen that the droplet volume is 20 μL, and the maximum flow velocity in the droplet is 0.34 mm/s, without taking the Marangoni effect into account. When the Marangoni effect is considered, the maximum flow velocity increases by almost 100 times. The Marangoni effect can cause the convection in the droplet to change direction, and the formation of the Marangoni flow may affect the temperature distribution within the droplet, thereby increasing the

 https://doi.org/10.3390/mi17010126

evaporation efficiency by 2.5%. The evaporation process will increase the velocity of the air that is close to the surface of the liquid, but the increase in air velocity close to the liquid surface is not sufficient to reinforce evaporation. There is a non-linear relationship between an increasing ambient temperature and increasing evaporation efficiency. For every 5 °C increase in the ambient temperature, the maximum increase in the rate of evaporation is approximately 22.7%.

Zhang et al. [Contribution 3] conduct an experimental study on the localized condensation frosting characteristics in the edge region of a cold plate. The edge effects on the water droplet condensation (WDC), water droplet frozen (WDF), and frost layer growth characteristics are quantitatively investigated. The results show that the number of droplets coalescing in the edge-affected regions is around 50% greater than in the unaffected regions. At the end of the WDC stages, the area-average equivalent contact diameter and coverage area ratio of water droplets in the edge-affected regions are 2.69 times and 11.6% greater than those in the unaffected regions under natural convection, and the corresponding values are 2.24 times and 9.9%, respectively, under forced convection. Compared with the unaffected regions, the WDF stage duration in the edge-affected regions decrease by 63.6% and 95.3% under natural and forced convection, respectively. Additionally, plate-type and feather-type frost crystals are, respectively, observed in natural and forced convection. The results of this study can help us to better understand the condensation frosting mechanism on a cold plate, which in turn provides guidelines for optimizing the design of heat exchanger structures and system control strategies facing frosting problems.

The underwater nonwetted state on a superhydrophobic surface is hardly maintained in flowing water because the entrapped gas dissolves into the water or is carried off by the flow. Therefore, a source gas is necessary to maintain a superhydrophobic state for its applications under realistic conditions. As detailed in this paper, based on the gas entrapped on a hydrophobic structured surface, the gas regeneration is experimentally achieved to replenish the losses of gas that are carried off by the flow and reduced through dissolution. Wang et al. [Contribution 4] investigate the mechanism of mass transfer at the liquid–gas interface by simulation. The results indicate that the water molecules at a liquid–gas interface escape to the entrapped gas when the water content does not reach saturation. This phenomenon could be due to the evaporation at the liquid–gas interface. With the increasing water content in the entrapped gas, the evaporation rate at the liquid–gas interface descends gradually. Under the action of flowing, the substances containing high concentrations of water molecule are washed away at the liquid–gas interface. Therefore, the low concentration of the water molecule at the liquid–gas interface is created. As a result, the equilibrium of water and gas at the liquid-gad interface is broken, and the evaporation continued to replenish the lost gas. Overall, the presented results in this study could be considered to be a promising candidate for replenishing the lost gas in hydrophobic structured surfaces by mass transfer at the liquid–gas interface.

Peng et al. [Contributions 13, 14] present a combined experimental and numerical investigation on the particle deposition in the vicinity of multiple film cooling holes to reveal the effect of interactions between cooling outflows on particle deposition. They experimentally investigate the transpiration cooling characteristics of a porous material: sintered wire mesh. Three samples with different porosities in a plain weave structure are tested with various blowing ratios in an open-loop wind tunnel with a heated mainstream flow. The temperature on the surface of the porous material is measured by an infrared camera to obtain the cooling efficiency. The measurements reveal nonuniform distributions of the surface temperature and cooling efficiency in both the flow direction and the transverse direction. The averaged cooling efficiency on the surface first decreases and then increases with the blowing ratio, but it increases and then decreases with the porosity

 https://doi.org/10.3390/mi17010126

of the material. The internal cooling by forced convection and its combination with the external film cooling from the transpiration cooling are considered to be attributed to those two cooling characteristics, respectively. Finally, they propose a modified blowing ratio to collapse the minima of the blowing ratio for all tested samples, providing a universal transition for the decreasing and increasing branches for all tested samples in the relationship between the averaged cooling efficiency and the blowing ratio.

Shang et al. [Contribution 16] experimentally and numerically study the dynamics of cavitation bubbles in a nozzle-shaped microfluidic channel with PZT (lead zirconate titanate) actuations. It is found that a cloud of bubbles can be generated near the center of the microfluidic channel when the actuation voltage is larger than the threshold of 1 kHz. After being generated, the bubbles under actuations oscillate radially with violent expansion and compression, and they simultaneously translate upstream towards the opening of the nozzle. Along with radial oscillation and translation, the bubbles undergo frequent and drastic coalescence and breakup, leading to vigorous churning of the surrounding liquids. The pressure variation and distribution in the microchannel are calculated by numerical simulation in Ansys Fluent, and results show that there is a low-pressure zone inside the microfluidic channel within each cycle of the actuation period, which is responsible for the observed bubble generation in the experiments. The method of bubble generation in this study is novel and can be applied for the enhancement of heat and mass transfer in microfluidic operations.

4. Applications for Single- and Multi-Phase Flows

Focused on design optimization and performance analysis for engineering systems, this segment includes a numerical comparison of fin shapes for microchannel evaporator efficiency [Contribution 6], a parametric study of supercritical CO_2 heat transfer in a spiral channel gas cooler [Contribution 9], the investigation of a reversed heat flux phenomenon in wire-wrapped nuclear fuel rods [Contribution 11], and experimental–analytical analysis of transient refrigerant behavior in microchannels [Contribution 12]. These works aim to solve practical thermal management challenges. This area addresses the performance characterization and operational limits of materials and systems involving phase change. One study measures and models the thermal conductivity of flexible composite phase-change materials [Contribution 7], while the other analytically determines the maximum heat load and temperature response of flat micro heat pipes with different working fluids [Contribution 10]. The findings offer vital data and guidelines for the design of reliable phase-change-based thermal systems.

Nguyen et al. [Contribution 6] numerically study the fin shape of a microchannel evaporator in a CO_2 air conditioning system. The study is performed at the inlet evaporative temperature of 10 °C and the vapor quality of 0.61. Two types of fin shapes are examined: straight fins and V-fins. The numerical results are verified by the experimental data. For the system under consideration and for the same heat-transfer area and the heat-transfer coefficient for the air side in the microchannel evaporator, the effect of the fin shape on the heat transfer is not different; however, the solution time and the physical memory for the straight fins are 1.3 and 1.45 times longer than those of the V-fins, respectively. Therefore, the V-fin shape should be used for numerical simulation to compare it with the straight fin shape. In this study, the evaporation of the refrigerant in the microchannel evaporator takes place in four passes. The normal heat flux from the air through the fins and tubes is almost reached at 1550 W/m^2, at the evaporative temperature of 10 °C. The obtained results from the experimental data are in good agreement with those obtained from the numerical results, with a deviation of less than 10%.

https://doi.org/10.3390/mi17010126

Jiang et al. [Contribution 9] study a heat exchanger with a spiral channel. ANSYS CFX software is used to analyze the flow and heat-transfer characteristics of the heat exchanger (single-plate model). The influences of the cooling pressure of CO_2, the mass flux of CO_2, the mass flux of water, and the channel radius of CO_2 are discussed. In this paper, the results show that the cooling pressure of CO_2, the mass flux of CO_2, and the channel radius of CO_2 all have a large influence on the local heat transfer coefficient: with an increase in the cooling pressure of CO_2, the peak value of the heat transfer coefficient of CO_2 decreases and the average heat transfer coefficient decreases; with an increase in the mass flux of CO_2, the peak value of the heat transfer coefficient of CO_2 increases and the average heat transfer coefficient increases; and with a decrease in the channel radius of CO_2, the peak value of the heat transfer coefficient of CO_2 increases. The water mass flux only has a small effect on the heat transfer, and the lower cooling pressure of CO_2 corresponds to a higher peak heat transfer coefficient, which can reach 27.5 kW m^{-2} K^{-1} at 9 MPa. A gas cooler is one of the important parts of a carbon dioxide (CO_2) heat pump water heater, and it must meet the needs of both pressurization and heat transfer. It is important to study gas coolers.

Tan et al. [Contribution 11] discover the existence of reversed heat flux from coolant to wrapped wire, which is contrary to our usual understanding. This phenomenon has not been reported in previous CFD calculations. Hence, a solid heat-conduction model is proposed to prove this phenomenon and analyze the heat-transfer process. The simulation results show that the wrapping wire embedding depth, the shape of the calculation domain, and the physical properties of all components have great influence on the magnitude of the reversed heat flux. The present findings will have a strong influence on the temperature field and maximum value of the fuel rod, as well as profound reference value for future flow calculation, especially in grid generation and treatment of the junction between the winding wire and fuel rod.

Mihai et al. [Contribution 12] conduct an analysis of the R134a (tetrafluoroetane) coolant's non-stationary behavior in rectangular microchannels with the help of a newly proposed miniature refrigerating machine of their own design and construction. The experimental device incorporates a condenser, a lamination tube and a vaporizer on the same plate, all of which integrate rectangular microchannels. The size of the rectangular microchannels is determined by laser profilometry. R-134a coolant vapors are pressurized using a small ASPEN rotary compressor. By using the variable soft spheres (VSS) model, the mean free path, Knudsen and Reynolds numbers, and the dimensionless velocity profile can be assessed analytically. In order to determine the average dimensionless temperature drop in the vaporizer's rectangular microchannels, in a non-stationary regime, an analytical solution for incompressible flow with slip at the walls, fully developed flow, and a laminar regime is used, with the aid of an integral transform approach. In the experimental study, the transitional distribution of temperature is tracked while modifying the R134a flow through the rectangular microchannels. Coolant flow is then maintained at a constant, while the amount of heat absorbed by the vaporizer was varied using multiple electric resistors. A comparative analysis of the analytical and experimental values is conducted.

Phase-change materials (PCMs) are widely used in energy storage and thermal management due to the large amount of latent heat in the phase-change process. As one of the most significant thermophysical properties of PCMs, the thermal conductivity has been extensively studied. Great attention has been paid to improving the thermal conductivities of PCMs; however, the studies on the thermal conductivities of flexible PCMs are relatively inadequate. Feng et al. [Contribution 7] use polyethylene glycol 1500 (PEG1500) as the base PCM, and expanded graphite (EG) and styrene–butadiene–styrene (SBS) are added to improve the thermal conductivity and flexibility of pure PCMs, respectively. A steady-state experimental test rig is built and verified with the measurement of the thermal conductivity

 https://doi.org/10.3390/mi17010126

of stainless steel and deionized water, and then the thermal conductivities of PCMs at different phases and qualitative temperatures are measured extensively. Compared to the PEG1500 with 5 wt.% EG, the addition of SBS sharply reduces the thermal conductivity, which is only 0.362 W/(m·K) at 12.5 °C when the addition ratio is 50%. This is approximately a 69% reduction compared with the composite PCMs without SBS. Furthermore, the theoretical thermal conductivities of the composite PCMs are calculated with six theoretical models of multiphase systems. The majority of the models provide a good prediction for the thermal conductivities of composite PCM with high SBS concentration, while the average deviation of the Agari–Uno model is only 20.5% with a different SBS concentration and, relatively, it agrees well with the experimental results.

Mihai et al. [Contribution 10] report a series of considerations for the occurrence of capillary boundaries in flat micro heat pipes (flat MHPs) and the conditions required for their stable operation in relation to the working circumstances and to the type of liquids inside the pipes. The particularities of heat transfer in a flat MHP are analyzed for situations with either excessive or deficient working liquid. Depending on the physical properties of the working liquids (acetone, methanol, and distilled water), the maximum rate of heat flow that can be applied to a flat MHP is determined analytically. The calculus is made with the assumption that constant vaporization of the liquid is ensured in the flat MHP's evaporator, with no overheating. The analytical models considered here allow for the evaluation of the liquid film thickness and the mass flow corresponding to the vaporization region. The temperature difference between the inner and outer walls of a flat MHP is found in the case of a transient regime and a variable thermal flow is applied in the evaporation region. The interior of the flat MHPs is modeled in MATLAB, using an FTCS (Forward Time Central Space) method, which is a finite difference method that is used to numerically solve the heat equation.

To conclude, we would like to acknowledge all the authors for their contributions to the success of this Special Issue on Heat and Mass Transfer in Microchannels, as well as the reviewers whose feedback helped to improve the quality of the published papers.

Author Contributions: Conceptualization, X.X., X.Z. and L.Z.; methodology, X.X., X.Z. and L.Z.; software, X.X., X.Z. and L.Z.; investigation, X.X., X.Z. and L.Z.; resources, X.X., X.Z. and L.Z.; writing—original draft preparation, X.Z. and L.Z.; writing—review and editing, X.X., X.Z. and L.Z.; visualization, X.Z. and L.Z.; supervision, X.X., X.Z. and L.Z.; project administration, X.X., X.Z. and L.Z.; funding acquisition, X.X., X.Z. and L.Z. All authors have read and agreed to the published version of the manuscript.

Funding: This research is funded by the National Natural Science Foundation of China (No. 52406007 and 52306003), Beijing Municipal Natural Science Foundation (No. 3244046 and 3232032), Key Laboratory of Icing and Anti/De-icing (No. IADL20230113 and IADL20220304), Young Elite Scientist Sponsorship Program by BAST (No. BYESS2023352).

Data Availability Statement: The raw data supporting the conclusions of this article will be made available by the authors on request.

Acknowledgments: The Guest Editors are very grateful to Scott Wang and Mars Tu from the Micromachines publishing office for their great guidance, assistance, help, and support that led to the success of this Special Issue.

Conflicts of Interest: The authors declare no conflicts of interest.

List of Contributions:

1. Liu, H.; Deng, J. Influence of Marangoni Effect on Heat and Mass Transfer during Evaporation of Sessile Microdroplets. *Micromachines* **2022**, *13*, 1968. https://doi.org/10.3390/mi13111968.

2. Gao, Y.; Huang, W.; Pratama, R.H.; Gong, H.; Wang, J. Investigation of Needle Motion Profile Effect on Diesel Spray in Near-Nozzle Field. *Micromachines* **2022**, *13*, 1944. https://doi.org/10.3390/mi13111944.
3. Zhang, L.; Song, M.; Chao, C.Y.H.; Dang, C.; Shen, J. Localized Characteristics of the First Three Typical Condensation Frosting Stages in the Edge Region of a Horizontal Cold Plate. *Micromachines* **2022**, *13*, 1906. https://doi.org/10.3390/mi13111906.
4. Wang, B.; Wang, C.; Weng, D.; Lazarus, M.; Yan, D.; Liu, X. Replenishment of the Gas in a Hydrophobically-Structured Surface by Mass Transfer at the Liquid-Gas Interface for Improving the Stability of Entrapped Gas. *Micromachines* **2022**, *13*, 1893. https://doi.org/10.3390/mi13111893.
5. Bafakeeh, O.T.; Ahmad, B.; Noor, S.; Abbas, T.; Khan, S.U.; Khan, M.I.; Elattar, S.; Eldin, S.M.; Oreijah, M.; Guedri, K. Nonlinear Thermal Diffusion and Radiative Stagnation Point Flow of Nanofluid with Viscous Dissipation and Slip Constrains: Keller Box Framework Applications to Micromachines. *Micromachines* **2022**, *13*, 1839. https://doi.org/10.3390/mi13111839.
6. Nguyen, T.; Dang, T. The Effect of Fin Shape on the Heat Transfer and the Solution Time of a Microchannel Evaporator in a CO_2 Air Conditioning System—A Numerical Investigation. *Micromachines* **2022**, *13*, 1648. https://doi.org/10.3390/mi13101648.
7. Feng, Z.; Xiao, X. Thermal Conductivity Measurement of Flexible Composite Phase-Change Materials Based on the Steady-State Method. *Micromachines* **2022**, *13*, 1582. https://doi.org/10.3390/mi13101582.
8. Sowmya, G.; Lashin, M.M.A.; Khan, M.I.; Kumar, R.S.V.; Jagadeesha, K.C.; Prasannakumara, B.C.; Guedri, K.; Bafakeeh, O.T.; Mohamed Tag-ElDin, E.S.; Galal, A.M. Significance of Convection and Internal Heat Generation on the Thermal Distribution of a Porous Dovetail Fin with Radiative Heat Transfer by Spectral Collocation Method. *Micromachines* **2022**, *13*, 1336. https://doi.org/10.3390/mi13081336.
9. Jiang, J.; Liang, S.; Ji, C.; Wang, L.; Guo, C. Study of New Mini-Channel Trans-Critical CO_2 Heat Pump Gas Cooler. *Micromachines* **2022**, *13*, 1206. https://doi.org/10.3390/mi13081206.
10. Mihai, I.; Suciu, C.; Picus, C.M. Considerations for the Maximum Heat Load and Its Influence on Temperature Variation of the Evaporator in Flat MHPs in Transient Regimes. *Micromachines* **2022**, *13*, 979. https://doi.org/10.3390/mi13060979.
11. Tan, X.; Wang, B.; Guo, Y.; Hu, M. Numerical Investigation of Special Heat Transfer Phenomenon in Wire-Wrapped Fuel Rod of SFR. *Micromachines* **2022**, *13*, 935. https://doi.org/10.3390/mi13060935.
12. Mihai, I.; Suciu, C.; Picus, C.M. Particularities of R134a Refrigerant Temperature Variations in a Transient Convective Regime during Vaporization in Rectangular Microchannels. *Micromachines* **2022**, *13*, 767. https://doi.org/10.3390/mi13050767.
13. Peng, Y.; Xu, G.; Luo, X.; He, J.; Liu, D. Particle Deposition in the Vicinity of Multiple Film Cooling Holes. *Micromachines* **2022**, *13*, 523. https://doi.org/10.3390/mi13040523.
14. Peng, Y.; Xu, G.; Luo, X.; He, J.; Liu, D. Experimental Investigation on the Transpiration Cooling Characteristics of Sintered Wire Mesh in Plain Weave. *Micromachines* **2022**, *13*, 450. https://doi.org/10.3390/mi13030450.
15. Ali, A.; Sajid, M.; Anjum, H.J.; Awais, M.; Nisar, K.S.; Saleel, C.A. Entropy Generation Analysis of Peristaltic Flow of Nanomaterial in a Rotating Medium through Generalized Complaint Walls of Micro-Channel with Radiation and Heat Flux Effects. *Micromachines* **2022**, *13*, 375. https://doi.org/10.3390/mi13030375.
16. Shang, X.; Huang, X. Investigation of the Dynamics of Cavitation Bubbles in a Microfluidic Channel with Actuations. *Micromachines* **2022**, *13*, 203. https://doi.org/10.3390/mi13020203.
17. Wu, T.; Wang, L.; Tang, Y.; Yin, C.; Li, X. Flow and Heat Transfer Performances of Liquid Metal Based Microchannel Heat Sinks under High Temperature Conditions. *Micromachines* **2022**, *13*, 95. https://doi.org/10.3390/mi13010095.

References

1. Li, R.; Zhang, J. Real-Time Heat Dissipation Model of Electronic Equipment for Determining the Dynamic Cooling Demand of Office Buildings. *J. Build. Eng.* **2022**, *45*, 103465. [CrossRef]
2. Lei, Y.; Chen, Z. Numerical Study of Condensation Heat Transfer in Curved Square and Triangle Microchannels. *Heat Transf. Eng.* **2020**, *41*, 768–778. [CrossRef]

3. Niu, Y.; Huang, P.; Pan, M. Study of Heat and Mass Transfer by Bionic Fractal Microchannel Plates. *Chem. Eng. Technol.* **2021**, *44*, 741–751. [CrossRef]
4. Xu, S.; Hu, G.; Qin, J.; Yang, Y. A Numerical Study of Fluid Flow and Heat Transfer in Different Microchannel Heat Sinks for Electronic Chip Cooling. *J. Mech. Sci. Technol.* **2012**, *26*, 1257–1263. [CrossRef]
5. Huang, S.-C.; Lee, G.-B.; Chien, F.-C.; Chen, S.-J.; Chen, W.-J.; Yang, M.-C. A Microfluidic System with Integrated Molecular Imprinting Polymer Films for Surface Plasmon Resonance Detection. *J. Micromech. Microeng.* **2006**, *16*, 1251–1257. [CrossRef]
6. Boogar, R.S.; Gheshlaghi, R.; Mahdavi, M.A. The Effects of Viscosity, Surface Tension, and Flow Rate on Gasoil-Water Flow Pattern in Microchannels. *Korean J. Chem. Eng.* **2013**, *30*, 45–49. [CrossRef]
7. Chen, S.; Lan, J.; Zhang, Y.; Guo, J.; Cao, Z.; Sha, Y. 3D Multiphase Flow Simulation of Marangoni Convection on Reactive Absorption of CO_2 by Monoethanolamine in Microchannel. *Chin. J. Chem. Eng.* **2022**, *43*, 370–377. [CrossRef]
8. Ichikawa, N.; Maeda, R. Interface Motion Driven by Capillary Action in Circular and Rectangular Microchannel. *Microscale Thermophys. Eng.* **2005**, *9*, 237–254. [CrossRef]
9. Chiriac, E.; Avram, M.; Balan, C. Investigation of Multiphase Flow in a Trifurcation Microchannel—A Benchmark Problem. *Micromachines* **2022**, *13*, 974. [CrossRef] [PubMed]
10. Wang, J.-X.; Wang, H.; Salmean, C.; Cui, B.; Zhong, M.-L.; Mao, Y.; Li, J.-X.; Yao, S. Sustained and Enhanced Nucleate Boiling Using Hierarchical Architectures at Large Superheats. *Exploration* **2025**, *5*, 20240137. [CrossRef] [PubMed]
11. Kuravi, S.; Du, J.; Chow, L.C. Encapsulated Phase Change Material Slurry Flow in Manifold Microchannels. *J. Thermophys. Heat Transf.* **2010**, *24*, 364–373. [CrossRef]
12. Tan, H.; Wu, L.; Wang, M.; Yang, Z.; Du, P. Heat Transfer Improvement in Microchannel Heat Sink by Topology Design and Optimization for High Heat Flux Chip Cooling. *Int. J. Heat Mass Transf.* **2019**, *129*, 681–689. [CrossRef]
13. Ikemoto, T.; Kokubun, Y. Fabrication of Microchannel with Thin Cover Layer for an Optical Waveguide MEMS Switch Based on Microfluidics. *IEICE Trans. Electron.* **2007**, *e90–c*, 78–86. [CrossRef]
14. Hua, Y.; He, L.; Zhang, L. Effective Transport of Passive Particles Induced by Chiral-Active Particles in Microchannel. *Chin. Phys. B* **2017**, *26*, 108–113. [CrossRef]
15. Yang, P.D.; Rizvi, A.H.; Messer, B.; Chmelka, B.F.; Whitesides, G.M.; Stucky, G.D. Patterning Porous Oxides within Microchannel Networks. *Adv. Mater.* **2001**, *13*, 427–431. [CrossRef]

 https://doi.org/10.3390/mi17010126

micromachines

Article

Influence of Marangoni Effect on Heat and Mass Transfer during Evaporation of Sessile Microdroplets

Haiting Liu and Jiewen Deng *

School of Energy and Power Engineering, Northeast Electric Power University, Jilin 132012, China
* Correspondence: 20182798@neepu.edu.cn

Abstract: Evaporative cooling is an important method for controlling the temperature of micro devices, and heat and mass transfer from the microdroplets in the evaporation process directly affect the cooling performance. In order to study the droplet heat and mass transfer law in the droplet evaporation process, this paper builds a coupled thermal mass model of droplet evaporation and tests the accuracy of the numerical model through theoretical results. In order to study the influence of the Marangoni effect on the droplet evaporation process and the effects of different initial droplet radius and ambient temperature on the temperature and flow, fields within the droplet are compared. From this result, it can be seen that the droplet volume is 20 μL, and the maximum flow velocity in the droplet is 0.34 mm/s, without taking into account the Marangoni effect. When the Marangoni effect is taken into account, the maximum flow velocity increases by almost 100 times. The Marangoni effect can cause the convection in the droplet to change direction, and the formation of the Marangoni flow may affect the temperature distribution within the droplet, thereby increasing the evaporation efficiency by 2.5%. The evaporation process will increase the velocity of the air close to the surface of the liquid, but the increase in air velocity close to the liquid surface is not sufficient to reinforce evaporation. There is a non-linear relationship between increasing ambient temperature and increasing evaporation efficiency. For every 5 °C increase in ambient temperature, the maximum increase in the rate of evaporation is approximately 22.7%.

Keywords: droplet evaporation; Marangoni effect; thermal mass coupling; enhanced evaporation; computational fluid dynamics

1. Introduction

Droplet evaporation is common in nature and human life. In practical engineering applications, many industrial technologies are closely related to droplet evaporation—such as inkjet printing, desalination of seawater, concentration of waste fluid, cooling of working fluids and medical diagnostics—and typically occur spontaneously when the vapor pressure near the liquid-gas interface of a liquid droplet is not saturated. This is a complex physical process accompanied by convective mass transfer and heat transfer accompanied by a phase change.

In 1805, following the landmark work on liquid evaporation by Young [1], one popular topic has become the process of evaporation-induced intra-drop changes. Researchers studying evaporation have uncovered many interesting phenomena caused by the process of evaporation, such as motion between liquid and solid during evaporation, heat transfer between droplet and substrate, and nanoparticle effect on evaporation. Marangoni found that changes in the concentration of local constituents on the surface of the liquid can lead to changes in surface tension, causing the fluid to flow out of a position with a lower surface tension and into a position with a higher surface tension [2]. This type of liquid flow due to gradients in surface tension has since been referred to as the "Marangoni effect". Gibbs [3] provided a more detailed and seminal theoretical contribution to this phenomenon, and thus, it is also called the "Gibbs-Marangoni effect". While research is ongoing, a large body of work has demonstrated that there is a link between liquid surface tension and droplet evaporation [4–6]. In order to test the effect of gravity on evaporation in the binary solution evaporation process, a mathematical model was proposed by Diddens [7] for predicting Rayleigh and Maranni convection phenomena. These results demonstrate that, while gravity may play a role in promoting evaporation, natural convection is the primary driving mechanism for the internal flow of binary droplets. The distribution law of the temperature and velocity fields under the Marangoni action was obtained by Girard [8]. By constructing averages of numerical simulations, he studied the influence of the Marangoni effect on evaporation and believed that heat conduction within the droplet was the major factor affecting the rate of evaporation.

The Marangoni effect is beneficial for improving the efficiency of micro-scale thermal mass transport and has an important influence on the component mass transfer process—not only in chemical processes, such as extraction and distillation [9], but also in heat transfer processes, such as evaporation [10] and condensation [11], which play an important role in increasing heat transfer within liquid droplets and films. Increased evaporation of droplets attached to the wall is also an important branch of research. The quasi-stationary Marangoni contraction of binary mixtures was studied by Stefan [12] through experiments and numerical analysis, and Stefan related it to the hydrodynamic model of droplet evaporation and compared the experimental results with numerical simulation. We propose that there is a general power–law relationship between the quasi-expansion contact angle and the relative saturation of the ambient gas phase. Marcus [13] distributed the fluid as an array of droplets on a super hydrophobic dose plate with 3×3 array pores, and the microheaters were screen printed on the back of the dose plate to study droplet evaporation on these pores; accurate measurement of the microdroplets is achieved. Ye [14] carried out an experimental study on the steady state evaporation of isopropanol droplets at different substrate temperatures. This study demonstrates that the height of the droplet can affect the evolution of the droplet flow pattern during the evaporation process, and increasing substrate temperature can cause droplet number of internal wrinkles. Wang [15] established a model for droplet evaporation on high-temperature substrates and droplet evaporation in suspension in air at elevated temperature and investigated the steady state flow process of Marangoni convection within droplets during evaporation. A numerical simulation was performed to investigate the effect of contact angle on Marangoni convection. Hu [16] constructed a relatively complete model of evaporation and investigated the characteristics of evaporation on both heated and unheated substrates, cooling effect. In addition, the theoretical model also considers buoyancy-driven convection and Marangoni convection. On the other hand, many simulations of evaporating droplet properties and inducing factors have been performed by many authors [17–20] in order to obtain the flow characteristics of Marangoni convection within the droplet on a tilted substrate, both in the oil phase and under laser irradiation, while also considering their influence on the heat and mass transfer process.

In general, the velocity of the liquid surface airflow is also a major factor affecting the evaporation efficiency, and the process of droplet evaporation increases the rate of liquid surface flow. It is, therefore, important to study whether droplet evaporation is a self-reinforcing process and to understand the effect of the Marangoni effect on evaporation; however, these studies have not discussed these issues but have mainly focused on the effects of the contact angle of the droplet and the temperature of the substrate on Marangoni convection.

The main purpose of this paper is to compare the heat and mass transfer differences in the wall droplet evaporation process with and without the Marangoni effect. The influence of ambient temperature and initial droplet size on evaporation under the influence of the Marangoni effect is also analyzed; the contribution of Marangoni convection during evaporation is analyzed in parallel.

2. Model Construction and Validation

2.1. Physical Model

In order to study the dynamic change process of the droplet surface during evaporation, this paper adopts the moving grid method to track the interface between the droplet and the air. This method realizes the simultaneous change of mass points and grid nodes at the moving interface so as to achieve the accurate capture of the location of the calculated interface by the deformed grid. The model boundary conditions and grid situation are shown in Figure 1. The initial radius of the droplet is R; the height of the stainless steel substrate is 1 mm; the width is $5R$; the radius of the air domain is $20R$; and the thickness of the infinite element domain is R. The minimum grid cell in the droplet region is 0.02 mm; the maximum grid cell is 0.08 mm; and the ambient temperature is T_{en}.

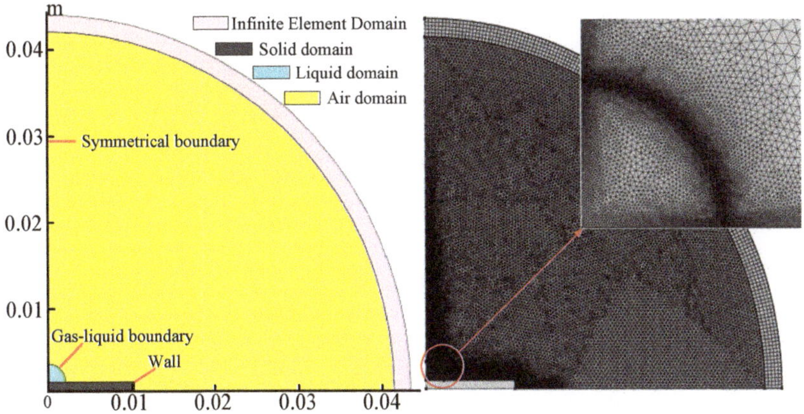

Figure 1. Model boundary conditions and mesh distribution.

2.2. Mathematical Model for the Internal Flow and Heat Transfer of Liquid Droplets

In this paper, we focus on the process of droplet evaporation on the surface of substrates. During evaporation, temperature gradients are generated within the droplet and surrounding air, which affect the convection process within the droplets and close to the liquid surface; hence, when describing the droplet evaporation process via numerical models, not only must the process of mass transfer of vapor and air around the droplet be considered but also the effect of convection caused by the temperature gradient during the evaporation process. In order to enhance the practicality of the numerical model, before constructing the numerical model in this paper, we make the following assumptions:

(1) Since some research findings show that the constant contact angle (CCA) [21] mode of evaporation accounts for over 50% of the evaporation process when the droplet is evaporating on a surface of constant heat [22,23], the process of evaporation is considered to be the CCA mode, and the change in contact angle is ignored.

(2) Droplet evaporation does not account for the change in moisture content of the initial environment, and to simplify the computational model, the environmental water vapor content is assumed to be 0 at the initial time, and water vapor in the environment only originates from the droplet during evaporation.

(3) Since the object of investigation in this paper is a micro-droplet and its size is small, the change in curvature of the liquid surface during evaporation is ignored. The effect of the curvature of the droplet interface on the concentration of vapor saturation can be neglected, according to the Kelvin equation [24]; thus, the vapor concentration at the droplet surface can be thought of as the saturated vapor concentration only relative to the local temperature, given by Equation (8).

(4) In the computational domain, the fluid is assumed to be incompressible due to the small velocity of the air flow around the droplet and the droplet.

Since this paper is investigating the process of droplet evaporation on a wall with a contact angle of 90°, as can be seen in Figure 2, the droplets can be equated to a spherical corona shape with the following volume equation:

$$V = \pi h\left(3r^2 + h^2\right)/6 \tag{1}$$

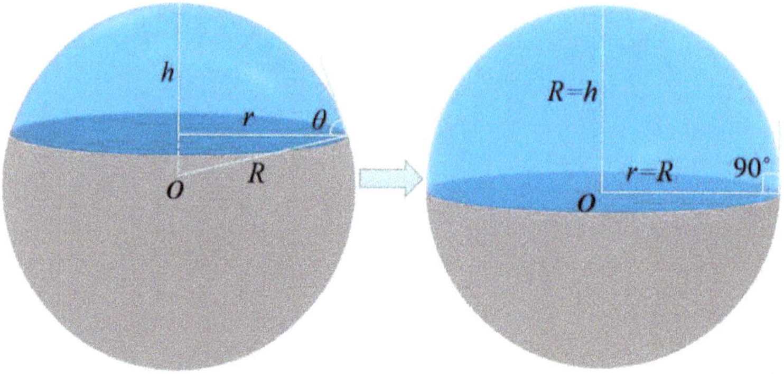

Figure 2. Schematic diagram of a spherical crown droplet.

The droplet contact angle is 90°, so the evaporation process does not take into account the change in contact angle. The volume of the droplet at which $R = r = h$ and at which the droplet is hemispherical is thus:

$$V = 2\pi R^3/3 \tag{2}$$

The velocity field $u(r, z)$, the vapor concentration field $c(r, z)$ and the temperature field $T(r, z)$ within the droplet and in the surrounding air during the evaporation of the droplet can be described by the mass conservation equation, the momentum conservation equation and the energy conservation equation, respectively, as follows:

$$\frac{\partial \rho}{\partial t} + \nabla\cdot(\rho u) = 0 \tag{3}$$

$$\frac{\partial(\rho u)}{\partial t} + \rho u\cdot\nabla u = -\nabla P - \rho g + \nabla\left[\mu\left(\nabla u + (\nabla u)^T\right) - \frac{2}{3}\mu(\nabla\cdot u)\right] \tag{4}$$

$$\frac{\partial(\rho c_p T)}{\partial t} + \nabla\cdot(\rho c_p u T) = \nabla\cdot(k\nabla T) \tag{5}$$

where u is the fluid velocity vector. P is the pressure; μ is the kinetic viscosity; and ρ is the fluid density, which is a temperature-dependent function, described as follows:

$$\begin{cases} \rho_{liquid} = \rho_0[1 + \alpha(T - T_0)] \\ \rho_{air} = \frac{P_{air}}{RT} M_{air} + \frac{M_{vapor} - M_{air}}{M_{vapor}} c \end{cases} \tag{6}$$

where P_{air} is the dry air pressure; R is the universal gas constant; M_{vapor} and M_{air} are the molar masses of water vapor and dry air, respectively; and c is the water vapor mass concentration.

The saturation vapor pressure is derived from the Antoine equation:

$$P_{sat} = 10^{[A - B/(C+T)]} \tag{7}$$

where A, B and C use the constants in the literature [25] and take the values 8.10765, 1750.286 and 235, respectively. The saturated vapor concentration is obtained from the ideal gas law as:

$$c_{sat} = \frac{M_{vapor} P_{sat}}{RT} \tag{8}$$

On the surface of the droplet, the evaporative mass flux of water vapor can be expressed as:

$$J = M_{vapor} \vec{n} \cdot (-D\nabla c) \tag{9}$$

$\vec{n}$ is the droplet surface normal, and D is the diffusion coefficient of water vapor in air. To simplify the calculation, in this paper, the diffusion coefficient of water vapor in air is assumed to be constant and taken as 2.85×10^{-5} m^2/s.

In general, the Marangoni effect occurs when there is a gradient in surface tension at the gas-liquid interface, and surface tension usually changes when the solute concentration, surfactant concentration and temperature along the interface change. In this paper, we focus on the evaporation of droplets under changing temperature conditions; therefore, the change in surface tension σ in the thermally induced Marangoni effect can be described as follows:

$$d\sigma = \frac{\partial \sigma}{\partial T} dT + \frac{\partial \sigma}{\partial c} dc \tag{10}$$

$$\left[\mu\left(\nabla u + (\nabla u)^T\right)\right] \cdot \vec{n} = \left|\frac{d\sigma}{dT}\right| \nabla_t T \tag{11}$$

2.3. Model Validation

The object of study in this article is a droplet of pure water, and under the assumption that evaporation is purely diffusion-driven, the rate of evaporation of a hemispherical droplet can be described as follows [26]:

$$\frac{dV}{dt} = -2\pi RD \frac{\rho_{sat} - \rho_{\infty}}{\rho_{liquid}} \tag{12}$$

where V is the droplet volume; ρ_{sat} is the saturated vapour density; $\rho_{\infty} = 0$ is the distant vapour density; and ρ_{liquid} is the droplet density.

A droplet contact angle of 90 degrees is used, and the law of variation of droplet diameter squared with evaporation time is illustrated in Figure 3. With a linear fit to the data and a correlation coefficient of $R^2 \approx 1$, we can see that the square of the droplet diameter decreases linearly with the evaporation time, and this finding is in agreement with the D^2 law of classical droplet evaporation [27]. As a further check on the reasonableness of the numerical model above, according to the theoretical droplet evaporation Equation (12) proposed in the literature [26], it is possible to obtain a trend of droplet mass change with time, which is compared to the numerical results in this paper. As can be seen in Figure 4, with continuous evaporation, the evaporation rate of the droplet gradually decreases, and the overall law of change is consistent. With or without the Marangoni effect, when the volume of the droplet is 20 μL with a dry evaporation time near 3870 s, it is near the theoretical evaporation time. On the other hand, when considering the Marangoni effect, the time taken to evaporate at the same droplet radius is shortened slightly, but the two are fundamentally consistent in the law of evaporation. The above numerical model can be seen to be reasonable for investigating the evaporation law of the CCA mode of microdroplets.

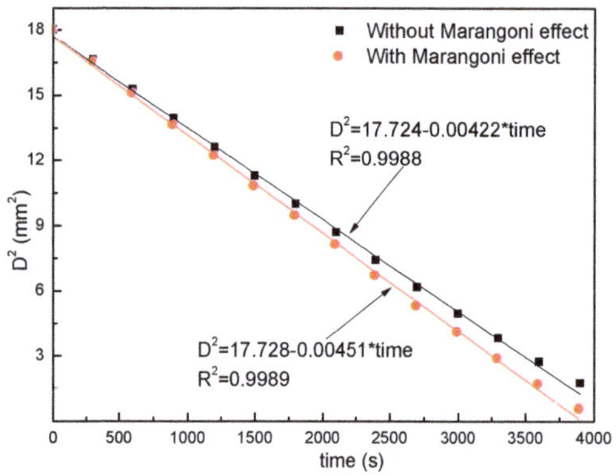

Figure 3. Variation of the squared droplet diameter with evaporation time.

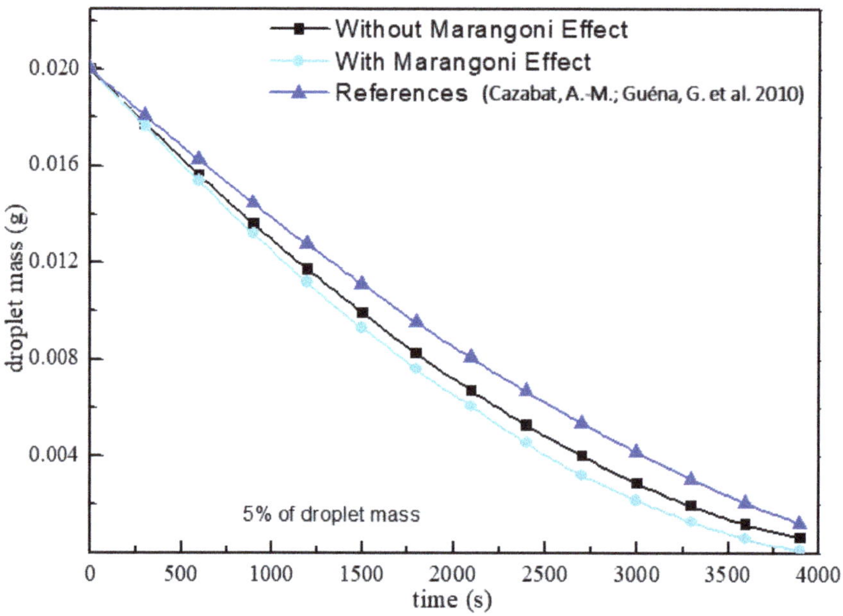

Figure 4. Comparison of simulated values with theoretical values from the literature (Reference data ia from Ref. [26]).

3. Numerical Results and Analysis

3.1. Effect of the Marangoni Effect on the Physical Field Inside the Droplet

Figure 5 shows the effect of the Marangoni effect on the physical field inside the droplet at an initial droplet volume of 20 μL and an ambient temperature of 297.15 K. Figure 5a shows the distribution of the velocity field inside the droplet. It can be seen that, when there is no Marangoni effect, the direction of flow within the droplet shows a pattern from the bottom centerline to the top of the droplet, which is primarily caused by the evaporative cooling effect at the droplet surface that causes the temperature at the bottom of the droplet to be higher than that at the top of the droplet. This generates a density difference, forming an upward buoyancy force and driving natural convection, and the surface of the droplet forms a center. Due to the small drop size, the temperature difference within the droplet is small, and the natural convection effect is small; thus, the maximum flow velocity is approximately 0.34 mm/s, and the change in flow velocity value as a function of evaporation time is not evident. The rule of thumb is similar to that of Ref. [8]; whereas, when the Marangoni effect is considered, the flow pattern within the droplet is opposite to that without the Marangoni effect. This phenomenon is rarely noticed in the relevant literature. The phenomenon is primarily due to the fact that the temperature at the bottom of the droplet is higher, leading to a lower surface tension, whereas a lower temperature at the top of the droplet leads to a greater surface tension. This leads to the formation of a gradient in surface tension, giving rise to Marangoni flow. Furthermore, the magnitude of the flow velocity increases by almost hundreds of times relative to the absence of the Marangoni effect, and the longer the evaporation time, the greater the value of the flow speed within the droplet. The temperature field distribution within the droplet is shown in Figure 5b. We have shown that, when there is no Marangoni effect, the temperature difference within the droplet is small, generating weak natural convection. That convective heat transfer to the internal heat transfer from the droplet is nearly negligible; thus, the internal temperature of the droplet is partitioned into the heat conduction and heat transfer characteristics of the droplet, and isotherms are near the horizontal and parallel distribution curves. The Marangoni effect reinforces convection within the droplet, causing the overall temperature of the droplet to rise. Simultaneously, the Marangoni effect causes a larger internal flow velocity, which reinforces convective heat transfer. From the formation of the central region, the temperature is low, and the temperature on the outer side is high.

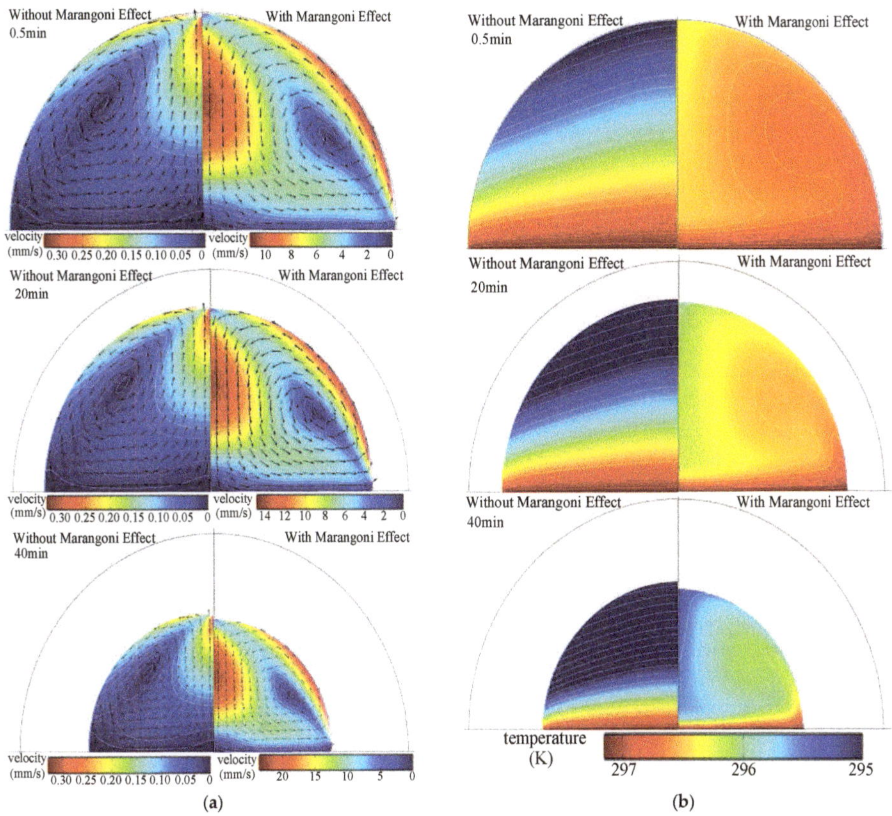

Figure 5. Differences in physical fields within droplets with and without the Marangoni effect. (**a**) Velocity field distribution within the droplet; (**b**) Temperature field distribution within the droplet.

Figures 6 and 7 show the flow velocity distribution at the boundary of the liquid surface and the centerline of the droplet, respectively, under the Marangoni effect. The change in flow velocity at the liquid surface boundary is shown in Figure 6. From these results, it can be seen that, under the Marangoni effect, the flow velocity from the top to the bottom of the droplet along the liquid surface boundary direction exhibits the law of first increasing and then decreasing. As the evaporation time is extended, the flow speed at the droplet boundary increases continuously, and the high-velocity surface is more concentrated. Figure 7 shows the change in flow velocity along the center of the droplet, which is influenced by the Marangoni effect.

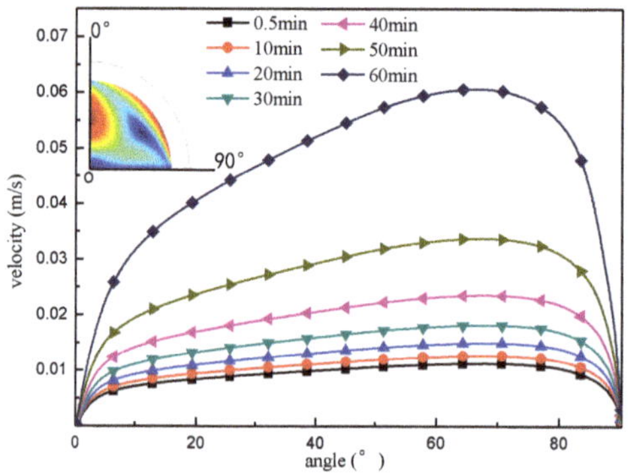

Figure 6. Droplet surface flow rate distribution.

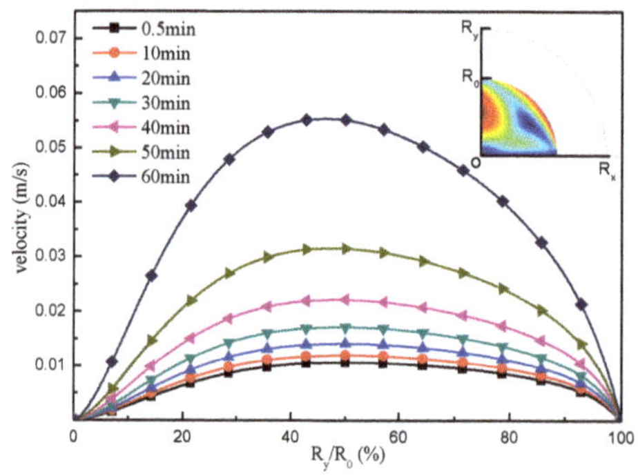

Figure 7. Flow velocity distribution on the droplet centerline.

The temperature distributions on the liquid surface boundary and droplet centerline, under the Marangoni effect, are shown in Figures 8 and 9, respectively. From the results in Figure 8 we see that, at the onset of evaporation, there is a relatively small difference in temperature between the top and bottom of the droplet. As evaporation progresses, the temperature of the top and the bottom of the droplet become a decreasing trend, and the temperature difference becomes progressively larger with a greater drop in temperature from the top of the droplet. The temperature distribution on the droplet centerline is shown in Figure 9, and the results demonstrate that the temperature of the droplet close to the substrate is higher, but along the centerline upward, the temperature remains fundamentally at a stable low temperature level. As evaporation develops, the temperature in the low-temperature region decreases in an increasingly obvious way.

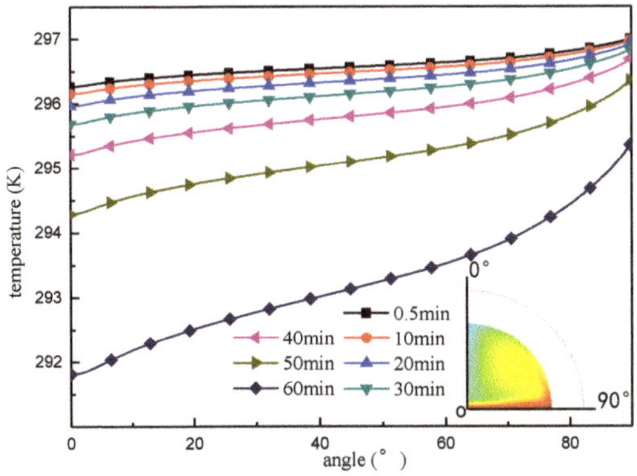

Figure 8. Droplet surface temperature distribution.

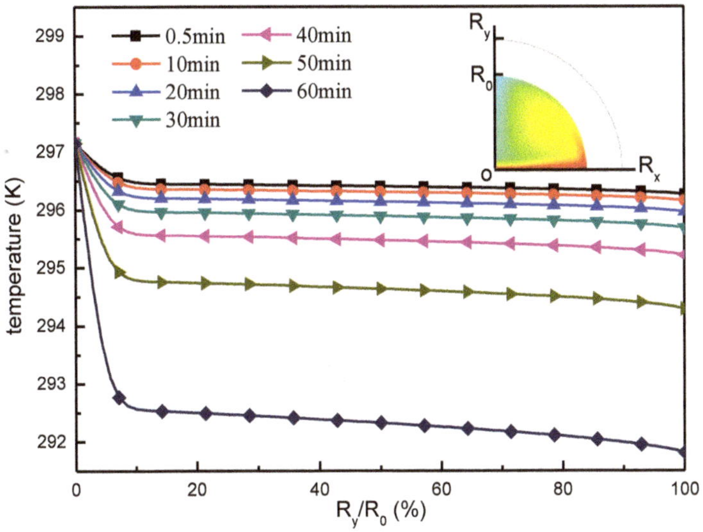

Figure 9. Temperature distribution on the centerline of the droplet.

As shown in Figure 5, without the Marangoni effect, convection within the droplet is weak, and the heat transfer is dominated by the heat conduction of the droplet. In order to further compare the contribution of the Marangoni effect to the efficiency of evaporation, in order to quantify and compare the two cases, we use the evaporation mass change of the droplet, as can be seen in Figure 10. The results show that, compared to the lack of the Marangoni effect, the effect can be used to improve the efficiency of evaporation, combined with the change in temperature and velocity fields between. The longer the duration of evaporation, the larger the temperature difference between the liquid at the top of the droplet and in the vicinity of the substrate, causing the Marangoni flow to be more intense and thus strengthening the convective heat exchange process of the droplets. The rate of evaporation is increased by almost 2.5%, compared to no Marangoni benefit. From Equation (9) one can see that, with the development of evaporation, the droplet surface area gradually decreases, and the evaporation rate is gradually reduced.

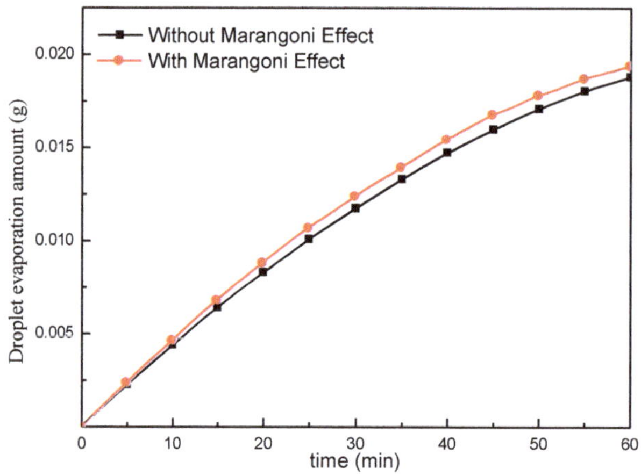

Figure 10. Effect of the presence or absence of Marangoni benefits on evaporation quality.

3.2. *Effect of Droplet Initial Radius on Droplet Evaporation*

The change in droplet radius and temperature and velocity distribution in the air domain close to the droplet during evaporation of the same droplet are shown in Figure 11. This shows that, as the vaporization process of water is a process of heat absorption, there is a decrease in air temperature near the droplet surface. In particular, the air temperature at the top of the droplet decreases in a more obvious way the greater the temperature decrease.

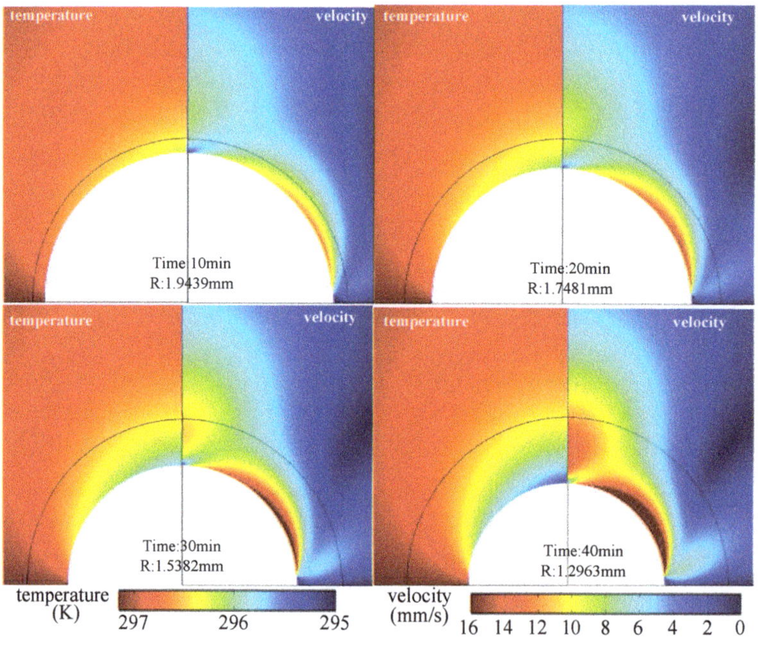

Figure 11. Distribution of temperature and velocity fields near the droplet surface as the droplet evaporates to different droplet radii.

The law of evaporation states that the velocity of the airflow on the droplet surface can also affect the evaporation process of the droplet. Figure 11 shows that the evaporation of droplets increases the air flow rate at the liquid surface, and as evaporation proceeds, the air flow rate near the liquid surface also increases; this means that this process is likely to be a self-reinforcing evaporation process. In order to analyze this question, this section compares the evaporation of droplets with different initial radii selected. Figure 11, at each instant corresponding to the initial radius of the droplet, provides a good comparison with the evaporation onset time for each droplet at the time corresponding to the chosen droplet radius. Because of the close variation in evaporation capacity during evaporation of droplets with different initial radii, to make the law more clear, we introduce the equivalent time as an abscissa. If the amount of evaporation is 0, the corresponding equivalent time corresponds to the initial evaporation time. Before this, the droplet is assumed not to evaporate. The results in Figure 12 show that the process of evaporation change of different initial radius of the droplet fundamentally coincides; essentially, the characteristic time corresponding to droplet diameter is the same. Thus, the initial radius of the droplet has little influence on the law of evaporation. This demonstrates that the initial droplet radius has little effect on the law of evaporation, and the increase in air velocity close to the liquid surface caused by evaporation does not have an obvious effect on the increased evaporation. The process of droplet evaporation is, thus, not a self-improving process.

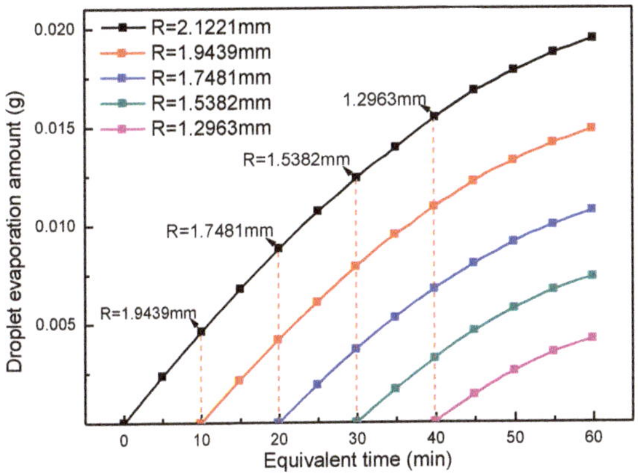

Figure 12. Evaporation at each moment for different initial droplet radii.

3.3. *Effect of Ambient Temperature on Droplet Evaporation*

One of the main factors affecting evaporation of liquids is the ambient temperature, whereas according to Equations (6)–(11), the density and surface tension of the droplet are related to the ambient temperature. To analyze the effect of the ambient temperature on the temperature and velocity distribution inside the droplet due to the Marangoni effect, in this section, the ambient temperatures of 297.15 K, 302.15 K, 307.15 K, 312.15 K, and, respectively, 317.15 K were compared to the changes in the flow velocity and temperature of the liquid surface during the evaporation of the 20 μL droplets, as shown in Figure 12. This shows that increasing the ambient temperature strengthens the Marangoni flow, and the surface liquid flow velocity increases. The increase in flow velocity is nonlinearly related to ambient temperature; the higher the temperature, the larger the increase in flow. So that the difference in the change in liquid surface temperature at different ambient temperatures could be compared, the ratio of the temperature to the ambient temperature at each part of the liquid surface will be used as an index of comparison. The results in Figure 13a show that, as the ambient temperature increases, the temperature difference at the liquid surface becomes more extreme. (The temperature at the top of the droplet is a very small value, and the temperature at the bottom edge of the liquid surface is an extreme value).

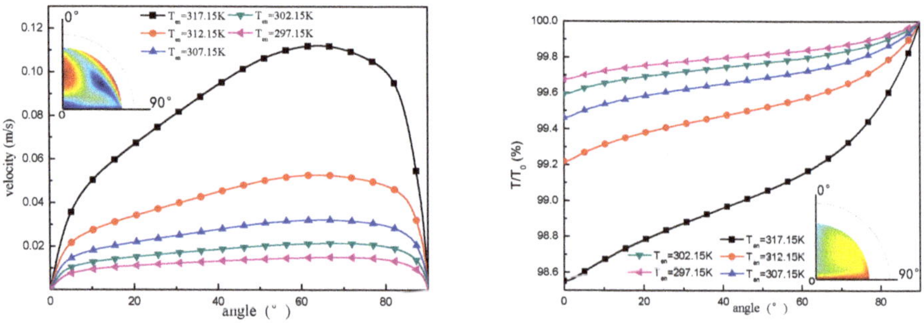

(a) (b)

Figure 13. Variation of the physical field on the liquid surface boundary line at different ambient temperatures. (**a**) Variation in droplet surface flow rate (20 min); (**b**) Change in droplet surface temperature (20 min).

To further analyze the liquid surface flow rate and temperature change process during the evaporation of droplets, a 20 µL droplet evaporated to a droplet radius of 1.9439 mm was chosen as the baseline for comparison in this section, and the results of the comparison are shown in Figure 14. As the temperature is increased linearly, the results show that the time taken to evaporate to the specified droplet radius exhibits nonlinear shortening, but the increase in flow velocity and temperature at each point on the liquid surface exhibits a quasi-linear increasing profile.

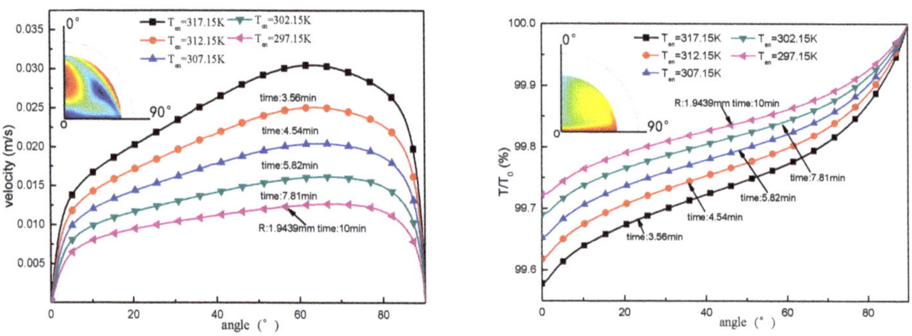

(a) (b)

Figure 14. Variation of the physical field on the liquid surface boundary line for different ambient temperature conditions when the droplets evaporate to the same radius. (**a**) Variation in droplet surface flow rate; (**b**) Change in droplet surface temperature.

Feargus [28] investigated the process of droplet evaporation on the thermally conducting substrate by constructing a coupling model and ignoring the Marangoni effect, emphasizing that the initial contact angle and substrate temperature will both affect the lifetime of the droplet. This paper considers the Marangoni effect and considers environmental temperature as a variable for extensive research, and the curve between the droplet lifetime and the environmental temperature is obtained as shown in Figure 15. It can be seen that, when the temperature is on the order of 297.15 K to 317.15 K, there is a nonlinear relationship between droplet lifetime (time required for evaporation) and ambient temperature; the higher the temperature, the shorter the droplet lifetime, but the higher the temperature, the lower the rate of droplet lifetime decay.

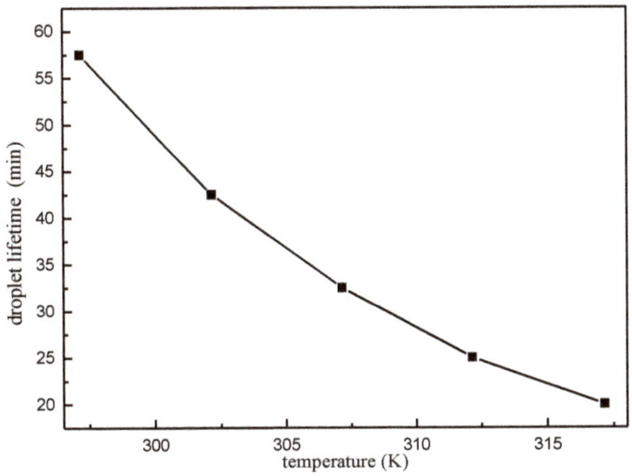

Figure 15. Droplet life versus ambient temperature variation curve.

The variation of droplet evaporative mass with time at different ambient temperatures is shown in Figure 16. This shows that an increase in the ambient temperature can significantly increase the rate of evaporation. At different ambient temperatures, the same radius of the droplet evaporates at the same time, and the relationship between increasing evaporation quality of droplets and increasing temperature is nonlinear. For each 5 °C increase in ambient temperature, the maximum increase in the rate of evaporation is approximately 22.7%.

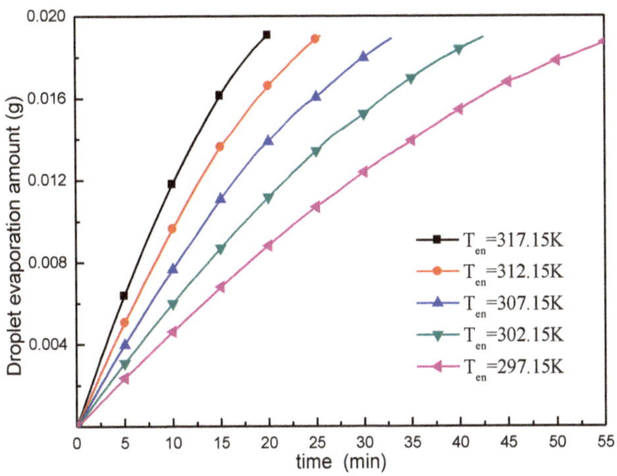

Figure 16. Variation in droplet evaporation mass at different ambient temperatures.

4. Conclusions

The influence of the Marangoni effect on droplet heat and mass transfer during droplet evaporation is primarily investigated in this paper. The next step will be to modulate the Marangoni effect by surfactant or by adding external physical fields, laying the foundation for achieving precise control of the evaporation process.

In this paper, through the construction of a numerical thermal mass coupling model for droplet evaporation and the verification of the accuracy of the model via theoretical values in the reference literature, we investigate the characteristics of the droplet flow field and the temperature field distribution at the gas–liquid interface during evaporation of tiny droplets on the wall. The main conclusions that can be drawn from the study are as follows:

(1) When considering the Marangoni effect, the evaporation process of the wall droplet leads to a slight increase in the bulk temperature of the droplet, and generation of Marangoni flow may increase the internal droplet velocity. In comparison to the lack of the Marangoni effect, the effect can increase the internal velocity of the droplet by almost 100 times and increase the rate of evaporation by approximately 2.5%.
(2) In the evaporative regime, the air temperature near the liquid surface exhibits a decreasing profile, and the flow velocity inside the air domain increases. However, increasing the rate of air flow at the liquid surface by evaporation is not enough to further increase droplet evaporation.
(3) Ambient temperatures can reinforce Marangoni flow and increase liquid surface velocity and temperature, shortening the droplet drying time; there is a non-linear variation between increasing the evaporation quality of the droplet and increasing the temperature of the droplet. For each 5 °C increase in ambient temperature, the maximum increase in the rate of evaporation is approximately 22.7%.

Author Contributions: Funding acquisition, J.D.; software, J.D.; Writing—original draft, J.D.; Investigation, H.L.; Project administration, H.L. All authors have read and agreed to the published version of the manuscript.

Funding: This research was supported by PhD research startup foundation of Northeast Electric Power University (BSJXM-2018208).

Conflicts of Interest: The authors declare no conflict of interest.

References

1. Young, T., III. An essay on the cohesion of fluids. *Philos. Trans. R. Soc. Lond.* **1805**, *95*, 65–87. [CrossRef]
2. Francescantonio, N.D.; Savino, R.; Abe, Y. New alcohol solutions for heat pipes: Marangoni effect and heat transfer enhancement. *Int. J. Heat Mass Transf.* **2008**, *51*, 6199–6207. [CrossRef]
3. Gibbs, J.W. On the Equilibrium of Heterogeneous Substances. *Trans. Connect. Acad. Arts Sci.* **1878**, *3*, 343–524. [CrossRef]
4. Su, X.; Shang, T.; Chao, Y. Evaporation characteristics of n-butanol aqueous solution droplets. *J. Therm. Sci. Technol.* **2021**, *1*, 22–27. [CrossRef]
5. Ryu, J.; Kim, J.; Park, J.; Kim, H. Analysis of vapor-driven solutal Marangoni flows inside a sessile droplet. *Int. J. Heat Mass Transf.* **2020**, *164*, 120499. [CrossRef]
6. Sun, H.-H.; Li, W.-B.; Ji, W.-J.; Dai, G.-L.; Huan, Y.; Wang, Y.-R.; Lan, D. Evaporation of saline colloidal droplet and deposition pattern. *Chin. Phys. B* **2019**, *29*, 014701. [CrossRef]
7. Diddens, C.; Li, Y.; Lohse, D. Competing Marangoni and Rayleigh convection in evaporating binary droplets. *J. Fluid Mech.* **2021**, *914*, 1–23. [CrossRef]
8. Girard, F.; Antoni, M.; Sefiane, K. On the Effect of Marangoni Flow on Evaporation Rates of Heated Water Drops. *Langmuir* **2008**, *24*, 9207–9210. [CrossRef]
9. Wang, Y.; Chen, J.; Wang, Z. Marangoni effect accompanying mass transfer processes in multiphase systems. *CIESC J.* **2013**, *64*, 124–132. [CrossRef]
10. Bozorgmehr, B.; Murray, B.T. Numerical Simulation of Evaporation of Ethanol–Water Mixture Droplets on Isothermal and Heated Substrates. *ACS Omega* **2021**, *6*, 12577–12590. [CrossRef]
11. Hu, S.; Yan, J.; Wang, J. Affecting factors of Marangoni condensation mode. *CIESC J.* **2011**, *62*, 3053–3059. [CrossRef]
12. Karpitschka, S.; Liebig, F.; Riegler, H. Marangoni Contraction of Evaporating Sessile Droplets of Binary Mixtures. *Langmuir* **2017**, *33*, 4682–4687. [CrossRef]
13. Hintermüller, M.A.; Offenzeller, C.; Knoll, M.; Tröls, A.; Jakoby, B. Parallel Droplet Deposition via a Superhydrophobic Plate with Integrated Heater and Temperature Sensors. *Micromachines* **2020**, *11*, 354. [CrossRef]
14. Ye, S.; Wu, C.-M.; Zhang, L.; Li, Y.-R.; Liu, Q.-S. Evolution of thermal patterns during steady state evaporation of sessile droplets. *Exp. Therm. Fluid Sci.* **2018**, *98*, 712–718. [CrossRef]
15. Wang, Z.; Guo, T.; Zhu, Z. Numerical study on transient flow in evaporating droplet. *J. Therm. Sci. Technol.* **2017**, *16*, 273–279.
16. Hu, D.; Wu, H. Numerical study and predictions of evolution behaviors of evaporating pinned drop-lets based on a comprehensive model. *Int. J. Therm. Sci.* **2015**, *96*, 149–159. [CrossRef]
17. Al-Sharafi, A.; Yilbas, B.S.; Ali, H.; Al-Qahtani, H. Adhesion of a water droplet on inclined hydrophilic surface and internal fluidity. *Int. J. Adhes. Adhes.* **2019**, *96*, 102464. [CrossRef]

18. Al-Sharafi, A.; Yilbas, B.S.; Hassan, G. Droplet on oil impregnated surface: Temperature and velocity fields. *Int. J. Therm. Sci.* **2019**, *146*, 106054. [CrossRef]
19. Tareie, Z.S.; Latifi, H.; Parchegani, S.; Soltanlou, K. Observation of water droplets oscillation due to laser induced Marangoni convection. *Opt. Laser Technol.* **2019**, *120*, 105749. [CrossRef]
20. Massing, J.; Mutschke, G.; Baczyzmalski, D.; Hossain, S.S.; Yang, X.; Eckert, K.; Cierpka, C. Thermocapillary convection during hydrogen evolution at microelectrodes. *Electrochim. Acta* **2018**, *297*, 929–940. [CrossRef]
21. Picknett, R.; Bexon, R. The evaporation of sessile or pendant drops in still air. *J. Colloid Interface Sci.* **1977**, *61*, 336–350. [CrossRef]
22. Gao, M.; Zhang, D.; Kong, P.; Zhang, L.-X. Experimental investigation of Marangoni convection in a sessile droplet at a constant heat flux condition. *Int. Commun. Heat Mass Transf.* **2020**, *115*, 104600. [CrossRef]
23. Nguyen, T.A.H.; Nguyen, A.V. On the Lifetime of Evaporating Sessile Droplets. *Langmuir* **2012**, *28*, 1924–1930. [CrossRef]
24. Semenov, S.; Starov, V.M.; Rubio, R.G. Evaporation of pinned sessile microdroplets of water on a highly heat-conductive substrate: Computer simulations. *Eur. Phys. J. Spec. Top.* **2013**, *219*, 143–154. [CrossRef]
25. Yao, Y. *Physical Chemistry Handbook*; Shanghai Science and Technology Press: Shanghai, China, 1985.
26. Cazabat, A.-M.; Guéna, G. Evaporation of macroscopic sessile droplets. *Soft Matter* **2010**, *6*, 2591–2612. [CrossRef]
27. Stephen, R.T. *An Introduction to Combustion: Concepts and Applications*; Tsinghua University Press: Beijing, China, 2009.
28. Schofield, F.; Pritchard, D.; Wilson, S.; Sefiane, K. The Lifetimes of Evaporating Sessile Droplets of Water Can Be Strongly Influenced by Thermal Effects. *Fluids* **2021**, *6*, 141. [CrossRef]

micromachines

MDPI

Article

Investigation of Needle Motion Profile Effect on Diesel Spray in Near-Nozzle Field

Ya Gao [1], Weidi Huang [2,*], Raditya Hendra Pratama [2], Huifeng Gong [2] and Jin Wang [1]

1 Advanced Photon Source, Argonne National Laboratory, 9700 S Cass Ave, Lemont, IL 60439, USA
2 Research Institute for Energy Conservation, National Institute of Advanced Industrial Science and Technology, Namiki 1-2-1, Tsukuba 305-8564, Japan
* Correspondence: wd.huang@aist.go.jp

Abstract: A variety of needle-motion profiles are used in diesel injectors. However, it is unclear what the underlying mechanism is to determine the needle-motion profiles and how they affect the spray dynamics. It has been of significant interest to examine how the spray dynamics will change if only altering the needle valve opening speed or closing speed while all other parameters are kept the same. The different needle-motion profiles were obtained using a piezo nozzle (Nozzle #P) and a solenoid nozzle (Nozzle #S), which have identical nozzle geometry. By utilizing the X-ray imaging technique, it was observed that the average needle valve speed of Nozzle #P was 51% higher at the opening stage but 17% lower at the closing stage than Nozzle #S. When the needle valve lift is low (approximately 200 μm), the needle valve opening speed has a crucial effect on spray dynamics. The faster needle valve opening of Nozzle #P results in a 42% larger spray spreading angle and 34% lower spray velocity at the downstream field. The spray dynamics may be controllable by properly designing the needle-motion profiles in the scenarios of the low needle lifts. However, when the needle valve is sufficiently open (approximately over 200 μm), almost identical spray characteristics were observed regardless of the needle-motion profiles.

Keywords: needle motion profiles; diesel spray; near-nozzle spray dynamics; x-ray imaging

1. Introduction

Spray characteristics are known to have a predominated effect on diesel engine performance and emissions. Many factors can affect the diesel spray characteristics, one of which refers to the injector nozzle. The injector nozzle injects fuel at a pressure of hundreds of MPa and imparts the resulting spray with specific properties. Therefore, the geometry of the injector nozzle needs to be designed properly, mainly referring to the shapes of the nozzle hole, nozzle sac, and needle valve. Studies have been extensively conducted to reveal the effect of the hole diameter [1,2], hole number [3,4], hole shape [5,6], and sac shape [7,8] on the nozzle internal flow and spray characteristics. However, it is still challenging to quantify the link between the nozzle geometry and spray characteristics because the detailed understanding of the nozzle internal flow is insufficient yet.

Another aspect regarding the injection nozzle effect on spray characteristics is the needle motion. In contrast to the geometry of the injector nozzle, the effect of needle motion on spray characteristics is more challenging to investigate. This is because, on the one hand, the transient dynamics must be considered when considering the needle motion effect. On the other hand, only limited techniques can measure the needle motion of the injector nozzle, mainly referring to the usage of optical nozzles [9,10] or optical sensors [11]. However, those methods are most suitable for low-pressure scenarios, which are different from the operating conditions of diesel nozzles. There have been some efforts to understand the needle motion effect in simulation [12–16]. However, experimental validation is vital for the simulation. In the early 2000s, X-ray diagnostics were introduced to examine the needle motion using a synchrotron X-ray source [17]. Later, by using the

X-ray imaging technique, Powell et al. [18] examined the correlations between eccentric motions of the injector needle valve and oscillations in the fuel density. Moon et al. [19] investigated the emerging jet flows and related flow breakups at different needle lifts. Viera et al. [20] reported an experimental analysis of the relationship between instantaneous partial needle lifts and the corresponding injection rate. Most recently, Raditya et al. [21] discussed the needle lift dependency of the near-nozzle spray dynamics, mainly including the injection velocity and spray spreading angle at the nozzle exit. Bae et al. [22] proposed an understanding of the fuel-temperature effect on fuel injection performance by measuring the transient needle motion and injection velocity.

Previous studies showed that the needle motion profiles are different in diesel nozzles. Two major categories of the needle motion profiles are commonly seen, including rectangle shape [18,20,23,24] and ramp shape [19,21,22,25,26]. It is also noted that the needle opening and closing speeds are varied among different nozzles. More specifically, some nozzles have almost the same needle valve opening and closing speed, but these speeds are different among the different nozzles, for instance, the results reported by [18] and [23]. As presented by [19,21], some nozzles have a higher needle valve opening speed relative to the closing speed, whereas some nozzles have the reverse trend, as seen in [22]. Despite a variety of needle motion profiles being used, it is unclear what the underlying mechanism is to determine the needle motion profiles and how they affect the spray dynamics. Therefore, it has been of significant interest to examine how the spray dynamics will change if only altering the needle valve rising speed or closing speed while all other parameters are kept the same.

In this work, the X-ray imaging technique was used to investigate the needle profile (needle valve opening speed) effect on the spray dynamics in the near-nozzle field. The different needle motion profiles were obtained using a piezo nozzle and a solenoid nozzle, which have identical nozzle holes and sac geometry. By conducting this study, it is expected to contribute to a comprehensive understanding of the needle motion effect on spray dynamics. Nevertheless, the current investigation could result in following investigations both in experiment or simulation to discuss the effect of more sophisticated needle motion profiles on spray dynamics, for instance, two-steps-like.

2. Methods

2.1. Needle Motion Measurements

The needle motion measurements were carried out using the X-ray phase-contrast imaging (XPCI) technique at the 7IDB station of the Advanced Photron Source (APS). First, a brief introduction is made to the XPCI technique. When an X-ray beam propagates through an object, X-ray energy attenuates due to the energy absorption, and the phase shift arises simultaneously due to the diffraction. To record the needle motion in a thick steel housing, the XPCI technique is unique because the phase shift of X-ray between the edges of the needle valve and fuel can provide several orders of magnitude larger contrast compared to the energy attenuation. This feature ensures a sensitive detection of the needle valve edges, which benefits the analysis of needle motion afterward. Meanwhile, the high contrast from the XPCI technique can also increase the shutter speed to avoid motion blur using the high-speed camera. Compared to the usage of optical nozzles or optical sensors, the XPCI technique can directly image the needle motion in the operating conditions of diesel nozzles, i.e., at injection pressures over 200 MPa. In addition, the image resolution using XPCI can be as high as a sub-micrometer because synchrotron X-ray is highly collimated. These imaging specifications ensure tracing the needle motion during injection well.

Based on the time-sequential images of the needle motion captured by the high-speed camera, the time traces of the needle motion can be obtained by the cross-analysis of the needle images during the injection and before the injection. More specifically, a region of interest covering the needle tip (base window) was set in a frame before the needle opened, representing the initial needle location. Second, a searching window having the same size as the base window was set in the frame to calculate (regarded as the calculation frame). Next, the correlation coefficient R between these two windows could be obtained using the two-dimensional cross-correlation calculation, as Equation (1).

$$R = \frac{\sum_m \sum_n (A_{mn} - \overline{A})(B_{mn} - \overline{B})}{\sqrt{\left(\sum_m \sum_n (A_{mn} - \overline{A})^2\right) \sum_m \sum_n (B_{mn} - \overline{B})^2}} \tag{1}$$

where A and B are the cropped images of the based window and the searching window, respectively. They have an image size of $m \times n$, and $\overline{A}$ and $\overline{B}$ are the averaged gray value of A and B. By incrementally shifting the searching window's location in the calculation frame and repeating the calculation using Equation (1), a series of R can be extracted corresponding to each searching window's location. Their maximum indicates where the searching window's cropped image perfectly matches the base window. Once the maximum of R is detected, needle lift can be known by counting the search window's displacement relative to the base window. The details of the needle motion analysis technique can be found in our work [26,27].

2.2. Near-Nozzle Spray Dynamics

The XPCI technique can also be used to measure the near-nozzle spray dynamics. There is an irregular beam mode at the APS, named the hybrid-fill mode. Figure 1 shows the pulse timing pattern of the hybrid-fill mode. It contains a single bunch with a 150 ps duration, and a 16 mA current isolated from the remaining 8 groups of 7 consecutive bunches (8 septets) by symmetrical 1.594 microseconds gaps. The group bunches have 11 mA current per group, a periodicity of 68 ns, with the total length of the bunch train being 500 ns. During the experiment, the single bunch, i.e., mode A was used for spray morphology investigation. This mode provided a limited pulse duration of 150 ps, which was beneficial for getting a still image of the spray.

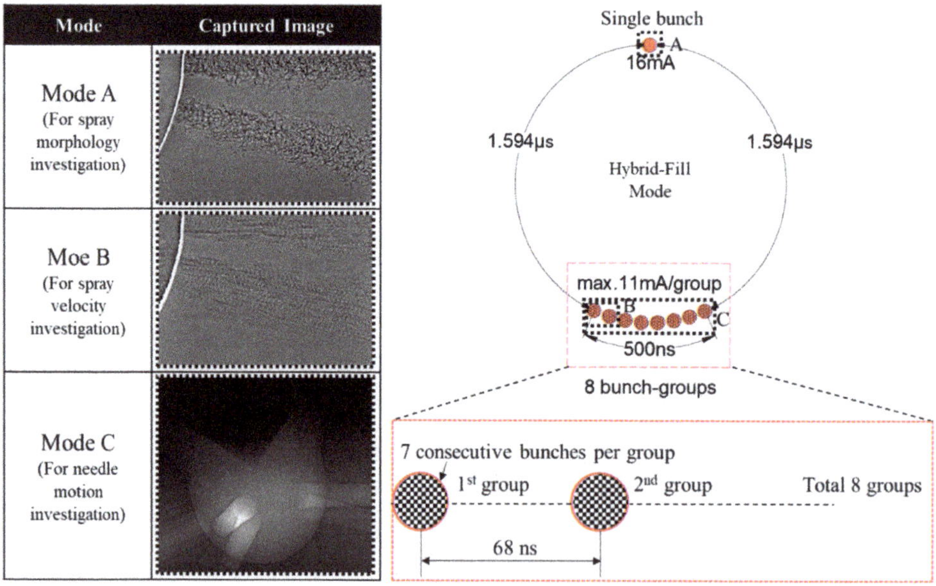

Figure 1. X-ray pulse timing patterns and modes used for single-exposure spray imaging, double-exposure spray imaging, and needle motion imaging.

A double-exposed spray image was obtained using two consecutive bunch groups, i.e., mode B in Figure 1. Based on an analysis of the double-exposed spray pattern on the image, the local velocity of the spray was known. More specifically, first, regions of interest (ROI) were selected at different locations along the spray axis of the double-exposed X-ray spray images. Then, an auto-correlation calculation was performed in the ROI at each location. The displacement vector of the imaged features during the 68 ns time interval can be obtained by detecting the relative location of the displacement peak concerning the center self-correlation peak. The details of the spray-velocity analysis technique can be found in our work [28,29].

Eight consecutive bunch groups, i.e., mode C, were used to get a bright image of needle motion inside the injector. To use different imaging modes, the pulse timing of the X-ray must be known. An imaging scan was done by shifting the camera shutter timing stepwise. Then, the pulse timing was known according to the intensity change in the sequence of images. Once the pulse timing was specified, different imaging purposes could be easily shifted using a digital delay generator.

2.3. Experiment Setup

The experiment setup is illustrated in Figure 2. Electrons circled in a storage ring under the guidance of insertion devices, i.e., different types of magnets. Whenever the electron-traveling path was changed, the electron emitted an X-ray beam along the tangent direction. The emitted X-ray beam was repositioned and refocused through a sophisticated setup of optic lenses and finally entered the experiment hutch for measuring uses. When the X-ray beam passed through the objective, a phase-contrast image was formed, which was converted using a scintillator crystal (LuAg:Ce) to visible lights. The synchrotron revolution frequency is 271 kHz. The higher the image rate, the smaller the image view field becomes. As a result, the high-speed camera used a reduced frequency signal as the triggering signal, i.e., once every four revolutions of the synchrotron. In addition, a mechanical chopper was placed upstream of the experiment system. This chopper enabled the X-ray to pass through only at the imaging instant. Otherwise, the scintillator crystal and camera could be damaged due to the heavy heat load of the X-ray. The converted visible-light image was recorded by the high-speed camera (Model SA-Z, Photron Inc., Tokyo, Japan). Several digital delay generators were used to synchronize the fuel injection and the timing of the incoming X-ray beam and the high-speed camera start-recording trigger. The view field of the images was 1.1 mm $\times$ 1.1 mm, and the spatial resolution of the images was 2.5 μm/pixel.

The details of the specification of the experimental apparatus are summarized in Table 1.

Table 1. Specification of the experimental apparatus.

Experimental Apparatus	Specification	Remark
Experimental X-rays	at APS 7IDB beamline	Special Operating Mode—hybrid fill, top-up [30]
Pressurizing fuel pump	HII 5L-DD-300	Maximum Outlet Pressure 3103 bar
Spray chamber window	Kapton film of 12 × 30 mm	Maximum Ambient Pressure11 bar
High-speed camera	FASTCAM SA-Z	1024 × 1000 pixels at 21000 fps
Scintillator	LuAG:Ce Crystals	10 mm Diameter 0.1 mm Thickness
Controller	Digital Delay Generator(DG 535)	Delay resolution of 5 psChannel Jitter of 50 ps

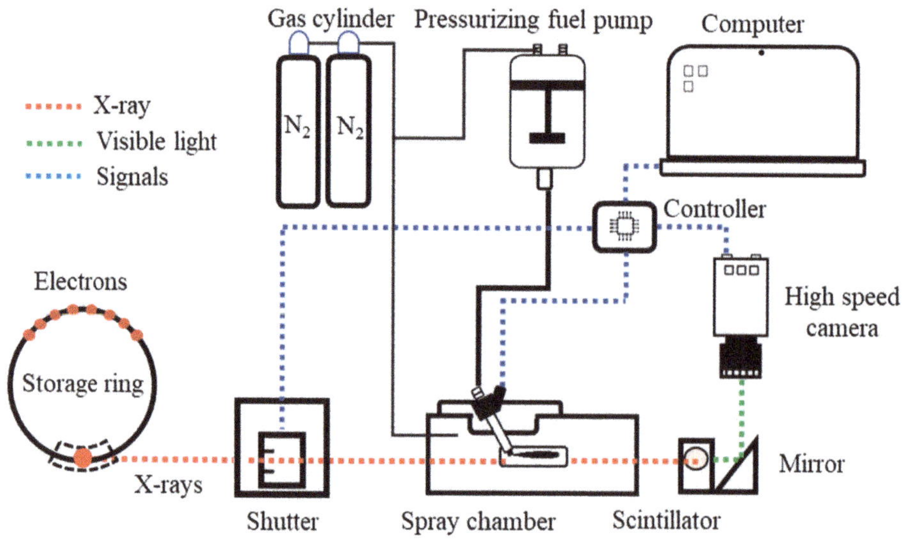

Figure 2. Schematic diagram of XPCI experiment.

2.4. Experimental Conditions

The experimental conditions are presented in Table 2. Two diesel injectors were used in this investigation, which are a 7-hole piezo injector and a 7-hole solenoid injector. These two nozzles have the same hole inlet and outlet diameters, hole length, and sac volume. A common-rail diesel injection system powered by an air-driven mechanical pump delivered fuel to the injectors. The fuel used in this investigation was US Diesel #2. Three different injection-pulse widths were applied, including 0.3, 0.5, and 1.0 ms. The varying pulse widths provided a wide range of needle motion. The results of needle motion will be presented in the next section. The initial conditions in the spray chamber were fixed to the ambient pressure and room temperature using N_2 as the ambient gas.

Table 2. Experimental conditions.

	Nozzle Geometries	
Injectors	Nozzle #P	Nozzle #S
Specification	7-hole piezo injector	7-hole solenoid injector
Hole diameters	Inlet: 0.14 mm	
	Outlet: 0.12 mm	
Hole length	0.80 mm	
Sac volume	0.22 mm^3	
	Initial conditions	
Fuels	Us Diesel #2	
Injection pressure	200 MPa (±5 MPa)	
Injection pulse widths	0.3, 0.5, 1.0 ms	
Ambient conditions	N_2 at 1.25 kg/m^3 and 25 °C	
Measuring repetitions	Ten shots	

Regarding the experimental uncertainty, the current experiment mainly used the needle motion, spreading angle, and spray velocity for a quantitative discussion based on image analysis. The image resolution, i.e., 2.5 μm in the current investigation, is the primary factor of the resulting uncertainty. To minimize the resulting trend's uncertainty, the ten-shot averaged results were used for the discussion, and the results' standard deviation (STD) was also presented in the figures. The experiment was conducted at a fixed injection pressure of 200 MPa, which may fluctuate ±5 MPa due to the pump operating. According to a theory calculation, the fluctuation in injection pressure may result in a 1.2% injection velocity fluctuation.

3. Results

3.1. Needle Motion Results

Figure 3 shows the needle motion results of the testing nozzles. Time ASOI in the figure stands for the time after the injection triggering signal. Two testing nozzles both have the ramp shape of needle motion. The longer the injection pulse width, the higher the needle can rise. First, it can be seen that the needle motion is highly repeatable, mostly having an STD of less than two μm. The needle valve of Nozzle #P opens earlier compared to Nozzle #S. This result is understandable because piezoelectric materials generally respond faster than solenoid coils when they are energized. Once energized, the needle valve keeps lifting until the injection pulse is off. The maximum opening of the needle valve appears shortly after the set pulse-width time. The existing delay is because of the inertia of the needle valve opening. The average needle valve speed of Nozzle #P is 51 % higher at the opening stage but 17% lower at the closing stage compared to Nozzle #S. It is challenging to clarify the specific mechanism causing the different needle valve speeds in the nozzles because the details of the injector upstream part are unknown. The possible reasons could rely on the characteristics of piezoelectric materials, the springs' stiffness, and other injector designs. However, the two testing nozzles having distinct needle motion profiles have been confirmed.

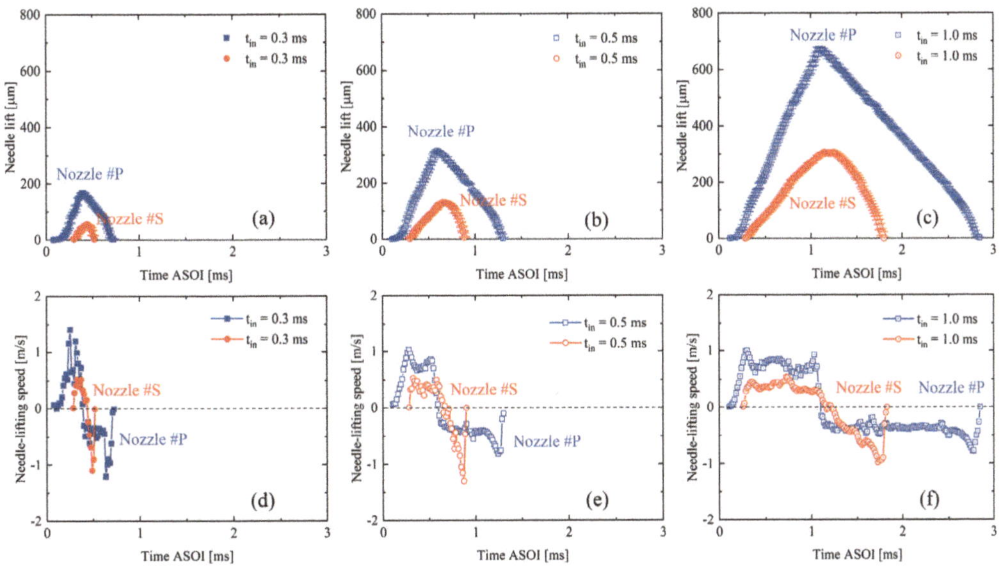

Figure 3. Results of needle motion: (**a–c**) needle lift, and (**d–f**) needle lifting speed.

3.2. Spray Center Velocity at the Nozzle Exit

The effect of needle motion profiles on the near-nozzle spray dynamics will be examined in this section. First, Figure 4 shows the time-sequence results of spray center velocities, measured at 1.5 mm from the nozzle-hole exit. The time-sequence velocity profiles are approximately rectangular. The spray velocities in Nozzle #P appear earlier than in Nozzle #S. This is understandable due to its earlier needle valve opening, as shown in Figure 3. The velocity amplitude slightly increases with the extending injection pulse width. This phenomenon is particularly apparent in Nozzle #S. The nozzle-exit spray velocity is well known to primarily rely on the pressure differential across the nozzle hole [31,32]. Since the ambient gas pressure was kept constant in this investigation, the higher the sac pressure, the larger the spray axial velocity can reach. Therefore, the results in Figure 4 indicate that the sac pressure in Nozzle #S is still increasing when the injection pulse sets 0.3 ms and 0.5 ms. Once the fuel flow entering the sac and leaving from the nozzle holes reaches balance, the sac pressure will be stabilized, i.e., reaching a steady state. As a result, the spray velocity at the nozzle exit should also keep constant. The results in Figure 4 reveal that the sac pressure in Nozzle #P reaches a steady state earlier than in Nozzle #S. After reaching the steady state, the spray center velocities of Nozzle #P and Nozzle #S are almost identical.

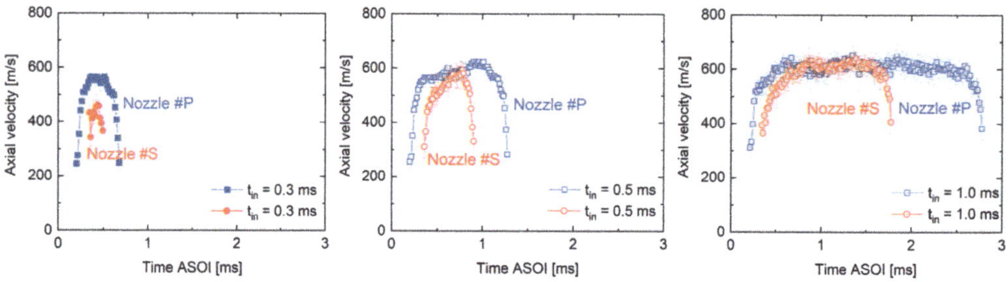

Figure 4. Time sequence results of spray axial velocities at the nozzle exit.

Figure 5 presents spray center velocities in the needle lift sequence. The spray center velocities of Nozzle #P and Nozzle #S are almost overlapped in the needle lift sequence, indicating a significant dependency of the spray center velocity on needle lift regardless of the nozzle. When the needle is partially opened, the spray center velocity increases rapidly with the lifting needle valve. At the same needle lift, Nozzle #P and Nozzle #S have a similar amplitude of spray center velocity.

The spray can step into a steady state after a specific needle lift since when the spray center velocity will barely change. This critical needle lift was found to be approximately 200 μm for these two nozzles. It is worth noting that the critical needle lift will possibly change in the different nozzle geometry. More specifically, the needle valve shape and sac volume are the other important influencing factors of the building sac pressure. On the one hand, the needle valve shape can affect the flow rate entering the sac, which changes the building sac pressure.

On the other hand, the sac pressure also relies on the sac volume because the smaller the sac volume, the faster the sac pressure can rise when the identical fuel mass enters the sac. It might also be noted that the spray velocities during the opening stage and closing stages are slightly varied even at the same needle lift. This phenomenon is likely explained as the dynamic flow lag or flow hysteresis [33,34]. In this investigation, the needle motion effect on the spray dynamics was discussed, focusing on the needle opening stage.

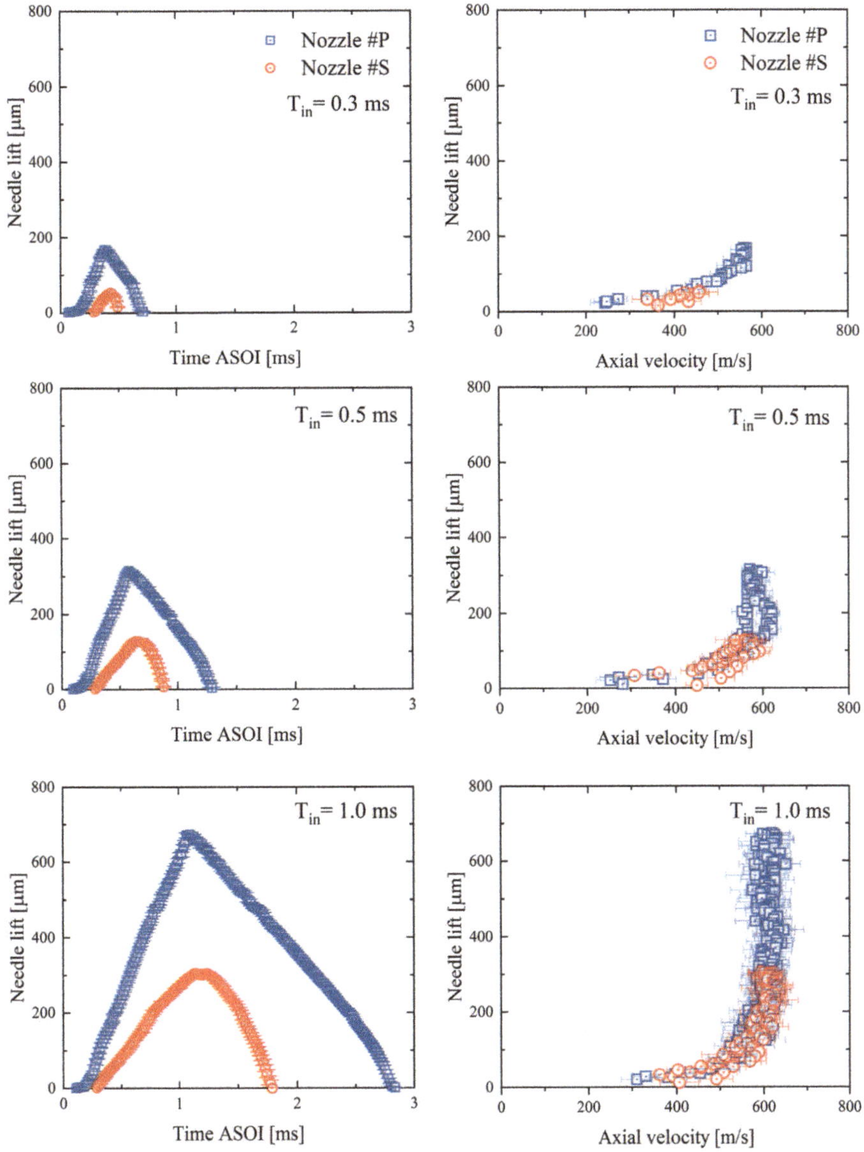

Figure 5. Needle lift sequence results of spray axial velocities at the nozzle exit.

3.3. Spray Spreading Angle at the Nozzle Exit

This section will discuss the spray spreading angle at the nozzle exit. First, Figure 6 shows spray morphology within 2 mm from the nozzle exit. The images of Nozzle #P and Nozzle #S captured at the same needle lifts were compared. At a needle lift of 50 μm, the sprays of both nozzles disperse widely, and the gap between two adjacent sprays almost disappears. A previous investigation [35] reported that the fuel flow would flush into the sac first rather than directly entering the nozzle hole under an extremely low needle lift. A vortex could form in this period, resulting in significant flow turbulence. Once this high-degree turbulent flow leaves the nozzle hole, it spreads immediately and widely. The current experimental results can be understood with the explanation above. The sprays get slimmer when the needle lift arrives at 130 μm and later at 200 μm. The spray of Nozzle #S seems to spread even less compared to Nozzle #P. When the needle lift reaches 300 μm, the sprays shrink obviously in both nozzles.

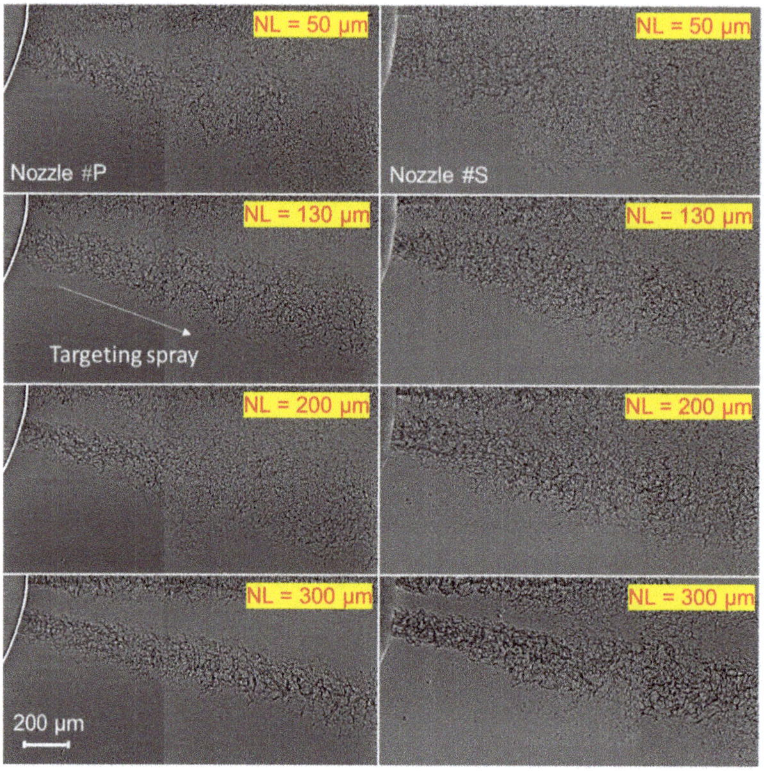

Figure 6. High-speed images of the injected sprays at the same needle lifts. (*NL* stands for the needle lift, and the targeting spray is inclined).

The spray spreading angle was further measured quantitatively based on a rectangle drawn from the nozzle exit until 1.5 mm, as shown in Figure 7a. More specifically, imaging processing codes were created to automatically detect the spray boundary at 1.5 mm from the nozzle exit. Then, the spray spreading angle was defined as the sum of the rectangle's upper (θ_1) and bottom parts angles (θ_2). There were some cases (mostly when the needle lift was less than 50 μm) where the spray spread so much that θ_1 was not recognizable. In such cases, the spray spreading angle was defined as twice θ_2 instead. This method should be reasonable because the large spreading angle is mainly due to the high-degree flow turbulence. The spray dynamics tend to distribute uniformly when the flow turbulence is high. It deserves mentioning that the conventional spray angles are measured in the downstream spray field (typically 50 mm from the nozzle exit or farther), where the aerodynamic effect has engaged in spray characteristics. In contrast, the spreading angles were derived adjacent to the nozzle exit, which should be more closely related to the nozzle's internal flow characteristics. As a result, the spreading angle and its STD are used as indicators of the internal flow turbulence.

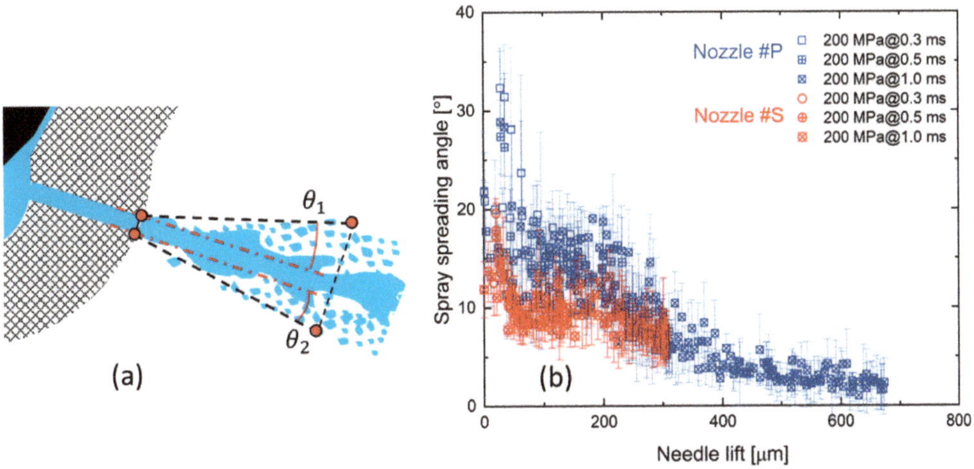

Figure 7. Comparison of the spray spreading angle between two nozzles: (**a**) the definition of spray spreading angle; (**b**) the results of spray spreading angle.

Figure 7b shows the results of the spray spreading angle. It is seen that the spreading angle of Nozzle #P is 42% larger than Nozzle #S within a needle lift ranging from 0 μm to 200 μm, approximately. When the needle lift exceeds 200 μm, the spreading angles of the two nozzles become identical. As mentioned above, the spray spreading angle mainly depends on the internal flow because the current spreading angle was measured at 1.5 mm from the nozzle exit. At such a distance, the aerodynamics should not have been effective yet. Therefore, the results shown in Figure 7b reveal that a higher degree of flow turbulence exists in Nozzle #P compared to Nozzle #S at the low needle lifts. However, the flow turbulence in both nozzles will become identical at the high needle lifts.

3.4. Spray Dynamic Evolution in the Near-Nozzle Field

The spray dynamic evolution in the near-nozzle field is further examined in this section. First, Figure 8 compares the spray velocity distributions in radial locations. The velocity distribution results were obtained at 1.5 mm from the nozzle exit under an injection pulse width of 1.0 ms. The velocity distribution at 1.5 mm from the nozzle exit is considered to represent the internal flow characteristics.

It is seen that the spray velocity distribution changes with the increasing needle lift. At the low needle lifts, i.e., 50 and 80 μm, the spray velocities are almost identical in the central region, whereas they decline rapidly in the periphery region. This distribution is typically called a top-hat shape of velocity distribution. The velocity distribution changes gradually from a top-hat shape to a parabolic shape with the lifting needle valve. This phenomenon is likely explained because the higher the flow turbulence, the more uniform the spray velocity distributes. The flow turbulence can also be known from the highly dispersed spray spreading angle. The spray becomes less turbulent at the high needle lift (200 and 300 μm), and the parabolic velocity distribution appears. Regarding the differences in the velocity distribution between the two nozzles, Nozzle #P has an overall lower velocity and flatter velocity distribution than Nozzle #S when the needle lifts are 50, 80, and 130 μm. However, as the needle lift increases, the velocity distribution of Nozzle #P and Nozzle #S becomes almost identical at the needle lift of 200 and 300 μm.

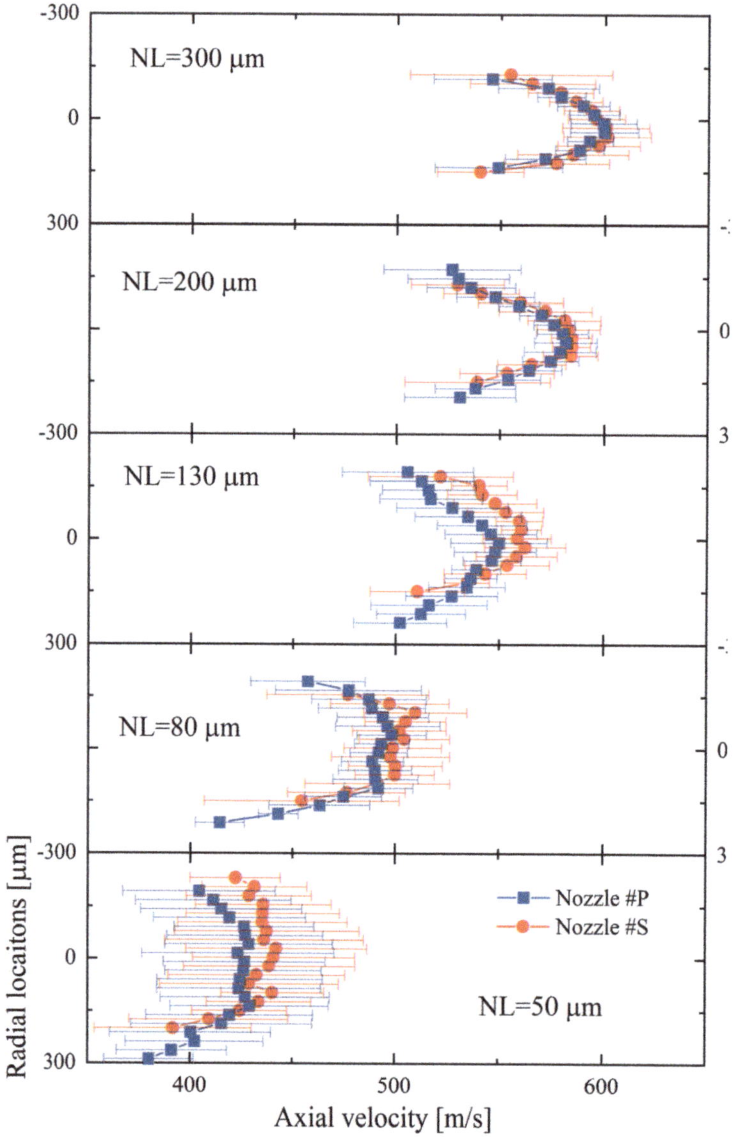

Figure 8. Nozzle−exit spray velocity distributions at different needle lifts.

In Figure 8, it is seen that the spray velocity gradually increases with the lifting needle valve. To compare the velocity distribution at different needle lifts, the spray velocities (U_r) were normalized by the spray center velocity ($U_{r=0}$) at each needle lift, as shown in Equation (2).

$$\overline{U_r} = U_r / U_{r=0} \tag{2}$$

where r is the radial location. The normalized nozzle-exit spray velocity distributions at different needle lifts are presented in Figure 9. It is evident that nozzle #S has a slightly narrower velocity distribution compared to Nozzle #P, especially at the low needle lifts. This result matches well with the observation in the spray spreading angle that the spreading angle of Nozzle #S is smaller than Nozzle #P when the needle lift is low. With the needle lift, the velocity distribution changes from top-hat to parabolic. At the 200 and 300 μm needle lifts, the velocity distributions are almost overlapped regardless of nozzle motion. It might also be noted that the fluctuation of the spray velocity, i.e., the error bar of spray velocity (ten shots were measured at each condition), is relatively larger than the results of needle motion. This is because the spray velocity, as well as the spray spreading angle, are also affected by flow turbulent instability. The STD of spray velocity decreases significantly with the lifting needle valve, indicating that the flow turbulence reduces at the high needle lifts.

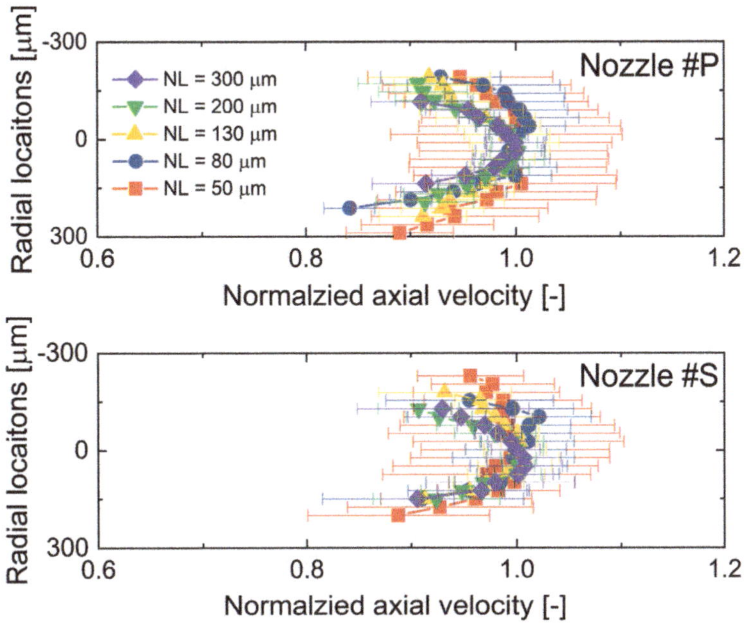

Figure 9. Normalized nozzle−exit spray velocity distributions at different needle lifts.

Figure 10 shows spray velocity evolutions along the spray axis. The results of the needle lifts of 80 and 300 μm under an injection pulse width of 1.0 ms were illustrated only as an example. The spray velocity of Nozzle #P declines rapidly along the spray axis at the 80 μm needle lift. In contrast, the spray velocities along the spray axis of Nozzle #P and Nozzle #S have become identical at the needle lift of 300 μm. This result is consistent with the results in the previous sections, showing that Nozzle #P and Nozzle #S have similar spray center velocities and spray spreading angles at the high needle lifts. Therefore, it is understandable that these two nozzles have an identical velocity distribution along the spray axis. It is worth noting that the spray velocity of Nozzle #S increases with the increasing needle lift, whereas the velocity declination rate along the spray axis is almost unchanged. Therefore, it is considered that Nozzle #S should have less flow turbulence inside the nozzle regardless of the needle lift because, as shown in Figure 7b, the spray spreading angles of Nozzle #S are similar at the needle lift of 80 μm and 300 μm. As such, the change in spray velocity is mainly related to the pressure increase in the nozzle sac.

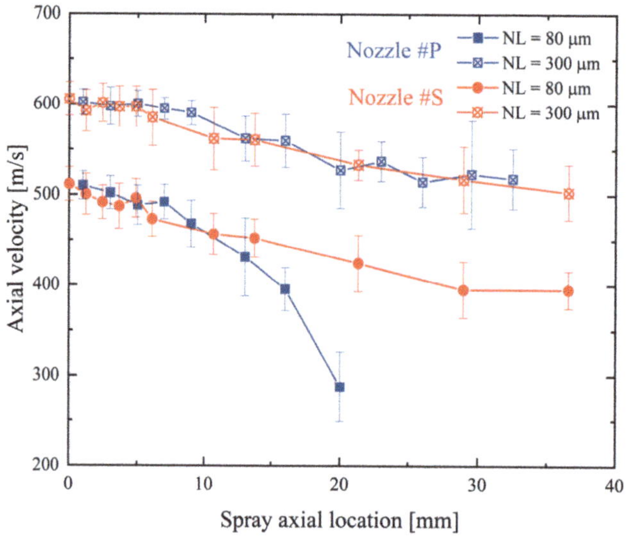

Figure 10. Spray velocity evolutions along the spray axis at the two different needle lifts.

4. Discussion

The spray dynamics of Nozzle #P and Nozzle #S can be summarized as shown in Figure 11. First, a faster needle-valve opening of Nozzle #P results in a larger spray spreading angle than Nozzle #S, even if the needle lift is the same. The larger spray spreading angle and its STD results likely indicate a higher degree of flow turbulence. As a result, faster needle valve opening was observed as well to have a slightly lower spray center velocity at the nozzle exit and a fast velocity decay along the spray axis. The injection pulse is increasingly shortened in modern diesel engines to achieve ultra-optimized combustion control. In such scenarios, the needle lifts are typically low, where the spray dynamics can be controlled by changing the needle valve opening speed according to the current investigation.

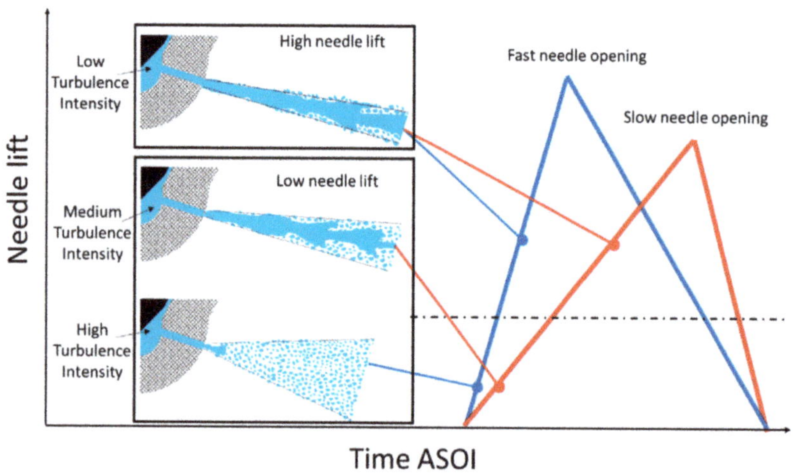

Figure 11. Needle motion profile effect on the spray dynamics.

With the needle valve further opening, a critical needle lift exists (200 μm in the current investigation) when the spray dynamics turn into a steady state. This change should relate to stabilizing sac pressure and nozzle internal flow. Two pieces of evidence are found to support this argument. One is that the spray spreading angle becomes smaller, and the other is that the STD of the spray spreading angle and spray velocity are largely reduced compared to the low needle lift. When the needle valve exceeds the critical lift, the spray dynamics will keep constant, i.e., being insensitive to the needle valve opening speed and the height of the needle lift. Almost identical spray characteristics were observed regardless of needle motion profiles, including the spray center velocities, spreading angles at the nozzle exit, and the spray dynamic evolution in the near-nozzle field.

Nozzle #P and Nozzle #S have identical nozzle holes and sac geometry but different needle motion profiles. More specifically, Nozzle #P has a faster needle opening and slower needle closing compared to Nozzle #S. Regarding the mechanism of why Nozzle #P, i.e., the faster needle valve opening tends to cause a larger spray spreading angle and faster velocity decay along the spray axis at the low needle lifts, a detailed simulation might be needed in the future. However, as a simplified analysis, the near-nozzle spray dynamics are typically regarded to rely on the nozzle internal flow characteristics. Second, the internal flow turbulence mainly relates to the flow momentum inside the nozzle sac, which further links to the time gradient of the sac pressure change [26,35]. Then, the time gradient of sac pressure change can be correlated with the needle lift using Equation (3),

$$\frac{dp}{dt} = \frac{dp/dl}{dt/dl} = V_{NL} \times \frac{dp}{dl} \tag{3}$$

where p and l are the sac pressure and needle lift, respectively. V_{NL} stands for the needle valve opening speed. It can be known from Equation (2) that the faster the needle valve opens, the larger the time gradient of the sac pressure change is. Therefore, it is reasonable that a faster needle opening can result in more flow turbulence. On the other hand, the spray velocity also depends on the absolute sac pressure. Once the sac pressure reaches a balance, i.e., being identical to the common rail pressure, the time gradient of the sac pressure will not change anymore. Therefore, the flow turbulence in the nozzle significantly reduces, making the spray dynamics irrelevant to the needle valve opening speed and the height of the needle lift.

5. Conclusions

A variety of needle motion profiles are extensively used in diesel injectors. However, it is unclear what the underlying mechanism is to determine the needle motion profiles and how they affect the spray dynamics. In this work, the effect of the needle motion profiles on diesel spray in the near-nozzle field was investigated. First, two needle motion profiles were obtained using a piezo nozzle (Nozzle #P) and a solenoid nozzle (Nozzle #S), which have almost identical nozzle holes and sac geometry. Then, by utilizing the X-ray imaging technique, the needle motion, spray center velocities, spreading angles at the nozzle exit, and the spray dynamic evolution in the near-nozzle field were quantitatively measured. The results were further discussed in detail. The key findings of this study can be summarized below.

1. Nozzle #P has a shorter opening delay compared to Nozzle #S. This result should rely on the fact that piezoelectric materials generally respond faster than solenoid coils when they are energized. The average needle valve speed of Nozzle #P was 51% higher at the opening stage but 17% lower than Nozzle #S at the closing stage.
2. When the needle valve lift is low (approximately 200 μm in the current investigation), the needle valve opening speed has a crucial effect on spray dynamics. The faster needle valve opening of Nozzle #P results in a 42% larger spray spreading angle and 34% lower spray velocity at the downstream field. These results reveal significant differences in the in-nozzle flow characteristics due to differences in needle valve opening speeds. When the needle valve is sufficiently open (approximately over

200 μm), almost identical spray characteristics were observed regardless of the needle motion profiles.
3. According to a simplified analytical analysis, the faster needle opening speed can result in a larger time gradient of sac pressure change, a more rapid change in the flow energy inside the nozzle, and a higher degree of flow turbulence. However, as the stabilizing sac pressure, the flow turbulence in the nozzle greatly reduces. Therefore, the spray dynamics become insensitive to the needle motion profiles at the high needle lifts.
4. The needle motion profile effect is found to be particularly evident at the low needle lifts. The injection pulse is increasingly shortened in modern diesel engines to achieve ultra-optimized combustion control. In such scenarios, the needle lifts are typically low, where the spray dynamics can be controlled by adequately designing the needle motion profiles.

Author Contributions: Conceptualization, Y.G., W.H., R.H.P. and J.W.; methodology, Y.G., W.H., R.H.P. and J.W.; software, Y.G.; validation, R.H.P. and H.G.; formal analysis, Y.G.; investigation, Y.G., W.H., R.H.P., and H.G.; resources, J.W.; data curation, Y.G., W.H. and R.H.P.; writing—original draft preparation, Y.G.; writing—review and editing, W.H.; visualization, Y.G. and R.H.P.; supervision, W.H. and J.W.; project administration, W.H. and J.W.; funding acquisition, W.H. All authors have read and agreed to the published version of the manuscript.

Funding: This research received no external funding.

Data Availability Statement: Not applicable.

Acknowledgments: We also thank Seoksu Moon of INHA University and other collaborators (Jeeyeon Jeon, Qing Zhang) for their assistance during the experiments.

Conflicts of Interest: The authors declare no conflict of interest.

References

1. Yao, C.; Geng, P.; Yin, Z.; Hu, J.; Chen, D.; Ju, Y. Impacts of nozzle geometry on spray combustion of high pressure common rail injectors in a constant volume combustion chamber. *Fuel* **2016**, *179*, 235–245. [CrossRef]
2. Nishida, K.; Zhu, J.; Leng, X.; He, Z. Effects of micro-hole nozzle and ultra-high injection pressure on air entrainment, liquid penetration, flame lift-off and soot formation of diesel spray flame. *Int. J. Engine Res.* **2017**, *18*, 51–65. [CrossRef]
3. Sayin, C.; Gumus, M.; Canakci, M. Influence of injector hole number on the performance and emissions of a di diesel engine fueled with biodiesel-diesel fuel blends. *Appl. Therm. Eng.* **2013**, *61*, 121–128. [CrossRef]
4. Moon, S.; Gao, Y.; Park, S.; Wang, J.; Kurimoto, N.; Nishijima, Y. Effect of the number and position of nozzle holes on in- and near-nozzle dynamic characteristics of diesel injection. *Fuel* **2015**, *150*, 112–122. [CrossRef]
5. Ning, W.; Reitz, R.D.; Diwakar, R.; Lippert, A.M. *A Numerical Investigation of Nozzle Geometry and Injection Condition Effects on Diesel Fuel Injector Flow Physics*; SAE Technical Papers; SAE International: Warrendale, PA, USA, 2008; pp. 776–790. [CrossRef]
6. Payri, R.; Salvador, F.J.; De la Morena, J.; Pagano, V. Using a one-dimensional spray model to improve liquid length and ignition delay estimations for diesel flames. *Appl. Therm. Eng.* **2017**, *124*, 1090–1102. [CrossRef]
7. Lee, W.G.; Reitz, R.D. A numerical investigation of transient flow and cavitation within minisac and VCO diesel injector nozzles. In Proceedings of the ASME 2009 Internal Combustion Engine Division Spring Technical Conference, Milwaukee, WI, USA, 3–6 May 2009; pp. 643–653. [CrossRef]
8. Salvador, F.J.; Carreres, M.; Jaramillo, D.; Martínez-López, J. Comparison of microsac and VCO diesel injector nozzles in terms of internal nozzle flow characteristics. *Energy Convers. Manag.* **2015**, *103*, 284–299. [CrossRef]
9. He, Z.; Zhang, Z.; Guo, G.; Wang, Q.; Leng, X.; Sun, S. Visual experiment of transient cavitating flow characteristics in the real-size diesel injector nozzle. *Int. Commun. Heat Mass Transf.* **2016**, *78*, 13–20. [CrossRef]
10. Xuan, T.; Sun, Z.; Lu, P.; Zhong, W.; Si, Z.; He, Z.; Wang, Q.; Chen, Z.; Guan, W. Optical study on needle lift and its effects on reacting diesel sprays of a single hole solenoid injector. *Therm. Sci.* **2021**, *25*, 3763–3773. [CrossRef]
11. Amirante, R.; Catalano, L.A.; Coratella, C. A new optical sensor for the measurement of the displacement of the needle in a common rail injector. In Proceedings of the 11th International Conference on Engines & Vehicles, Naples, Italy, 15–19 September 2013; SAE Technical Paper. SAE International: Warrendale, PA, USA, 2013; Volume 6. [CrossRef]
12. Gu, F.; Ball, A.D. Diesel injector dynamic modelling and estimation of injection parameters from impact response Part 1: Modelling and analysis of injector impacts. *Proc. Inst. Mech. Eng. Part D J. Automob. Eng.* **1996**, *210*, 293–302. [CrossRef]

13. Arcoumanis, C.; Gavaises, M.; Nouri, J.M.; Abdul-Wahab, E.; Horrocks, R.W. Analysis of the flow in the nozzle of a vertical multi-hole diesel engine injector. In Proceedings of the International Congress & Exposition, Detroit, MI, USA, 23–26 February 1998; SAE International. Warrendale, PA, USA, 1998. [CrossRef]
14. Kouremenos, D.A.; Hountalas, D.T.; Kouremenos, A.D. Development and validation of a detailed fuel injection system simulation model for diesel engines. In Proceedings of the International Congress & Exposition, Detroit, MI, USA, 1–4 March 1999; SAE Technical Paper. SAE International: Warrendale, PA, USA, 1999; Volume 108, pp. 694–702. [CrossRef]
15. Chiavola, O.; Palmieri, F. Modeling needle motion influence on nozzle flow in high pressure injection system. In Proceedings of the SAE World Congress & Exhibition, Detroit, MI, USA, 16–19 April 2007; SAE Technical Paper. SAE International: Warrendale, PA, USA, 2007. [CrossRef]
16. Payri, R.; Gimeno, J.; Martí-Aldaraví, P.; Alarcón, M. Numerical simulation of needle movement nozzle flow coupled with spray for a diesel injector using an Eulerian spray atomization model. *J. Braz. Soc. Mech. Sci. Eng.* **2017**, *39*, 2585–2592. [CrossRef]
17. Fezzaa, K.; Lee, W.K.; Cheong, S.K.; Powell, C.F.; Wang, J.; Lai, M.C.; Powell, C.F.; Lai, M.C.; Fezzaa, K.; Lee, W.K.; et al. Ultrafast X-ray Phase-Enhanced Microimaging for Visualizing Fuel Injection Process. In Proceedings of the 7th International Conference on Engines for Automobile, Capri Naples, Italy, 11–16 September 2005; SAE Technical Paper. SAE International: Warrendale, PA, USA, 2005; pp. 0–4.
18. Powell, C.F.; Kastengren, A.L.; Liu, Z.; Fezzaa, K. The effects of diesel injector needle motion on spray structure. In Proceedings of the ASME 2009 Internal Combustion Engine Division Fall Technical Conference, Lucerne, Switzerland, 27–30 September 2009; Volume 133, pp. 207–217. [CrossRef]
19. Moon, S.; Zhang, X.; Gao, J.; Fezzaa, K.; Durfresne, E.; Wang, J.; Xie, X.; Wang, F.; Lai, M.-C. Morphological Exploration of Emerging Jet Flows From Multi-Hole Diesel Injectors At Different Needle Lifts. *At. Sprays* **2015**, *25*, 375–396. [CrossRef]
20. Viera, J.P.; Payri, R.; Swantek, A.B.; Duke, D.J.; Sovis, N.; Kastengren, A.L.; Powell, C.F. Linking instantaneous rate of injection to X-ray needle lift measurements for a direct-acting piezoelectric injector. *Energy Convers. Manag.* **2016**, *112*, 350–358. [CrossRef]
21. Pratama, R.H.; Huang, W.; Moon, S. Unveiling needle lift dependence on near-nozzle spray dynamics of diesel injector. *Fuel* **2021**, *285*, 119088. [CrossRef]
22. Bae, G.; Choi, S.; Lee, S.; Moon, S.; Wang, J.; Kim, K.; Lee, J.; Kang, J. Experimental investigation of fuel temperature effects on transient needle motion and injection velocity of solenoid type diesel injector. *Int. J. Heat Mass Transf.* **2021**, *181*, 121838. [CrossRef]
23. Huang, W.; Moon, S.; Gao, Y.; Wang, J.; Ozawa, D.; Matsumoto, A. Hole number effect on spray dynamics of multi-hole diesel nozzles: An observation from three- to nine-hole nozzles. *Exp. Therm. Fluid Sci.* **2019**, *102*, 387–396. [CrossRef]
24. Huang, W.; Moon, S.; Gao, Y.; Li, Z.; Wang, J. Eccentric needle motion effect on near-nozzle dynamics of diesel spray. *Fuel* **2017**, *206*, 409–419. [CrossRef]
25. Battistoni, M.; Xue, Q.; Som, S.; Pomraning, E. Effect of off-axis needle motion on internal nozzle and near exit flow in a multi-hole diesel injector. *SAE Int. J. Fuels Lubr.* **2014**, *7*, 167–182. [CrossRef]
26. Huang, W.; Moon, S.; Ohsawa, K. Near-nozzle dynamics of diesel spray under varied needle lifts and its prediction using analytical model. *Fuel* **2016**, *180*, 292–300. [CrossRef]
27. Huang, W.; Moon, S.; Jeon, J.; Gao, Y.; Li, Z.; Wang, J. A Model for Predicting the Transient Spray Dynamics of Engine Multiple Injections. *Proc. Korean Soc. Liq. Fine Chem.* **2017**, *1*, 72–75.
28. Pratama, R.H.; Huang, W.; Moon, S.; Wang, J.; Murayama, K.; Taniguchi, H.; Arima, T.; Sasaki, Y.; Arioka, A. Hydraulic flip in a gasoline direct injection injector and its effect on injected spray. *Fuel* **2022**, *310*, 122303. [CrossRef]
29. Gong, H.; Huang, W.; Gao, Y.; Wang, J.; Arioka, A.; Sasaki, Y. End-of-injection fuel dribbling dynamics of multi-hole GDI injector. *Fuel* **2022**, *317*, 123406. [CrossRef]
30. Storage Ring Operation Modes. Available online: https://ops.aps.anl.gov/SRparameters/node5.html (accessed on 9 July 2020).
31. Payri, R.; García, J.M.; Salvador, F.J.; Gimeno, J. Using spray momentum flux measurements to understand the influence of diesel nozzle geometry on spray characteristics. *Fuel* **2005**, *84*, 551–561. [CrossRef]
32. Monsalve-Serrano, J.; Belgiorno, G.; Di Blasio, G.; Guzmán-Mendoza, M. 1D simulation and experimental analysis on the effects of the injection parameters in methane–diesel dual-fuel combustion. *Energies* **2020**, *13*, 3734. [CrossRef]
33. Payri, R.; Salvador, F.J.; Gimeno, J.; de la Morena, J. Study of cavitation phenomena based on a technique for visualizing bubbles in a liquid pressurized chamber. *Int. J. Heat Fluid Flow* **2009**, *30*, 768–777. [CrossRef]
34. Margot, X.; Hoyas, S.; Fajardo, P.; Patouna, S. CFD study of needle motion influence on the exit flowconditions of single-hole injectors. *At. Sprays* **2011**, *21*, 31–40. [CrossRef]
35. Moon, S.; Huang, W.; Wang, J. First observation and characterization of vortex flow in steel micronozzles for high-pressure diesel injection. *Exp. Therm. Fluid Sci.* **2019**, *105*, 342–348. [CrossRef]

micromachines

Article

Localized Characteristics of the First Three Typical Condensation Frosting Stages in the Edge Region of a Horizontal Cold Plate

Long Zhang [1], Mengjie Song [1,*], Christopher Yu Hang Chao [2], Chaobin Dang [3] and Jun Shen [1]

1 Department of Energy and Power Engineering, School of Mechanical Engineering, Beijing Institute of Technology, Beijing 100081, China
2 Department of Building Environment and Energy Engineering & Department of Mechanical Engineering, The Hong Kong Polytechnic University, Hong Kong 999077, China
3 Graduate School of Engineering, University of Fukui, 3-9-1 Bunkyo, Fukui 910-8507, Japan
* Correspondence: mengjie.song@gmail.com; Tel.: +86-188-1172-6830

Abstract: Condensation frosting usually causes a negative influence on heat exchangers employed in engineering fields. As the relationships among the first three typical condensation frosting stages in the edge regions of cold plates are still unclear, an experimental study on the localized condensation frosting characteristics in the edge region of a cold plate was conducted. The edge effects on the water droplet condensation (WDC), water droplet frozen (WDF) and frost layer growth characteristics were quantitatively investigated. The results showed that the number of droplets coalescing in the edge-affected regions was around 50% greater than in the unaffected regions. At the end of the WDC stages, the area-average equivalent contact diameter and coverage area ratio of water droplets in the edge-affected regions were 2.69 times and 11.6% greater than those in the unaffected regions under natural convection, and the corresponding values were 2.24 times and 9.9% under forced convection. Compared with the unaffected regions, the WDF stage duration in the edge-affected regions decreased by 63.6% and 95.3% under natural and forced convection, respectively. Additionally, plate-type and feather-type frost crystals were, respectively, observed in natural and forced convection. The results of this study can help in the better understanding of the condensation frosting mechanism on a cold plate, which provides guidelines for optimizing the design of heat exchanger structures and system control strategies facing frosting problems.

Keywords: condensation frosting; edge effect; droplet size and distribution; freezing wave propagation; frost layer surface roughness

1. Introduction

As a type of efficient heat transfer device, heat exchangers have been widely applied in many engineering applications, such as air source heat pumps, refrigerators, and liquid natural gas air vaporizers [1,2]. However, frost may accumulate on the surfaces of heat exchangers when their surface temperatures are lower than 0 °C and in a humid ambient.

Frost may facilitate heat transfer at the early frosting stage of the heat exchanger due to the increased heat transfer area and turbulent flow caused by frost [3].

On the other hand, the gradually accumulated frost may block the air passages of a heat exchanger and increase the thermal resistance between the air and the heat exchanger. As a result, the heat transfer performance of the heat exchanger deteriorates as frosting progresses [4]. Accordingly, the operational efficiency of the related system decreases notably. For example, the heating capacity and coefficient of performance of an air source heat pump unit decreased by 45.1% and 42.5%, respectively, after a 60 min frosting duration [5]. To resolve the frosting problem of heat exchangers, a considerable amount of literature has been published over previous decades.

In general, studies on the frosting characteristics of heat exchangers have mainly focused on the following two aspects. The first is the thermal hydraulic performances of heat exchangers with different fin types under frosting conditions. Liu et al. [6] found that the total heat transfer rate and coefficient of a heat exchanger with perforated fins were 38.9% and 31.8% higher than those with plain fins, but the frost mass of the former was 46.7% higher than that of the latter. Silva et al. [7] demonstrated that frost morphology on the heat exchanger at -5 °C was quite different from that at -10 °C, and the air side pressure dropped significantly depending on frost morphology. Additionally, the air face velocity of the heat exchanger increased as frosting progressed, which may increase the heat transfer coefficient [8]. Moallem et al. [9] developed a semi-empirical correlation of the j factor for a microchannel heat exchanger. The experimental results also indicated that the air side heat transfer coefficient increased at the early frosting stage and decreased at the later frosting stage. The second notable aspect of the studies regarded the frost growth and distribution characteristics on heat exchangers. In most of the related studies, the phenomena of uneven frost distribution along the air flow direction due to the leading-edge effect have been mentioned [10–12]. Recently, Zhang et al. [13] suggested that frost on a finned tube heat exchanger could be divided into frost on the edge of windward fins, and the surfaces of fins and tubes; after that, the effects of air temperature, relative humidity, and velocity on these two parts of the frost layer were further quantitatively investigated [14]. A frosting model of a finned tube heat exchanger based on the uneven frost distribution characteristic was also developed [15]. Regarding the complex structures of heat exchangers, it is hard to reveal frosting mechanisms in terms of frosting studies on heat exchangers.

To better understand frosting mechanisms, more attention has been paid to frosting on simple geometries, especially on a flat plate [16]. In terms of the physical process of the early frosting stage, frosting can be divided into desublimation frosting and condensation frosting. Desublimation frosting indicates that water vapor desublimates directly as ice crystals, while condensation frosting indicates that water vapor condenses as water droplets first, and then solidifies [17]. The critical conditions for desublimation and condensation frosting mainly depend on the cold surface contact angle, surface temperature, and water vapor partial pressure [18]. Generally, desublimation frosting usually occurs when the surface temperature [19] or the water vapor partial pressure [20] is extremely low. Since the conditions favorable for condensation frosting are much more accessible than those for desublimation frosting, frosting phenomena in practical engineering are mostly condensation frosting. In terms of the physical process of condensation frosting, it can be divided into four stages. The first is the water droplet condensation (WDC) stage, during which water vapor nucleates as tiny water droplets on the cold surface and keeps growing [21]. The second is the water droplet frozen (WDF) stage, during which water droplets solidify and interdroplet freezing occurs [22]. The third is the frost layer growth (FLG) stage, during which frost crystals grow on the surface of solidified water droplets in three dimensions and form a porous layer [23]. The last is the frost layer full growth (FLFG) stage, during which the frost layer surface melts and freezes periodically, creating a more compact frost layer [24].

In practical engineering, since the defrosting operation usually begins before frosting enters the final stage, the related studies have mainly focused on the first three stages. The WDC stage characteristics of embryo formation [21], droplet growth and coalescence [25], jumping [26], self-propulsion [27], and distribution density [28] have been further investigated. The WDF stage begins with the solidification of one of these water droplets on a cold plate. A complete solidification process of a single water droplet includes subcooling, recalescence, freezing, and subcooling stages [29,30]. Additionally, a notable phenomenon of freezing wave propagation can be observed once one of these droplets solidifies. In 2013, Guadarrama-Cetina first proposed that the freezing wave was mainly propagated by ice bridges, which grew from frozen droplets to liquid droplets [22]. Apart from ice bridges, frost halo [31] and freezing rain [32] have also been demonstrated as other types of propagation mechanisms. The velocity of freezing wave propagation among water droplets was much higher than that of the freezing front of a single water droplet. For the FLG stage, numerical and experimental studies on spatial and temporal characteristics of the frost layer, such as frost thickness, density, and surface roughness, were conducted under different frosting conditions [33,34]. Apart from intelligent models [35], most related frosting models required initial frost thickness and density [36], which are strongly associated with the characteristic parameters at the end of the WDF stage. In particular, FLG characteristics in the edge region are quite different from those in the interior region [37]. However, it can be noted that all the published studies on WDC and WDF stages focused on the interior region of cold plates. As a result, the edge effects on WDC and WDF characteristics are still unclear, as well as the relationships among WDC, WDF, and FLG stages in the edge regions of cold plates.

Therefore, a comprehensive study on the localized condensation frosting characteristics in the edge region of a horizontal cold plate is carried out in this paper. The edge effects on the WDC, WDF and FLG characteristics and their interplay were quantitatively investigated. The specifications of the experimental set-up, experimental procedures, and conditions, as well as data reductions and error analysis, are firstly presented. This is followed by the detailed results and discussion of localized WDC, WDF, and FLG characteristics. Finally, conclusions are given. The results of this paper reveal not only the edge effect on the condensation frosting characteristics, but also the internal relationships among WDC, WDF, and FLG stages in the edge regions of cold plates. Accordingly, the results of this study can help in better understanding of the condensation frosting mechanism on a cold plate, which provides guidelines for optimizing the design of heat exchanger structures and system control strategies facing frosting problems.

2. Methodology

2.1. Experimental Set-Up

A specific experimental set-up was constructed to investigate the localized condensation frosting characteristics in the edge region of a horizontal cold plate. The photograph of the experimental set-up and its schematics are illustrated in Figure 1a,b, respectively. The experimental set-up comprised the air conditioning section and the test section. In the air conditioning section, a constant flow of air with specific temperature, relative humidity, and velocity was processed by a typical air conditioning unit. The air temperature, relative humidity, and velocity were measured by thermocouples (T-TT-36) with an accuracy of ±0.15 °C, humidity sensor (Rotronic HF320) with an accuracy of ±2%, and air velocity sensor (EE650) with an accuracy of ±0.2 m/s, respectively. The air temperature, air relative humidity and air velocity sensors were installed at the inlet of the test section, around 100 mm away from the experimental plate, as illustrated in Figure 1a,b. Using PID controller A, the controlling accuracy of the air temperature, relative humidity, and velocity were ±0.2 °C, ±3.5%, and ±0.2 m/s, respectively. In the test section, the experimental cold plate was horizontally attached to a thermoelectric module and a micro heat exchanger by thermal grease. The dimension of each component is presented in Figure 1c. The static contact angle of the cold plate was 74.6° ± 2.5°. Since the material of the cold plate was copper and its thickness was 2 mm, the difference in temperature between the upper and bottom surfaces of the cold plate was less than 0.1 °C due to little thermal resistance. Therefore, the upper surface temperature of the experimental cold plate could be evaluated by four uniformly distributed superfine thermocouples that were installed at the bottom of the plate. By using PID controller B, the controlling accuracy of the cold plate surface temperature was ±0.2 °C. In general, the air temperature, air relative humidity, air velocity and cold plate temperature could be maintained in the range of 5.0~25.0 °C (±0.2 °C), 30~90% (±3.5%), 0~3 m/s (±0.2 m/s), and −30.0~30.0 °C (±0.2 °C), respectively.

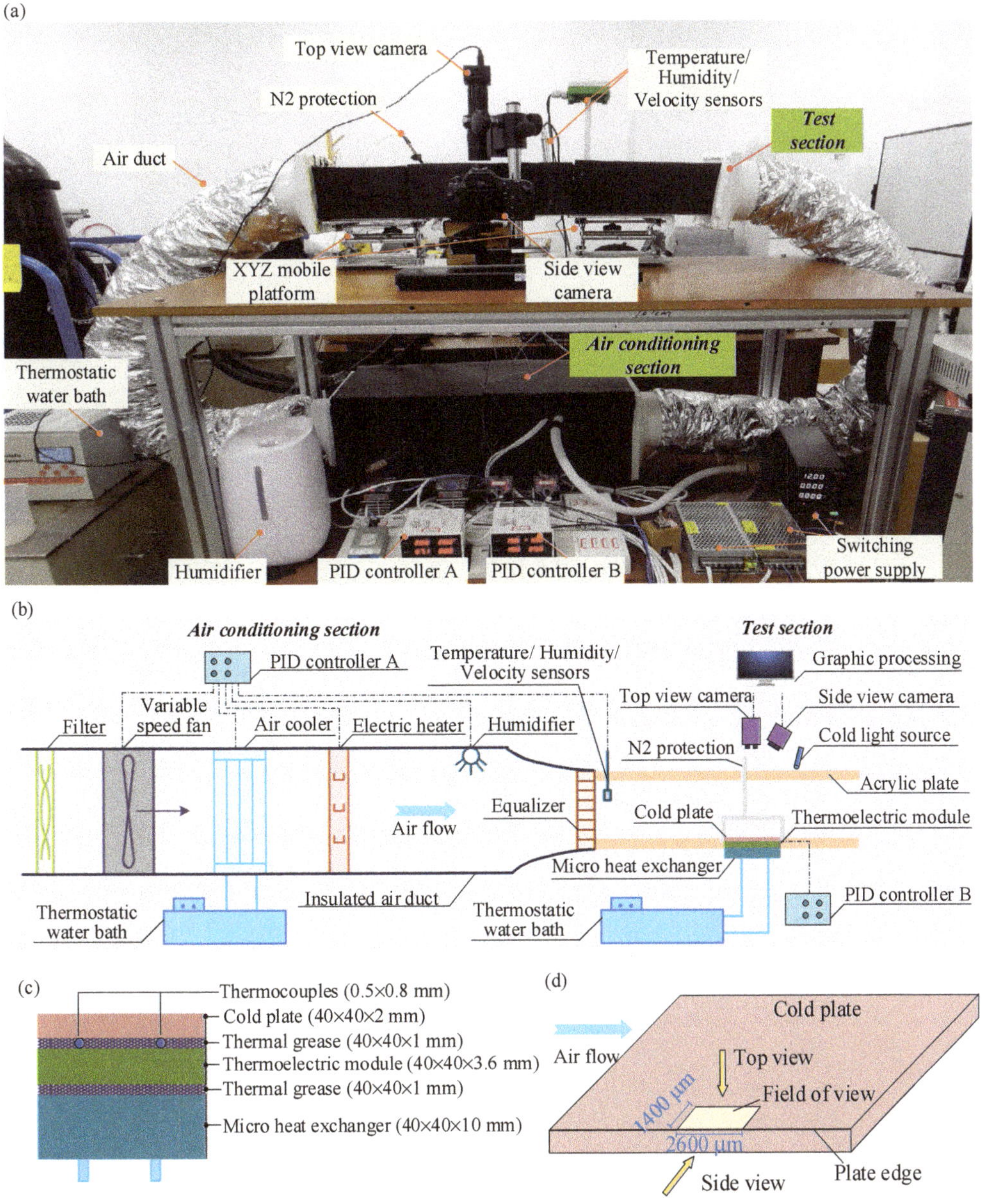

Figure 1. (**a**) photograph of the experimental set-up, (**b**) schematics of the experimental set-up, (**c**) surface temperature measurement, and (**d**) field of view of the experimental cold plate.

To avoid undesired condensation on the cold plate at the preparation stage of each experiment, an N_2 protection unit was employed. It was composed of an N_2 pipe and N_2 mask, which separated the cold plate from the air at the preparation stages. To better investigate the localized frosting characteristics in the edge region of the cold plate, a top view camera (CCD AO-508U) and a side view camera (Canon-60D) were simultaneously employed. The field of view of the cold plate was around 2600 × 1400 µm, as shown in Figure 1d. The resolutions of the top and side view cameras were 2592 × 1944 and 1920 × 1080, respectively.

2.2. Experimental Procedure and Conditions

The complete experiment comprised three stages: the preparation stage, the frosting experiment stage, and the defrosting experiment stage. The detailed experimental procedure in each stage was as follows:

- Preparation: (1) adjust the air temperature, relative humidity, and velocity to the designated values, on the air conditioning unit; (2) place the N_2 mask on the cold plate to avoid undesired condensation; (3) adjust the cold plate surface temperature to the designated value by the thermoelectric module;
- Frosting experiment: (1) remove the N_2 mask to allow the processed humid air to directly contact with the cold plate; (2) maintain the air temperature, relative humidity and velocity, and cold plate surface temperature at the designated values;
- Defrosting experiment: (1) turn off the air conditioning unit when the frosting time reached the set value; (2) heat the cold plate by the thermoelectric module to melt frost; (3) wipe away the melted water on the cold plate and clean the cold plate with absolute ethyl alcohol.

To increase the reliability of the experimental results, each experiment was repeated at least three times. In this study, two cases were designed in terms of the typical experimental conditions in related references. Cases 1 and 2 corresponded to natural and forced convection, and the detailed experimental conditions are presented in Table 1.

Table 1. Experimental conditions.

No.	Item	Unit	Case 1	Case 2
1	Air velocity	m/s	0.1	2.5
2	Air temperature	°C	20.0	20.0
3	Air relative humidity	-	50%	50%
4	Cold plate surface temperature	°C	−15.0	−15.0
5	Frosting time	s	1200	1200

2.3. Data Reductions and Error Analysis

In this study, the localized characteristics of the first three stages of a condensation frosting process in the edge region of the experimental cold plate were comprehensively investigated. Based on the visualized data, the schematics of the temporal and spatial divisions of condensation frosting stages in the edge region of a cold plate are illustrated in Figure 2. As seen, the condensation frosting process can be divided into WDC, WDF, and FLG stages, in terms of time sequence. The WDC stage started when humid air contacted the cold plate and a large amount of water vapor nucleated on the cold plate surface. These water droplets gradually grew, and droplet coalescence occurred when two droplets contacted. The WDF stage started when one of the water droplets in the field of view solidified, and interdroplet freezing, also known as freezing wave propagation, could be observed as time passed. The FLG stage started when all the water droplets in the field of view solidified. At that time, frost crystal embryos grew on some early frozen water droplets. As frosting progressed, frost crystals grew in three dimensions and covered each other, creating a porous frost layer. On the other hand, a notable difference in the condensation frosting characteristics was observed in different zones of the cold plate. As a result, the condensation frosting process on the cold plate was further divided into frosting in the edge-affected zone and the unaffected zone regarding its spatial distribution characteristic.

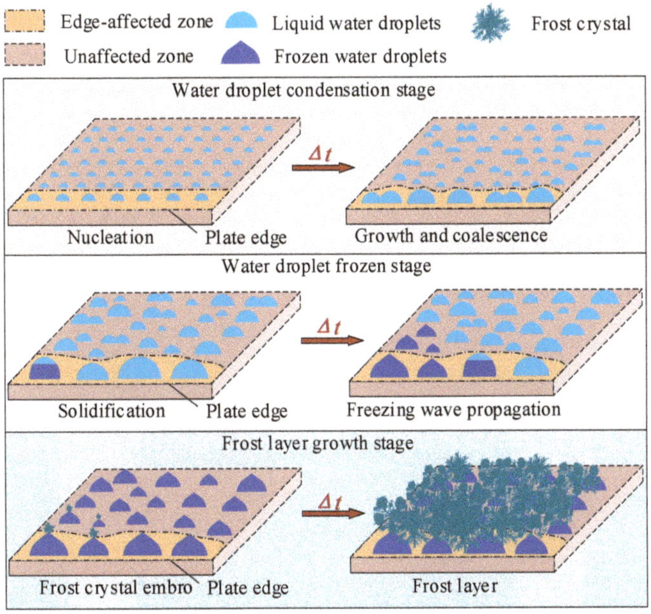

Figure 2. Schematics of the temporal and spatial divisions of condensation frosting stages in the edge region of a cold plate.

The corresponding characteristic parameters are presented in this section to better understand the temporal and spatial condensation frosting characteristics. Firstly, the contact area of a water droplet can be expressed as:

$$A_{\mathrm{w}} = N_{\mathrm{p,in}} \times A_{\mathrm{p}} \tag{1}$$

where $N_{\mathrm{p,in}}$ is the number of pixels covered by a droplet, and A_{p} is the area of a pixel.

The equivalent contact diameter of a water droplet can be expressed as:

$$D_{\mathrm{w}} = \sqrt{\frac{4 \times A_{\mathrm{w}}}{\pi}} \tag{2}$$

The area-average equivalent contact diameter of water droplets can be expressed as:

$$\overline{D}_{\mathrm{w}} = \sqrt{\frac{4 \sum_{i=1}^{N_{\mathrm{w}}} A_{\mathrm{w},i}}{\pi N_{\mathrm{w}}}} \tag{3}$$

where N_{w} is the number of water droplets.

The coverage area ratio of water droplets to the corresponding region can be expressed as:

$$R_{\mathrm{A}} = \frac{\sum_{i=1}^{N_{\mathrm{w}}} A_{\mathrm{w},i}}{A_{\mathrm{s}}} \tag{4}$$

where A_{s} is the area of the corresponding region.

During the droplet frozen stage, the average freezing wave propagation velocity can be expressed as:

$$v_{\mathrm{fw}} = \frac{L}{\Delta t} \tag{5}$$

where L is the equivalent distance in the corresponding region, and Δt is the WDF stage duration in the corresponding region.

Based on the photos captured by the side view cameras, the average water droplet height during WDC and WDF stages can be expressed as:

$$\overline{H}_{\mathrm{w}} = \sqrt{\frac{\sum_{i=1}^{N_{\mathrm{w}}} H_{\mathrm{w},i}}{N_{\mathrm{w}}}} \quad (6)$$

where H_w is the height of a water droplet.

The average frost layer thickness at the FLG stage can be expressed as:

$$\overline{\delta}_{\mathrm{f}} = \frac{A_{\mathrm{f}}}{L'} \quad (7)$$

where A_f is the area of the frost layer measured by the side view camera, and L' is the length of the edge region.

Based on the average frost layer thickness, the frost layer growth rate can be evaluated by:

$$g_{\mathrm{f}} = \frac{\overline{\delta}_{\mathrm{f}}(t + \Delta t') - \overline{\delta}_{\mathrm{f}}(t)}{\Delta t'} \quad (8)$$

where $\Delta t'$ is the time interval used for measuring frost thickness.

The frost layer surface roughness can be evaluated by:

$$RMS_{\mathrm{f}} = \sqrt{\frac{\sum_{i=1}^{N_{\mathrm{sp}}} (\delta_{\mathrm{f},i} - \overline{\delta}_{\mathrm{f}})^2}{N_{\mathrm{sp}}}} \quad (9)$$

where N_{sp} is the number of sampling points, and $\delta_{\mathrm{f},i}$ is the local frost layer thickness of the sampling points.

The uncertainties of the parameters above can be evaluated using the error propagation formula presented by Rasul et al. [38]. For example, the uncertainty of the frost layer surface roughness can be evaluated by:

$$u_{RMS_{\mathrm{f}}} = \frac{\sqrt{\sum_{i=1}^{N_{\mathrm{sp}}} \left(\frac{\partial RMS_{\mathrm{f}}}{\partial(\delta_{\mathrm{f},i} - \overline{\delta}_{\mathrm{f}})} u_{(\delta_{\mathrm{f},i} - \overline{\delta}_{\mathrm{f}})} \times (\delta_{\mathrm{f},i} - \overline{\delta}_{\mathrm{f}}) \right)^2}}{RMS_{\mathrm{f}}} \quad (10)$$

In general, the uncertainties of the characteristic parameters are presented in Table 2.

Table 2. Uncertainties of the characteristic parameters.

Characteristic Parameter	Uncertainty
Equivalent contact diameter of a water droplet	±0.9%
Area-average equivalent contact diameter of water droplets	±2.4%
Coverage area ratio of water droplets	±1.1%
Average height of water droplets	±2.8%
Frost layer thickness	±3.7%
Frost layer growth rate	±2.9%
Frost layer surface roughness	±2.2%

3. Results and Discussion

3.1. Localized Water Droplet Condensation Characteristics in Edge-Affected and Unaffected Regions

In this study, the WDC stage started when air with high temperature and humidity contacted the cold plate, and the time point was set as 0 s. Figure 3 compares the top-view images of the localized WDC characteristics in the edge region of the experimental cold plate in Cases 1 and 2. As seen, a large number of nanoscale water droplets nucleated on the cold plate surface at the initial WDC stage and kept growing and coalescing as time passed. A similar phenomenon can also be found in many other related studies [21,25,28]. However, water droplet jumping and self-propulsion were not observed in our study, which was different from [26] and [27], as our experimental cold plate surface was not super-hydrophobic or at a gradient. Additionally, it can be noted that significant differences in water droplet size and distribution characteristics were observed. In general, a single row of water droplets close to the plate edge was much larger and sparser than others, due to the edge effect. Therefore, the region covered by these water droplets was defined as the edge-affected region in this study, as circled by the magenta dash line. The plate edge mainly affected the WDC characteristics through the following three aspects. Firstly, more defects existed in the plate edge, leading to a lower nucleation barrier. Secondly, the plate edge was a joint between the upper and side surfaces of the cold plate, which provided more probability of coalescence of water droplets. Finally, the air side heat and mass transfer rate around the plate edge was much higher due to its geometric singularity. Additionally, it can also be noted that the growth rate of water droplets in Case 2 was much faster than in Case 1 due to the enhanced air side heat and mass transfer rates caused by the air velocity. For example, the water droplets in Case 1 at 210 s were slightly smaller than those in Case 2 at 40 s.

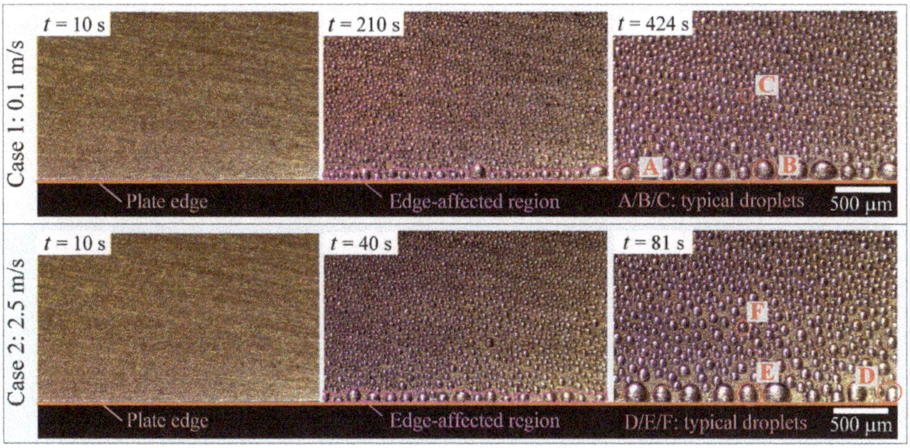

Figure 3. Top-view images of water droplet condensation characteristics.

In particular, the water droplets on the cold plate surface were subcooled at the WDC stage and could solidify at a certain time as they gradually grew up, which was also the indicator of the termination of the WDC stage. Since the plate edge contained more defects, ice nucleation usually occurred first at water droplets close to the plate edge, such as Droplet A in Case 1 at 424 s and Droplet D in Case 2 at 81 s. As a result, the WDC stage duration in Cases 1 and 2 was 424 s and 81 s, respectively. It can be observed that the WDC stage duration was notably shortened by 80.9% when the air velocity increased from 0.1 to 2.5 m/s. To better analyze the droplet dynamics at the WDC stage, four typical water droplets named Droplets B, C, E, and F were selected, as circled by the red line in Figure 3. Droplets B and E represented water droplets in the edge-affected regions in Cases

1 and 2, and Droplets C and F represented those in the unaffected regions in Cases 1 and 2, respectively. The characteristics of growth and coalescence of Droplets B, C, E, and F are further compared in Figure 4.

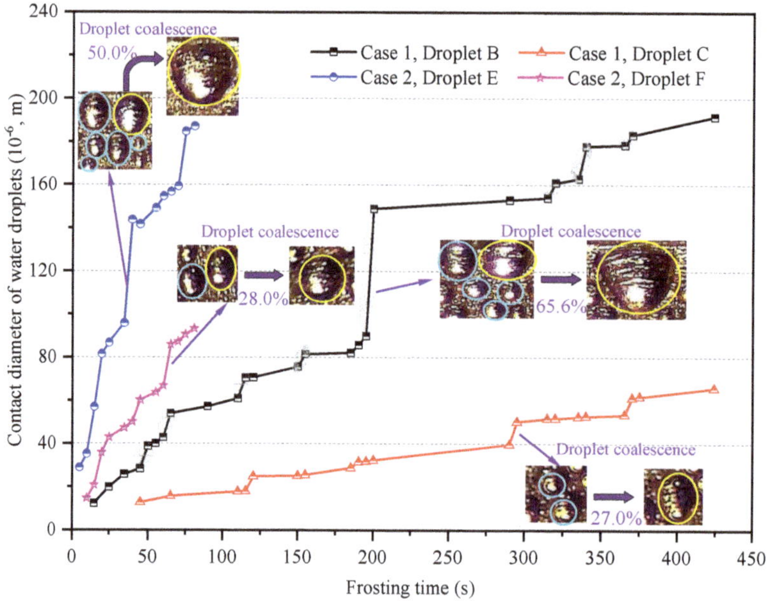

Figure 4. Characteristics of growth and coalescence of water droplets B, C, E, and F.

As shown in Figure 4, the growing processes of Droplets B, C, E, and F were all coupled with droplet coalescence, as indicated by the shadows in the corresponding polygonal lines. For a specific water droplet, its contact diameter notably increased when droplet coalescence occurred. For example, droplet coalescence occurred in the periods of 195 to 200 s, 290 to 295 s, 35 to 40 s, and 60 to 65 s, which increased the contact diameters of Droplets B, C, E, and F, circled by the yellow lines, by 65.6%, 27.0%, 50.0%, and 28.0%, respectively. In addition, the number of droplet coalescences for Droplets B and E were eight and six, which increased by four and three when compared with those for Droplets C and F, respectively. In general, it can be found that the number of droplet coalescences for water droplets in the edge-affected region was around 50% greater than in the unaffected region. Considering the droplet coalescence was a type of rapid droplet growth, the average growth rates of droplet contact diameter for Droplets B and E were 0.45×10^{-6} and 2.31×10^{-6} m/s, and 31.7% and 100.4% greater than those for Droplets C and F, respectively. To better understand the difference in droplet size and characteristics between edge-affected and unaffected regions, and those between nature and forced convection, the temporal and spatial area-average equivalent contact diameter, and coverage area ratio of water droplets are presented in Figures 5 and 6.

Figure 5 presents the area-average equivalent contact diameter of water droplets in edge-affected and unaffected regions in Cases 1 and 2. As seen, the area-average equivalent contact diameters of water droplets in edge-affected regions were much greater than those in unaffected regions. With respect to Case 1, the area-average equivalent contact diameters of water droplets in edge-affected regions at 60 and 424 s were 39.6×10^{-6} and 146.0×10^{-6} m, which were 2.92 and 2.69 times greater than those in unaffected regions at the same time points, respectively. With respect to Case 2, the area-average equivalent contact diameters of water droplets in edge-affected regions at 15 and 81 s were 53.4×10^{-6} and 164.1×10^{-6} m, which were 3.16 and 2.24 times greater than those in unaffected regions at the same time points, respectively. On the other hand, the area-average equivalent contact diameters of water droplets in Case 2 were notably greater than those in Case 1. For example, the area-average equivalent contact diameters of water droplets in edge-affected and unaffected regions in Case 2 at 60 s were 3.24 and 4.38 times greater than those in Case 1 at the same time. However, the WDC stage duration in Case 1 was notably longer than in Case 2. As a result, the area-average equivalent contact diameters of water droplets in edge-affected and unaffected regions in Case 2 were 8.9% and 34.9% greater than those in Case 1 at the end of their respective WDC stages.

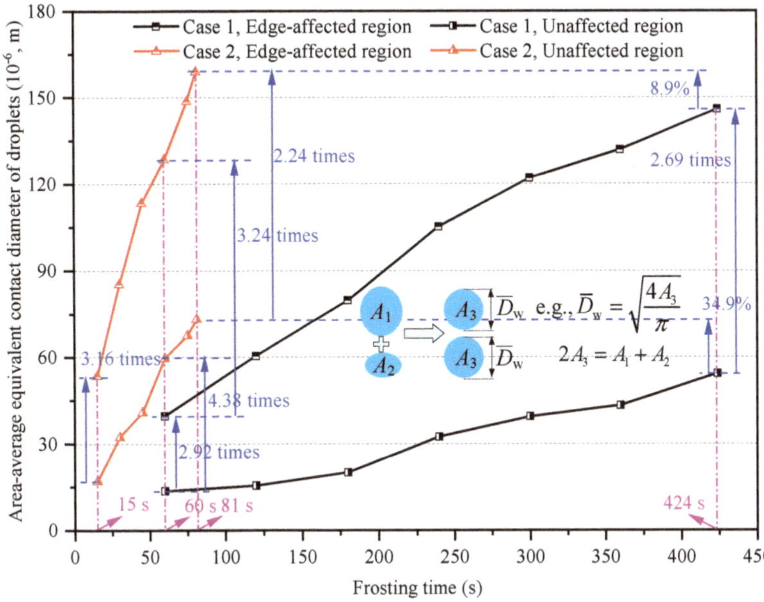

Figure 5. The area-average equivalent contact diameter of water droplets in edge-affected and unaffected regions.

Similar to the area-average equivalent contact diameter of water droplets, their coverage area ratio in edge-affected regions was also much greater than that in unaffected regions, as illustrated in Figure 6. At the end of the WDC stages in Cases 1 and 2, the coverage area ratios in edge-affected regions were 2.69 and 2.24 times larger than the corresponding values in unaffected regions, respectively. Additionally, the coverage area ratio of water droplets in Case 2 was notably greater than in Case 1 due to the increased heat and mass transfer rate caused by the increased air velocity. For example, the coverage area ratio of water droplets in edge-affected and unaffected regions in Case 2 at 60 s were 26.4% and 19.6% greater than those in Case 1. However, although the coverage area ratio of water droplets in Case 2 was notably greater than in Case 1 at the same time point, an opposite trend was observed at the end of their respective WDC stages. The coverage area ratios of water droplets in edge-affected and unaffected regions in Case 2 were, in turn, 4.9% and 3.1% smaller than those in Case 1 at the end of their respective WDC stages. The main reason for this was that the WDC stage duration in Case 1 was notably longer than that in Case 2, and the coverage area ratio of water droplets consistently increased as time passed. Therefore, it was found that the final coverage area ratio of water droplets at the end of the WDC stage was not only dependent on the growth rate of water droplets, but also on the WDC stage duration.

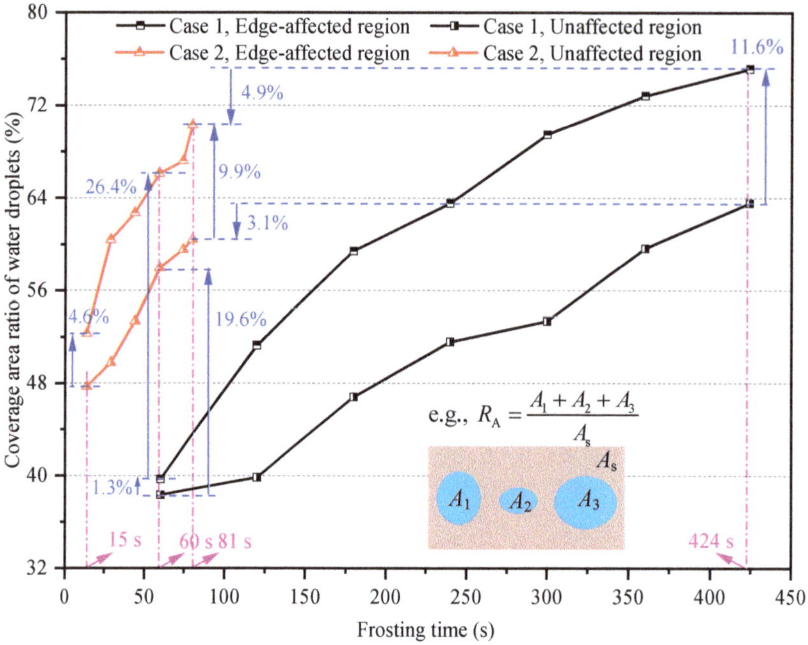

Figure 6. The coverage area ratio of water droplets in edge-affected and unaffected regions.

Considering the area of the edge-affected region was closely related to the length of the field of view and its shape was not regular, a parameter defined as the equivalent width of the edge-affected region was employed, as presented in Figure 7. The equivalent width of the edge-affected region was used to represent the effective range impacted by the edge effect. As seen, the equivalent width of the edge-affected region consistently increased as time passed and was notably affected by air velocity. At 60 s, the equivalent width of the edge-affected region in Case 2 was 131.9×10^{-6} m/s, which was 1.51 times greater than in Case 1. As time passed, the equivalent width of the edge-affected region in Cases 1 and 2 were 145.5×10^{-6} and 173.6×10^{-6} m/s at 424 and 81 s, respectively. It can be found that the equivalent width of the edge-affected region in Case 2 was only 19.3% greater than in Case 1 at the end of their respective WDC stages because the WDC stage duration in Case 1 was notably longer than in Case 2.

On the other hand, the average height of the water droplets in the edge-affected regions was evaluated by images captured by the side view camera, which was also presented in Figure 7. The variation trend in the average height of the water droplets was similar to the equivalent width of the edge-affected region. In general, the average height of the water droplets in the edge-affected regions consistently increased as time passed. At the end of the WDC stages in Cases 1 and 2, their final values were 80.7×10^{-6} and 105.3×10^{-6} m, respectively. Although the WDC stage duration in Case 1 was 80.9% longer than in Case 2, the final average height of the water droplets in the edge-affected region in Case 1 was 30.5% smaller than in Case 2, due to the lower heat and mass transfer rate. Similar conclusions can also be found in the study carried out by Zhang et al. [39].

In general, the water droplet size and distribution characteristics at the WDC stage were notably impacted by the plate edge. At the end of their respective WDC stages, the area-average equivalent contact diameter and coverage area ratio of water droplets in edge-affected regions were 2.69 times and 11.6% greater than those in unaffected regions under natural convection, and the corresponding values were 2.24 times and 9.9% under forced convection. Additionally, the differences in the area-average equivalent contact diameter and coverage area ratio of water droplets in edge-affected and unaffected regions, the equivalent width of the edge-affected region, and the average height of water droplets in edge-affected regions, were highly significant between natural and forced convection at the same time point. However, the differences were notably decreased at the end of their respective WDC stages because the WDC stage duration under natural convection was much longer than under forced convection.

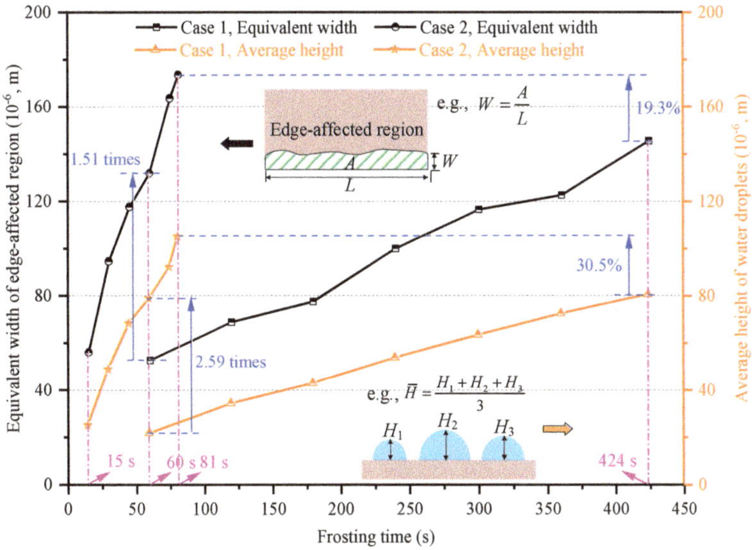

Figure 7. The equivalent width of the edge-affected region and average height of water droplets in the edge-affected region.

3.2. Localized Water Droplet Frozen Characteristics in Edge-Affected and Unaffected Regions

As mentioned in Section 2.3, the WDF stage started when the ice nucleation occurred for the first time. Figure 8 compares the top-view images of the localized WDF characteristics in the edge region of the experimental cold plate in Cases 1 and 2. It was not difficult to distinguish the solidified water droplets from liquid water droplets due to their different refractive index. As seen, ice nucleation firstly occurred at water droplets in the corner of the edge-affected regions. After that, significant water droplet freezing wave propagation was observed, indicated by the irregular red lines and arrows. It was noted that the water droplet freezing wave propagation velocities in edge-affected regions were much faster than those in unaffected regions. As a result, the WDF stage duration in edge-affected regions was much shorter than in unaffected regions. In general, the difference in freezing wave propagation between edge-affected and unaffected regions was mainly caused by the following three reasons. The first was because the contact diameter and coverage ratio of water droplets in edge-affected regions were greater than those in unaffected regions, which decreased the number of times of freezing wave propagation. The second reason was because there existed a secondary freezing propagation path for the water droplets in the edge-affected region, through the water droplets on the side surface of the cold plate. The third reason was because the heat and mass transfer rates in the edge-affected region were much higher than in the unaffected region, which was also the main reason why the freezing wave propagation in Case 2 was much faster than in Case 1. More quantified data on the WDF stage duration, average freezing wave velocity, water droplet size, and distribution characteristics are presented in Figures 9–11.

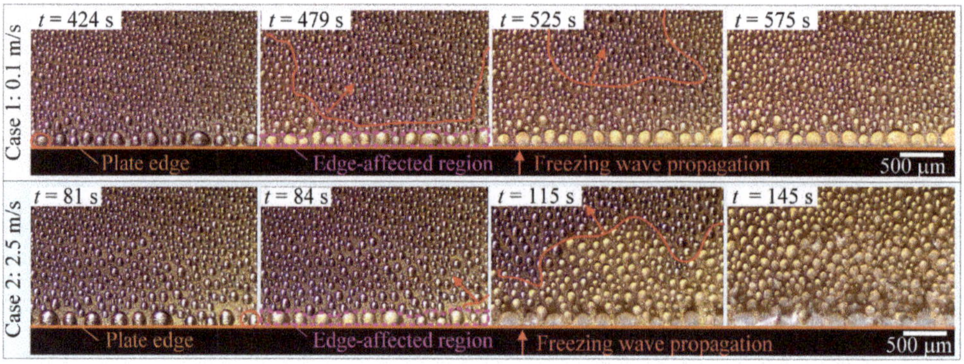

Figure 8. Top-view images of water droplet frozen characteristics.

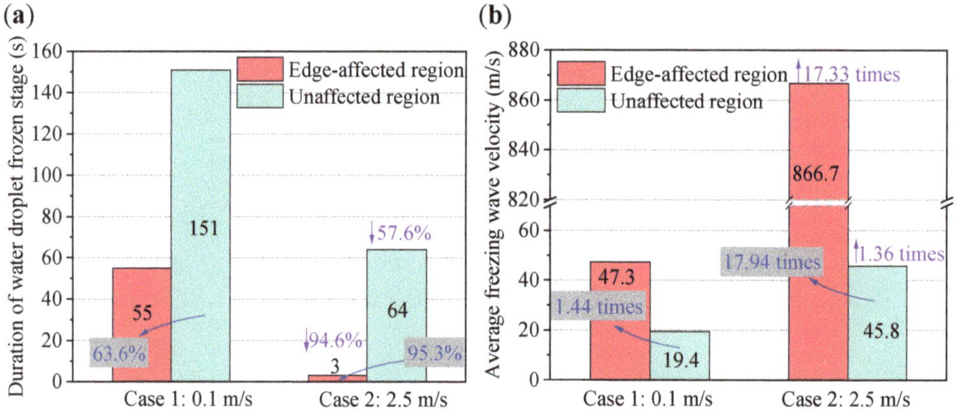

Figure 9. (**a**) Duration of water droplet frozen stage, and (**b**) average freezing wave velocity.

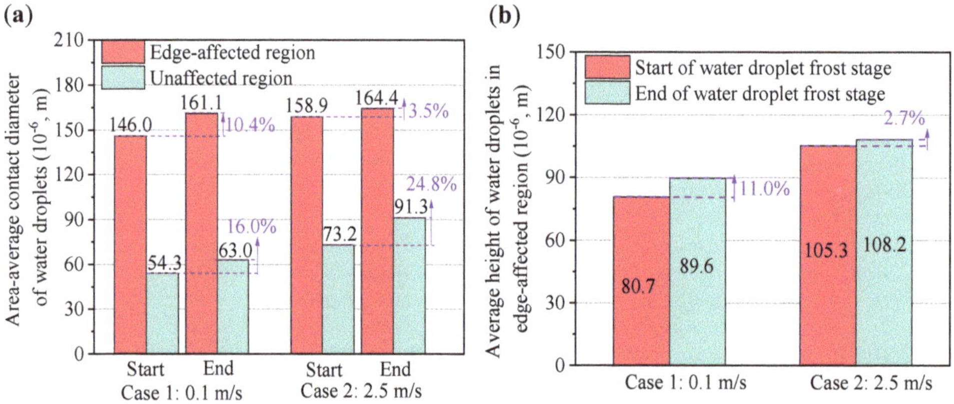

Figure 10. Water droplet size characteristics: (**a**) area-average contact diameter, and (**b**) average height of water droplets.

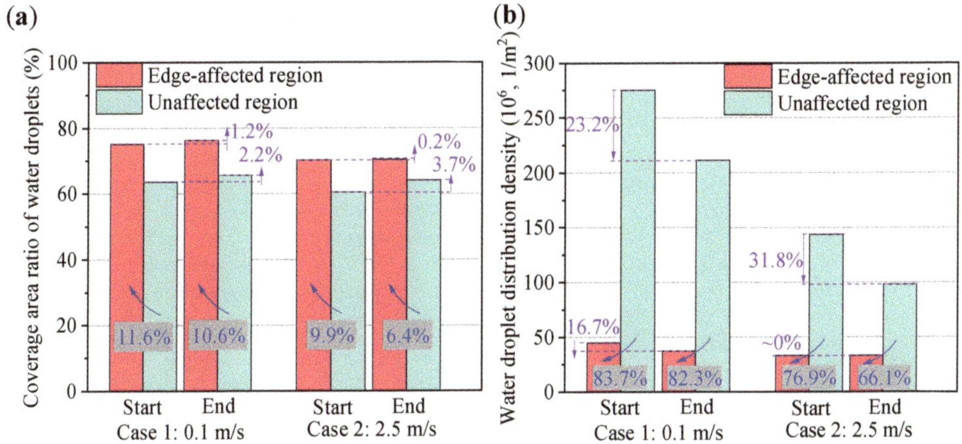

Figure 11. Water droplet distribution characteristics: (**a**) coverage area ratio, and (**b**) distribution density of water droplets.

Figure 9a,b providethe WDF stage duration and average freezing wave velocity in edge-affected and unaffected regions, respectively. With respect to Case 1, the WDF stage duration in edge-affected and unaffected regions was 55 and 151 s, respectively. With respect to Case 2, the WDF stage duration in edge-affected and unaffected regions was 3 and 64 s, respectively. Compared with the unaffected region, the WDF stage duration in edge-affected regions decreased by 63.6% in Case 1, and 95.3% in Case 2, respectively. On the other hand, the average freezing wave propagation velocity in edge-affected and unaffected regions in Case 1 was 47.3×10^{-6} and 19.4×10^{-6} m/s, and in Case 2 was 866.7×10^{-6} and 45.8×10^{-6} m/s, respectively. The magnitudes of the average freezing wave propagation velocities in unaffected regions in Cases 1 and 2 were the same as reported in [22], from 10×10^{-6} to 100×10^{-6} m/s. However, compared with the unaffected region, the average freezing wave propagation velocities in edge-affected regions increased by 1.44 and 17.94 times in Cases 1 and 2, respectively. In general, both the plate edge and air velocity positively impacted the propagation of the freezing wave at the WDF stage. As a result, it was noted that the WDF stage duration in edge-affected and unaffected regions in Case 2 decreased by 94.6% and 57.6%, respectively, when compared with Case 1. In contrast, the average freezing wave propagation velocity in edge-affected and unaffected regions in Case 2 increased by 17.33 and 1.36 times, respectively.

The area-average contact diameter and average height of water droplets at the start and end of the WDF stages in Cases 1 and 2 are illustrated in Figure 10a,b, respectively. As seen, the area-average contact diameters of water droplets in edge-affected and unaffected regions were 161.1×10^{-6} and 63.0×10^{-6} m at the end of the WDF stage in Case 1, increased by 10.4% and 16.0% when compared with those at the start of the WDF stage. The area-average contact diameters of water droplets in edge-affected and unaffected regions were 164.4×10^{-6} and 91.3×10^{-6} m at the end of the WDF stage, and increased by 3.5% and 24.8% when compared with those at the start of the WDF stage in Case 2. Additionally, it was also found that the difference in area-average contact diameter of water droplets between Cases 1 and 2 in edge-affected regions was much smaller than that in unaffected regions. The former was around 2.0%, while the latter was 31.0% at the end of the WDF stage. On the other hand, the average heights of water droplets in edge-affected regions at the start and end of the WDF stage in Case 1 were 80.7×10^{-6} and 89.6×10^{-6} m, and those in Case 2 were 105.3×10^{-6} and 108.2×10^{-6} m/s, respectively. In other words, the average height of water droplets in edge-affected regions in Cases 1 and 2 at the end of their WDF stages increased by 11.0% and 2.7%, when compared with those at the start of the WDF stages, respectively.

The coverage area ratio and distribution density of water droplets at the start and end of the WDF stages in Cases 1 and 2 are compared in Figure 11a,b. As seen, the coverage area ratios of water droplets in edge-affected and unaffected regions were 76.3% and 65.7% at the end of the WDF stage in Case 1, which increased by 1.2% and 2.2% when compared with those at the start of the WDF stage. For Case 2, the corresponding increase rates in edge-affected and unaffected regions were 0.2% and 3.7%. Additionally, the coverage area ratio of water droplets in edge-affected region was greater than in the unaffected region because the growth rate of water droplets of the former was faster than that of the latter. The coverage area ratio of water droplets in edge-affected regions in Case 1 at the start and end of the WDF stages increased by 11.6% and 10.6%, when compared with those in unaffected regions, respectively. Similarly, the corresponding increase rates were 9.9% and 6.4% in Case 2. The difference in the coverage area ratios of water droplets between edge-affected and unaffected regions decreased as air velocity increased from 0.1 m/s (Case 1) to 2.5 m/s (Case 2).

The distribution density of water droplets was another water droplet distribution characteristic. As seen in Figure 11b, the distribution densities of water droplets in edge-affected and unaffected regions were 37.4×10^6 and 211.1×10^6 $1/m^2$ at the end of the WDF stage in Case 1, and decreased by 16.7% and 23.2% when compared with those at the start of the WDF stage. The corresponding decrease rates were ~0% and 31.8% in Case 2. Additionally, the distribution density of water droplets in edge-affected regions was more than 60% smaller than in unaffected regions due to the faster growth and greater coalescence of water droplets. It was also found that the difference in the distribution density of water droplets between the edge-affected and unaffected regions decreased as air velocity decreased from 0.1 m/s (Case 1) to 2.5 m/s (Case 2).

3.3. Localized Frost Layer Growth Characteristics in Edge-Affected Region

The FLG stage started when all the water droplets solidified in the field of view. Since the frost layer in the unaffected region was covered by that in the edge-affected region in terms of the side-view images, only the frost layer growth characteristics in the edge-affected regions are presented in this section. Figure 12 presents the average frost layer thickness in the edge-affected regions at the FLG stages in Cases 1 and 2. As seen, the onset time of the FLG stage in Cases 1 and 2 were 479 and 84 s, respectively. Thereafter, frost crystals rapidly grew on the surface of solidified water droplets due to the large temperature difference between air and water droplet surface. As a result, the average frost layer thickness increased as time passed. It can be noted that the average frost layer thickness continued to increase during the remaining frosting time, even though local reverse melting occurred for part of each frost crystal at the later FLG stage. For example, many shining points can be observed in the images of the frost layer at 720 and 1200 s in Case 1. These bright points indicated the local reverse melting of frost crystals. At 1200 s, the average frost layer thickness in edge-affected regions in Cases 1 and 2 was 342.5×10^{-6} and 1442.2×10^{-6} m, respectively. In other words, the average frost layer thickness increased by 3.21 times after a 1200 s frosting period when the air velocity increased from 0.1 to 2.5 m/s.

The frost layer growth rate in the edge-affected regions at the FLG stages in Cases 1 and 2 are compared in Figure 13. As seen, the general variation trend in the frost layer growth rate in Case 1 first increased, and then decreased, as time passed, which was similar to the results reported in [24,34]. At the initial FLG stage in Case 1, the frost layer growth rate was not the maximum, although the frost crystal growth rate later reached the maximum value. The main reason was that many gaps existed among these solidified water droplets, and the frost crystal growth rate in Case 1 was not big enough to balance out these initial gaps. As a result, the average frost layer thickness increased slowly at the initial FLG stage, as demonstrated in Figure 12. The overall frost layer growth rate increased when most initial gaps were filled with frost crystals. At the later FLG stage, the frost layer growth rate decreased as time passed as the surface temperature of the frost layer gradually increased. By contrast, the initial frost crystal growth rate in Case 2 was big enough to balance out the initial gaps among solidified water droplets. Therefore, the frost layer growth rate was maximum at the initial FLG stage and then gradually decreased as time passed. The maximum frost layer growth rate in Case 2 was 3.46×10^{-6} m/s at 84 s, which was 2.93 times larger than the maximum frost layer growth rate of 0.52×10^{-6} m/s in Case 1 at 720 s. Additionally, the frost layer growth rates in edge-affected regions in Cases 1 and 2 were 0.29×10^{-6} and 0.50×10^{-6} m/s at the end of the frosting period, respectively. In other words, the frost layer growth rate increased significantly by 71.2% at 1200 s when the air velocity increased from 0.1 to 2.5 m/s.

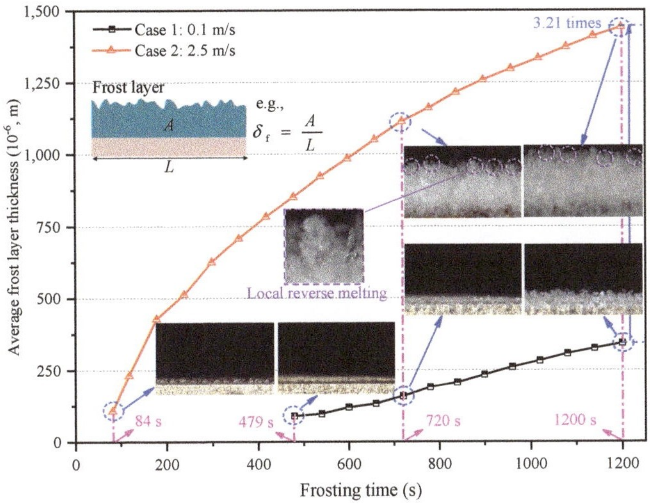

Figure 12. Average frost layer thickness in the edge-affected region.

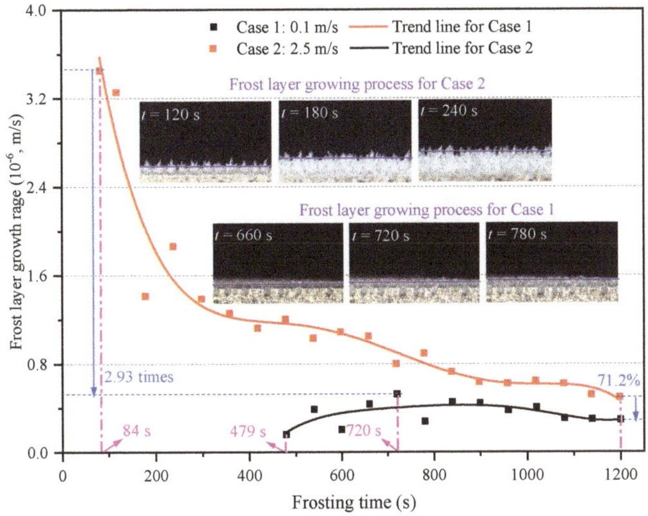

Figure 13. Frost layer growth rate in the edge-affected region.

Figure 14 shows the frost layer surface roughness in the edge-affected regions at the FLG stages in Cases 1 and 2. As seen, the surface roughness of the frost layer in Case 2 fluctuated more significantly than in Case 1. One reason for this was that the growth rate of frost crystals in Case 2 was much higher than in Case 1, which may have facilitated the uneven growth of each frost crystal. Another reason was that frost crystal collapse occurred more frequently in Case 2 than in Case 1. As illustrated in the images of Figure 14, frost crystal collapse occurring in the period of 176 to 177 s in Case 1 significantly impacted frost layer surface roughness. Additionally, it was also noted that the frost layer surface roughness in Case 2 was much larger than in Case 1. This was mainly because the frost layer thickness in Case 2 was notably larger than in Case 1. At the FLG stage, frost layer surface roughness increased with an increase in frost layer thickness in general. Additionally, the frost crystal in Case 2 was feather-type, which differed from the plate-type frost crystal in Case 1. Feather-type frost crystals also led to a larger frost layer surface roughness when compared with plate-type frost crystals. The frost layer surface roughness in edge-affected regions in Cases 1 and 2 was 29.5×10^{-6} and 48.3×10^{-6} m/s at the end of the frosting period, respectively. In other words, the frost layer surface roughness increased by 63.8% at 1200 s when the air velocity increased from 0.1 to 2.5 m/s.

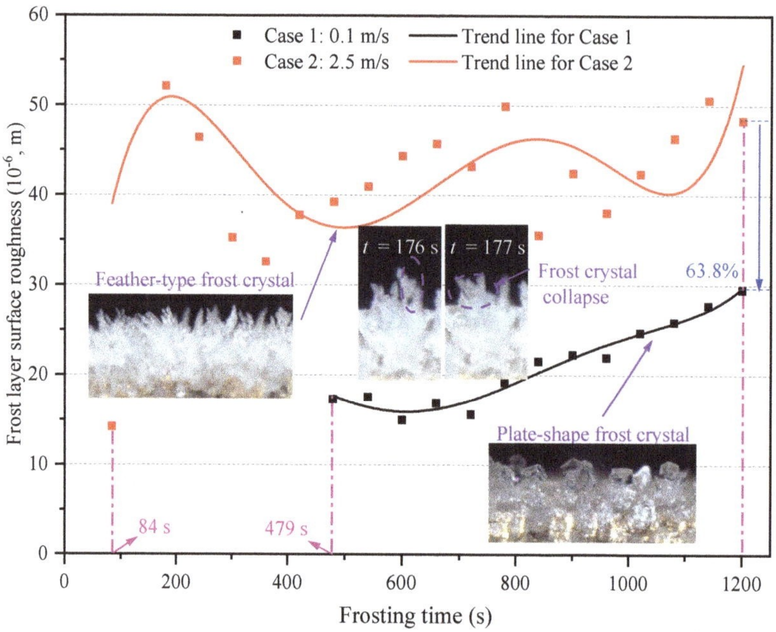

Figure 14. Frost layer surface roughness in the edge-affected region.

4. Conclusions

This paper comprehensively studied the localized condensation frosting characteristics in the edge region of a horizontal cold plate. Regarding the characterized condensation frosting stages, the main conclusions can be drawn as follows:

(a) Forced convection may decrease the WDC stage duration and the influence area of the plate edge when compared with natural convection. The WDC stage duration was notably shortened from 424 to 81 s with a decrease of 80.9% when the air velocity increased from 0.1 to 2.5 m/s. The equivalent width of the edge-affected region under forced convection was 19.3% greater than under natural convection, at the end of their respective WDC stages;

(b) The plate edge may facilitate the growth rate of water droplets, which increases their size and coverage area ratio at the early frosting stage. The number of droplet coalescence for water droplets in edge-affected regions was around 50% greater than in unaffected regions. At the end of their respective WDC stages, the area-average equivalent contact diameter and coverage area ratio of water droplets in edge-affected regions were 2.69 times and 11.6% greater than those in unaffected regions under natural convection, and the corresponding values were 2.24 times and 9.9% under forced convection;

(c) Both the air velocity and plate edge positively impacted the propagation of the freezing wave at the WDF stage. Compared with the unaffected region, the WDF stage duration in edge-affected regions decreased by 63.6% under natural convection and 95.3% under forced convection, respectively. The average freezing wave propagation velocities in edge-affected and unaffected regions under natural convection were 47.3×10^{-6} and 19.4×10^{-6} m/s, and those under forced convection increased by 17.33 and 1.36 times, respectively. Compared with the unaffected region, the average freezing wave propagation velocities in edge-affected regions increased by 1.44 and 17.94 times under natural and forced convection, respectively.

(d) The average frost layer thickness continued to increase at the later FLG stage, even though local reverse melting occurred for part of each frost crystal. The general variation trend in frost layer growth rate under natural convection first increased, and then decreased, while under forced convection it continued to decrease as time passed. The maximum frost layer growth rate under forced convection was 3.46×10^{-6} m/s at 84 s, which was 2.93 times larger than the maximum frost layer growth rate of 0.52×10^{-6} m/s under natural convection at 720 s. Additionally, the frost layer surface roughness under forced convection fluctuated more significantly than under natural convection, and feather-type and plate-type frost crystals were, respectively, observed in the former and latter.

In general, this paper mainly focused on the localized condensation frosting characteristics in the edge region of a horizontal cold plate under designated experimental conditions, and those for other cold plates with different surface treatments and installation sets and frosting conditions should be further investigated.

Author Contributions: Conceptualization, L.Z. and M.S.; methodology, L.Z.; investigation, L.Z. and M.S.; resources, M.S.; data curation, L.Z.; writing—original draft preparation, L.Z.; writing—review and editing, M.S., C.Y.H.C. and C.D.; supervision, M.S. and C.Y.H.C.; funding acquisition, M.S. and J.S. All authors have read and agreed to the published version of the manuscript.

Funding: This research was funded by the National Natural Science Foundation of China (grant no. 52076013), National Foreign Expert Project (grant no. G2022178023L), Beijing Municipal Science and Technology Commission (grant no. 3212024), Open Fund of Key Laboratory of Icing and Anti/De-icing (grant no. IADL20200104), CAS Key Laboratory of Cryogenics, Technical Institute of Physics and Chemistry, China (grant no. CRYO202001, grant no. CRYO202104), and National Natural Science Foundation of China (grant no. 51925605).

Institutional Review Board Statement: Not applicable.

Informed Consent Statement: Not applicable.

Data Availability Statement: The data presented in this study are available on request from the corresponding author. The data are not publicly available due to restrictions.

Acknowledgments: The authors would like to acknowledge the Frost Lab, Beijing Institute of Technology, China, for equipment access and technical support.

Conflicts of Interest: The authors declare no conflict of interest.

Nomenclatures

A	area (m^2)
D	equivalent diameter (m)
$\overline{D}$	area-average equivalent diameter (m)
FLG	frost layer growth
FLFG	frost layer fully growth
g	growth rate (m/s)
H	water droplet height (m)
$\overline{H}$	average water droplet height (m)
L	equivalent distance in the corresponding region (m)
L'	length of the edged region (m)
N	number
R_A	coverage area ratio of droplets (%)
RMS	root-mean-square roughness (m)
u	uncertainty (%)
v	velocity (m/s)
W	equivalent width of the edge-affected region
WDC	water droplet condensation
WDF	water droplet frozen
Greek symbols	
δ	thickness (m)
$\overline{\delta}$	average thickness (m)
Δt	water droplet frozen stage duration (s)
$\Delta t'$	time interval used for measuring frost thickness (s)
Subscripts	
f	frost
fw	freezing wave
p	pixel
p,in	pixels inside a water droplet
sp	sampling points
s	surface of the corresponding region
w	water

References

1. Azeez Mohammed Hussein, H.; Zulkifli, R.; Faizal Bin Wan Mahmood, W.M.; Ajeel, R.K. Structure parameters and designs and their impact on performance of different heat exchangers: A review. *Renew. Sustain. Energy Rev.* **2022**, *154*, 111842. [CrossRef]
2. Moradkhani, M.A.; Hosseini, S.H.; Song, M.J. Robust and general predictive models for condensation heat transfer inside conventional and mini/micro channel heat exchangers. *Appl. Therm. Eng.* **2022**, *201*, 117737. [CrossRef]
3. Lee, K.; Kim, W. The effects of design and operating factors on the frost growth and thermal performance of a flat plate fin-tube heat exchanger under the frosting condition. *KSME Int. J.* **1999**, *13*, 973–981. [CrossRef]
4. Sadeghianjahromi, A.; Wang, C. Heat transfer enhancement in fin-and-tube heat exchangers—A review on different mechanisms. *Renew. Sustain. Energy Rev.* **2021**, *137*, 110470. [CrossRef]
5. Hu, W.J.; Fan, J.; Song, M.J.; Jia, P.; Gao, Y. An experimental study on the frosting characteristic and performance of a micro-channel evaporator in an air source heat pump unit. *Energy Build.* **2020**, *224*, 110254. [CrossRef]
6. Liu, X.; Yu, J.; Yan, G. An experimental study on the air side heat transfer performance of the perforated fin-tube heat exchangers under the frosting conditions. *Appl. Therm. Eng.* **2020**, *166*, 114634. [CrossRef]
7. Da Silva, D.L.; Melo, C.; Hermes, C.J.L. Effect of frost morphology on the thermal-hydraulic performance of fan-supplied tube-fin evaporators. *Appl. Therm. Eng.* **2017**, *111*, 1060–1068. [CrossRef]
8. Zhang, L.; Jiang, Y.Q.; Dong, J.K.; Yao, Y.; Deng, S.M. A comparative study of frosting behavior on finned tube heat exchanger under different fan control modes. *Appl. Therm. Eng.* **2019**, *160*, 114063. [CrossRef]
9. Moallem, E.; Cremaschi, L.; Fisher, D.E.; Hong, T. Effects of frost growth on louvered folded fins of microchannel heat exchangers on the time-dependent air side convective heat transfer coefficient. *Exp. Therm. Fluid Sci.* **2017**, *88*, 326–335. [CrossRef]
10. Kim, K.; Lee, K.S. Frosting and defrosting characteristics of surface-treated louvered-fin heat exchangers: Effects of fin pitch and experimental conditions. *Int. J. Heat Mass Transf.* **2013**, *60*, 505–511. [CrossRef]
11. Kim, K.; Kim, M.H.; Kim, D.R.; Lee, K.S. Thermal performance of microchannel heat exchangers according to the design parameters under the frosting conditions. *Int. J. Heat Mass Transf.* **2014**, *71*, 626–632. [CrossRef]

12. Kim, M.H.; Kim, H.; Kim, D.R.; Lee, K.S. A novel louvered fin design to enhance thermal and drainage performances during periodic frosting/defrosting conditions. *Energy Convers. Manag.* **2016**, *110*, 494–500. [CrossRef]
13. Zhang, L.; Jiang, Y.Q.; Dong, J.K.; Yao, Y.; Deng, S.M. An experimental study of frost distribution and growth on finned tube heat exchangers used in air source heat pump units. *Appl. Therm. Eng.* **2018**, *132*, 38–51. [CrossRef]
14. Zhang, L.; Jiang, Y.Q.; Dong, J.K.; Yao, Y.; Deng, S.M. An experimental study on the effects of frosting conditions on frost distribution and growth on finned tube heat exchangers. *Int. J. Heat Mass Transf.* **2019**, *128*, 748–761. [CrossRef]
15. Zhang, L.; Song, M.; Mao, N.; Dong, J. Temporal and spatial frost growth prediction of a tube-finned heat exchanger considering frost distribution characteristics. *Int. J. Heat Mass Transf.* **2022**, *183*, 122192. [CrossRef]
16. Zhang, L.; Song, M.; Deng, S.; Shen, J.; Dang, C. Frosting mechanism and behaviors on surfaces with simple geometries: A state-of-the-art literature review. *Appl. Therm. Eng.* **2022**, *215*, 118984. [CrossRef]
17. Choi, C.; Mittal, K.L. *Ice Adhesion: Mechanism, Measurement, and Mitigation*; John Wiley & Sons: Hoboken, NJ, USA, 2020.
18. Walker, C.; Lerch, S.; Reininger, M.; Eghlidi, H.; Milionis, A.; Schutzius, T.M.; Poulikakos, D. Desublimation Frosting on Nanoengineered Surfaces. *ACS Nano* **2018**, *12*, 8288–8296. [CrossRef]
19. Byun, S.; Jeong, H.; Hyeon, S.; Lee, K. Quantitative analysis of frosting characteristics at ultra-low temperatures under forced convection conditions. *Energy Build.* **2021**, *248*, 111186. [CrossRef]
20. Qu, K.; Xu, X.; Duan, Z. Experimental study on behavior of initial frost crystal formation under lower water vapor pressures. *Int. Commun. Heat Mass* **2006**, *33*, 819–826. [CrossRef]
21. Piucco, R.O.; Hermes, C.J.L.; Melo, C.; Barbosa, J.R. A study of frost nucleation on flat surfaces. *Exp. Therm. Fluid Sci.* **2008**, *32*, 1710–1715. [CrossRef]
22. Guadarrama-Cetina, J.; Mongruel, A.; González-Viñas, W.; Beysens, D. Percolation-induced frost formation. *Europhys. Lett.* **2013**, *101*, 16009. [CrossRef]
23. Wang, Y.; Cheng, Y. Early stage condensation frosting characteristics on plain and nano Al_2O_3-epoxy mixture-coated brass. *Appl. Therm. Eng.* **2019**, *160*, 113971. [CrossRef]
24. Mengjie, S.; Shangwen, L.; Hosseini, S.H.; Xiaoyan, L.; Zhihua, W. An experimental study on the effect of horizontal cold plate surface temperature on frosting characteristics under natural convection. *Appl. Therm. Eng.* **2022**, *211*, 118416. [CrossRef]
25. Sheng, W.; Pei, Y.; Li, X.; Ming, P.; Zhao, W. Effect of surface characteristics on condensate droplets growth. *Appl. Therm. Eng.* **2020**, *173*, 115260. [CrossRef]
26. Boreyko, J.B.; Collier, C.P. Delayed Frost Growth on Jumping-Drop Superhydrophobic Surfaces. *ACS Nano* **2013**, *7*, 1618–1627. [CrossRef] [PubMed]
27. Boreyko, J.B.; Chen, C. Self-propelled dropwise condensate on superhydrophobic surfaces. *Phys. Rev. Lett.* **2009**, *103*, 184501. [CrossRef]
28. Shen, Y.; Zou, H.; Wang, S. Condensation Frosting on Micropillar Surfaces—Effect of Microscale Roughness on Ice Propagation. *Langmuir* **2020**, *36*, 13563–13574. [CrossRef]
29. Lu, M.; Song, M.; Pang, X.; Dang, C.; Zhang, L. Modeling study on sessile water droplet during freezing with the consideration of gravity, supercooling, and volume expansion effects. *Int. J. Multiph. Flow* **2022**, *147*, 103909. [CrossRef]
30. Dang, Q.; Song, M.J.; Dang, C.B.; Zhan, T.Z.; Zhang, L. Experimental study on solidification characteristics of sessile urine droplets on a horizontal cold plate surface under natural convection. *Langmuir* **2022**, *38*, 7846–7857. [CrossRef]
31. Jung, S.; Tiwari, M.K.; Poulikakos, D. Frost halos from supercooled water droplets. *Proc. Natl. Acad. Sci. USA* **2012**, *109*, 16073–16078. [CrossRef]
32. Graeber, G.; Dolder, V.; Schutzius, T.M.; Poulikakos, D. Cascade Freezing of Supercooled Water Droplet Collectives. *ACS Nano* **2018**, *12*, 11274–11281. [CrossRef] [PubMed]
33. Léoni, A.; Mondot, M.; Durier, F.; Revellin, R.; Haberschill, P. State-of-the-art review of frost deposition on flat surfaces. *Int. J. Refrig.* **2016**, *68*, 198–217. [CrossRef]
34. Lei, S.W.; Song, M.J.; Dang, C.B.; Xu, Y.J.; Shao, K.K. Experimental study on the effect of surface temperature on the frost characteristics of an inverted cold plate under natural convection. *Appl. Therm. Eng.* **2022**, *211*, 118470. [CrossRef]
35. Moradkhani, M.A.; Hosseini, S.H.; Shangwen, L.; Mengjie, S. Intelligent computing approaches to forecast thickness and surface roughness of frost layer on horizontal plates under natural convection. *Appl. Therm. Eng.* **2022**, *217*, 119258. [CrossRef]
36. Brèque, F.; Nemer, M. Frosting modeling on a cold flat plate: Comparison of the different assumptions and impacts on frost growth predictions. *Int. J. Refrig.* **2016**, *69*, 340–360. [CrossRef]
37. Zhang, L.; Song, M.; Hosseini, S.H.; Shen, J.; Jiang, Y. A modeling study of spatial and temporal frost growth on the edge of windward fins for a tube-finned heat exchanger. *Int. J. Heat Mass Transf.* **2022**, *183*, 122093. [CrossRef]
38. Rasul, S.B.; Kajal, A.M.; Khan, A.H. Quantifying Uncertainty in Analytical Measurements. *J. Bangladesh Acad. Sci.* **2018**, *41*, 145–163. [CrossRef]
39. Zhang, L.; Song, M.; Chao, C.Y.H.; Shen, J. An experimental study on the dynamic frosting characteristics on the edge zone of a horizontal copper plate under forced convection. *Int. J. Heat Mass Transf.* **2023**, *200*, 123541. [CrossRef]

micromachines

MDPI

Article

Replenishment of the Gas in a Hydrophobically-Structured Surface by Mass Transfer at the Liquid-Gas Interface for Improving the Stability of Entrapped Gas

Bao Wang [1,2], Caihua Wang [1], Ding Weng [2], Marisa Lazarus [3], Dayun Yan [4] and Xiaoyan Liu [1,*]

1 Department of Mechanical Engineering, Northeast Petroleum University, Daqing 163318, China
2 State Key Laboratory of Tribology, Tsinghua University, Beijing 100084, China
3 Department of Physics, George Washington University, Washington, DC 20052, USA
4 Department of Mechanical and Aerospace Engineering, George Washington University, Washington, DC 20052, USA
* Correspondence: liu_xydq@163.com

Abstract: The underwater nonwetted state on a superhydrophobic surface is hardly maintained in flowing water because the entrapped gas dissolves into the water or is carried off by flow. Therefore, a source gas is necessary to maintain a superhydrophobic state for its applications under realistic conditions. As detailed in this paper, based on the gas entrapped on a hydrophobic structured surface, the gas regeneration was experimentally achieved to replenish the losses of gas carried off by the flowing and reduced through dissolution. Furthermore, the mechanism of mass transfer at the liquid-gas interface was investigated by simulation. The results indicated that water molecules at a liquid-gas interface should escape to entrapped gas when water content didn't reach saturation. This phenomenon could be due to the evaporation at the liquid-gas interface. With the increasing water content in the entrapped gas, the evaporation rate at the liquid-gas interface descended gradually. Under the action of flowing, the substances containing high concentrations of water molecule was washed away at the liquid-gas interface. Therefore, the low concentration of the water molecule at the liquid-gas interface was created. As a result, the equilibrium of water and gas at the liquid-gad interface was broken, and the evaporation continued to replenish the lost gas. Overall, the presented results in this study could be considered a promising candidate for replenishing the lost gas in hydrophobic structured surfaces by mass transfer at the liquid-gas interface.

Keywords: hydrophobic; flow; evaporation; micro-grooved surface; entrapped gas

1. Introduction

Recently, the wettability of solid surfaces, particularly superhydrophobic surfaces with a big contact angle (>150°), has attracted extensive attention. When submerged in water, these superhydrophobic surfaces can sustain gas between their surface microstructures, resulting in a layer of gas found at the solid-liquid interface. Therefore, these surfaces are understood to contain a superior alternative gas-sustaining phase that achieves an obvious slip with a length ranging from several micrometers to hundreds [1–5]. Experiments utilizing these superhydrophobic surfaces have achieved drag-reduction rates of 5–30% [6–8]. These common superhydrophobic surfaces can hold air pockets (in rough structures) and have demonstrated an effective slip for underwater drag-reduction [9]. However, it is still difficult to sustain an air-pocket stability, particularly under conditions where the liquid is flowing over the surface at a high speed or under a liquid pressure [10–16]. In these realistic conditions, no superhydrophobic surface has shown nonwetting properties underwater [17] because this requires first that the surface maintains a no-wetted state. Therefore, drag reduction in a superhydrophobic surface has never been achieved under realistic conditions [14]. In flowing water, higher flow velocities result in fast surface air layer removal rates [11,15]. Aljallis et al., experimentally indicated that the air was dissipated from the superhydrophobic surface in shear flow [18]. Furthermore, the air layer between surface structures becomes unstable immediately by the liquid pressure or over time by gas diffusion, leading to the collapse of the gas layer and the filling in of the surface structures with the surrounding water [13]. Theoretically, the collapse pressure at which the wetting transition occurs is inversely proportional to the pitch of surface structures [13]. In agreement with theoretical prediction, it was shown that the wetting transition occurred at varying liquid pressure ranging from 5 to 30 kPa above the ambient, depending on microstructures [13].

Once the gas layer is disrupted, the system becomes ineffective, and the gas loss at the liquid-solid interface is irreversible. Without any energy or gas-providing device, the loss of trapped gas to the surrounding liquid by diffusion is inevitable [13]. To counteract this, a lubricating gas film can be generated by vaporizing liquid at the surface at a temperature above the Leidenfrost point or by producing bubbles at the interface [19–22]. However, these methods require continuous energy input or a gas-providing device, both of which are active methods, limiting their application in practical conditions.

As passive methods, both evolution and evaporation of dissolved gas may generate gas underwater. During the evolution of dissolved gas, the time required for a gas bubble to grow ten times in size is several hundred seconds, even when the ratio of the dissolved gas concentration to the dissolved gas concentration for a saturated solution is 1.25 [23]. The growth rate of gas by evolution is too slow to achieve an effective gas layer on a solid surface underwater. Therefore, the evolution of dissolved gas has never been used to provide gas. On the other hand, evaporation may provide a feasible method to generate gas. Evaporation rate is inversely proportional to the concentrations of water molecules in the gas. Generally, concentrations of water molecules increase as the liquid vaporizes until it reaches a water-saturated state. However, suppose the concentrations of water molecules in the entrapped gas near the liquid-gas interface could be kept low. In that case, evaporation should be a relatively more violent and effective phenomenon to generate gas.

As detailed in this paper, water flow was considered essential to wash away the gas with high-concentration of the water molecule. This should achieve the sustained low concentration of water molecule needed to enable rapid evaporation at the liquid-gas interface. The mechanism of mass transfer at the liquid-gas interface was investigated by numerical simulation, and the experiments were conducted to verify the proposed approach. The state on the hydrophobic structured surface was directly observed under the flowing water condition. These results have surpassed prior limitations of the evaporation method at the liquid-gas interface for gas generation in flowing water.

2. Materials and Methods

2.1. Theoretical Evaporation Rate

Evaporation at the liquid-gas interface is a process of mass transfer. This phenomenon should be considered an important means of replenishing the gas for immersed hydrophobic surfaces [24]. Hertz conducted evaporation experiments and was the first to state an upper limit to the maximum rate of evaporation (Hertz [24], as cited by Eames et al.):

$$\mathrm{n} = \frac{1}{4} N\overline{C}, \tag{1}$$

where n is the number of impacts per unit area per unit time ($\mathrm{kmol \cdot m^{-2} \cdot s^{-1}}$), N is the number of molecules per volume ($\mathrm{kmol \cdot m^{-3}}$), and $\overline{C}$ is the arithmetic mean of the velocities of the molecules.

From Maxwell's law of velocity distribution of molecules

$$\overline{C} = \left(\frac{8k_B T_s}{\pi m} \right)^{1/2}, \tag{2}$$

where K_B is Boltzmann' constant 1.38×10^{-23} JK^{-1}, T_S is liquid surface temperature, and m is the mass of each molecule.

And from the Idea Gas Laws [25]:

$$P_S V = m k_B T_S, \tag{3}$$

where P_s is gas pressure, and V is gas volume. And in a unit volume of gas:

$$N = \frac{P_s}{k_B T_s}, \tag{4}$$

Substituting for N and $\overline{C}$ in Equation (1)

$$n = \frac{P_s}{(2\pi m k_B T_s)^{1/2}}, \tag{5}$$

When some water molecules escaped from the liquid-gas interface, others in gas entered the interface; effectively, evaporation and condensation happened simultaneously. Therefore, the evaporation rate should be modified to give Equation (6).

$$n = \frac{1}{(2\pi m k_B)^{1/2}} \left(\frac{P_s}{\sqrt{T_s}} - \frac{P_v}{\sqrt{T_v}} \right), \tag{6}$$

The evaporation coefficient of water has been the subject of many theoretical and experimental papers and is still an area of dispute. Part of the reason for this intense interest is that some liquids, water included, appear to evaporate at a rate far less than the theoretical rate calculated using Knudsen's equation. However, this approach can be used to analyze the factors contributing to evaporation at the liquid-gas interface.

2.2. Preparation of the Hydrophobic Structured Surface

To obtain entrapped gas in structures for enough time to provide the necessary premise for the evaporation at liquid-gas interface, the gas must not be completely carried away by flowing water [26]. Therefore, a hydrophobic transverse microgrooved (HTM) structure was designed to achieve our goal of sustaining an amount of gas in gaps [27]. Furthermore, when water perpendicularly flows over our designed HTM surface, gases are blocked by ridges and cannot escape easily.

In addition to its hydrophobicity, the physical surface characteristics of our design contributed to sustaining gas in the gaps. In our experiments, the turning method fabricated a series of transverse microgrooves, and the surface topographies were measured (Figure 1). The outer diameter of the tube sample was 39 mm, and the turning length was 325 mm. The profile of the microgroove had a cross-section with an isosceles trapezoidal shape with a height (depth: D) of 12 μm), groove width (W) of 10 μm, and pitch (L) between two neighboring grooves of 30 μm. The roughness of the platform between two neighboring microgrooves was the same as that of the smooth surface, which was used as a comparison (roughness was smaller than 0.8 μm).

The original smooth surface had a contact angle of 65°, which was hydrophilic, prior to treatment. To improve the water repellency of the surface, a coating of a low-surface-energy material named fluoroalkyl silane (FAS-17) [27,28] was applied. Based on contact-angle measurements of the corresponding profile of water droplets (5 μL) on the smooth samples modified by this low-surface-energy material, we ascertained that the FAS-17 coating can achieve a contact angle of 115 ± 2° on the smooth surface and an angle of about 130° on the hydrophobic micro-grooved surface.

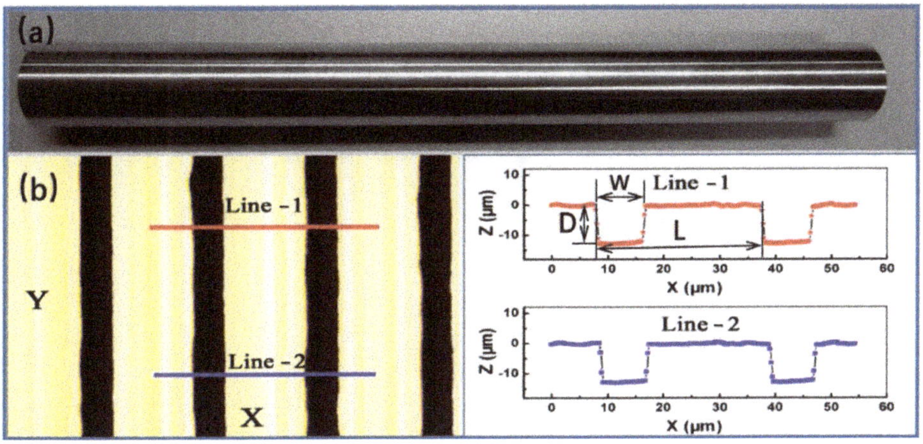

Figure 1. (color online) (**a**) Sample with HTMs with an outer diameter of 39 mm and length of 325 mm, (**b**) surface topography measurement of microgrooved surface at the position of line-1 and line-2, the profile of the microgroove had a cross section with an isosceles trapezoidal shape with a height (depth: D) of 12 μm, groove width (W) of 10 μm, and pitch (L) between two neighboring grooves of 30 μm.

According to the previous investigation, the intruding angle β (the angle between the tangent of a three-phase junction and the horizontal line), which was an acritical parameter, was proposed for the entrapped gas in the surface profile underwater [29]. To achieve a Cassies state, the intruding angle should be less than the maximum asperity slope angle. Based on the profile information of a water droplet on the surface shown in Table 1, the intruding angle for a droplet is described as:

$$\beta = \pi - \theta_0 + \arcsin\left\{\frac{f_{LG}}{\lambda}\frac{1}{\sigma}\left[\frac{2\sigma}{R_D} + \rho g R_D(1-\cos\theta)\right]\right\}, \quad (7)$$

where, θ_0 is the true contact angle; f_{LG} is the ratio of the projected area of the liquid-gas interface to the apparent contact area under the droplet; λ is the contact line density, the length of the contact line over the entrapped gas per unit apparent contact area; $2\sigma/R_D$ is pressure induced by the surface tension at the apex of droplet; R_D is the radius of the droplet; and θ is the apparent contact angle. The true contact angle can be considered as the contact angle on the original smooth surface coated with FAS-17, which is 115°. The apparent contact angle, 128°, is the contact angle on the hydrophobic grooved surface. f_{LG} is 33.3%; λ is 6.67 × 104 1/m; σ is 0.0727 N/m, liquid density is 1003 kg/m^3; R_D is 1.2 mm. Thus, the intruding angle of, 65.5°, was less than the maximum asperity slope angle (π/2−α). Thus, the three-phase lines of contact didn't enter the grooves (If the calculated intruding angle exceeds the maximum, the liquid-gas interface will not keep balance, which will result in the entrance of liquid into grooves). Additionally, according to the literature [27], initial gas can be entrapped in the above microgrooves on the surface, which was proved by optical measurements [27].

Table 1. Results of hydrophobic treatment.

Surface Properties	Images	Contact Angle
smooth surface		65°
smooth surface with FAS film	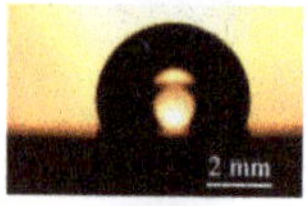	115 ± 2°
microgrooved surface with FAS-17 film	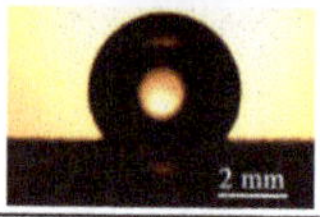	130°

2.3. Experimental Methods

An optical observation was performed to evaluate the effectiveness of our approach when water flowed over the HTM surface in a water tunnel with a closed circulation system. Visualizations of gas on samples were conducted in a test section of the water tunnel, as shown in Figure 2a, which was transparent (Plexiglas®, Beijing, China, with an inner diameter of 120 mm and thickness of 15 mm) for observation. As shown in Figure 2, three positions (A, B and C) were selected to observe. When the superhydrophobic surface (sustaining the gas between structures) was immersed in water and viewed at a glancing angle, it appeared to have high brightness [30–32]. According to the optical property, there was a distance (h) between the camera and the sample's center. Therefore, a glancing angle (φ) was achieved. Based on the geometrical calculations, the glancing angle was equal to another angle γ, which indicates that the height (h) determines the glancing angle. Therefore, we can obtain an appropriate glancing angle for our observations by only moving the camera in the vertical direction.

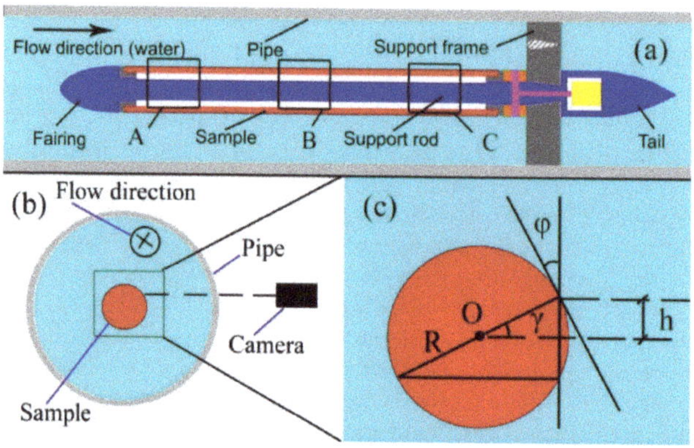

Figure 2. (color online) (**a**) Schematic diagrams of the water tunnel used for optical measurements of gas, A, B, and C are three positions for observation, (**b**) the position of camera, which is a sectional view of (**a**,**c**) schematic diagram of the glancing angle, R is the radius of the sample, h is the offset distance from the camera.to the center line of sample, φ is a glancing angle was equal to another angle γ.

During the measurements, water flowed over the surface in the test section. Therefore, the intensity of white light reflected from the submerged surface was measured as a description of gas. The water used in the experiment was tap water, and its temperature in the tunnel was 20 ± 2 °C.

2.4. Numerical Simulation

Here, a two-dimensional model was established to simulate the shape of the gas-liquid interface under dynamic conditions. The flow on the grooved surface is fast-strained, and the unsteady state-based renormalized group (RNG) k-ε model was used to improve the accuracy of numerical simulation. The implicit solution scheme combined algebraic multigrid methods to achieve faster convergence. The discretization scheme of all equations adopted the second-order upwind scheme to obtain higher result accuracy. The velocity-pressure coupling was established through the pressure-velocity correlation using the SIMPLE (semi-implicit method for pressure-linked equation) algorithm. Residuals were continuously monitored for continuity, x-velocity, y-velocity, k, and ε. The calculation time step was set to 5×10^{-8} s, and the total calculation time was greater than 0.5 ms. According to the geometric characteristics of the designed grooved surface, a series of regularly distributed grooves was selected as the research object, and the periodic boundary condition was set to simulate the actual situation. In this model, the designed surface and the smooth surface were located on the upper and the lower walls of the channel. The no-slip condition was adopted for the smooth surface so that the drag on the smooth surface and the hydrophobic grooved surface under the same conditions can be calculated at the same time. The height of the calculation domain was H = 450 μm, which was proven to be high enough to ensure the boundary layers of the upper and lower walls did not affect each other. Sufficient grid density was ensured in the area near the wall, and unstructured grids were used to densify the area near the groove wall to ensure calculation accuracy. The total number of grids used was 71,046. After the grid independence verification, the grid was obtained.

3. Results and Discussion

Observation of Gases

When the hydrophobic structured surface was immersed in static water, the partial pressure of water (Pv) in the entrapped gas was lower than the saturated vapor pressure (Ps), because the entrapped gas was carried from the atmosphere. According to Equation (5), evaporation at the liquid-gas interface occurred. Accompanied by the evaporation, the gas ascended until the partial water pressure reached the saturated vapor pressure. As shown in Figure 3, the gas on the submerged hydrophobic surface grew within 300 s (Figure 3a–c), but then it stopped growing (from Figure 3d–h). The results indicate that evaporation at the liquid-gas interface was transient. Generally, the phenomenon ends when the partial pressure of water reaches the saturated vapor pressure.

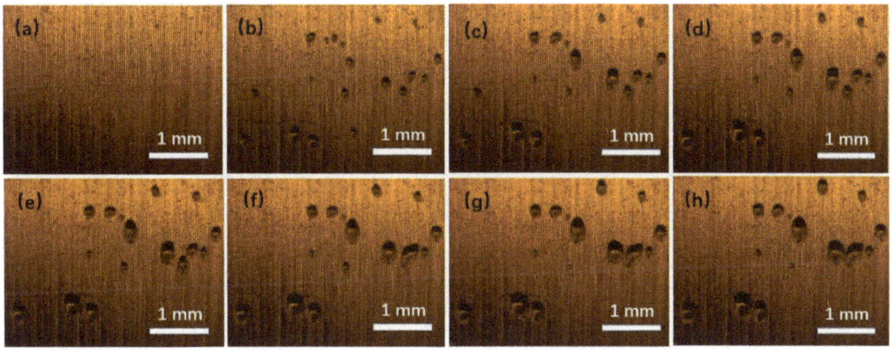

Figure 3. (color online) The state of gas at the submerged hydrophobic structured surface under static conditions at different time; (**a**) 0 s; (**b**) 150 s; (**c**) 300 s; (**d**) 450 s; (**e**) 600 s; (**f**) 750 s; (**g**) 900 s; (**h**) 1050 s.

As the results in Equation (5) and Figure 3 show, the evaporation rate declined gradually with increasing water concentration in the entrapped gas. As shown in Figure 4a, when the concentration of water in the trapped gas reached the saturated state, it achieved an equilibrium between evaporation and condensation at the liquid-gas interface. At equilibrium, the evaporation effect cannot be used to provide extra gas to sustain the nonwetted state of a submerged hydrophobic surface. However, according to Equation (5), if the partial pressure of water (Pv) in the entrapped gas stays lower than saturated vapor pressure (Ps), the equation rate is greater than zero, which indicates that the evaporation could go on. As shown in Figure 4b, once the gas layer reached high concentrations, gas was carried away by water flow, and the entrapped gas with a low concentration of water was achieved. Therefore, the partial pressure of water (Pv) in the trapped gas could remain low for a long time. Thus, under the action of flowing water, the concentration of water in entrapped gas cannot reach the saturated state, resulting in continuous evaporation.

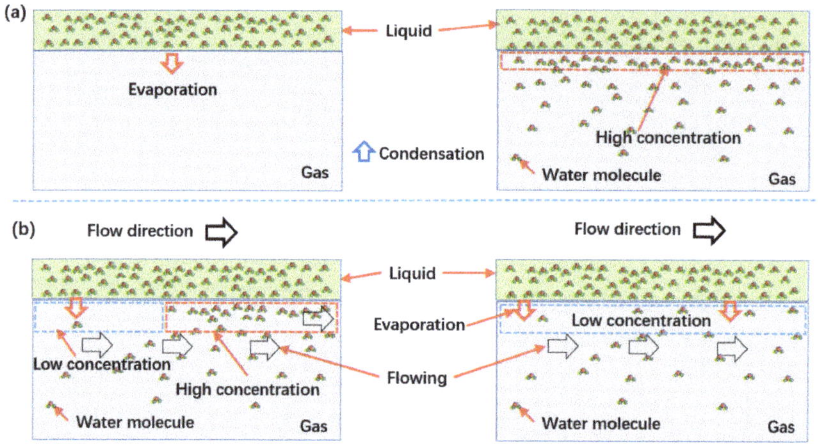

Figure 4. Schematic of evaporation at the liquid-gas interface, (**a**) for generation condition, the evaporation occurred when there was no water molecule in gas; the evaporation would stop due to the high concentration of water reaching to statured state; (**b**) under the action of flow, the water molecules in gas were washed away, the molecule with high concentration moved with flowing water; the low concentration could be sustained; therefore, the evaporation could keep happening.

To verify the influence of flowing water on the liquid-gas interface, a numerical simulation was carried out to investigate the flow field. As shown in Figure 5a, when water flows over the hydrophobic structured surface, the entrapped gas is sustained in the gap. According to the velocity distribution, the results indicate that the gas layer near the liquid-gas interface has velocity, which means that the gas layer with a high-water concentration was washed away by the flowing water. Therefore, the flowing water influenced the effects of the partial pressure of water on the evaporation rate. Because the field with a high concentration of water molecule was continually washed away by flowing water, the concentration of water was kept at a low level (Figure 5b). As a result, the partial pressure of water in entrapped gas could be considered a constant, and the evaporation rate didn't descend with time.

Experiments have verified this proposed approach for gas replenishment by evaporation at the liquid-gas interface. As shown in Figure 6, when water perpendicularly flowed over the HTM surface, the result indicates that the phase structure corresponding to the bright area exists at the solid-liquid interface underwater [30–32], which confirms that hydrophobic grooves can hold gases within the microstructures in flowing water. Furthermore, the gas on the HTM surface was not static and instead fluctuated in the flowing water. The alternating variation of bright (caused by liquid-gas interface) and dark lines was observed (Figure 6), which indicates that the gas entrapped in HTMs was continuously carried away and then replenished by renewal gas without any energy or gas-providing device to supply it [32]. To measure this gas generation effect, as shown in Figure 6, a particular area (10 mm × 5 mm) was selected to quantifiably illustrate the gas coverage on the designed HTM surface, and then these chosen images were processed by a brightness thresholding algorithm to identify wetted (corresponding to the dark area) and nonwetted (the light area caused by gas) regions. To minimize the error caused by the definition of the threshold value, the OTSU algorithm (which depends on the image itself) was used to calculate this gray value. When the gray value of a pixel exceeded this threshold (141), calculated by the OTSU algorithm [33], the pixel was identified as nonwetted; otherwise, the pixel was identified as wetted.

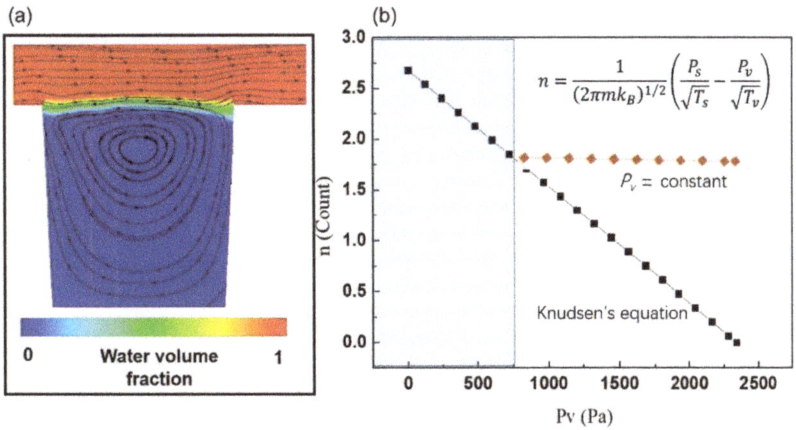

Figure 5. Influence of flowing water on the evaporation rate at the liquid-gas interface; (**a**) the flow field of hydrophobic structured surface; (**b**) the evaporation rate under the action of flowing water, when the concentration of water was kept at a low level, the partial pressure of water in entrapped gas could be considered a constant, and the evaporation rate didn't descend with time.

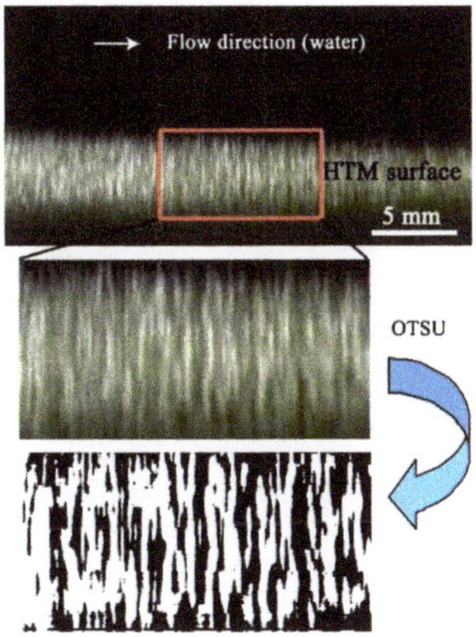

Figure 6. (color online) Image of gas on the HTM surface in flowing water, the selected part used for quantitative analysis of the growth rate and reduction rate of gas, and the image processed by the OTSU threshold method to differentiate the nonwetted area on the HTM surface.

As shown in Figure 7, the results indicated that the gas coverage on the HTM surface was averagely stable, with immersion time in flowing water. In Figure 7, the fluctuation of the gas coverage with time showed that the gas was continually carried off (corresponding to the areas changing from light to dark) and was replenished by the renewal gas (corresponding to the regions changing from dark to light) in flowing water. The fluctuation of gas coverage was induced by the successive alternation of gas generation and gas loss [32].

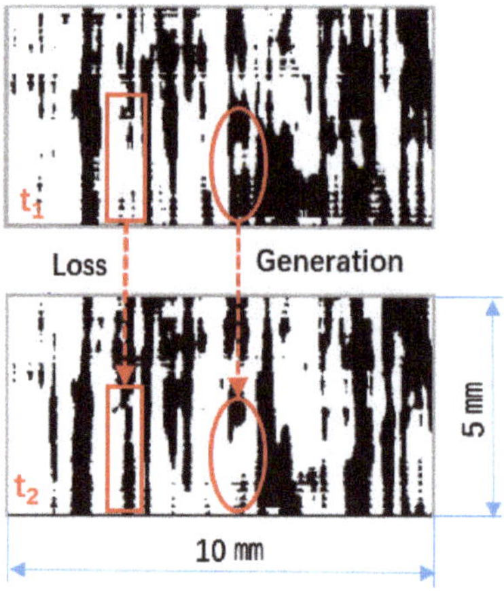

Figure 7. (color online) The gas coverage achieved by the threshold method on HTM surface in flowing water at different times, t_1 and t_2 is the different time point, the time interval between t_1 and t_2 is 0.125 s.

The entrapped gas within the surface was essential for gas generation, and the flow velocity was also a key factor for gas generation. In the experiments, the fluctuation of gas coverage was induced by the successive alternation of gas generation and gas loss. Therefore, the gas automatic generation rate can be gauged by the sum of the new light pixels corresponding to the light area transformed from the dark during a unit time and it can be described as flow:

$$G\% = \frac{Q_1 + Q_2 + Q_3 + \ldots\ldots + Q_n}{ntQ_{total}} \times 100\%, \quad (8)$$

where G is the gas growth rate; n is the count of pairs of images; t is the interval time between two close images; Q_i is the count of pixels becoming light from the i-th image to $(i + 1)$-th one; Q_{total} is the total count of pixels in the selected image. Therefore, the reduction rate of gas (R) can be described as flow:

$$R\% = \frac{q_1 + q_2 + q_3 + \ldots\ldots + q_n}{ntQ_{total}} \times 100\%, \quad (9)$$

where, q_i the count of pixels becoming dark from the i-th image to $(i + 1)$-th one.

In the statistics, the time interval between each pair of images was set as 0.125 s, which is short enough to capture the progressive increase in a light area; 16 pairs of images were selected, which is enough to describe the fluctuation of the gas coverage. As shown in Figure 8, the gas generation rate is approximately equal to the gas loss rate, indicating that the loss of gas in flow was supplemented by the generation of gas. The gas coverage at three positions (positions A, B and C shown in Figure 2) was (55.89 ± 1.42)%, (55.95 ± 1.98)%, and (57.28 ± 1.65)%, respectively.

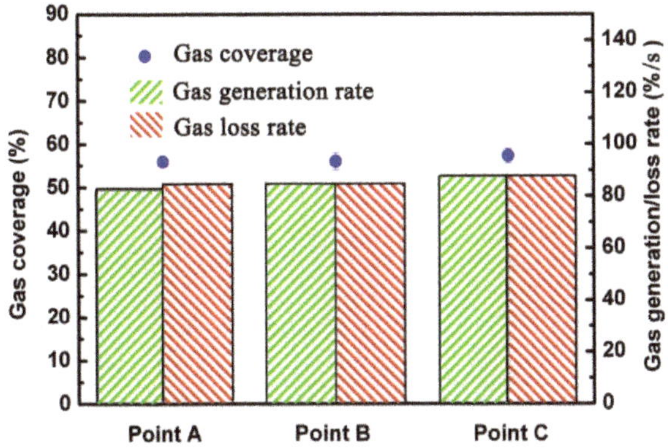

Figure 8. (color online) The gas coverage, gas generation rate, and gas loss rate at different sample positions with HTMs in flowing water at 5 m/s.

The entrapped gas within the surface was essential for gas generation, and the flow velocity was also a key factor for gas generation. Although the higher flow velocities resulted in faster removal rates of gas [13] on the HTM surface, the gas-generation rate was enhanced with the flow velocity increase. This observation indicates that the accelerating generation of gas supplemented the increasing loss of gas with the increase in flow velocity, as shown in Figure 9, which finally results in a durable gas film on the HTM surface.

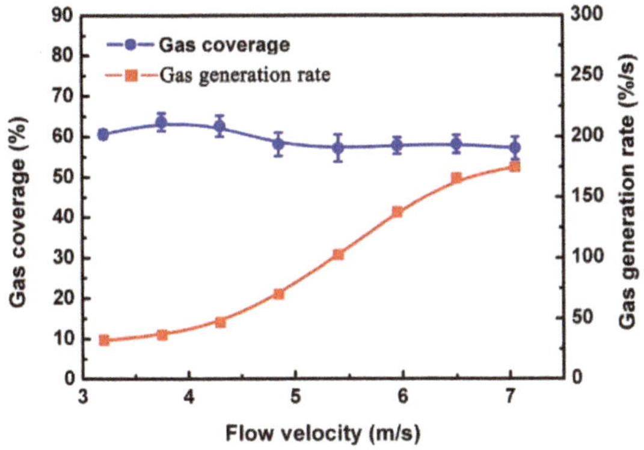

Figure 9. (color online) The gas coverage and generation rate on the HTM surface at different flow velocities.

When there was no gas-providing device or energy supply, the evaporation at the liquid-gas interfaces above the microgrooves and the evolution of dissolved gas might cause the gas generation underwater. However, considering the evolution of dissolved gas, from the previous investigation, the time required for a gas bubble to grow ten times in size was several hundred seconds, even when the ratio of the dissolved gas concentration to the dissolved gas concentration for a statured solution was 1.25 [22]. Therefore, the growth rate of gas by evolution was too slow to achieve this phenomenon in our experiment. Therefore, gas generation in flowing water should primarily be attributed to the evaporation at the liquid-gas interface. To clarify this point further, the observations of gas on the HTM in the water with different gas contents were performed.

Here, ambient air was chosen as the solute gas, the gas concentration of a volume of water was accurately controlled by adequately exposing (72 h) the volume to a constant gas pressure for equilibration before the experiment. The steady-state gas concentration of liquids was described in terms of ambient pressure conditions through Henry's Law [34]:

$$p = k_H c \tag{10}$$

where p is the partial pressure of gas component above the liquid, k_H is the Henry's constant (material property) for a particular liquid and gas, and c is the steady-state concentration of gas within the liquid. The states of HTM surface under different gas concentrations (60 kPa, 101 kPa, 200 kPa) were investigated. The gas concentration of 60 kPa, 101 kPa, and 200 kPa corresponded to undersaturation, saturation, and supersaturation, respectively. According to the optical test results, the statistical results were shown in Table 2.

Table 2. The effects of air content on the gas state on the HTM surface.

No.	Air Content (kPa)	State on HTM	Gas Coverage (%)	Gas Generation Rate (%)
1	60 (Undersaturation)		60.68 ± 3.39	92.48 ± 3.04
2	101 (Saturation)		57.12 ± 3.33	102.48 ± 6.39
3	200 (Supersaturation)		63.44 ± 4.75	95.78 ± 1.90

From the statistical results, both the gas coverage and the gas generation rate were not directly related to the air content in water, which conflicts with the precipitation of dissolved gas. Additionally, the entrapped gas (liquid-gas interface) was essential for regenerating gas, and the flow velocity significantly influenced the gas generation. Although the mechanism of gas generation needs to be fully characterized in future investigations, based on the results in this work, we surmise that the phenomenon of gas generation without any gas or energy supply should be attributed to the evaporation induced by flow at the liquid-gas interface.

In previous investigation [18], when the surface was just taken out from the water, the surface was covered by discrete water layers which were sticky to the surface. It suggested that the entrapped air was mostly removed. After the flow test, our experiment immediately removed the HTM surface from the water, and surface-wetting conditions were examined again. The result in Figure 10 indicates that the sticky water residues on the HTM surface were much less, and, after the test, the surface was still hydrophobic. The HTM surface still sustained a noticeable amount of gas within the surface, even in fully turbulent flow for effective drag reduction under water. Although the entrapped gas might not be durable enough for several days or longer due to the dissolution [34], this HTM surface improved the longevity of trapped gas in turbulent flow due to the evaporation at the liquid-gas interface.

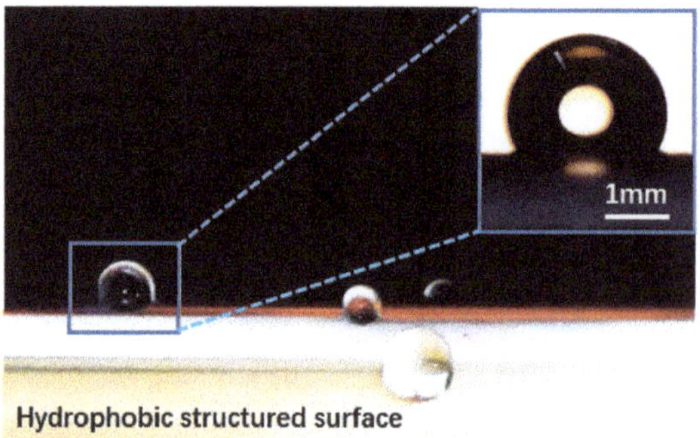

Figure 10. (color online) Images of a water droplet (5 μL) on the surfaces after the water tunnel experiments.

4. Conclusions

In summary, hydrophobic transverse microgrooves have been designed and fabricated to sustain air pockets in the gaps. The designed structures presented here successfully entrapped gas. Renewable gases were experimentally observed to form lubricating films at solid-liquid interfaces in flowing water. The mechanism of the gas generation should be attributed to evaporation at the liquid-gas interface induced by the flow. The results indicated that water molecules at the liquid-gas interface should escape to entrapped gas. When water content did not reach saturation levels, this phenomenon could be considered evaporation at the liquid-gas interface. With increasing water content in trapped gas, the evaporation rate at the liquid-gas interface descended gradually. Under the action of flowing, the substances containing high concentrations of water molecules at the liquid-gas interface were washed away, creating the low concentration of water molecules at the liquid-gas interfaces while evaporation continued to replenish the lost gas. Although further experimental and theoretical studies are needed to fully characterize the generation of gas and understand its underlying mechanism, this study can lead us down a new path toward achieving an effective underwater drag reduction approach and may further guide the development of related speed-enhancing and energy-saving technologies for underwater vehicles.

Author Contributions: Conceptualization, B.W. and X.L.; methodology, B.W.; software, B.W. and C.W.; validation, D.W., M.L. and D.Y.; formal analysis, B.W.; resources, B.W. and X.L.; data curation, B.W.; writing—original draft preparation, B.W.; writing—review and editing, B.W., D.Y., M.L. and X.L.; supervision, B.W., D.Y. and X.L. All authors have read and agreed to the published version of the manuscript.

Funding: This research was funded by the National Natural Science Foundation of China (Grant Nos. 52076036, 51775296, 51703116, and 51375253), the Northeast Petroleum University Youth Science Foundation of China (Grant No. 2019QNQ-01), and Tribology Science Fund of State Key Laboratory of Tribology (No. SKLTKF19B05).

Data Availability Statement: Not applicable.

Acknowledgments: This work was supported by the National Natural Science Foundation of China (Grant Nos. 52076036, 51775296, 51703116, and 51375253), the Northeast Petroleum University Youth Science Foundation of China (Grant No. 2019QNQ-01), and Tribology Science Fund of State Key Laboratory of Tribology (No. SKLTKF19B05).

Conflicts of Interest: The authors declare no conflict of interest.

References

1. Lee, C.; Choi, C.H. Structured surfaces for a giant liquid slip. *Phys. Rew. Lett.* **2008**, *101*, 064501. [CrossRef] [PubMed]
2. Lee, C.; Kim, C.J. Maximizing the giant liquid slip on superhydrophobic microstructures by nanostructuring their sidewalls. *Langmuir* **2009**, *25*, 12812–12818. [CrossRef] [PubMed]
3. Choi, C.-H.; Ulmanella, U.; Kim, J. Effective slip and friction reduction in nanograted superhydrophobic microchannels. *Phys. Fluids* **2006**, *18*, 087105. [CrossRef]
4. Zhu, Z.; Li, J.; Peng, H.; Liu, D. Nature-inspired structures applied in heat transfer enhancement and drag reduction. *Micromachines* **2021**, *12*, 656. [CrossRef]
5. Qin, L.; Hafezi, M.; Yang, H.; Dong, G.; Zhang, Y. Constructing a dual-function surface by microcasting and nanospraying for efficient drag reduction and potential antifouling capabilities. *Micromachines* **2019**, *10*, 490. [CrossRef]
6. Yao, J.; Teo, C.J. Effect of the liquid–gas interface curvature for a superhydrophobic surface with longitudinal grooves in turbulent flows. *Phys. Fluids* **2021**, *33*, 075116. [CrossRef]
7. Rajappan, A.; Golovin, K.; Tobelmann, B.; Pillutla, V.; Abhijeet; Tuteja, A.; McKinley, G.H. Influence of textural statistics on drag reduction by scalable, randomly rough superhydrophobic surfaces in turbulent flow. *Phys. Fluids* **2019**, *31*, 042107. [CrossRef]
8. Moaven, K.; Rad, M.; Taeibi-Rahni, M. Experimental investigation of viscous drag reduction of superhydrophobic nano-coating in laminar and turbulent flows. *Exp. Therm. Fluid Sci.* **2013**, *51*, 239–243. [CrossRef]
9. McHale, G.; Flynn, M.R.; Newton, M.I. Plastron induced drag reduction and increased slip on a superhydrophobic sphere. *Soft Matter.* **2011**, *7*, 10100–10107. [CrossRef]
10. Govardhan, R.N.; Srinivas, G.S.; Asthana, A.; Bobji, M.S. Time dependence of effective slip on textured hydrophobic surfaces. *Phys. Fluids* **2009**, *21*, 052001. [CrossRef]

11. Sakai, M.; Nakajima, A.; Fujishima, A. Removing an Air Layer from a Superhydrophobic Surface in Flowing Water. *Chem. Lett.* **2010**, *39*, 482–484. [CrossRef]
12. Poetes, R.; Holtzmann, K.; Franze, K.; Steiner, U. Metastable underwater superhydrophobicity. *Phys. Rev. Lett.* **2010**, *105*, 166104. [CrossRef] [PubMed]
13. Forsberg, P.; Nikolajeff, F.; Karlsson, M. Cassie–Wenzel and Wenzel–Cassie transitions on immersed superhydrophobic surfaces under hydrostatic pressure. *Soft Matter.* **2011**, *7*, 104–109. [CrossRef]
14. Xu, M.; Grabowski, A.; Yu, N.; Kerezyte, G.; Lee, J.W.; Pfeifer, B.R. Superhydrophobic drag reduction for turbulent flows in open water. *Phys. Rev. Appl.* **2020**, *13*, 034056. [CrossRef]
15. Samaha, M.A.; Tafreshi, H.V.; Gad-el-Hak, M. Influence of flow on longevity of superhydrophobic coatings. *Langmuir* **2012**, *28*, 9759–9766. [CrossRef]
16. Xue, Y.; Chu, S.; Lv, P.; Duan, H. Importance of hierarchical structures in wetting stability on submersed superhydrophobic surfaces. *Langmuir* **2012**, *28*, 9440–9450. [CrossRef]
17. Wu, H.; Yang, Z.; Cao, B.; Zhang, Z.; Zhu, K.; Wu, B.; Chai, G. Wetting and dewetting transitions on submerged superhydrophobic surfaces with hierarchical structures. *Langmuir* **2017**, *33*, 407–416. [CrossRef]
18. Aljallis, E.; Sarshar, M.A.; Datla, R.; Sikka, V.; Jones, A.; Choi, C.H. Experimental study of skin friction drag reduction on superhydrophobic flat plates in high Reynolds number boundary layer flow. *Phys. Fluids* **2013**, *25*, 025103. [CrossRef]
19. Jacob, B.; Olivieri, A.; Miozzi, M.; Campana, E.F.; Piva, R. Drag reduction by microbubbles in a turbulent boundary layer. *Phys. Fluids* **2010**, *22*, 115104. [CrossRef]
20. Vakarelski, I.U.; Marston, J.O.; Chan, D.Y.C.; Thoroddsen, S.T. Drag Reduction by Leidenfrost Vapor Layers. *Phys. Rev. Lett.* **2011**, *106*, 214501. [CrossRef]
21. Ceccio, S.L. Friction drag reduction of external flows with bubble and gas injection. *Annu. Rev. Fluid Mech.* **2010**, *42*, 183–203. [CrossRef]
22. Lee, C.; Kim, C.J. Underwater Restoration and Retention of Gases on Superhydrophobic Surfaces for Drag Reduction. *Phys. Rev. Lett.* **2011**, *106*, 014502. [CrossRef] [PubMed]
23. Epstein, P.S.; Plesset, M.S. On the Stability of Gas Bubbles in Liquid-Gas Solutions. *J. Chem. Phys.* **1950**, *18*, 1505–1509. [CrossRef]
24. Eames, I.W.; Marr, N.J.; Sabir, H. The evaporation coefficient of water: A review. *Int. J. Heat Mass Transf.* **1997**, *40*, 2963–2973. [CrossRef]
25. Delchar, T.A. *Vacuum Physics and Techniques*; Chapter 1; Chapman and Hall: London, UK, 1993.
26. Hrubes, J.D. High-speed imaging of supercavitating underwater projectiles. *Exp. Fluids* **2001**, *30*, 57–64. [CrossRef]
27. Wang, B.; Wang, J.; Dou, Z.; Chen, D. Investigation of retention of gases in transverse hydrophobic microgrooved surfaces for drag reduction. *Ocean Eng.* **2014**, *79*, 58–63. [CrossRef]
28. Brassard, J.D.; Sarkar, D.K.; Perron, J. Fluorine based superhydrophobic coatings. *Appl. Sci.* **2012**, *2*, 453–464. [CrossRef]
29. Wang, J.D.; Chen, D.R. Criteria for Entrapped Gas under a Drop on an Ultrahydrophobic Surface. *Langmuir* **2008**, *24*, 10174–10180. [CrossRef]
30. Larmour, I.A.; Bell, S.E.J.; Saunders, G.C. Remarkably simple fabrication of superhydrophobic surfaces using electroless galvanic deposition. *Angew. Chem.* **2007**, *119*, 1740–1742. [CrossRef]
31. Sakai, M.; Yanagisawa, T.; Nakajima, A.; Kameshima, Y.; Okada, K. Effect of Surface Structure on the Sustainability of an Air Layer on Superhydrophobic Coatings in a Water−Ethanol Mixture. *Langmuir* **2009**, *25*, 13–16. [CrossRef]
32. Wang, B.; Wang, J.; Chen, D. Continual Automatic Generation of Gases on Hydrophobic Transverse Microrooved Surface. *Chem. Lett.* **2014**, *43*, 646–648. [CrossRef]
33. Otsu, N. An automatic threshold selection method based on discriminate and least squares criteria. *Denshi. Tsushin Gakkai Ronbunshi* **1979**, *63*, 349–356.
34. Sun, W.Y.; Kim, C.J. The role of dissolved gas in longevity of cassie states for immersed superhydrophobic surfaces. In *Micro Electro Mechanical Systems (MEMS)*; IEEE: Taipei, Taiwan, 2013; pp. 397–400.

micromachines

MDPI

Article

Nonlinear Thermal Diffusion and Radiative Stagnation Point Flow of Nanofluid with Viscous Dissipation and Slip Constrains: Keller Box Framework Applications to Micromachines

Omar T. Bafakeeh [1], Bilal Ahmad [2], Skeena Noor [2], Tasawar Abbas [2], Sami Ullah Khan [3], Muhammad Ijaz Khan [4,5,*], Samia Elattar [6], Sayed M. Eldin [7], Mowffaq Oreijah [8] and Kamel Guedri [8,9]

1 Department of Industrial Engineering, Jazan University, Jazan 82822, Saudi Arabia
2 Department of Mathematics, University of Wah, Wah Cantt 47040, Pakistan
3 Department of Mathematics, COMSATS University Islamabad, Sahiwal 57000, Pakistan
4 Department of Mathematics and Statistics, Riphah International University I-14, Islamabad 44000, Pakistan
5 Department of Mechanical Engineering, Lebanese American University, Beirut 1102, Lebanon
6 Department of Industrial & Systems Engineering, College of Engineering, Princess Nourah bint Abdulrahman University, P.O. Box 84428, Riyadh 11671, Saudi Arabia
7 Center of Research, Faculty of Engineering, Future University in Egypt, New Cairo 11835, Egypt
8 Mechanical Engineering Department, College of Engineering and Islamic Architecture, Umm Al-Qura University, Makkah 21955, Saudi Arabia
9 Research Unity: Materials, Energy and Renewable Energies, Faculty of Science of Gafsa, University of Gafsa, Gafsa 2100, Tunisia
* Correspondence: muhammadijaz.khan@lau.edu.lb

Abstract: The radiated flow of magnetized viscous fluid subject to the viscous dissipation phenomenon is numerically studied. The radiative phenomenon is addressed with nonlinear relations. Further, analysis is performed by using the slip effects and convective thermal flow constraints. The transformed problem is numerically evaluated using the Keller Box method. The physical parameter effects, such as the magnetic parameter for the velocity profile, Prandtl number, Brownian motion parameter and Biot number for the energy profile and Lewis number, and the thermophoresis parameter for the concentration profile are discussed. The obtained results suggest applications in enhancing the heat transfer phenomenon, thermal system, energy generation, heat transmission devices, power generation, chemical reactions, etc.

Keywords: thermal flow; nanofluid; stagnation point flow; thermal radiation; viscous dissipation; porous medium

1. Introduction

The interaction of magnetic forces has significance in engineering applications, such as machine cooling, the lubrication industry, oil pumping, etc. The concept is associated with the impact of electromagnetic body forces on fluid flows. The primary role of magnetic force is in the suction phenomenon, wall motion and blowing. The flow of electrically conducted fluids is usually considered to make the process more effective. The impact of an externally applied magnetic field of a certain strength causes electrical conductivity in the flow of the considered fluids. This is the traditional type of MHD flow mechanism. In sea water, an externally applied magnetic field has much lower thermal conductivity, so an external electrical field is applied to overcome this problem to control the flow. An intersected electric and magnetic field causes the production of a wall-parallel Lorentz force with the objective of varying the procedure of the pressure gradient driven in the boundary layer. Bhatti et al. [1] evaluated the magnetized flow of nanofluids with the enhancement of the solar energy framework. Sabu et al. [2] discussed the particles' shapes for MHD flow in a rotating flow with the passive control phenomenon. The Williamson nanofluid flow with MHD effects was examined by Song et al. [3]. Jie et al. [4] reported sinusoidal wave flows due to oscillations under a magnetic force impact. Wang et al. [5] performed a bioconvective investigation with the slip phenomenon created by Maxwell liquids.

Nanotechnology is extensively used in industry because nanometer-sized particles provide a number of physical and chemical resources. The term nanofluid was first introduced by Choi [6] in 1995. "Nanofluids" are beneficial for reducing abrasion and reducing in operating components, including pumps and compressors, and result in gasoline savings of more than 6%. It is quite possible that great financial savings can ultimately be made. The appearance of nanofluids is the focal point of research on the sliding of nanofluids in the presence of nanoparticles. Nanofluids consist of particles that are nanometer-sized that slide inside the base fluids. Nanofluids are artificially created colloids containing a base fluid and nanoparticles. Nanoparticles are widely used in improving heat transfer systems, thermal energy, new fuels, hybrid engines, prescription drugs, electricity production and fuel reduction and most cancer remedies. Convection heat transfer fluids, as well as mixtures of oil, ethylene glycol and water, have thermal conductivity, which is important for the constant switching between heat transfer and surface heat transfer. This latest form of liquid was first discovered by increases in thermal conductivity in conventional liquids. The heat transfer mechanism in nanofluid is imprecise, but predictable, as it clearly relies on various additional structures, such as surfactant effects, particle size and particle dispersion of dispersed debris. In addition to progress, nanofluids are regularly used in biomedicine for the identification of tumor cells using nano-scale preparation methods and delivery systems, and breaking blockages in the circulatory system in supply pathways in thallium research. Rasheed et al. [7] modeled the curved surface flow problem with Prandtl nanofluid via convective constraints. Zeeshan [8] reported the entropy-generation significance for viscous nanomaterials with a nonlinear radiation impact. Rasheed et al. [9] evaluated Jeffrey nanofluid due to a vertically moving cylinder with dissipation impact. Zeeshan et al. [10] predicted a rotating surface with a thin film of nanoparticles. Hybrid nanofluid analysis with ethylene glycol as the base liquid was explored by Zeeshan et al. [11].

The radiative phenomenon is an interesting mode of heat transfer consisting of the transfer of energy as electromagnetic waves. At high temperatures, the radiation effects are important and have different applications in thermal systems, heat transfer devices, chemical processes, fusion and fission reactions, atomic blasts, nuclear systems, etc. In solar power, gas turbines, aircrafts, space vehicles and missile technology, the application of the radiative phenomenon is also observed. Zhao et al. [12] addressed the features of thermal radiation associated with street canyons. Ramesh et al. [13] presented the decomposition of the thermophoretic phenomenon for carbon nanotubes with radiative analysis. Famakinwa et al. [14] used nonlinear thermal radiation for examining the parallel plate flow of a hybrid nanomaterial with variable features. Ketchatel et al. [15] observed the onset of nonlinear thermal radiation for aluminum oxide and water-based nanoparticles in porous channels. Some modern developments regarding the fluid flow and nanoparticle concepts are listed in Refs. [16–21].

The aim of the current work is to report the stagnation point flow of nanofluid due to a moving surface with magnetic force and porous medium effects. The heat transfer phenomenon is observed by considering the viscous dissipation and radiative phenomenon. The nonlinear thermal radiation relationships are followed, instead of the traditional linear radiation phenomenon. The flow is subject to slip effects. Further, thermal analysis is performed using the convective boundary conditions. For the formulated flow system, the Keller Box numerical scheme is implemented to compute the numerical simulations. It is noted that the Keller Box scheme is widely used in the flow problem; however, numerical simulations for the nanofluid problem with these flow features have not yet been reported. The heat and mass transfer rate and wall shear stress are assessed using a useful resource of plotting the Sherwood Number and Nusselt number relative to several physical elements.

2. Mathematical Modeling

A two-dimensional, steady boundary layer incompressible flow of nanofluid is considered over a stretching sheet in the presence of solar radiation (thermal radiation). Flow occurs due to the linear velocity of a surface, which is mathematically written as $u_w(x) = ax$ and is considered through a porous medium. The y-axis of the coordinate system should be perpendicular to the stretching sheet denoted as the x-axis. Assume that B_0 is the uniform magnetic field strength, which is applied perpendicularly to the stretching sheet. T represents the fluid temperature and C represents the concentration. T_w is the fluid's reference temperature, while T_∞ is the fluid's ambient temperature. The effects of viscous dissipation are also incorporated during flow. Under these assumptions, the governing equations after using conservation law are modeled [8,15].

$$\frac{\partial u}{\partial x} + \frac{\partial v}{\partial y} = 0, \tag{1}$$

$$u\frac{\partial u}{\partial x} + v\frac{\partial v}{\partial y} = u_\infty \frac{\partial u_\infty}{\partial x} + v_f \frac{\partial^2 u}{\partial y^2} - \frac{\sigma_e {B_0}^2}{\rho_f}(u - u_\infty) - \frac{\nu}{k_p}(u_\infty - u), \tag{2}$$

$$\begin{aligned} u\frac{\partial T}{\partial x} + v\frac{\partial T}{\partial y} &= \alpha\frac{\partial^2 T}{\partial y^2} + \frac{v_f}{C_f}\left(\frac{\partial u}{\partial y}\right)^2 - \frac{1}{(\rho C)_f}\left(\frac{\partial q_r}{\partial y}\right) + \frac{\sigma_e B_0^2}{(\rho C)_f}(u_\infty - u)^2 \\ &+ \tau\left[D_B\frac{\partial T}{\partial y}\frac{\partial C}{\partial y} + \frac{D_T}{T_\infty}\left(\frac{\partial T}{\partial y}\right)^2\right] \end{aligned} \tag{3}$$

$$u\frac{\partial C}{\partial x} + v\frac{\partial C}{\partial y} = D_B\frac{\partial^2 C}{\partial y^2} + \frac{D_T}{T_\infty}\frac{\partial^2 T}{\partial y^2}, \tag{4}$$

With boundary conditions given as:

$$\begin{gathered} y = 0 : u = u_w(x) + L\frac{\partial u}{\partial y} = ax + L\frac{\partial u}{\partial y}, \; v = 0, -k\frac{\partial T}{\partial y} = h(T - T_w), \; C = C_w, \\ y \to \infty : u \to u_\infty(x) = bx, \; T \to T_\infty, \; C \to C_\infty. \end{gathered} \tag{5}$$

The mathematical expression for radiative heat flux is given as follows [13,14]:

$$q_r = -\frac{4\sigma^*}{3k^*}\frac{\partial T^4}{\partial y} = -\frac{16\sigma^* T^3}{3k^*}\frac{\partial T}{\partial y}, \tag{6}$$

Therefore, Equation (3) becomes:

$$\begin{aligned} u\frac{\partial T}{\partial x} + v\frac{\partial T}{\partial y} &= \frac{\partial}{\partial y}\left[\left(\alpha + \frac{16\sigma^* T^3}{3(\rho C)_f k^*}\right)\frac{\partial T}{\partial y}\right] + \frac{v_f}{C_f}\left(\frac{\partial u}{\partial y}\right)^2 + \frac{\sigma_e B_0^2}{(\rho C)_f}(u_\infty - u)^2 \\ &+ \tau\left[D_B\frac{\partial T}{\partial y}\frac{\partial C}{\partial y} + \frac{D_T}{T_\infty}\left(\frac{\partial T}{\partial y}\right)^2\right] \end{aligned} \tag{7}$$

In Equation (7) on the R.H.S., the first term is expressed as [14,15]:

$$\alpha\frac{\partial}{\partial y}\left[\left(1 + R_d(1 + (\theta_w - 1)\theta)^3\right)\frac{\partial T}{\partial y}\right] \tag{8}$$

where $Rd = \frac{16\sigma^* T_\infty^3}{3kk^*}$ is the radiation parameter. Equation (7) then becomes:

$$\begin{aligned} u\frac{\partial T}{\partial x} + v\frac{\partial T}{\partial y} &= \alpha\frac{\partial}{\partial y}\left[\left(1 + R_d(1 + (\theta_w - 1)\theta)^3\right)\frac{\partial T}{\partial y}\right] + \frac{v_f}{C_f}\left(\frac{\partial u}{\partial y}\right)^2 \\ &+ \frac{\sigma_e B_0^2}{(\rho C)_f}(u_\infty - u)^2 + \tau\left[D_B\frac{\partial T}{\partial y}\frac{\partial C}{\partial y} + \frac{D_T}{T_\infty}\left(\frac{\partial T}{\partial y}\right)^2\right] \end{aligned}$$

By using the similarity transformation

$$\eta = \sqrt{\frac{a}{v}}y,\ u = axf'(\eta),\ v = -\sqrt{av_f}f,\ \theta(\eta) = \frac{T-T_\infty}{T_w-T_\infty},\ \varphi(\eta) = \frac{C-C_\infty}{C_w-C_\infty}, \quad (9)$$

the continuity equation will be identically satisfied and Equations (2)–(4) can be written as:

$$f''' + ff'' - \{f'\}^2 + \lambda^2 + M(\lambda - f') - Kf' = 0 \quad (10)$$

$$\frac{1}{Pr}\left[\theta''\left(1 + R_d(1+(\theta_w-1)\theta)^3\right) + 3\theta'^2(\theta_w-1)R_d(1+(\theta_w-1)\theta)^2\right] + f\theta' + E_C^* f''^2 + ME_C^*(\lambda - f')^2 + N_b\theta'\varphi' + N_t\theta'^2 = 0 \quad (11)$$

$$\phi'' + Lef\varphi' + \frac{N_t}{N_b}\theta'' = 0 \quad (12)$$

With the thermal slip boundary conditions:

$$f(\eta) = 0,\ f'(\eta) = 1 + \beta f''(\eta),\ \theta'(\eta) = -Bi(1-\theta(\eta)),\ \varphi(\eta) = 1 \quad \text{as } \eta = 0 \quad (13)$$

$$\eta \to \infty : f'(\eta) \to \lambda, \theta(\eta) \to 0, \varphi(\eta) \to 0 \quad (14)$$

The involved physical parameters are defined as:
where $M = \frac{\sigma B_0}{\rho_f a}$ is the magnetic parameter, $K = \frac{v}{k_p}$ is the porosity parameter, $\beta = L\sqrt{\frac{a}{v_f}}$ represents the slip parameter, $\lambda = \frac{b}{a}$ represents the stretching parameter, $Bi = \frac{h}{k}\sqrt{\frac{v_f}{a}}$ is the Biot number, $Nb = \frac{\tau D_B(C_w - C_\infty)}{v_f}$ is the Brownian motion parameter, $Nt = \frac{\tau D_T(T_w - T_\infty)}{T_\infty v_f}$ is the thermophoresis parameter, $E_c^* = \frac{U_W^2}{C_p(T_w - T_\infty)}$ is the local Eckert number, $Le = \frac{v}{D_B}$ is the Lewis number and $Pr = \frac{v_f}{a}$ is the Prandtl number.

3. Keller Box Scheme

The nonlinear coupled system justified via Equations (10)–(12) is numerically solved with the Keller Box scheme. This scheme was used due to its high accuracy. First, the governing problem is reduced into the first-order equations

$$f' = u \quad (15)$$

$$u' = v \quad (16)$$

$$\theta' = p \quad (17)$$

$$\varphi' = g \quad (18)$$

Equations (1)–(3) become:

$$v' + fv - u^2 + \lambda^2 + M(\lambda - u) - Ku = 0 \quad (19)$$

$$\frac{1}{Pr}\left[p'\left(1 + Rd(1+(\theta_w-1)\theta)^3\right) + 3p^2(\theta_w-1)Rd(1+(\theta_w-1)\theta)^2\right] + fp + N_b pg + N_t p^2 + Ec^* v^2 + MEc^*(\lambda - u)^2 = 0 \quad (20)$$

$$g' + Lefg + \frac{N_t}{N_b}p' = 0 \quad (21)$$

Therefore, BC becomes:

$$f(0) = 0,\ \ u(0) = 1 + \beta v,\ \ p(0) = -B_i[1-\theta(0)],\ \ \varphi(0) = 1 \quad \text{at } \eta = 0 \quad (22)$$

$$u(\infty) \to \lambda,\ \ \theta(\infty) \to 0,\ \ \phi(\infty) \to 0 \ \text{ at } \eta \to \infty \quad (23)$$

By applying the Keller Box method:

$$\frac{f_j - f_{j-1}}{h_j} = \frac{u_j + u_{j-1}}{2}$$

$$f_j - f_{j-1} - \frac{h_j}{2}(u_j + u_{j-1}) = 0 \tag{24}$$

$$\frac{u_j - u_{j-1}}{h_j} = \frac{v_j + v_{j-1}}{2}$$

$$u_j - u_{j-1} - \frac{h_j}{2}(v_j + v_{j-1}) = 0 \tag{25}$$

$$\frac{\theta_j - \theta_{j-1}}{h_j} = \frac{p_j + p_{j-1}}{2}$$

$$\theta_j - \theta_{j-1} - \frac{h_j}{2}(p_j + p_{j-1}) = 0 \tag{26}$$

$$\frac{\varphi_j - \varphi_{j-1}}{h_j} = \frac{g_j + g_{j-1}}{2}$$

$$\varphi_j - \varphi_{j-1} - \frac{h_j}{2}(g_j + g_{j-1}) = 0 \tag{27}$$

Using Equations (10)–(13) in Equations (5)–(7), we get:

$$v_j - v_{j-1} + h_j[\frac{1}{4}(f_j + f_{j-1})(v_j + v_{j-1}) - \left(\frac{u_j + u_{j-1}}{2}\right)^2 + \lambda^2 + M\left\{\lambda - \left(\frac{u_j + u_{j-1}}{2}\right)\right\} - \frac{K}{2}\{u_j + u_{j-1}\}] = 0 \tag{28}$$

$$\begin{aligned} &\tfrac{1}{Pr}\left[\left(\tfrac{p_j - p_{j-1}}{h_j}\right)\left(1 + Rd\left(1 + (\theta_w - 1)\left(\tfrac{\theta_j + \theta_{j-1}}{2}\right)\right)^3\right) + 3\left(\tfrac{p_j + p_{j-1}}{2}\right)^2 (\theta_w - 1) Rd\left(1 + (\theta_w - 1)\left(\tfrac{\theta_j + \theta_{j-1}}{2}\right)\right)^2\right] + \left(\tfrac{f_j + f_{j-1}}{2}\right)\left(\tfrac{p_j + p_{j-1}}{2}\right) \\ &\quad + N_b\left(\tfrac{p_j + p_{j-1}}{2}\right)\left(\tfrac{g_j + g_{j-1}}{2}\right) + N_t\left(\tfrac{p_j + p_{j-1}}{2}\right)^2 + Ec^*\left(\tfrac{v_j + v_{j-1}}{2}\right)^2 + MEc^*\left[\left(\lambda - \left(\tfrac{u_j + u_{j-1}}{2}\right)\right)^2\right] = 0 \end{aligned} \tag{29}$$

$$\begin{aligned} &p_j - p_{j-1} + \frac{h_j}{\left(1 + Rd\left(1 + (\theta_w - 1)\left(\frac{\theta_j + \theta_{j-1}}{2}\right)\right)^3\right)}\left[\tfrac{\Pr}{4}\left\{(f_j + f_{j-1})(p_j + p_{j-1}) + N_b(p_j + p_{j-1})(g_j + g_{j-1}) + N_t(p_j + p_{j-1})^2\right.\right. \\ &\quad \left.\left. + Ec^*(v_j + v_{j-1})^2 + MEc^*\left[(2\lambda - (u_j + u_{j-1}))^2\right]\right\} + 3\left(\tfrac{p_j + p_{j-1}}{2}\right)^2 (\theta_w - 1) Rd\left(1 + (\theta_w - 1)\left(\tfrac{\theta_j + \theta_{j-1}}{2}\right)\right)^2\right] = 0 \end{aligned} \tag{30}$$

$$g_j - g_{j-1} + \frac{h_j}{4} Le(f_j + f_{j-1})(g_j + g_{j-1}) + \frac{N_t}{N_b}(p_j - p_{j-1}) = 0 \tag{31}$$

4. Newton Method for Linearization

Linearize the system of nonlinear equations:

$$f_j^{(i+1)} = f_j^{(i)} + \delta f_j^{(i)} \tag{32}$$

$$u_j^{(i+1)} = u_j^{(i)} + \delta u_j^{(i)} \tag{33}$$

$$v_j^{(i+1)} = v_j^{(i)} + \delta v_j^{(i)} \tag{34}$$

$$p_j^{(i+1)} = p_j^{(i)} + \delta p_j^{(i)} \tag{35}$$

$$\theta_j^{(i+1)} = \theta_j^{(i)} + \delta \theta_j^{(i)} \tag{36}$$

$$\varphi_j^{(i+1)} = \varphi_j^{(i)} + \delta \varphi_j^{(i)} \tag{37}$$

Therefore, Equation (18) becomes:

$$\delta f_j - \delta f_{j-1} - \frac{h_j}{2}\left[\delta u_j + \delta u_{j-1}\right] = (r_1)_j \tag{38}$$

$$\delta u_j - \delta u_{j-1} - \frac{h_j}{2}\left[\delta v_j + \delta v_{j-1}\right] = (r_2)_j \tag{39}$$

$$\delta \theta_j - \delta \theta_{j-1} - \frac{h_j}{2}\left[\delta p_j + \delta p_{j-1}\right] = (r_3)_j \tag{40}$$

$$\delta \varphi_j - \delta \varphi_{j-1} - \frac{h_j}{2}\left[\delta g_j + \delta g_{j-1}\right] = (r_4)_j \tag{41}$$

where

$$\begin{aligned}(\boldsymbol{r_1})_j &= f_{j-1} - f_j + h_j u_{j-\frac{1}{2}} \\ (\boldsymbol{r_2})_j &= u_{j-1} - u_j + h_j v_{j-\frac{1}{2}} \\ (\boldsymbol{r_3})_j &= \theta_{j-1} - \theta_j + h_j p_{j-\frac{1}{2}} \\ (\boldsymbol{r_4})_j &= p_{j-1} - p_j + h_j g_{j-\frac{1}{2}}\end{aligned}$$

Equation (14) becomes:

$$\delta v_j[A_1] + \delta v_{j-1}[A_2] + \delta f_j[A_3] + \delta f_{j-1}[A_4] + \delta u_j[A_5] + \delta u_{j-1}[A_6] = \boldsymbol{r_5} \tag{42}$$

Equation (15) becomes:

$$\delta p_j[B_1] + \delta p_{j-1}[B_2] + \delta f_j[B_3] + \delta f_{j-1}[B_4] + \delta g_j[B_5] + \delta g_{j-1}[B_6] + \delta v_j[B_7] + \delta v_{j-1}[B_8] = \boldsymbol{r_6} \tag{43}$$

Equation (17) becomes:

$$\delta g_j[C_1] + \delta g_{j-1}[C_2] + \delta f_j[C_3] + \delta f_{j-1}[C_4] + \delta p_j[C_5] + \delta p_{j-1}[C_6] = \boldsymbol{r_7} \tag{44}$$

where

$$\boldsymbol{A_1} = 1 + \frac{h_j}{2} f_{j-\frac{1}{2}},\ \boldsymbol{A_2} = -1 + \frac{h_j}{2} f_{j-\frac{1}{2}},\boldsymbol{A_3} = \boldsymbol{A_4} = \frac{h_j}{2} v_{j-\frac{1}{2}},\ \boldsymbol{A_5} = \boldsymbol{A_6} = -\frac{1}{2} + \frac{M}{2} - \lambda - \frac{K}{2}$$

$$\boldsymbol{B_1} = 1 + \frac{1}{2} f_{j-\frac{1}{2}} + \frac{N_b}{2} g_{j-\frac{1}{2}} + N_t p_{j-\frac{1}{2}},\ \boldsymbol{B_2} = -1 + \frac{1}{2} f_{j-\frac{1}{2}} + \frac{N_b}{2} g_{j-\frac{1}{2}} + N_t p_{j-\frac{1}{2}}$$

$$\boldsymbol{B_3} = \boldsymbol{B_4} = \frac{1}{2} p_{j-\frac{1}{2}} + M E_c^*\left(f_{j-\frac{1}{2}} + \lambda\right),\ \boldsymbol{B_5} = \boldsymbol{B_6} = \frac{N_b}{2} p_{j-\frac{1}{2}},\ \boldsymbol{B_7} = \boldsymbol{B_8} = E_c^* p_{v-\frac{1}{2}}$$

$$\boldsymbol{C_1} = 1 + \frac{h_j}{2} Le f_{j-\frac{1}{2}},\ \boldsymbol{C_2} = 1 - \frac{h_j}{2} Le f_{j-\frac{1}{2}},\ \boldsymbol{C_3} = \boldsymbol{C_4} = \frac{h_j}{2} Le g_{j-\frac{1}{2}},\ \boldsymbol{C_5} = \boldsymbol{C_6} = \frac{N_t}{N_b}$$

and

$$\boldsymbol{r_5} = v_{j-1} - v_j - h_j[\left(f_{j-\frac{1}{2}}\right)\left(v_{j-\frac{1}{2}}\right) - \left(u_{j-\frac{1}{2}}\right)^2 - \lambda^2 - M\left\{\lambda^2 + \left(u_{j-\frac{1}{2}}\right)^2 - \lambda(u_j + u_{j-1})\right\} + \frac{K}{2}(u_j + u_{j-1})$$

$$\begin{aligned}\boldsymbol{r_6} = p_{j-1} - p_j + \frac{Pr}{1+Rd\left(1+(\theta_w-1)\left(\frac{\theta_j+\theta_{j-1}}{2}\right)\right)^3}&\left[\left(p_{j-\frac{1}{2}}\right)\left(f_{j-\frac{1}{2}}\right) + E_c^*\left(v_{j-\frac{1}{2}}\right)^2 + M E_c^*\left\{\lambda^2 + \left(f_{j-\frac{1}{2}}\right)^2 + \lambda(f_j + f_{j-1})\right\}\right. \\ &\left. + \tfrac{N_b}{2}\left(f_{j-\frac{1}{2}}\right)\left(g_{j-\frac{1}{2}}\right) + N_t\left(p_{j-\frac{1}{2}}\right)^2\right] + 3(\theta_w - 1) Rd\left(1 + (\theta_w - 1)\left(\tfrac{\theta_j+\theta_{j-1}}{2}\right)\right)^2\left(p_{j-\frac{1}{2}}\right)^2\end{aligned}$$

$$\boldsymbol{r_7} = g_{j-1} - g_j - \frac{h_j}{2} Le\left(g_{j-\frac{1}{2}}\right)\left(f_{j-\frac{1}{2}}\right) - \frac{N_t}{N_b}(p_j - p_{j-1})$$

$$\begin{gathered} f(0) = 0,\ \ u(0) = 1 + \beta v,\ \ p(0) = -B_i[1 - \theta(0)],\ \ \varphi(0) = 1 \quad \text{at } \eta = 0 \\ u(\infty) \to \lambda,\ \ \theta(\infty) \to 0,\ \ \phi(\infty) \to 0 \quad \text{at } \eta \to \infty \end{gathered} \tag{45}$$

5. Computational Method

In this section, the arising PDEs that are nonlinear along with the BCs are solved by using the Keller Box method. For this, non-linear equations are first converted into ordinary differential equations of the first order. The conversion process is carried out by using the following transformation:

$$(y_1, y_2, y_3, y_4, y_5, y_6, y_7, y_8, y_9, y_{10}) = (f, f', f'', f'''\theta, \theta', \theta'', \varnothing, \varnothing', \varnothing'')$$

$$\begin{pmatrix} y'_1 \\ y'_2 \\ y'_3 \\ y'_4 \\ y'_5 \\ y'_6 \\ y'_7 \\ y'_8 \end{pmatrix} = \begin{pmatrix} 1 \\ y_3 \\ y_4 \\ \left[y_3^2 - y_2y_4 - \lambda^2 - M(\lambda - y_3) - Ky_4\right] \\ y_6 \\ \frac{-1}{1+Rd(1+zy_4)^3}\left[\Pr\left\{y_1y_5 + \mathrm{Nb}y_5y_6 + \mathrm{Nt}(y_5)^2 + Ec^*y_3^2 + MEc(\lambda - y_2)^2\right\} + 3y_5^2zRd(1+zy_4)^2\right] \\ y_8 \\ -\left(Ley_2y_8 + \frac{N_t}{N_b}y'_6\right) \end{pmatrix}$$

$$\begin{pmatrix} y_1(0) \\ y_2(0) \\ y_3(0) \\ y_4(0) \\ y_5(0) \\ y_6(0) \\ y_7(0) \end{pmatrix} = \begin{pmatrix} u_1 \\ 1 + \beta f'' \\ \lambda \\ u_2 \\ -Bi(1 - y_4) \\ 1 \\ u_3 \end{pmatrix}. \quad (46)$$

The above system of first-order ODEs and corresponding ICs is resolved using the Keller Box method. The corresponding values of the mysterious ICs u_1, u_2 and u_3 are considered by Newton's approach to BCs $f'(\eta) \to \lambda, \theta(\eta) \to 0, \phi(\eta) \to 0$ as $n \to \infty$ and the boundary conditions are satisfied. The calculations were conducted using the mathematical software MATLAB.

6. Results and Discussion

A two-dimensional, steady boundary layer flow of MHD nanofluid through an extending sheet with the occurrence of thermal radiation was analyzed. We examined the graphical results of the problem from an actual perspective. The variations in speed, temperature and nanoparticles concentration with governing parameters, for example N_b, the parameter for Brownian motion; the Prandtl number Pr, N_t, the parameter for thermophoresis; the Biot number B_i and the magnetic parameter (M), were determined using the Keller Box technique and the mathematical calculations were processed utilizing the MATLAB program. The outcomes are presented in Tables and the diagrams displayed in Figures 1–12.

Figure 1 shows that the velocity in the boundary area was reduced by the slip parameter. Figure 2 shows that, by increasing the values of K, the velocity profile declined, i.e., with an increase in the value of K, the boundary layer thickness increased, so the velocity decreased with the increase in the value of K because the effect of the porous medium, which would act against the flow, also increased and enhanced the deceleration of the flow.

The results of magnetic field parameter M over the dimensionless velocity and temperature are displayed in Figures 3 and 4, respectively. Figure 4 shows the effect of M on the velocity profile; it caused the velocity profile to decrease with a decrease in force, which resulted in a reduction in speed. In addition, higher values of M decreased the boundary layer thickness. In the Lorentz force, expansion was caused by the magnetic field; subsequently, the highly restricting power made it difficult for the liquid to stream without any hindrance. The magnetic field acted against the vehicle cycle. Figure 4 shows the impact of the magnetic field on the temperature profiles, which increased with an increase in M. The Lorentz force tended to increase the temperature in nanofluid movement. Thus, the thermal boundary layer thickness increased for a more grounded magnetic field. Figure 5 illustrates the impact of stretching boundary λ on the speed profile. It is apparent that expanding the upper side resulted in an increase in the speed field. However, expanding the upper side decreased the thickness of the force and limited the layers for both the non-Newtonian and Newtonian liquid cases.

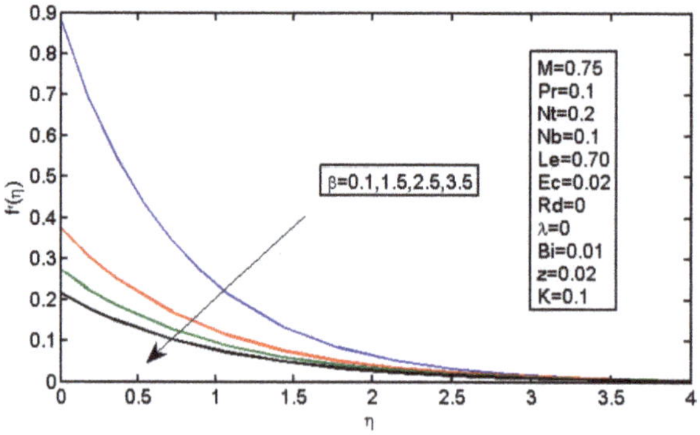

Figure 1. Impact of β on the velocity profile.

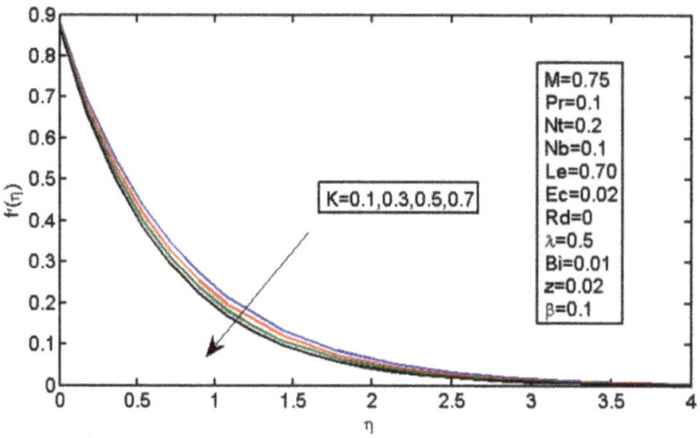

Figure 2. Impact of K on the velocity profile.

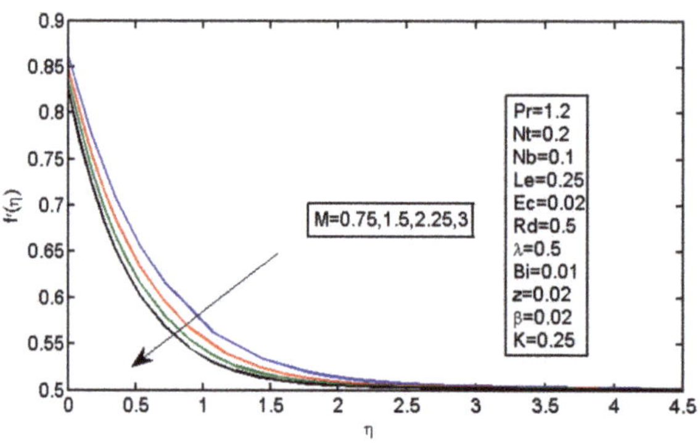

Figure 3. Impact of M on the velocity profile.

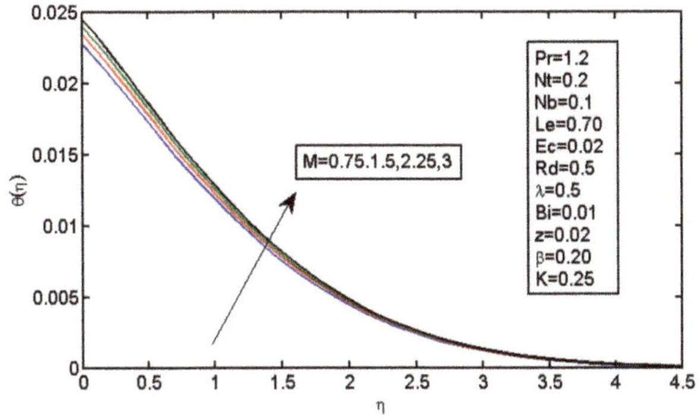

Figure 4. Impact of M on the temperature profile.

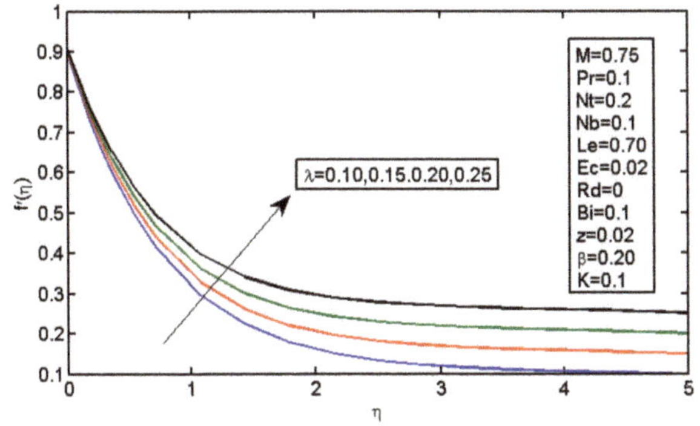

Figure 5. Impact of λ on the velocity profile.

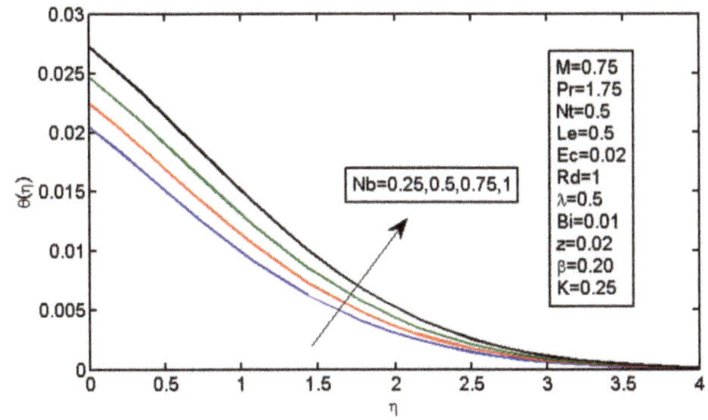

Figure 6. Impact of N_b on the temperature profile.

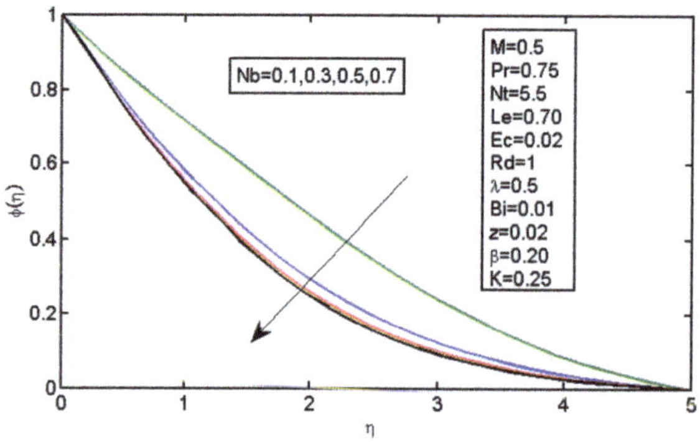

Figure 7. Impact of N_b on the concentration profile.

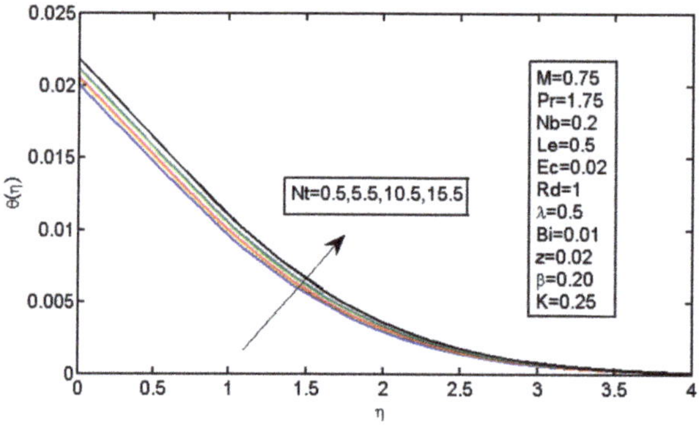

Figure 8. Impact of N_t on the temperature profile.

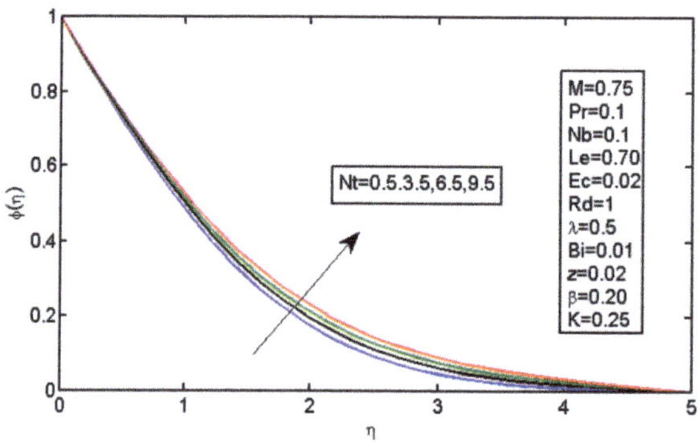

Figure 9. Impact of N_t on the concentration profile.

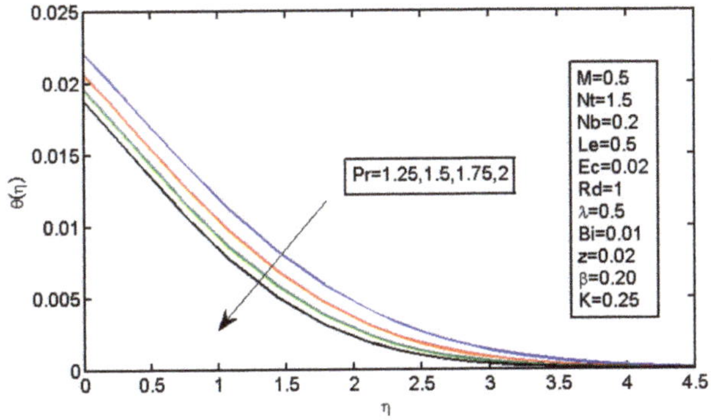

Figure 10. Impact of Pr on the temperature profile.

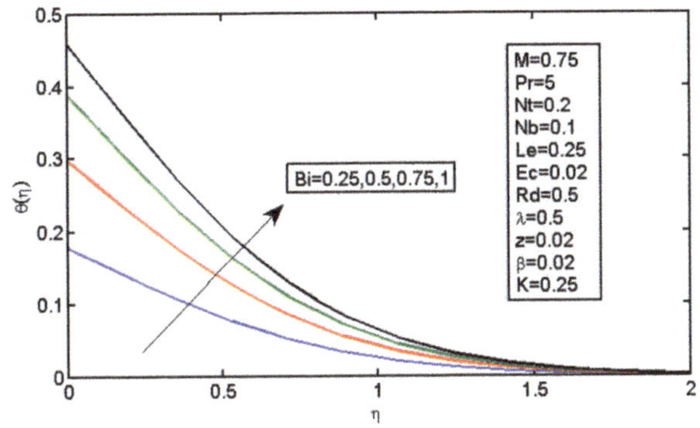

Figure 11. Impact of B_i on the temperature profile.

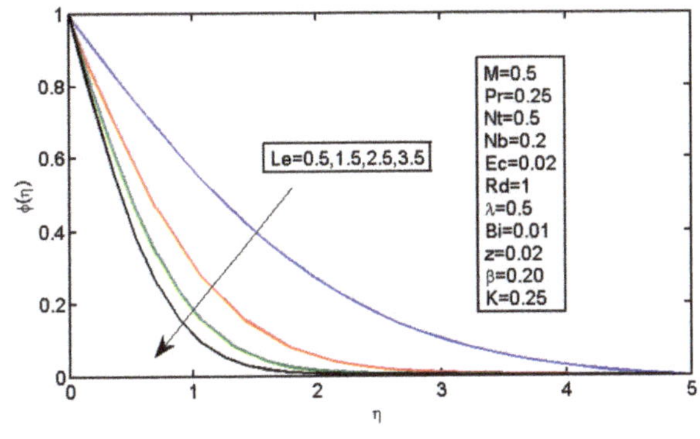

Figure 12. Impact of Le on the concentration profile.

Figures 6 and 7 illustrate the impact of Brownian motion parameter N_b on the temperature and nanoparticle volume concentration profile, respectively. Figure 6 presents the effects of different values of N_b on the thermal boundary layer. At this phase, Brownian motion occurred due to the size of the nanoparticles in the nanofluids. In the transfer of heat, the motion of the particles in the fluid played a significant role. In the flow, the nanoparticles' motion was affected by an increase in N_b. The temperature of the nanofluid increased when the kinetic energy of the nanoparticles was increased by the strength of this chaotic motion. Figure 7 illustrates the variations in the nanoparticle concentration with N_b. For stronger Brownian motion, a slight decrease in φ occurred accounted away from the sheet. The impact of thermophoresis parameter N_t on the temperature profile and the concentration profile is presented in Figures 8 and 9. Figure 8 illustrates the impact of the thermophoresis parameter of nanoparticles and thermal radiation on the thermal boundary layer. Additionally, it shows that the expansion in the thermophoresis boundary resulted in an increase in the thermophoretic force that transported nanoparticles from warm to cool spaces; however, the temperature profile increased due to this. Figure 9 shows the impact of N_t on the nanoparticle concentration profile. With an increase in N_t, there was an increase in the wideness of the concentration boundary and overshoot formed near the wall. Figure 10 shows the impact of the Prandtl number on the nanofluid's thermal boundaries. Practically, if Pr increased, there would be a decrease in the thermal diffusivity, and this principle would diminish the ability to transfer energy, decreasing the thermal boundary layer. Figure 11 depicts the impact of B_i on the temperature profile. As the Biot number increased, the thermal boundary increased. Figure 12 illustrates the impact of Le on the nanoparticle concentration profile. At higher values of Le, the concentration of the nanoparticles greatly decreased. As Le increased, the thermal boundary layer thickness also increased; however, the thickness of the concentration boundary layer decreased.

Table 1 indicates that, for a weaker Brownian movement, $N_b = 0.1$, the Nusselt number decreased with an increase in N_t. As the Brownian motion increased (for example, N_b changed from 0.1 to 0.4), the decrease in the Nusselt number became critical with an expansion in N_t. This is reasonable as the strengthened Brownian movement was comparable to the instant movement of nanoparticles that stretched the wall to the dormant fluid. The heat movement rate at the wall $\theta'(0)$ was an increasing function of the radiation boundary. It was additionally seen that, for a given N_b value, the decrease in Nusselt number with N_t was greater in the case of $R_d = 0$ when compared with $R_d = 1$. These observations were obtained from the results of Zeeshan [8]. Table 2 provides information for the decreased Sherwood number. The Sherwood number decreased when N_t increased. This was because of the way that stronger thermophoresis impacts would generally move the nanoparticle volume fraction away from the surface, as the hot surface repulsed the sub-micron-sized particles, subsequently forming a moderately molecule-free layer close to the surface. For weaker Brownian movements and stronger Brownian movements, this decrease was constant.

Table 1. Values of the decreased Nusselt number $-\theta'(0)$ when $Pr = 7, M = 0.5, \lambda = 0.5, Ec = 0.2, \theta_w = 1.5, Le = 10$ and $B_i = 0.1$.

N_b	R_d	N_t			
		0.1	0.2	0.3	0.4
0.1	0	0.07543	0.07418	0.07271	0.07095
	1	0.3597	0.3584	0.3570	0.3556
0.2	0	0.06623	0.06365	0.06042	0.05623
	1	0.3486	0.3469	0.3452	0.3434
0.3	0	0.05249	0.04712	0.04013	0.03023
	1	0.3362	0.3341	0.3319	0.3296
0.4	0	0.03307	0.02328	0.01018	−0.006327
	1	0.3225	0.3200	0.3173	0.3145

Table 2. Values of the decreased Sherwood number $-\varphi'(0)$ when $Pr = 7, M = 0.5, \lambda = 0.5, Ec = 0.2, \theta_w = 1.5, Le = 10$ and $B_i = 0.1$.

N_b	R_d	N_t			
		0.1	0.2	0.3	0.4
0.1	0	2.43	2.535	2.65	2.783
	1	2.37	2.396	2.424	2.456
0.2	0	2.417	2.5	2.597	2.714
	1	2.364	2.384	2.406	2.429
0.3	0	2.418	2.504	2.609	2.737
	1	2.363	2.381	2.401	2.422
0.4	0	2.424	2.518	2.630	2.756
	1	2.362	2.380	2.399	2.419

7. Conclusions

This research is about the numerical examination of the impacts of thermal radiation on the MHD nanofluid flow when an extending sheet that is not linear is saturated in a permeable medium. The Keller Box method was used to obtain a numerical solution for governing non-linear equations after transformation by using similarity transformation. The mass and heat transfer rates, shear stress and other factors, such as the nanoparticle concentration profile, temperature and velocity, were computed numerically by using MATLAB, and their graphical representation is also provided. The MHD nanofluid flow was used to investigate the effects of slip with velocity when thermal radiation is applied. Some of the observations of interest in this study are

- The increase in the stretching sheet parameter suggests an increase in the velocity profile;
- The increase in the parameter of velocity slip decreased the velocity profile;
- The increase in the porosity parameter reduced the velocity profile;
- When the Brownian motion parameter increased, the nanoparticle concentration decreased, whereas the temperature increased;
- The nanoparticle concentration and temperature both increased with an increase in the thermophoresis parameter;
- By increasing the Prandtl number, the temperature profile decreased;
- The increase in the Biot number indicated an increase in the temperature profile;
- As *Le* increased, the concentration decreased;
- By increasing the Brownian motion parameter, the Sherwood number increased in contrast to the increase in the thermophoresis parameter;
- These results can be further extended for different non-Newtonian fluid models, such as Jeffrey fluid, viscoelastic fluid, couple stress fluid, Maxwell fluid, Oldroyd-B fluid, etc. Moreover, some other features can be discussed for current flow problems, such as entropy generation, heat source, activation energy, bioconvection, etc.

Author Contributions: All authors are equally contributed in the mathematical modeling, data analysis and final drafting. All authors have read and agreed to the published version of the manuscript.

Funding: The authors would like to thank the Deanship of Scientific Research at Umm Al-Qura University for supporting this work by Grant Code: 22UQU4340474DSR15. Princess Nourah bint Abdulrahman University Researchers Supporting Project number (PNURSP2022R163), Princess Nourah bint Abdulrahman University, Riyadh, Saudi Arabia.

Institutional Review Board Statement: Not applicable.

Informed Consent Statement: Not applicable.

Data Availability Statement: All the data are fully available in the manuscript.

Acknowledgments: The authors would like to thank the Deanship of Scientific Research at Umm Al-Qura University for supporting this work by Grant Code: 22UQU4340474DSR15. Princess Nourah bint Abdulrahman University Researchers Supporting Project number (PNURSP2022R163), Princess Nourah bint Abdulrahman University, Riyadh, Saudi Arabia.

Conflicts of Interest: The authors declare no conflict of interest.

Nomenclature

a Stretching rate	T Fluid temperature
C Concentration	T_w Surface temperature
T_∞ Free stream temperature	u_∞ Free stream velocity
(u, v) Velocity components	(x, y) Cartesian coordinates
σ_e Electrically conductivity	B_0 Magnetic force strength
T Fluid temperature	C Fluid concentration
D_B Brownian diffusion	q_r Radiative flux
v_f Fluid viscosity	α Thermal conductivity
L Slip factor	M Magnetic parameter
Rd Radiation parameter	β Slip parameter
Bi Biot number	Nb Brownian motion parameter
K Porosity parameter	λ Stretching parameter
Nt Thermophoresis parameter	Le Lewis number and
E_c^* Local Eckert number	Pr Prandtl number

References

1. Bhatti, M.M.; Bég, O.A.; Abdelsalam, S.I. Computational framework of magnetized MgO-Ni/water-based stagnation nanoflow past an elastic stretching surface: Application in solar energy coatings. *Nanomaterials* **2022**, *12*, 1049. [CrossRef] [PubMed]
2. Sabu, A.S.; Wakif, A.; Areekara, S.; Mathew, A.; Shah, N.A. Significance of nanoparticles' shape and thermo-hydrodynamic slip constraints on MHD alumina-water nanoliquid flows over a rotating heated disk: The passive control approach. *Int. Commun. Heat Mass Transf.* **2021**, *129*, 105711. [CrossRef]
3. Song, Y.-Q.; Hamid, A.; Sun, T.-C.; Khan, M.I.; Qayyum, S.; Kumar, R.N.; Prasannakumara, B.C.; Khan, S.U.; Chinram, R. Unsteady mixed convection flow of magneto-Williamson nanofluid due to stretched cylinder with significant non-uniform heat source/sink features. *Alexandria Eng. J.* **2022**, *61*, 195–206. [CrossRef]
4. Jie, Z.; Khan, M.I.; Al-Khaled, K.; El-Zahar, E.R.; Acharya, N.; Raza, A.; Khan, S.U.; Xia, W.-F.; Tao, N.-X. Thermal transport model for Brinkman type nanofluid containing carbon nanotubes with sinusoidal oscillations conditions: A fractional derivative concept. *Waves Random Complex Media* **2022**, *32*, 1–20. [CrossRef]
5. Wang, F.; Ahmad, S.; Al Mdallal, Q.; Alammari, M.; Khan, M.N.; Rehman, A. Natural bio-convective flow of Maxwell nanofluid over an exponentially stretching surface with slip effect and convective boundary condition. *Sci. Rep.* **2022**, *12*, 2220. [CrossRef]
6. Choi, S.U.S. Nanofluids: A new field of scientific research and innovative applications. *Heat Transf. Eng.* **2008**, *29*, 429–431. [CrossRef]
7. Rasheed, H.U.; Islam, S.; Zeeshan; Khan, J.; Abbas, T.; Mohmand, M.I. Numerical solution of chemically reactive and thermally radiative MHD Prandtl nanofluid over a curved surface with convective boundary conditions. *Z. Angew. Math. Mech.* **2021**, e202100125. [CrossRef]
8. Zeeshan. Second law and entropy generation analysis of magnetized viscous fluid flow over a permeable expandable sheet with nonlinear thermal radiation: Brownian and thermophoresis effect. *Adv. Mech. Eng.* **2022**, *14*, 1–9. [CrossRef]
9. Rasheed, H.U.; Al-Zubaidi, A.; Islam, S.; Saleem, S.; Khan, Z.; Khan, W. Effects of Joule heating and viscous dissipation on magnetohydrodynamic boundary layer flow of jeffrey nanofluid over a vertically stretching cylinder. *Coatings* **2021**, *11*, 353. [CrossRef]
10. Zeeshan; Rasheed, H.U.; Khan, W.; Khan, I.; Alshammari, N.; Hamadneh, N. Numerical computation of 3D Brownian motion of thin film nanofluid flow of convective heat transfer over a stretchable rotating surface. *Sci. Rep.* **2022**, *12*, 2708. [CrossRef]
11. Zeeshan; Khan, I.; Weera, W.; Mohamed, A. Heat transfer analysis of Cu and Al_2O_3 dispersed in ethylene glycol as a base fluid over a stretchable permeable sheet of MHD thin-film flow. *Sci. Rep.* **2022**, *12*, 8878. [CrossRef] [PubMed]
12. Zhao, H.; Xu, G.; Shi, Y.; Li, J.; Zhang, Y. The characteristics of dynamic and non-uniform thermal radiation experienced by pedestrians in a street canyon. *Build. Environ.* **2022**, *222*, 109361. [CrossRef]
13. Ramesh, G.; Madhukesh, J.; Shah, N.A.; Yook, S.-J. Flow of hybrid CNTs past a rotating sphere subjected to thermal radiation and thermophoretic particle deposition. *Alexandria Eng. J.* 2022, *in press*. [CrossRef]
14. Famakinwa, O.A.; Koriko, O.K.; Adegbie, K.S. Effects of viscous dissipation and thermal radiation on time dependent incompressible squeezing flow of CuO-Al_2O_3water hybrid nanofluid between two parallel plates with variable viscosity. *J. Comput. Math. Data Sci.* **2022**, *5*, 100062. [CrossRef]

15. Ketchate, C.G.N.; Kapen, P.T.; Fokwa, D.; Tchuen, G. Stability analysis of mixed convection in a porous horizontal channel filled with a Newtonian Al2O3/Water nanofluid in presence of magnetic field and thermal radiation. *Chin. J. Phys.* **2022**, *79*, 514–530. [CrossRef]
16. Abbasi, A.; Farooq, W.; Tag-ElDin, E.S.M.; Khan, S.U.; Khan, M.I.; Guedri, K.; Elattar, S.; Waqas, M.; Galal, A.M. Heat transport exploration for hybrid nanoparticle (Cu, Fe_3O_4)-based blood flow via tapered complex wavy curved channel with slip features. *Micromachines* **2022**, *13*, 1415. [CrossRef]
17. Waqas, H.; Oreijah, M.; Guedri, K.; Khan, S.U.; Yang, S.; Yasmin, S.; Khan, M.I.; Bafakeeh, O.T.; Tag-ElDin, E.S.M.; Galal, A.M. Gyrotactic motile microorganisms impact on pseudoplastic nanofluid flow over a moving Riga surface with exponential heat flux. *Crystals* **2022**, *12*, 1308. [CrossRef]
18. Shahid, M.; Javed, H.M.A.; Ahmad, M.I.; Qureshi, A.A.; Khan, M.I.; Alnuwaiser, M.A.; Ahmed, A.; Khan, M.A.; Tag-ElDin, E.S.M.; Shahid, A.; et al. A brief assessment on recent developments in efficient electrocatalytic Nitrogen reduction with 2D non-metallic nanomaterials. *Nanomaterials* **2022**, *12*, 3413. [CrossRef]
19. Manzoor, N.; Qasim, I.; Khan, M.I.; Ahmed, M.W.; Guedri, K.; Bafakeeh, O.T.; Tag-Eldin, E.S.M.; Galal, A.M. Antibacterial applications of low pressure plasma on degradation of multidrug resistant V. Cholera. *Appl. Sci.* **2022**, *12*, 9737. [CrossRef]
20. Mamatha, S.U.; Devi, R.L.V.R.; Ahammad, N.A.; Shah, N.A.; Rao, B.M.; Raju, C.S.K.; Khan, M.I.; Guedri, K. Multi-linear regression of triple diffusive convectively heated boundary layer flow with suction and injection: Lie group transformations. *Int. J. Mod. Phys. B* **2022**, *in press*. [CrossRef]
21. Kiranakumar, H.V.; Thejas, R.; Naveen, C.S.; Khan, M.I.; Prasanna, G.D.; Reddy, S.; Oreijah, M.; Guedri, K.; Bafakeeh, O.T.; Jameel, M. A review on electrical and gas-sensing properties of reduced graphene oxide-metal oxide nanocomposites. *Biomass Convers. Biorefinery* 2022, *in press*. [CrossRef]

micromachines

MDPI

Article

The Effect of Fin Shape on the Heat Transfer and the Solution Time of a Microchannel Evaporator in a CO_2 Air Conditioning System—A Numerical Investigation

Tronghieu Nguyen and Thanhtrung Dang *

Department of Thermal Engineering, HCMC University of Technology and Education, Ho Chi Minh City 71307, Vietnam
* Correspondence: trungdang@hcmute.edu.vn; Tel.: +84-913-606-261

Abstract: Numerical simulations on the fin shape of a microchannel evaporator in a CO_2 air conditioning system were performed at the inlet evaporative temperature of 10 °C and the vapor quality of 0.61. Two types of fin shapes were dealt with: the straight fins and V-fins. The numerical results were verified by the experimental data. For the system under consideration and for the same heat transfer area and the heat transfer coefficient for the air side in the microchannel evaporator, the effect of the fin shape on the heat transfer was not different; however, the solution time and the physical memory for the straight fins were 1.3 and 1.45 times compared with the V-fins, respectively. Therefore, the V-fin shape should be used for numerical simulation to compare it with the straight fin shape. In this study, the evaporation of the refrigerant in the microchannel evaporator took place in four passes. The normal heat flux from the air through the fins and tubes was almost reached at 1550 W/m^2 at the evaporative temperature of 10 °C. The results obtained from the experimental data were in good agreement with those obtained from the numerical results, with a deviation of less than 10%.

Keywords: fin shape; CO_2; microchannel evaporator; heat flux; temperature profile

1. Introduction

Heat exchangers, especially fin tube heat exchangers, are widely used in industrial and domestic applications, such as air conditioners, heat pumps, refrigeration systems, cooling systems, etc. In the fin tube heat exchangers, the geometries of the fin, including the pitch, thickness, shape and angle, are important parameters, which strongly affect their capacities. Recently, many types of research related to the effect of the fin shape on the performance of the exchanger have been reported [1–6]. For instance, Hsieh et al. [5] optimized the louver finned-tube heat exchangers using the Taguchi method. The results indicated that three main factors, including fin collar outside diameter, transverse tube pitch and fin pitch, significantly influence the thermal hydraulic performance of the heat exchanger [5].

Recently, the application of CO_2 as a refrigerant in the fin tube heat exchangers has attracted particular attention due to its non-toxicity, safety and environmental friendliness. Unfortunately, the previous report showed that the use of CO_2 in an air conditioning system led to its low performance [7]. Thus, to increase the efficiency of heat exchange in the CO_2 systems and reduce their size as well, lately, the microchannel or minichannel heat exchangers have been installed in the systems [8–11]. Moreover, to develop the new CO_2 microchannel system in a short time and at a lower cost, the numerical simulation has been extensively used [9,12–15]. For example, based on the finite volume method, Jin et al. [11] developed and predicted the performance of a CO_2 transcritical system. The model had a mean deviation with the experimental cooling capacities of 1.9% and with refrigerant-side pressure drops of 12.3%. Using numerical analysis, Yun et al. [12] developed a microchannel evaporator for a CO_2 air conditioning system. The performance of the developed model can be improved by changing the mass flow rate and varying the fin pitch to expand the two-phase region. Furthermore, also applying the finite volume method, Kim and Bullard [13] developed a two-slab microchannel evaporator for a CO_2 air conditioning system. However, the results in Refs [7–15] did not mention the solution time for numerical simulation in the microchannel evaporator.

To verify the theoretical calculation and the numerical simulation or to find the physical rules for the models, the corresponding experimental results are needed for developing the fin tube heat exchangers. Patino et al. [16] used experimental data to demonstrate heat transfer correlations in a CO_2 evaporator model, which was developed using a finite volume method. For both numerical and experimental methods, Yang and Ning [17] studied a CO_2 double pipe evaporator to find out the effects of the tube diameters and the operation parameters on its performance. The experimental data yielded a good correlation with the evaporator heat transfer and pressure drop, verified by the relative errors of 5.21% and 3.78%, respectively.

From the literature above, it is found that numerical simulation is a strong method for developing a fin tube microchannel evaporator using CO_2 as a refrigerant. Moreover, the corresponding experiments are needed to enhance the system's reliability. The review results did not deal with the fin shape and solution time in more detail. Herein, we numerically investigate the effect of the fin shape and the solution time on the heat transfer of a microchannel evaporator in a CO_2 air conditioning system. There are two types of fin shapes: V-shaped fins and straight fins. Moreover, experiments are also carried out to demonstrate the numerical simulation results.

2. Methodology

2.1. Design of Evaporator Model

The flat tube evaporator was designed with a heat transfer area of 2.5 m^2 and a cooling capacity of 2.6 kW. Figure 1 displays the dimensions of the microchannel evaporator. The evaporator is composed of 29 flat tubes and divided into 6 passes. There are 3, 4, 5, 6, 6 and 5 flat tubes in the first, second, third, fourth, fifth and sixth pass, respectively. Each flat tube has 10 rectangular microchannels with dimensions of 1.2 mm × 0.6 mm.

There are two fin shapes being investigated with the same heat transfer area and the heat transfer coefficient for the air side. The straight fins have a height of 4.1 mm, and the V-fins have a height of 4.05 mm, as shown in Figure 1. Table 1 presents the detailed evaporator specifications.

Table 1. Detailed geometries of the microchannel evaporator with the straight fins and the V-fins.

Name	Specifications of Straight Fin	Specifications of V-Fin
Heat transfer area of evaporator (m^2)	2.5	2.5
Evaporator Size (L × H × W) (mm)	330 × 285 × 16	330 × 285 × 16
Flat tube size (mm)	1.3 × 300 × 16	1.3 × 300 × 16
Number of microchannels in a flat tube	10	10
Microchannel size (mm)	0.6 × 1.2	0.6 × 1.2
Fin size (mm)	4.1 × 0.1 × 16	4.05 × 0.1 × 16
Fin pitch (mm)	1.1	1.1
Angle of fin (deg)	0°	13.8°
Number of fins per flat tube	536	536
Heat transfer area of a flat tube (m^2)	81.3×10^{-3}	81.3×10^{-3}
Number of flat tubes in each pass	3-4-5-6-6-5	3-4-5-6-6-5

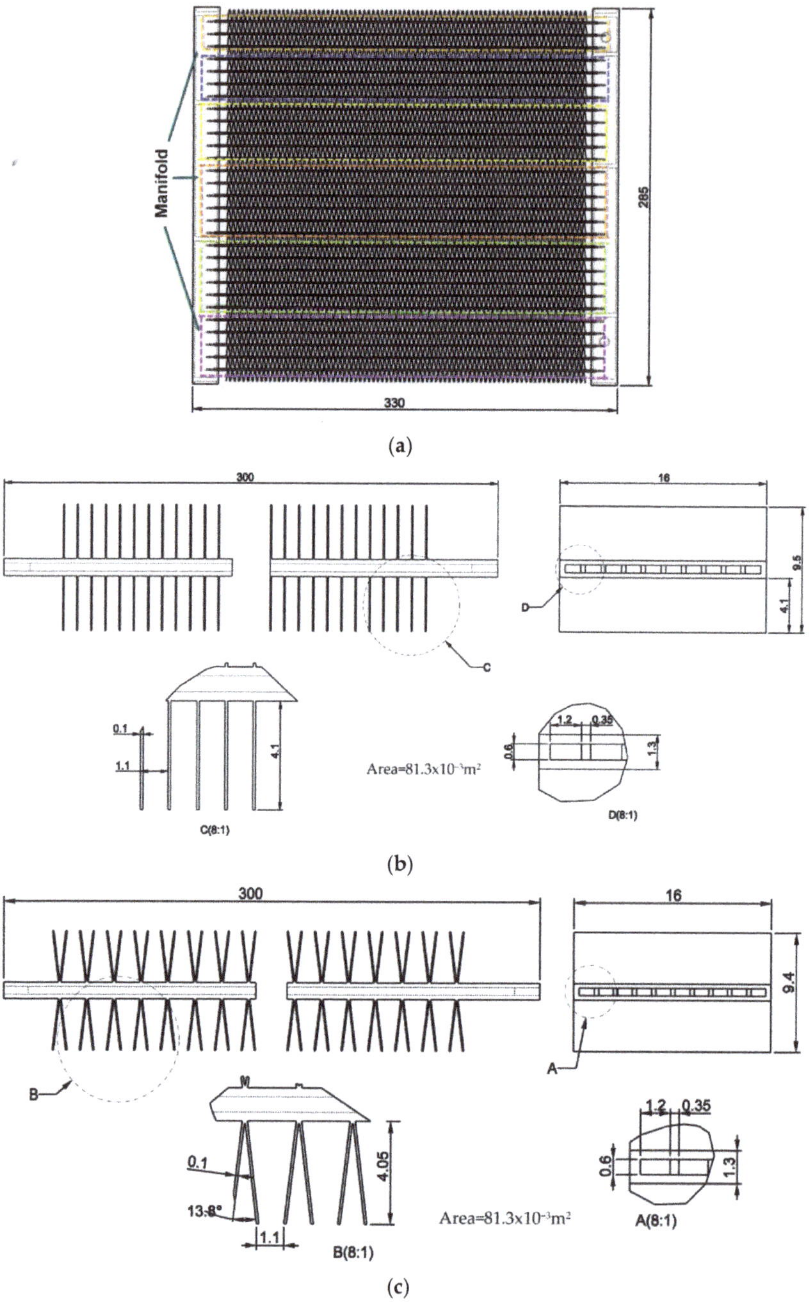

Figure 1. Dimensions of the microchannel evaporator: (**a**) Front view; (**b**) A flat tube drawing with the straight fins; (**c**) A flat tube drawing with the V-fins.

2.2. Mathematical Model

In this study, a turbulent model was used for numerical simulation. The governing equations are shown as follows [18].

The momentum formulation is as follows:

$$\rho\mathbf{u}\cdot\nabla(\mathbf{u}) = \nabla\cdot\left[-p + (\mu+\mu_T)\left(\nabla\mathbf{u} + (\nabla\mathbf{u})^T\right) - \frac{2}{3}(\mu+\mu_T)(\nabla\cdot\mathbf{u})\mathbf{I}\right] + F \tag{1}$$

The conservation of mass equation

$$\nabla\cdot(\rho\mathbf{u}) = 0 \tag{2}$$

Heat transfer equations

$$\rho C_p\mathbf{u}\cdot\nabla T + \nabla\cdot\mathbf{q} = \mathbf{Q} \tag{3}$$

$$\mathbf{q} = -k\nabla T \tag{4}$$

where T is the absolute temperature (K), p is the pressure (Pa), **u** is the fluid velocity (m/s), μ_T is the turbulence viscosity (Pa.s), F is the external forces applied to the fluid (N/m^3), ρ is the density (kg/m^3), Cp is the specific heat capacity at constant pressure (J/(kg·K)), Q contains additional heat sources (W/m^3), and q is the heat flux (W/m^2).

The phase change equations are as follows:

$$\rho = \theta_1\rho_1 + \theta_2\rho_2 \tag{5}$$

The specific heat capacity is

$$C_p = \frac{1}{\rho}\left(\theta_1\rho_1 C_{p,1} + \theta_2\rho_2 C_{p,2}\right) + L_{1-2}\frac{\partial\alpha_m}{\partial T} \tag{6}$$

The mass fraction is

$$\alpha_m = \frac{1}{2}\frac{\theta_2\rho_2 - \theta_1\rho_1}{\theta_1\rho_1 + \theta_2\rho_2} \tag{7}$$

The effective thermal conductivity is as follows:

$$k = \theta_1 k_1 + \theta_2 k_2 \tag{8}$$

The velocity of the refrigerant in the microchannel is

$$u_{ch} = \frac{\dot{m}}{n\rho A_{ch}} \tag{9}$$

The relationship between the vapor quality and the phase indicator is

$$x = \theta_2 \text{ and } 1 - x = \theta_1 \tag{10}$$

$$\theta_2 = \frac{h - h_1}{L_{1-2}} \tag{11}$$

where indices 1 and 2 indicate a refrigerant in phase 1 (liquid) and in phase 2 (vapor), respectively, h is the enthalpy (kJ/kg), $L_{1\text{-}2}$ is the latent heat (J/kg), x is the vapor quality, and θ is the phase indicator.

In addition, a general k—ε model was selected for the turbulent flow in the present simulation. The equations of the k—ε model are shown below.

The turbulent viscosity is modeled as

$$\mu_T = \rho C_\mu \frac{k_T^2}{\varepsilon} \tag{12}$$

The transport equation for k reads

$$\rho(\mathbf{u}\cdot\nabla)k = \nabla\cdot\left[\left(\mu + \frac{\mu_T}{\sigma_k}\right)\nabla k\right] + p_k - \rho\varepsilon \tag{13}$$

where the production term p_k is

$$p_k = \mu_T\left[\nabla\mathbf{u} : \left(\nabla\mathbf{u} + (\nabla\mathbf{u})^T\right) - \frac{2}{3}(\nabla\cdot\mathbf{u})^2\right] - \frac{2}{3}\rho_k\nabla\cdot\mathbf{u} \tag{14}$$

The transport equation for ε is as follows:

$$\rho(\mathbf{u}\cdot\nabla)\varepsilon = \nabla\cdot\left[\left(\mu + \frac{\mu_T}{\sigma_\varepsilon}\right)\nabla\varepsilon\right] + C_{\varepsilon 1}\frac{\varepsilon}{k_T}p_k - C_{\varepsilon 2}\frac{\varepsilon^2}{k_T} \tag{15}$$

where k_T is the turbulent viscosity. For the superheated passes, the phase indicator θ_1 (liquid) equals zero. The steady-state method was used to solve the CO_2 microchannel evaporator.

2.3. Numerical Simulation

It is very difficult and complex to simulate the microchannel evaporator at one time because the evaporator has thin fins and microchannels. Therefore, there are six simulation times corresponding to six passes of the evaporator. This approach is used to reduce the computing resources. The input conditions for the first pass are the parameters in Table 2. The input conditions for the second pass are the numerical simulation results of the first pass and so on, until the numerical simulation of the sixth pass is completed. Table 2 shows some inlet parameters and outlet parameters. The inlet parameters include some initial values. The unknown parameters were calculated using Engineering Equation Solver (EES) [19], REFPROP [20] and DORIN [21] software.

Table 2. Inlet parameters and measured parameters.

Pass Number	Inlet Parameter	Outlet Parameter
1	T_{1_i} = 10 °C, p_{1_i} = 45 bar; T_a = 25 °C, h_a = 110 W/(m^{2}K) x_{1_i} = 0.61, m_{p_1} = 30/3 = 10 g/s	T_{1_o}, p_{1_o}, x_{1_o}
2	T_{1_o}, p_{1_o} T_a = 25 °C, h_a = 110 W/(m^{2}K) $x_{2\_i} = x_{1\_o}$, m_{p_2} = 30/4 = 7.5 g/s	T_{2_o}, p_{2_o}, x_{2_o}
3	T_{2_o}, p_{2_o} T_a = 25 °C, h_a = 110 W/(m^{2}K) $x_{3\_i} = x_{2\_o}$, m_{p_3} = 30/5 = 6 g/s	T_{3_o}, p_{3_o}, x_{3_o}
4	T_{3_o}, p_{3_o} T_a = 25 °C, h_a = 110 W/(m^{2}K) $x_{4\_i} = x_{3\_o}$, m_{p_3} = 30/6 = 5 g/s	T_{4_o}, p_{4_o}, x_{4_o}

(where indices 1, 2, 3 and 4 indicate a refrigerant in pass 1, 2, 3 and 4, respectively. Indices i and o indicate an inlet and outlet of the flat tube, respectively. T_a is the air room temperature, m is the CO_2 mass flow rate, x is the vapor quality, and h_a is the air convective heat transfer coefficient).

Figure 2 shows two flat tube models of the CO_2 microchannel evaporator. The CO_2 flows inside the flat tube, and the air is outside with the convective heat transfer. To determine the variations of the vapor quality at the evaporator inlet, the EES software [19] was used based on some initial conditions and the pressure–enthalpy diagram of R744.

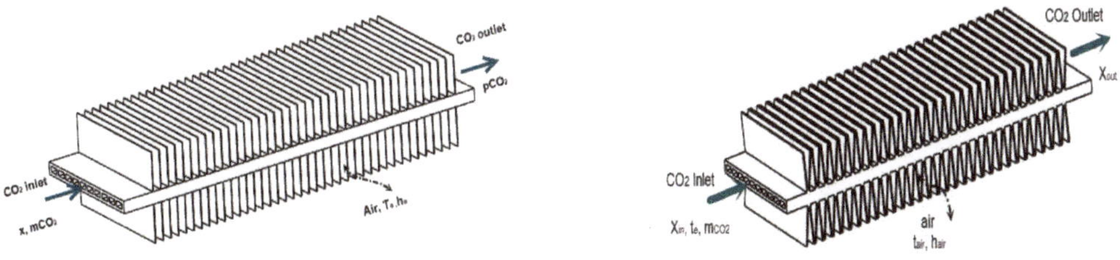

(a) (b)

Figure 2. Two flat tube models of the CO_2 microchannel evaporator: (**a**) a model of straight finned flat tube; (**b**) a model of V-finned flat tube.

In Figure 3, the finite elements in the grid meshes of the straight fin model were partitioned to be triangular. For the fluid flow, the boundary layer was divided into 5 layers to investigate the characteristics of the wall, as shown in Figure 3a. For important parts, the finer mesh level was used, i.e., the edges of the corners in the microchannels were surveyed on 4 different levels (1, 10, 20, 30), as shown in Figure 3b. For the V-fin model, the method and the meshing level are the same as in the straight fin model. The model consists of approximately 610,000 domain elements and 122,000 boundary elements. The average number of degrees of freedom (DOF) was 11,300,000 (plus 134,000 internal DOFs). The parallel sparse direct solver (PARDISO) algorithm is a solver that was developed based on the linear equation system. The relative tolerance was set at 10^{-6}. For numerical simulation in this study, the COMSOL Multiphysics software-version 6.0 was used with the PARDISO solver.

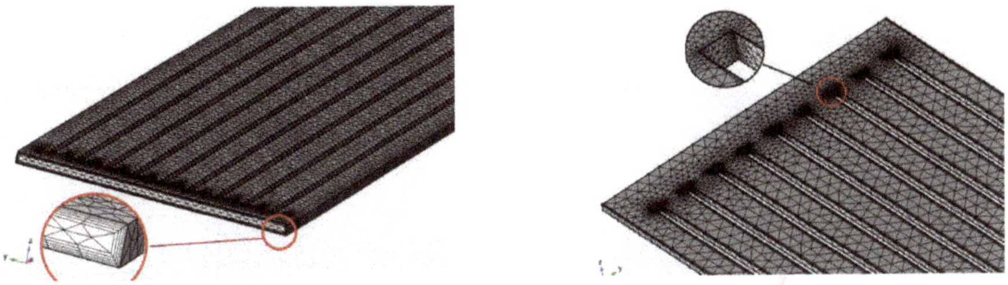

(a) (b)

Figure 3. Grid mesh diagram of the straight fin model. (**a**) The boundary layer; (**b**) The edge of the corners.

2.4. Experimental Setup

Figure 4 shows a schematic diagram of the test apparatus used in the study. It consists of a CO_2 DORIN compressor, a throttle valve, a gas cooler, a V-fin tube evaporator and an internal heat exchanger. The pressure sensors and the thermocouples were installed at the inlet and the outlet of each component to measure the pressure and temperature. Their ranges and accuracies of measurement are displayed in Table 3. The mass flow rate of the refrigerant was measured by a digital volumetric flow rate meter or determined by the DORIN software based on the suction temperature, the evaporative temperature and the gas cooler pressure. The experimental data of the microchannel evaporator with a V-finned flat tube will be used to validate the numerical simulation results for the flat tube with straight fins and V-fins. From Figure 5, the experimental data were imported to the EES software; the vapor quality was determined by $x = 0.61$; the value was used to import the numerical simulation. The uncertainty values of several parameters, such as the vapor quality and the normal heat flux, were 3.17% and 6.2%, respectively. The evaporator with the V-finned flat tube for the CO_2 air conditioning system is shown in Figure 6.

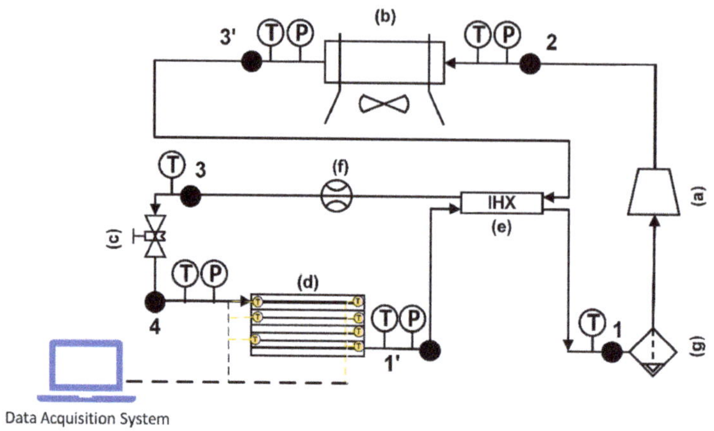

Figure 4. Schematic of the test loop for CO_2 air conditioning system: (a) a CO_2 compressor; (b) a gas cooler; (c) a throttle valve; (d) a microchannel evaporator; (e) an internal heat exchanger; (f) a flow meter; (g) a gas–liquid separator.

Table 3. Accuracies and ranges of testing devices.

Testing Apparatus	Accuracy	Range
Infrared thermal camera, Fluke Ti9	2%	−20~250 °C
Thermometer, Extech 421,509	0.75% of rdg	−20~250 °C
Thermocouples, T—Type	±0.1 °C	0~100 °C
Digital volumetric flow rate meter	±0.5% FS	400 to 5000 l/h
Pressure sensor, SENSYS—Korea	±0.5 FS	0~100 bar

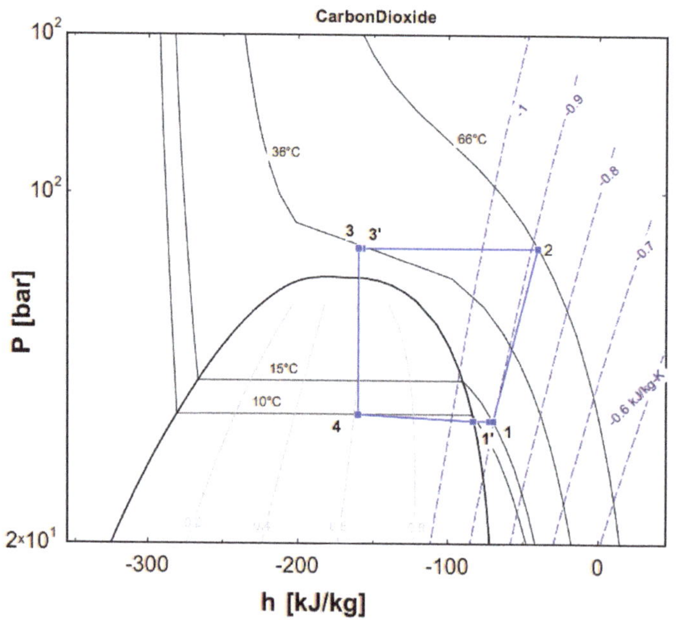

Figure 5. The experimental data presented on a p–h diagram.

Figure 6. A V-finned tube evaporator for the CO_2 air conditioning system.

3. Results and Discussion

The experimental data obtained from the test loop were under the ambient (outdoor) temperature of 33 °C. The inlet evaporative temperature and inlet vapor quality were fixed at 10 °C and 0.61, respectively. The temperature slice of the straight finned flat tube is shown in Figure 7. The inlet and outlet refrigeration temperatures of the straight finned flat tube were the same in the first pass of the microchannel evaporator; this result is consistent with the theory of evaporation. The temperature of the fins was about 1 °C higher than that of the CO_2 refrigerant.

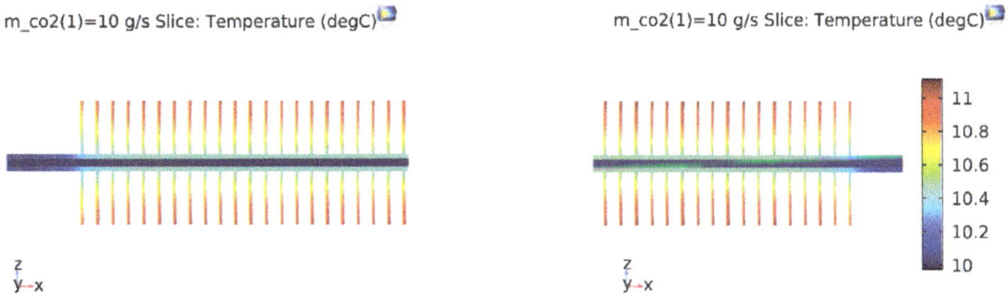

(a) (b)

Figure 7. The temperature slice of the straight finned flat tube in the first pass. (**a**) Inlet; (**b**) Outlet.

With the same initial simulation conditions, the temperature slice on the V-finned flat tube was the same as that obtained from the straight finned flat tube, as shown in Figure 8. The temperature profile obtained from the straight finned flat tube was also the same as that obtained from the V-finned flat tube, as shown in Figures 9 and 10. However, the arrows of the heat fluxes were different for these two types. The conductive heat flux of the V-finned flat tube model was more concentrated in the connection location between the fins and the flat tube.

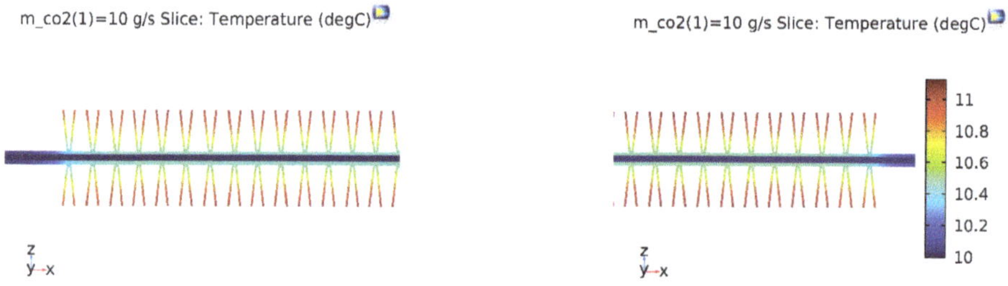

(a) (b)

Figure 8. The temperature slice of the V-finned flat tube in the first pass. (**a**) Inlet; (**b**) Outlet.

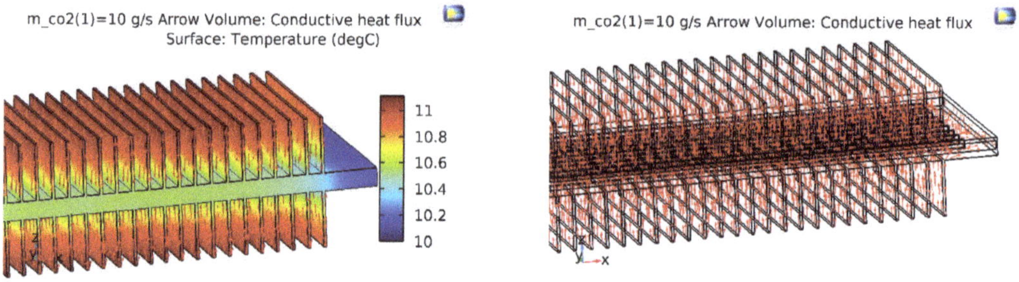

(a) (b)

Figure 9. The temperature and conductive heat flux of the straight finned flat tube of the first pass. (**a**) Temperature profile; (**b**) Conductive heat flux.

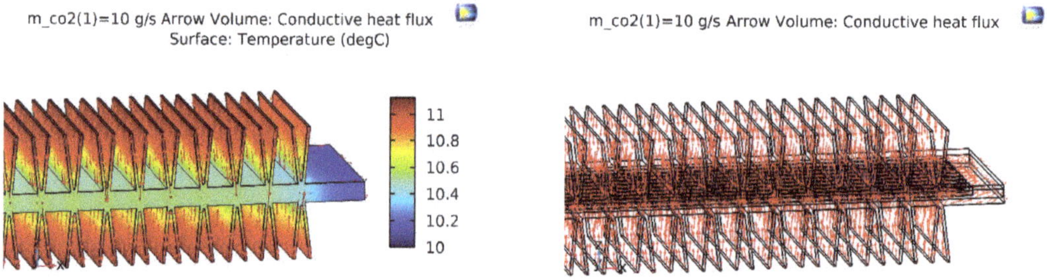

(a) (b)

Figure 10. The temperature and conductive heat flux of the V-finned flat tube of the first pass. (**a**) Temperature profile; (**b**) Conductive heat flux.

Figure 11 shows the vapor quality of the straight finned flat tube evaporator model and the V-finned flat tube evaporator model at the evaporative temperature of 10 °C. Due to the same heat transfer area (2.5 m^2), the straight fins had a height of 4.1 mm, and the V-shaped fins had a height of 4.05 mm (Figure 1). Therefore, the simulated vapor qualities in these two cases were the same. Each microchannel pass of the evaporator had a length of 300 mm. The evaporation of the refrigerant took place in four passes, so the normal heat flux from the air through the fins and tubes was almost reached at 1550 W/m^2. It was observed that the saturated vapor state (dry state) was achieved at the outlet of the fourth pass. When the CO_2 was superheated, the normal heat flux rapidly declined from 1550 to 700 W/m^2. From Figures 7–11, with the same heat transfer area and the heat transfer coefficient for the air side, the effect of the fin shape on the heat transfer of a microchannel evaporator was no different between the straight fin model and the V-fin model.

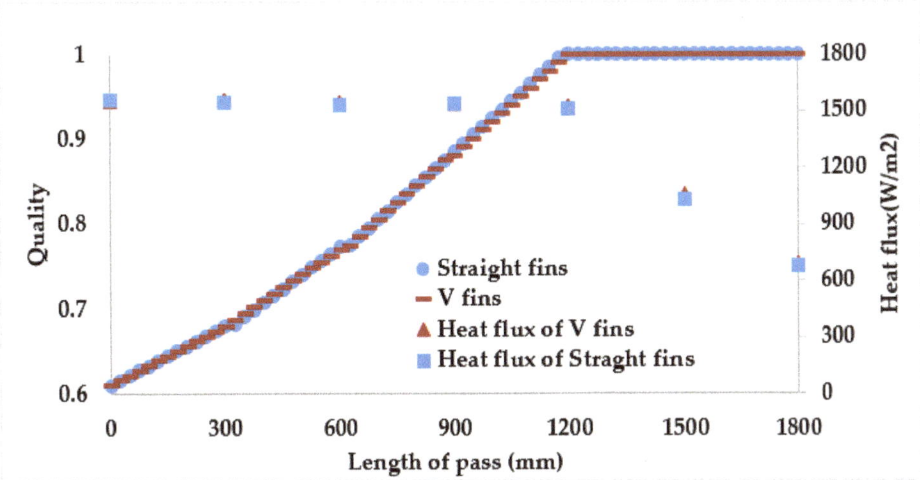

Figure 11. The vapor quality and the heat flux along the length of the flat tube.

Figure 12 shows the velocity field of the refrigerant along the length of the flat tube at the evaporative temperature of 10 °C. At the inlet, the velocity rose sharply because the refrigerant from the manifold entered the microchannel with a smaller cross-section. The velocity increased gradually in the microchannel; this was due to the mixture density decreasing. At the outlet, the velocity fell rapidly due to the refrigerant leaving the microchannel to enter the manifold, which saw the cross-section increase suddenly.

The comparison between the numerical simulation results and the experimental data is shown in Figure 13 at the evaporative temperature of 10 °C. The experimental temperature profile was captured by an infrared thermal camera and then compared with the numerical simulation results. The experimental temperatures were also obtained from the thermocouples. It could be observed that they had similarities at the first four passes, and the refrigerant was superheated at the fifth and sixth passes. From the figure, the deviation between the numerical results and the experimental data was less than 10% at the hottest spot.

Table 4 shows the results on the number of elements, the solution time and the physical memory between the straight finned flat tube model and the V-finned flat tube model. For the same method and meshing level, the V-fin tube model had more elements but less solution time and physical memory compared with those obtained from the straight fin tube model. The results showed that the fin shape affected the solution time and the memory. The solution time and the physical memory for the straight fins were 1.3 and 1.45 times those of the V-fins, respectively. This could be due to the grid meshes and the PARDISO solver in the V-fin tube model being more suitable than those obtained from the straight fin tube model. Therefore, the V-fin tube model should be used for numerical simulation to save the computing resources. These results will make additional contributions to numerical simulation studies in microchannel evaporators, especially with a CO_2 refrigerant.

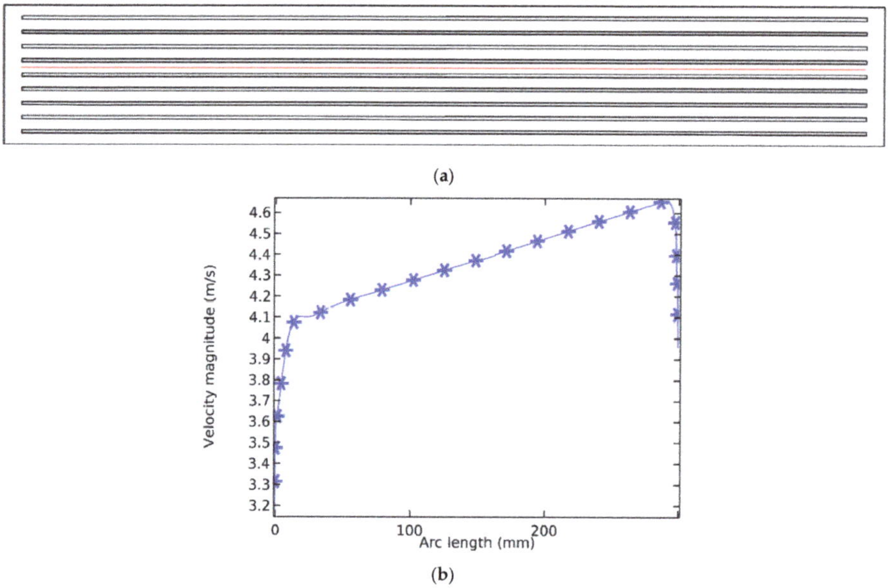

Figure 12. The refrigerant velocity in the microchannel for the straight fin tube model. (**a**) Location of the velocity discussed; (**b**) The velocity field in a microchannel.

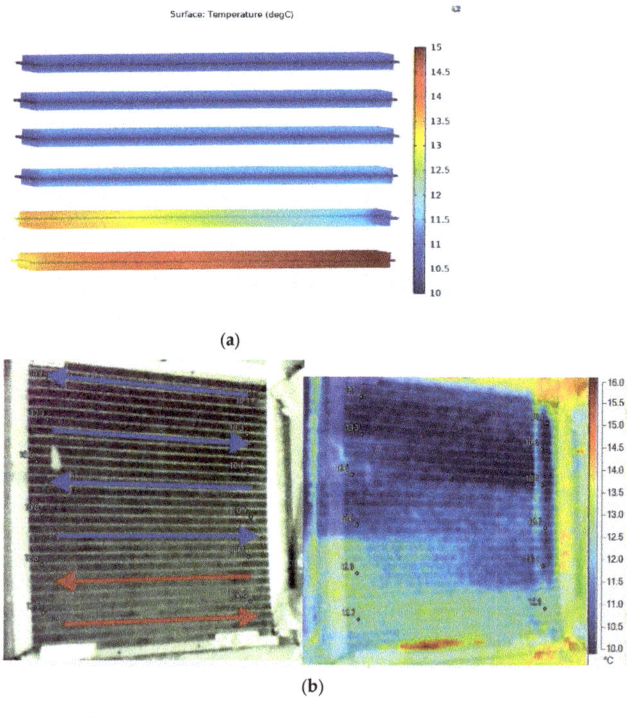

Figure 13. A comparison between the numerical simulation and experimental data. (**a**) The numerical results of the straight finned flat tube model; (**b**) The experimental temperature profile of the evaporator was captured by the thermal camera.

Table 4. The solution time and the physical memory.

Types	All Domains (Elements)	All Boundaries (Elements)	Solution Time (s)	Physical Memory (GB)
Straight fins	601,833	121,428	1653	9.7
V-fins	618,006	122,592	1268	6.69

4. Conclusions

A numerical study of the effect of the fin shape on the heat transfer and the solution time of a microchannel evaporator in a CO_2 air conditioning system was completed at the evaporative temperature of 10 °C. The two types of fin shapes were mentioned: V-shaped fins and straight fins. The experimental data were used to verify the numerical results. Some important contributions can be expressed as follows:

- The mathematical model, the boundary conditions, the meshing method and the PARDISO solver were applied to numerically simulate a microchannel evaporator model. The numerical results were in good agreement with those obtained from the experimental results, with an error of less than 10%.
- For the system under consideration and for the same heat transfer area and the heat transfer coefficient for the air side, the effect of the fin shape on the heat transfer of a microchannel evaporator was not different. However, the solution time and the physical memory for the straight fins were 1.3 and 1.45 times those of the V-fins, respectively. Under the same conditions, the V-fin shape should be used for numerical simulation and not the straight fin shape.
- The evaporation of the refrigerant in the microchannel evaporator took place in four passes. The normal heat flux from the air through the fins and tubes was almost reached at 1550 W/m^2 at the evaporative temperature of 10 °C.

Author Contributions: Conceptualization, T.N. and T.D.; methodology, T.D.; software, T.N.; validation, T.N. and T.D.; formal analysis, T.N.; investigation, T.N.; resources, T.D.; data curation, T.D.; writing—original draft preparation, T.N.; writing—review and editing, T.D.; visualization, T.N.; supervision, T.D.; project administration, T.D.; funding acquisition, T.D. All authors have read and agreed to the published version of the manuscript.

Funding: Sponsored by the Vietnam Ministry of Education and Training.

Acknowledgments: The funding for this work from project No. B2020-SPK-04 (sponsored by the Vietnam Ministry of Education and Training) is deeply appreciated.

Conflicts of Interest: The authors declare no conflict of interest.

References

1. Kim, M.H.; Bullard, C.W. Air-side performance of brazed aluminum heat exchangers under dehumidifying conditions. *Int. J. Refrig.* **2002**, *25*, 924–934. [CrossRef]
2. Kim, M.; Bullard, C.W. Air-side thermal hydraulic performance of multi-louvered fin aluminum heat exchangers. *Int. J. Refrig.* **2002**, *25*, 390–400. [CrossRef]
3. Okbaz, A.; Pınarbaşı, A.; Olcay, A.B. Experimental investigation of effect of different tube row-numbers, fin pitches and operating conditions on thermal and hydraulic performances of louvered and wavy finned heat exchangers. *Int. J. Therm. Sci.* **2020**, *151*, 106256. [CrossRef]
4. Ayad, F.; Benelmir, R.; Idris, M. Thermal-hydraulic experimental study of louvered fin-and-flat-tube heat exchanger under wet conditions with variation of inlet humidity ratio. *Appl. Therm. Eng.* **2021**, *183*, 116218. [CrossRef]
5. Hsieh, C.T.; Jang, J.Y. Parametric study and optimization of louver finned-tube heat exchangers by Taguchi method. *Appl. Therm. Eng.* **2012**, *42*, 101–110. [CrossRef]
6. Jang, J.Y.; Chen, C.C. Optimization of louvered-fin heat exchanger with variable louver angles. *Appl. Therm. Eng.* **2015**, *91*, 138–150. [CrossRef]
7. Kim, M.H.; Pettersen, J.; Bullard, C.W. Fundamental process and system design issues in CO_2 vapor compression systems. *Prog. Energy Combust. Sci.* **2004**, *30*, 119–174. [CrossRef]
8. Choi, K.I.; Pamitran, A.S.; Oh, J.T. Two-phase flow heat transfer of CO_2 vaporization in smooth horizontal minichannels. *Int. J. Refrig.* **2007**, *30*, 767–777. [CrossRef]

9. Brix, W.; Kærn, M.R.; Elmegaard, B. Modelling distribution of evaporating CO_2 in parallel minichannels. *Int. J. Refrig.* **2010**, *33*, 1086–1094. [CrossRef]
10. Nguyen, T.; Dang, T.; Vo, K. Experimental Investigation on a Microchannel Evaporator of CO_2 Air Conditioning System with an Internal Heat Exchanger. *Int. J. Emerg. Res. Manag. Technol.* **2017**, *6*, 40–45. [CrossRef]
11. Jin, J.; Chen, J.; Chen, Z. Development and validation of a microchannel evaporator model for a CO_2 air-conditioning system. *Appl. Therm. Eng.* **2011**, *31*, 137–146. [CrossRef]
12. Yun, R.; Kim, Y.; Park, C. Numerical analysis on a microchannel evaporator designed for CO_2 air-conditioning systems. *Appl. Therm. Eng.* **2007**, *27*, 1320–1326. [CrossRef]
13. Kim, M.H.; Bullard, C.W. Development of a microchannel evaporator model for a CO_2 air-conditioning system. *Energy* **2001**, *26*, 931–948. [CrossRef]
14. Dang, T.; Teng, J.T. Comparison on the heat transfer and pressure drop of the microchannel and minichannel heat exchangers. *Heat Mass Transf.* **2011**, *47*, 1311–1322. [CrossRef]
15. Dang, T.; Teng, J.; Chu, J. A study on the simulation and experiment of a microchannel counter-flow heat exchanger. *Appl. Therm. Eng.* **2010**, *30*, 2163–2172. [CrossRef]
16. Patiño, J.; Llopis, R.; Sánchez, D.; Sanz-Kock, C.; Cabello, R.; Torrella, E. A comparative analysis of a CO_2 evaporator model using experimental heat transfer correlations and a flow pattern map. *Int. J. Heat Mass Transf.* **2014**, *71*, 361–375. [CrossRef]
17. Yang, J.; Ning, S. Experimental and Numerical Study of Double-Pipe Evaporators Designed for CO_2 Transcritical Systems. *Processes* **2019**, *7*, 547. [CrossRef]
18. Comsol Software, M. Heat Transfer Module, Version 6.0, 2021, Inc. 100 District Avenue Burlington, MA 01803, USA. Available online: https://www.comsol.com/heat-transfer-module (accessed on 30 June 2022).
19. EES. *Engineering Equation Solver*; F-Chart MdthaSoftware Inc.: Madison, WI, USA, 2005; Available online: https://fchartsoftware.com/ees/ (accessed on 30 June 2022).
20. REFPROP. *REFPROP. NIST Refrigerant Properties Database 23*; Version 9; Gaithersburg, MD, USA, 2013. Available online: https://www.nist.gov/srd/refprop (accessed on 30 June 2022).
21. Dorin. Mario Dorin innovation, Compiobbi (FI), Italy, Version 19.1. [Online]. Available online: https://www.dorin.com/en/Software/ (accessed on 30 June 2022).

micromachines

Article

Thermal Conductivity Measurement of Flexible Composite Phase-Change Materials Based on the Steady-State Method

Ze Feng [1] and Xin Xiao [1,2,*]

[1] School of Environmental Science and Engineering, Donghua University, Shanghai 201620, China
[2] Yunnan Provincial Rural Energy Engineering Key Laboratory, Kunming 650550, China
* Correspondence: xin.xiao@dhu.edu.cn

Abstract: Phase-change materials (PCMs) are widely used in energy storage and thermal management due to the large latent heat in the phase-change process. As one of the most significantly thermo-physical properties of PCMs, the thermal conductivity has been extensively studied. Great attention has been paid to improving the thermal conductivities of PCMs; however, the studies on the thermal conductivities of flexible PCMs are relatively inadequate. In this study, polyethylene glycol 1500 (PEG1500) was used as the base PCM, and expanded graphite (EG) and styrene–butadiene–styrene (SBS) were added to improve the thermal conductivity and flexibility of pure PCMs, respectively. A steady-state experimental test rig was built and verified with the measurement of the thermal conductivity of stainless steel and deionized water, and then the thermal conductivities of PCMs at different phases and qualitative temperatures were measured extensively. Compared to the PEG1500 with 5 wt.% EG, the addition of SBS sharply reduces the thermal conductivity, which is only 0.362 W/(m·K) at 12.5 °C when the addition ratio is 50%. This is approximately a 69% reduction compared with the composite PCMs without SBS. Furthermore, the theoretical thermal conductivities of the composite PCMs were calculated with six theoretical models of multiphase systems. The majority of the models provide a good prediction of thermal conductivities of composite PCM with high SBS concentration, while the average deviation of Agari-Uno model is only 20.5% with different SBS concentration and relatively agrees well with the experimental results.

Keywords: phase-change materials; enhancement; flexibility; thermal conductivity

1. Introduction

Currently, massive fossil energy is consumed by automobiles with a large amount of waste gas exhaust, and this causes serious damage to the ecological environment. Therefore, as the preferred type of transportation for environmental protection and sustainable development, the demand and output of new energy vehicles have increased rapidly in recent years, along with lithium batteries. This is because lithium batteries are the power source of electric vehicles, and the loss rate is directly determined by the working temperature [1]. Jin et al. [2] indicated that the optimal working temperature of lithium battery was in the range of 20 to 40 °C, and the maximum local temperature difference was 5 °C. This could cause irreversible capacity loss to the battery if the working temperature exceeds the range or if the local temperature difference is overly large. In addition, there is a certain risk of heat accumulation at high temperature, inducing the battery burning or becoming thermally out of control [3,4]. Therefore, for the thermal management of a lithium battery, a method of passive temperature control of battery through phase-change materials (PCMs) to ensure the battery optimal lifespan has been proposed. Due to the poor heat dissipation of batteries by the PCMs of low thermal conductivity, the enhancement and accurate measurement of the thermal conductivity are becoming the current research hotspots.

The traditional method is to integrate carbon materials with high thermal conductivities, such as expanded graphite (EG) [5], graphene [6], carbon nanotubes [7] and nano-metal particles, such as aluminum [8] and aluminum oxide [9], with pure PCM to significantly enhance the thermal conductivity. Zou et al. [10] used graphene and multi-walled carbon nanotubes (MWCNTs) to enhance the thermal conductivity of paraffin. The thermal conductivities of the paraffin/graphene and paraffin/MWCNTs composite PCMs were 0.75 and 0.61 W/(m·K), respectively, with an impregnation ratio of 1.5%, and these were measured using the transient plane heat source technique. Ling et al. [11] adsorbed sodium acetate trihydrate-urea (SAT-Urea) with EG, and the thermal conductivities of the composite PCMs increased from 0.61 to 4.19 W/(m·K) with the addition of 20 wt.% EG, which were also measured using the transient plane heat source technique.

Krishna et al. [12] measured the thermal conductivities of tricosane-based composite PCMs by the hot-wire method, and the increase reached 32% by adding 2% Al_2O_3 nanoparticles. Mehrabi-Kermani et al. [13] impregnated paraffin into copper foam with the porosity of 0.903 and pore size of 20 PPI to make the heat dissipation faster of the lithium-ion battery, which was further enhanced by combining with forced-air convection caused by fans. The thermal conductivity of the composite PCM was 11.98 W/(m·K) measured by hot-wire method, which was 57 times than that of pure paraffin.

Hussain et al. [6] attached a graphene coating to the surface of nickel foam, which had the porosity of 0.9 and pore size of 12.7 PPI and infiltrated with paraffin, improving the thermal conductivities of pure paraffin from 0.19 to 4.6 W/(m·K) measured by the laser flash method. Lin et al. [14] prepared a graphene skeleton by reducing graphene oxide and then impregnated paraffin into the porous structure under vacuum condition so that its axial and lateral thermal conductivities increased to 2.58 and 1.78 W/(m·K), respectively, which were also measured by laser flash method at room temperature. The increased ratios of the thermal conductivities reached 1260% and 840%, respectively.

However, there are limitations in the measurements of thermal conductivities by transient plane heat source technique [11], hot wire method [13] and laser flash method [14]. Since these three methods do not distinguish the thermal contact resistance (TCR) of the sample surface from its own thermal resistance, the results obtained are the summation of both, indicating that the roughness of the sample surface would greatly affect the thermal conductivity [15].

Liu et al. [16] found that TCR played a dominant role in the total thermal resistance and was clearly influenced by changing the temperature, pressure and roughness of the contact. Ahadi et al. [17] analyzed the effective thermal resistance measured by conventional transient plane heat source technique and concluded that the proportion of TCR in total thermal resistance decreased with the increase in the thickness of sample, while the ignored TCR lead to the thermal conductivities having a 66.7% deviation from those measured with the steady-state method. As for the hot wire method, it was proposed to measure the samples with large size and smooth surface to reduce the uncertainty and the thermal conductivity of the measured samples is limited to 15 W/(m·K) [17,18]. The laser flash method is performed by heating one side of the sample with the pulse of energy, which results in the temperature on the back of sample rising to determine the thermal diffusivity. Due to this, TCR is regularly ignored, and the thickness of the sample is severely restricted by the timescales of the energy pulse and detection, and the specific heat must be known [19]. Xiao et al. [20] found that the uncertainty of thermal conductivity measured by the laser flash method can be as large as 40% when the TCR on both sides of the measured sample is ignored by numerical investigations. Additionally, these methods are not suitable for measuring the thermal diffusivity of thermal insulation materials [21].

In contrast, the steady-state method is more widely used with higher reliability. The heat flux and temperature gradient of the sample is kept stable in the steady state to obtain an accurate thermal resistance. For at least two samples with different thicknesses are measured, not only the effective thermal conductivities but the TCR values are obtained by the evolution of temperature in the direction of heat flux [22].

For instance, Xiao et al. [15] measured the thermal conductivities of copper foam with different porosities after paraffin adsorption using the steady-state method and found that the TCR accounts for 15%~50% of the total thermal resistance. Wang et al. [23] measured the thermal conductivities of the composite PCMs composed of paraffin, HDPE and EG with steady-state heat flux meter. The results showed that the thermal conductivities increased linearly with temperature in solid and liquid states, respectively, and the uncertainty was only 4.72%. Feng et al. [24] measured the thermal conductivity of paraffin by both transient plane heat source technique and steady-state method, and the experimental results showed that the thermal conductivity deviation between them could be less than 7% while the influence of TCR was considered.

According to the differences of the shape, size, proportion, dispersion and interaction of fillers, different theoretical models were proposed to predict the theoretical thermal conductivities of the composite PCMs and then compared with the experimental results measured by above-mentioned methods [25–27]. In these models, the isotropic fillers with high thermal conductivity were considered as the dispersed phase and as dispersed in the polymer matrix, which was the continuous phase. Subsequently, several studies were reported for modifications, e.g., Wang et al. [28] proposed the influence of air with low thermal conductivity on porous matrix, and then Agari-Uno model was modified and effectively predicted the theoretical thermal conductivities of composite PCM, with a deviation of less than 9.6% from the experimental value. Xu et al. [29] introduced the contact resistance among fillers into the classical Maxwell model, and the predicted value by reconstruction of the Maxwell model agreed firmly well with the experimental results over a wide range of fillers concentrations and diameters.

Another shortcoming for composite PCMs in practical applications is the insufficient flexibility, which restricts the reduction of the large TCR between PCM and battery and phase separation occurs during the recycling of PCM. In order to address the issues, the integration of copolymer with memory shape, such as styrene–butadiene–styrene (SBS), is proposed to have a good effect. By blending with the PCM to form a film on its surface, not only the flexibility is improved, but the PCM is prevented from separating out during circulation [30].

In the present study, polyethylene glycol (PEG1500) was used as PCM for the characteristics of being difficult to burn, while EG and SBS were used for synergistic packaging. A new type of flexible composite PCM was prepared. More importantly, a thermal conductivity determination system was built to accurately measure the thermal conductivity at different temperatures and additive mass fractions with steady-state method. In this way, the flexible PCM with suitable thermal conductivity was chosen, and its heat dissipation effects in the battery thermal management system are the main concern of future works. Additionally, the theoretical thermal conductivities of samples with different SBS concentrations were predicted by well-known semi-empirical models from the literature and compared with the experimental results.

2. Experimental Setup and Thermal Conductivity Measurement

2.1. Preparation of PEG1500/EG/SBS Composite PCMs

In the present study, PEG1500 was used as the initial PCM, and EG was used to adsorb PEG1500 and improve the thermal conductivity of the composite PCM [31]. According to previous research [32], the adsorption effects of low EG content may not meet the requirement of preventing the leakage of the PCM together with the limited improvement of thermal conductivity, and it was observed that the shape of PCM is powdered with the low latent heat when EG content is overly large, which may be not suitable for battery thermal management. Thus, composite PCM with a suitable EG mass fraction was strongly proposed for battery thermal management [33]. Therefore, PEG1500/EG composite PCM was prepared with 95 wt.% PEG1500 and 5 wt.% EG as the matrix, and SBS was chosen to improve the flexibility and prevent leakage of PEG1500 by forming a film on the surface of EG [34]. The organic solvent dichloromethane (CH_2Cl_2) was used to dissolve SBS at room temperature due to its low toxicity and easy decomposition under light, which would not remain in the composite.

Firstly, after the proportion of SBS (Huizhou LCY Elastomers Co., Ltd., Huizhou, China) was determined, the SBS particles were dissolved into a gel-like solution by CH_2Cl_2 (Shanghai Gaoxinhuabo Instrument Co., Ltd., Shanghai, China) and the PEG1500 (Sinopharm Chemical Reagent Co., Ltd., Shanghai, China) particles were placed in water bath at a temperature of 70 °C until it was melted. EG (Nantong Yifan Graphite Co., Ltd., Nantong, China) was prepared by placing expandable graphite in a muffle furnace at 850 °C for about 10 mins, which was then added into the molten PEG1500. Secondly, the above-mentioned composite was stirred with a glass rod for about 15 mins until it was fully mixed and solidified at room temperature, which was subsequently crushed and ground into powder. Then, the PEG1500/EG powder was added into the solution of CH_2Cl_2 containing SBS and stirred for about 15 mins to become effectively wrapped by the SBS, whose contents were 10 wt.%, 20 wt.%, 30 wt.%, 40 wt.% and 50 wt.%, respectively. Finally, the flexible composite PCM powder was placed in a ventilated place for 4 hours to completely volatilize CH_2Cl_2 and then pressed into PCM blocks with the diameter of 50 mm and thicknesses of about 15, 30 and 45 mm, respectively, under a constant pressure of 6 MPa by a pelletizer (769YP-15A, Tianjin Keqigaoxin Technology Co., Ltd., Tianjin, China). The SBS content and the detailed values of these samples are listed in Table 1, and the addition of SBS was used to improve the flexibility of materials.

Table 1. The components and properties of flexible composite PCMs.

Samples	PEG1500/g	EG/g	SBS/g	Height/mm	Mass/g	Average Density/(kg/m^3)
PES0	95	5	0	10.24 32.46 45.84	24.51 77.9 109.72	1220.10
PES10	85.5	4.5	10	15.28 20.33 46.16	34.62 46.1 105.19	1156.46
PES20	76	4	20	13.84 31.36 42.38	29.74 65.21 94.25	1095.35
PES30	66.5	3.5	30	15.8 30.7 45.66	30.41 61.55 92.48	1010.95
PES40	57	3	40	11.2 30.65 43.71	21.64 59.3 88.69	1000.93
PES50	47.5	2.5	50	13.27 17.41 31.17	22.48 30.35 52.02	866.86

2.2. Experimental Device

In the present study, a steady-state system with constant heat sources was used to test the thermal conductivities of cylindrical samples with different thicknesses. The structural diagram of the experimental system is shown in Figure 1. The upper constant heat source was controlled by the water circulation of the thermostatic bath (DC2010, Shanghai Hengping Instrument Co., Ltd., Shanghai, China), and the lower cold source at the bottom was provided by the ethylene glycol/water mixture (i.e., 1:3) circulating in another thermostatic bath (DC2010, Henan Xinling Instrument Co., Ltd., Henan, China).

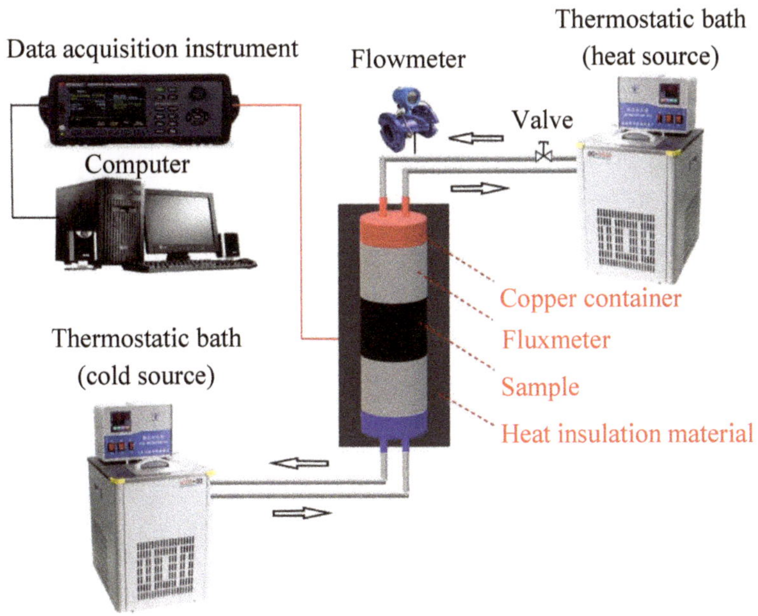

Figure 1. Schematic diagram of the experimental system.

For the power difference of circulating pump in thermostatic bath, a valve with an electronic flowmeter was installed between the heat source and thermostatic bath to control the flow rate, which was set to 0.7 L/min during the experiment. The heat flow was conducted from the heat source to the cold source through upper fluxmeter, samples and lower fluxmeter then adjusted to be stable after it reached a steady state.

Figure 2 shows the detailed parameters of the test section and its schematic diagram. The test section consists of two copper containers with diameters of 50 mm as heat and cold sources, upper and lower fluxmeters as the heat transfer media and the temperature measurement section, where the fluxmeters made of 304 stainless steel are with the diameter of 50 mm. Each fluxmeter was perforated with four holes with a diameter of 2 mm up to the axis as shown in Figure 2.

In this way, eight platinum resistance thermometers of PT100 (i.e., T1–T8) coated by thermal conductive silicone grease were placed into holes to obtain the temperature of the fluxmeters with a data acquisition instrument (DAQ970A, Keysight Technology Co., Ltd., Penang, Malaysia) for subsequent thermal conductivity calculations.

During the experiment, the upper and lower thermostatic baths were set to different temperatures to measure the thermal conductivities of solid samples at different temperatures. As for the liquid or molten samples at high temperature, a custom-made glass container was proposed as shown in Figure 3. The container body was made of glass with the upper and lower parts sealed by copper plates of the thickness of 2 mm, and two nylon rubber rings with the thickness of 2 mm were attached to the edge of the copper plate to fix the container, which was done to ensure that the fluxmeter was consistent with the sample in the vertical direction and the heat flux flowed vertically.

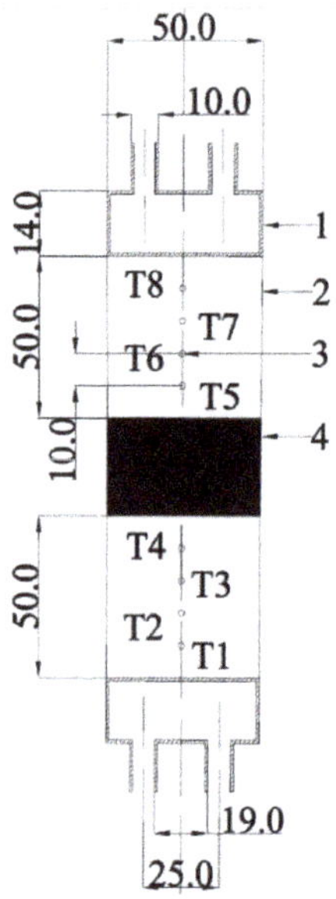

Figure 2. Schematic diagram and parameters of the test section. (1) Copper container, (2) fluxmeter, (3) platinum resistance thermometer and (4) sample to be tested (unit: mm).

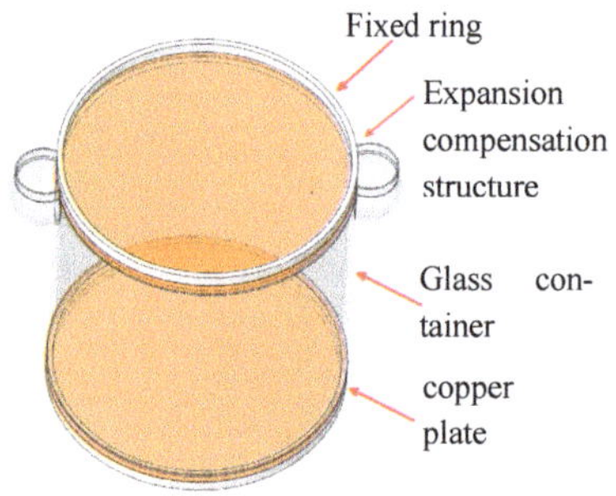

(**a**)

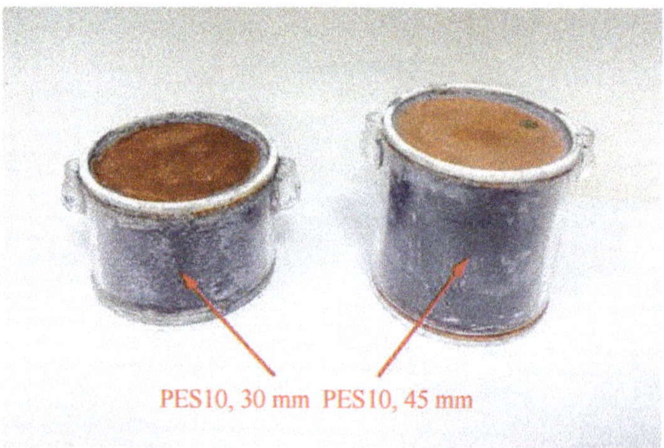

(**b**)

Figure 3. Schematic diagram of the sample containers in the liquid state. (**a**) Glass container and (**b**) layout of samples.

In addition, the container was also designed with two expansion compensation structures due to the volume change when the phase of PCM changes. The liquid in the compensation structures flowed back into the container to ensure that the solution was in close contact with the top copper plate without any gap when the volume of the solution decreased. While the volume of the solution expanded, the liquid was squeezed to compensate the rise of the liquid level in the structure, keeping the pressure balanced in the container.

It should be noted that the contact surfaces of copper pieces, fluxmeters and samples were coated with thermal conductive silicone grease to decrease the TCR ratio regarding the total thermal resistance. The experimental device was tightly coated with several layers of thermal insulation materials, which is silica aerogel pad with a thermal conductivity of 0.019 W/(m·K) and a rubber–plastic sponge with a thermal conductivity of 0.034 W/(m·K), to prevent the influence of external temperature on the heat flow, while the heat leakages through the platinum resistance thermometers were estimated to be 1.16% compared to the axial heat flux. The temperature evolutions of platinum resistance thermometers were verified to be less than 0.1 °C when both thermostatic baths were set as 20 °C as shown in Figure 4. As of the tolerance of the thermostatic bath is the cold source, the fluctuations of temperatures of T8 and T7 were within 0.01 °C, which can be ignored in the experiments.

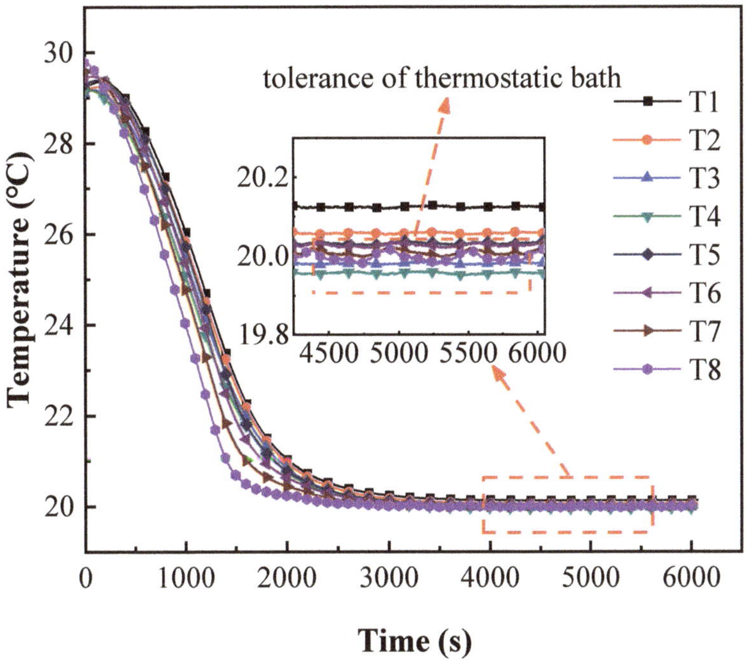

Figure 4. Pre-calibration of the platinum resistance thermometers at 20 °C.

During the experiment, the heat flow from the upper copper container reached the lower copper container through fluxmeters and samples. Since the temperatures of the cold and heat sources were constant, the system quickly reached a steady state within 60 mins and then was maintained for 30 mins to obtain the averaged temperatures of T1–T8, which were obtained by connecting platinum resistance thermometers with a data acquisition instrument.

2.3. Experimental Method

Since the diameter of fluxmeters are the same as those of the samples to be measured with a uniform wall temperature provided by cold and heat sources and the thermal insulation measures are taken in the radial direction, the whole process is considered as an axial one-dimensional heat transfer process. The heat flux of each interface is almost the same and can be given by Fourier's Law (1):

$$q = -\lambda_{st} \times \frac{dt}{d\delta} \tag{1}$$

$$\lambda_{st} = 11.05 + 0.0525 \times T_{qua} \tag{2}$$

where λ_{st} is the thermal conductivity of the stainless steel fluxmeter, and it can be seen from Equation (2) that the temperature deviation dt is measured by the platinum resistance during the experiment. The distance $d\delta$ is determined by the test rig. T_{qua} in Equation (2) indicates the qualitative temperature of stainless steel fluxmeter at the specified position, which is generally averaged from the temperatures of adjacent measuring points.

The temperature difference between the top and bottom of a sample can be deduced from the temperature distribution of the fluxmeter. The average heat flux of the fluxmeter is approximately regarded as that passing through the sample. As a result, the total thermal resistance of the samples to be measured is deduced by Equation (3).

$$R_{all} = \frac{\Delta T_s}{\overline{q}} = TCR + R_s = TCR + \frac{\delta}{\lambda} \tag{3}$$

where R_{all} is the total thermal resistance of the sample, including the thermal resistance R_s, and the TCR, δ and λ are the thickness and the thermal conductivity of the sample respectively. ΔT_s is the temperature difference between the upper and lower interfaces of the sample, and $\overline{q}$ is the average heat flux of the fluxmeter.

The total thermal resistance can be obtained by the experiments, while TCR is considered to be a constant for same types of samples as the surface of the sample is covered with thermal conductive silicone grease. Therefore, the thermal conductivity of the sample can be determined by the total thermal resistance at different thicknesses, and to further reduce the tolerance of the results, the total thermal resistance of each thickness is obtained by averaging three independent experimental results, where the properties of the samples are listed in Table 1.

2.4. Experimental Uncertainty Analysis

During the measurement of thermal conductivity in the present study, the uncertainty analysis of related parameters is involved in Equation (4):

$$R_s = f(q,\ \Delta T,\ h,\ A) \tag{4}$$

The total uncertainty of thermal resistance is obtained by the tolerance propagation law, which depends on the uncertainty of heat flux Δq, temperature difference Δt, thickness difference Δh and the contact area difference ΔA. The typical sample PES0 with the thickness of 45.84 mm under the condition of heat and cold sources temperature of 35 and −10 °C, respectively, is taken as an example, and the uncertainty of each parameter with detailed calculation processes are shown in Table 2. The tolerances of testing instruments are considered to be the reason for experimental uncertainty, such as the platinum resistance thermometer, data acquisition instrument, vernier caliper and thermostatic bath. In addition, the testing section can not be completely heat insulated, leading part of heat flow transferred from the radial direction to the outside, which changes the radial temperature distribution.

Table 2. Uncertainty analyses of the parameters.

Parameter	Uncertainty Analysis
Measurement parameters	
Diameter, d (mm)	0.50 (mm)
Thickness, h (mm)	0.02 (mm)
Temperature, T (°C)	0.10 (°C)
Export parameters	
Thickness, $\frac{\delta h}{h}$ (%)	$\delta h = 0.02$ (mm) $\frac{\delta h}{h} = \frac{0.02}{45.84} = 0.04\%$
Contact area, $\frac{\delta A}{A}$ (%)	$A = \frac{\pi}{4}d^2$ $\delta A = (0.0505^2 - 0.05^2)\frac{\pi}{4} = 3.95 \times 10^{-5}$ (m^2) $\frac{\delta A}{A} = \frac{3.95\times10^{-5}}{\frac{\pi}{4}\times0.05^2} = 2.01\%$
Difference in temperature, $\frac{\delta\Delta T}{\Delta T}$ (%)	$\delta\Delta T = 0.10$ (°C) $\frac{\delta\Delta T}{\Delta T} = \frac{0.10}{2.53} = 3.95\%$
Heat flow, $\frac{\delta Q}{Q}$ (%)	$\delta Q_1 = \sqrt{(\lambda_{st}\frac{\Delta t}{\Delta x}\delta A)^2 + (A\frac{\Delta t}{\Delta x}\Delta\lambda_{st})^2 + (\lambda_{st}\frac{\Delta A}{\Delta x}\delta\Delta t)^2 + (\lambda_{st}A\frac{\Delta t}{(\Delta x)^2}\delta\Delta x)^2}$ $= \sqrt{(10.69 \times \frac{2.53}{0.02} \times 0.0000395)^2 + (0.001963 \times \frac{2.53}{0.02} \times 0.10)^2 + (10.69 \times \frac{0.001963}{0.02} \times 0.10)^2 + (10.69 \times 0.001963 \times \frac{2.53}{(0.02)^2} \times 0.001)^2}$ $= 0.1791$ (W) $\frac{\delta Q}{Q} = \sqrt{(\frac{\delta Q_1}{Q_1})^2 + (\frac{\delta Q_2}{Q_2})^2}$ $= \sqrt{(\frac{0.1791}{2.7034})^2 + (0.0116)^2} = 6.73\%$
Effective thermal conductivity, $\frac{\delta\lambda_e}{\lambda_e}$ (%)	$\frac{\delta\lambda_e}{\lambda_e} = \sqrt{(\frac{\delta Q}{Q})^2 + (\frac{\delta\Delta T}{\Delta T})^2 + (\frac{\delta h}{h})^2 + (\frac{\delta A}{A})^2}$ $= \sqrt{(0.0673)^2 + (0.0395)^2 + (0.0005)^2 + (0.0201)^2}$ $= 8.06\%$

3. Experimental Results and Analysis

3.1. Verification of Experimental System

The verification of thermal conductivity measurements by the system includes the experiments of the solid and liquid samples. The former was verified by measuring the thermal conductivities of 304 stainless steel at the temperatures of 12.5, 22.5, 32.5, 42.5, and 52.5 °C, in comparison with Ref. [15], while the latter was verified by measuring the thermal conductivities of deionized water at the temperatures of 20 and 40 °C, in comparison with Ref. [35].

The temperature variations of stainless steel during the experiment are shown in Figure 5, where the qualitative temperature is 12.5 °C as the example. After the temperature of the heat and cold sources is set as 25/0 °C, the circulating water temperature in the thermostatic bath begins to change rapidly, leading the heat flow to transfer from top to bottom until reaching a steady state after about 5000 s as shown in sections ab in Figure 5, which is kept for approximately 2000 s to estimate the actual thermal resistance through the stable temperature gradient.

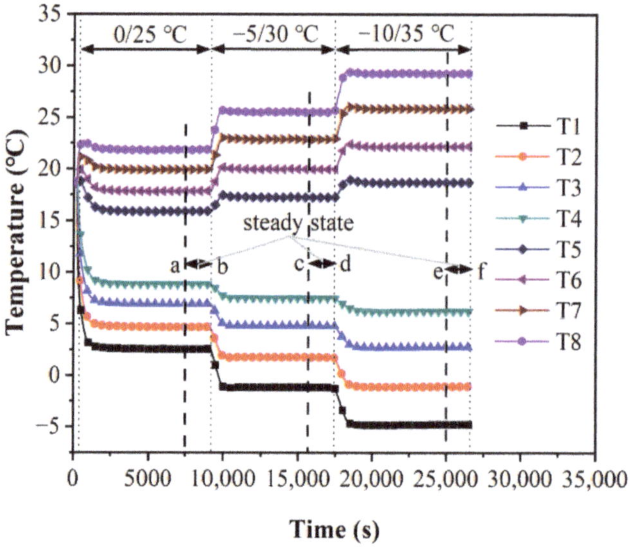

(**a**)

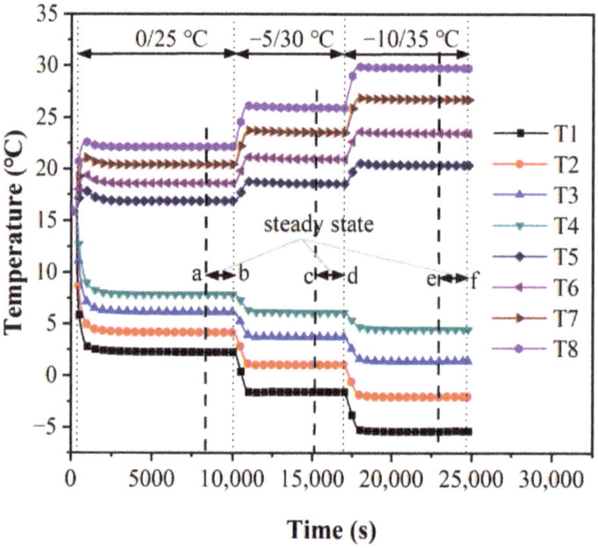

(**b**)

Figure 5. *Cont.*

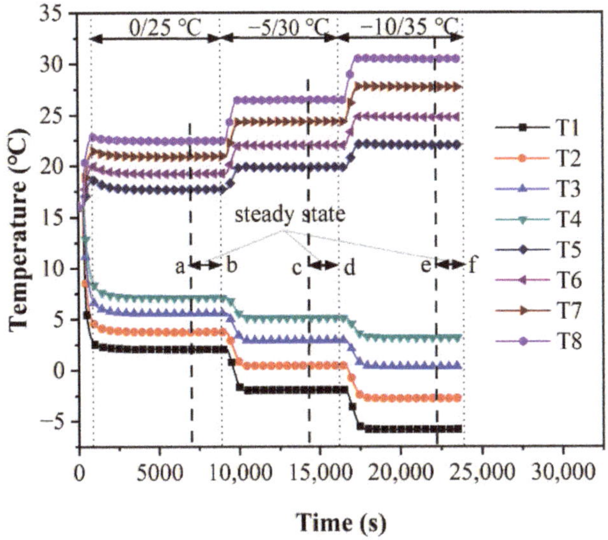

(c)

Figure 5. Experimental results of 304 stainless steel with the qualitative temperature of 12.5 °C. (**a**) h = 10.22 mm, (**b**) h = 25.06 mm and (**c**) h = 40.02 mm.

Subsequently, the temperature of the thermostatic baths is set as 30/−5 and 35/−10 °C, respectively, to obtain the thermal resistance in the other two steady states represented by sections cd and ef in Figure 5. During the whole process, the other two sets of data are measured for stainless steel samples with the qualitative temperature at 12.5 °C to eliminate the tolerances and by changing the temperatures of heat and cold sources. Those of stainless samples at different qualitative temperatures are obtained by the experimental test rig, and the experimental results are shown in Figure 6, including the values of Ref. [15]. As the maximum deviation is 3.6%, the experimental results agree reasonably well with those of the reference, indicating the high efficiency with reliability of the experimental system.

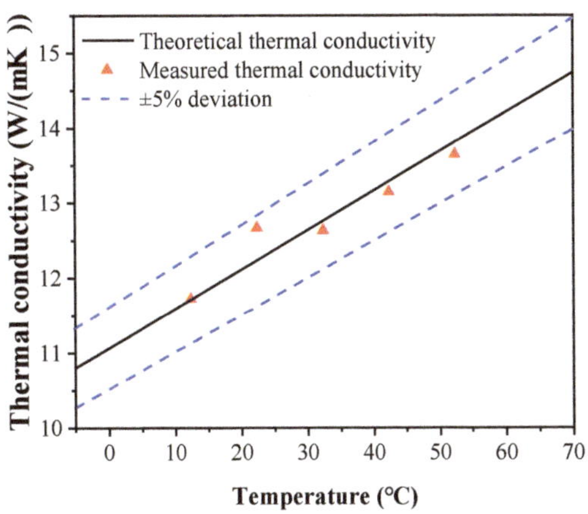

Figure 6. Comparisons between the experimental results and literature data [15] of 304 stainless steel.

As for the liquid samples, due to the inner diameter of the glass container being larger than the fluxmeter, the heat from heat source is almost conducted through specimens to the cold source and little influence is caused by the axial heat flux transfer. However, it should be noted that the main experimental uncertainty is from the expansion compensation structure of glass container. It changes the cylindrical shape of the liquid samples and causes local heat loss. Therefore, it is designed to be small enough to reduce the experimental deviation. The temperature curves of deionized water are shown in Figure 7, where the qualitative temperature is 40 °C. Similar to the experimental results of stainless steel, the steady-state stage can be quickly obtained after the temperatures of cold and heat sources are set, which shows the high efficiency of the experimental system designed in the present study.

As shown in the enlarged diagram in Figure 7, the temperatures of the fluxmeters are firmly stable in the steady state as well as the interval from T1 to T8, indicating the uniform distribution of heat flow. However, it is clear that the temperature evolution between these platinum resistance thermometers in the steady-state stages of water is greatly reduced. Although the temperature difference between the hot and cold sources reaches 32 °C, that of two measured points is about 0.3 °C, which is far less than the 2.5 °C of the stainless steel samples.

It is consistent with the inference of Fourier law that the thermal conductivities of the water are much lower than those of stainless steel with the larger thermal resistance and reduces the heat transfer from the upper fluxmeter to the lower one to lead a smaller temperature gradient. Interestingly, the temperature difference between the upper and lower fluxmeters is also apparent and causes a slight deviation of the thermal resistance. Therefore, the axial heat flux is averaged to calculate the effective thermal conductivity. The thermal conductivities of deionized water at temperatures of 20 and 40 °C are 0.653 and 0.689 W/(m·K), while those in Ref [35] are 0.599 and 0.635 W/(m·K), respectively. The deviations of 8.3% and 7.8% indicate that the method is reliable for measuring liquid samples.

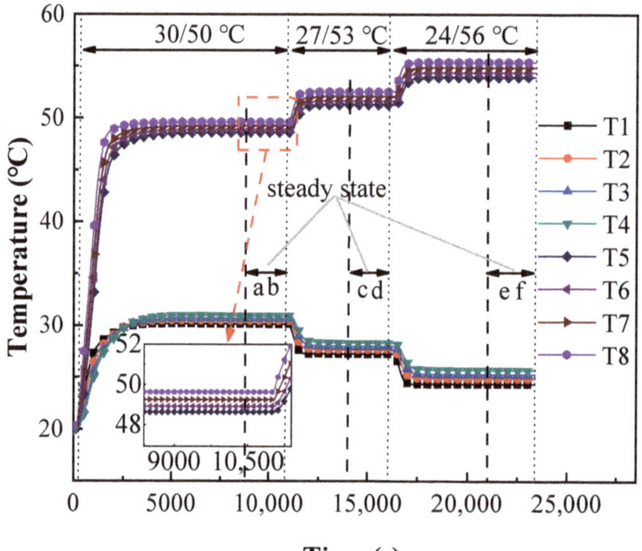

(a)

Figure 7. *Cont.*

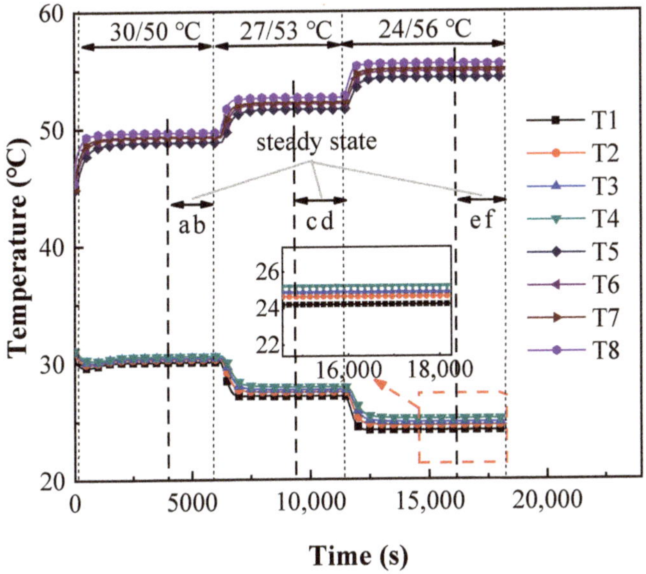

(b)

Figure 7. The experimental results of deionized water with the qualitative temperature of 40 °C as the example. (**a**) $h = 30$ mm and (**b**) $h = 45$ mm.

3.2. Effective Thermal Conductivity of Flexible Composite PCM

The thermal conductivity of pure SBS was measured first. The total thermal resistance of SBS blocks with thicknesses of 12.89, 18.77 and 26.03 mm were tested and linear fitted, indicating that the thermal conductivities of pure SBS were 0.182 and 0.188 W/(m·K) at the temperatures of 12.5 and 30 °C, respectively, which increased 3.3% with the higher temperature. Although the thermal conductivities of SBS have not been measured in the previous study, it was considered to be close to those of paraffin (0.23 W/(m·K)) [30], which showed agreement with the present results. Furthermore, those of pure PEG1500 were measured to be 0.346 and 0.353 W/(m·K) at 12.5 and 27 °C respectively, which agreed well with the value of 0.31 W/(m·K) in Ref. [36].

Each thermal conductivity is obtained by the linear dependence of the total thermal resistance on three different thicknesses. We mentioned above that the total thermal resistance is measured at three steady states to obtain the average value and the experimental deviation is extremely small as shown in Figure 8. Considering PES 10, the maximum deviation of total thermal resistance is only 4.1% with the average deviation of 1.37%, which indicates the reliability of the experimental system in the present study.

Figure 9a shows the experimental results of the thermal conductivities of the solid flexible composite PCMs at different temperatures. Then, the experimental results at 12.5 °C are discussed as an example. First, the thermal conductivity of PEG1500/EG is as large as 1.970 W/(m·K), while that of pure PEG1500 is 0.346 W/(m·K) [36] due to the effective paths for heat conduction formed by EG particles. After being integrated with 10 wt.% SBS, the value drops sharply to 1.197 W/(m·K), and the other values are 0.980, 0.725, 0.541 and 0.362 W/(m·K), when the addition ratios of SBS are 20 wt.%, 30 wt.%, 40 wt.% and 50 wt.%, respectively.

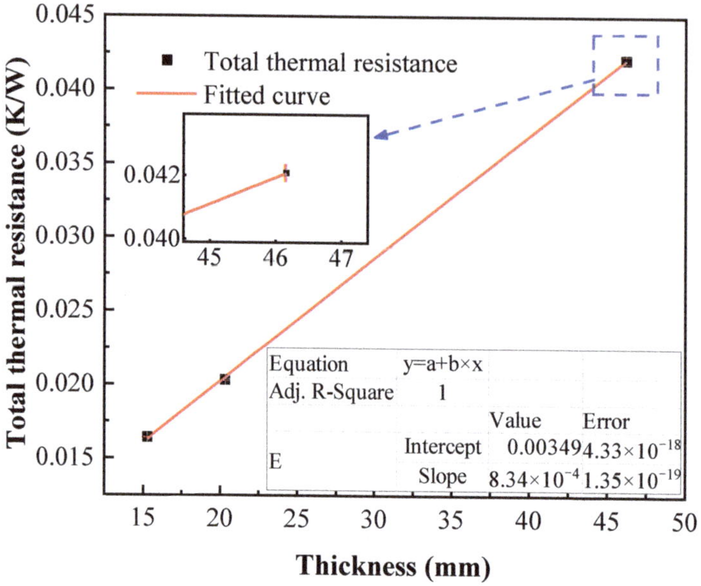

(**a**)

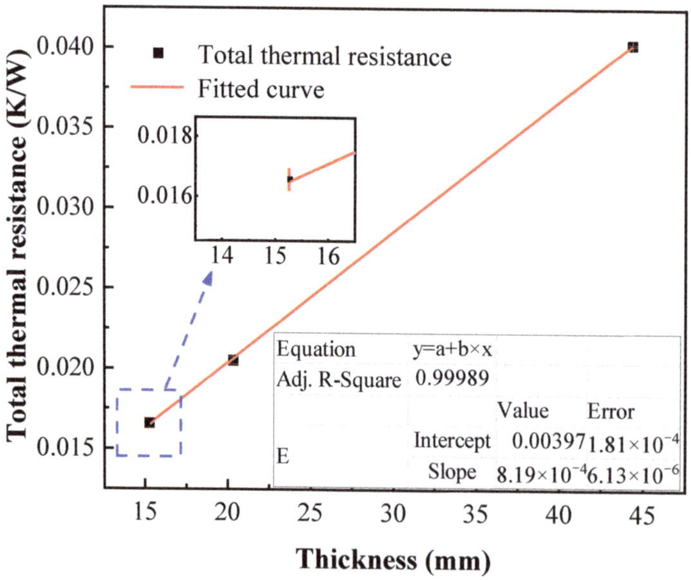

(**b**)

Figure 8. The total thermal resistances of SEP10 with different thickness. (**a**) Qualitative temperature of 12.5 °C and (**b**) qualitative temperature of 27 °C.

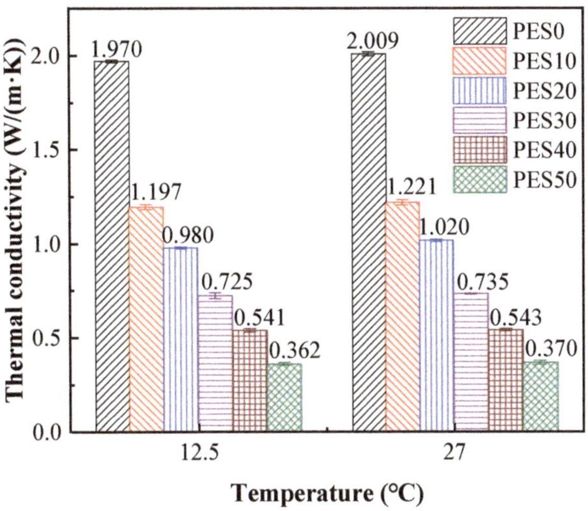

(**a**)

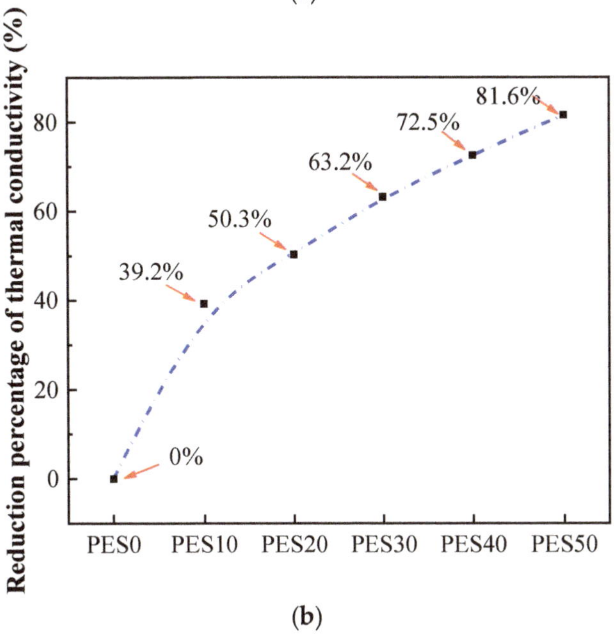

(**b**)

Figure 9. The thermal conductivities and reduction percentages of solid samples. (**a**) Samples of different SBS concentrations at 12.5 and 27 °C. The error bars are the standard deviation of the measurements (n = 3). (**b**) the reduction percentages of thermal conductivities compared to PES0 at 12.5 °C.

Those results are the average values of different heat flows at the same qualitative temperature. Additionally, the error bars show the high repeatability of the experiments, considering the maximum deviation of thermal conductivities is less than 4.3%. As shown in the Figure 9b, the initial addition of 10 wt.% SBS reduces the thermal conductivity of PCM by 39.2% and the latter 10 wt.% SBS further reduces it by about 10%, and the slope of the curve noticeably decreases.

We conclude that the excellent paths for heat conduction of EG are destroyed when its surface is covered with a film of SBS, which restricts the transfer of heat. In this way, although the thermal conductivity of composite PCM keeps decreasing with the increase in the SBS content, the decreasing range is not as apparent as the surface of EG is almost covered by SBS.

Since the slow accumulation of heat in the process of battery thermal management may be caused due to the low thermal conductivity of PCMs, the PES 20 with thermal conductivity of 0.980 W/(m·K) is taken as the acceptable one of the flexible PCMs. In addition, the trend of the thermal conductivities with the temperature is raised to be the same. When the temperature varies from 12.5 to 27 °C, the thermal conductivities of different samples increase accordingly, although this is limited. It should be noted that the increased percentages of PES0 to PES50 are 2.0%, 2.0%, 4.1%, 1.4%, 0.4% and 2.8%, which are irregular with respect to different SBS concentrations. The possible reason is that the increases in the thermal conductivity of SBS and PES0 with temperature are similar, 3.3% and 2.0%, respectively. However, when the temperature is higher than the phase-change temperature of the composite flexible PCMs, the thermal conductivities in the present study decrease in relatively large percentages as shown in Figure 10.

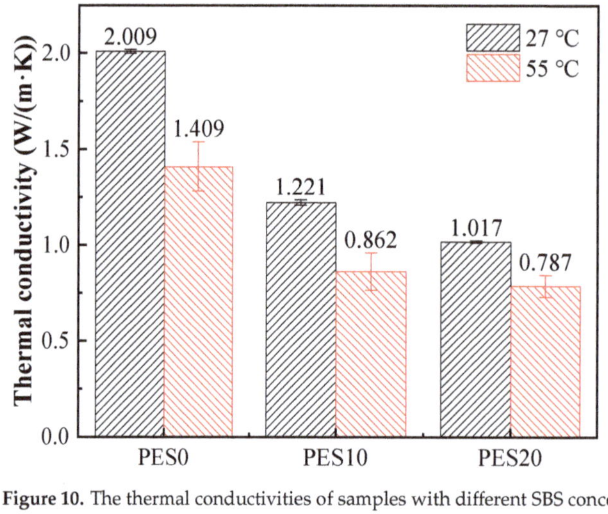

Figure 10. The thermal conductivities of samples with different SBS concentrations at 27 and 55 °C. The error bars are the standard deviation of the measurements (n = 3).

The thermal conductivities of the flexible materials PES10 and PES20 and the substrate PES0, which have high thermal conductivities, and the prospect of battery thermal management are measured at high temperature. The thermal conductivities of PES0, PES10 and PES20 decrease from 2.009, 1.221 and 1.017 W/(m·K) to 1.409, 0.862 and 0.787 W/(m·K), respectively, reducing by approximately 29.9%, 29.4% and 22.6%.

This may be due to the inherent low thermal conductivity of the molten PEG1500; however, the addition of SBS appears to reduce the declining trend of the thermal conductivities. It can be predicted that the encapsulation effect of SBS with stable thermal conductivities reduces the leakage of PEG1500 impregnated with EG. Due to this, the heat accumulation at high temperatures in battery thermal management using the flexible PCMs can be reduced in an effective way.

3.3. Theoretical Prediction of Thermal Conductivity

In the present study, the thermal conductivities of the composite flexible PCMs at 27 °C were theoretically predicted. During the preparation process of the composite PCMs, different raw materials were physically mixed, indicating that the dispersed phase particle

EG was uniformly dispersed in the PCM, and thus it was considered as a composite material with dispersed high thermal conductivity fillers in the continuous polymer matrix. Based on factors, such as the shape, size, proportion, dispersion and interaction of fillers, its effective thermal conductivities were predicted by well-known theoretical models for multiphase composite as shown in Table 3, and the rationality and theoretical reasons for its change in EG content were verified by comparison with the experimental results.

Table 3. Predictions for the effective thermal conductivity of the multiphase composite using theoretical models from the literature.

Items	Theoretical Predictions	Equations
Russell model [25]	$\lambda_{eff} = \lambda_m \frac{\phi^{\frac{2}{3}} + \frac{\lambda_m}{\lambda_f}(1-\phi^{\frac{2}{3}})}{\phi^{\frac{2}{3}} - \phi + \frac{\lambda_m}{\lambda_f}(1+\phi-\phi^{\frac{2}{3}})}$	(5)
Cheng-Vachon model [26]	$\frac{1}{\lambda_{eff}} = \frac{1-A}{\lambda_m} + \frac{\ln \frac{\{\lambda_m + A(\lambda_f-\lambda_m)\}^{\frac{1}{2}} + \frac{A}{2}\{B(\lambda_f-\lambda_m)\}^{\frac{1}{2}}}{\{\lambda_m + A(\lambda_f-\lambda_m)\}^{\frac{1}{2}} - \frac{A}{2}\{B(\lambda_f-\lambda_m)\}^{\frac{1}{2}}}}{[B(\lambda_f-\lambda_m)\{\lambda_m + A(\lambda_f-\lambda_m)\}]^{\frac{1}{2}}} A = (\frac{3\phi}{2})^{\frac{1}{2}} B = (\frac{2}{3\phi})^{\frac{1}{2}}$	(6)
Agari-Uno model [27]	$\lambda_{eff} = (\frac{\lambda_f}{C_p\lambda_m})^{\phi C_f} \times C_p\lambda_m$	(7)
Meredith-Tobias model [37]	$\lambda_{eff} = \lambda_m \left[\frac{2(2+\frac{\lambda_f}{\lambda_m})-2\phi(1-\frac{\lambda_f}{\lambda_m})}{2(2+\frac{\lambda_f}{\lambda_m})+\phi(1-\frac{\lambda_f}{\lambda_m})}\right] \times \left[\frac{(2-\phi)(2+\frac{\lambda_f}{\lambda_m})-2\phi(1-\frac{\lambda_f}{\lambda_m})}{(2-\phi)(2+\frac{\lambda_f}{\lambda_m})+\phi(1-\frac{\lambda_f}{\lambda_m})}\right]$	(8)
Maxwel-Eucken model [38]	$\lambda_{eff} = \lambda_m \frac{2\lambda_m+\lambda_f+2\phi(\lambda_f-\lambda_m)}{2\lambda_m+\lambda_f-\phi(\lambda_f-\lambda_m)}$	(9)
Ling model [39]	$\lambda_{eff} = \frac{\lambda_f \rho_{composite} \varphi_{EG}}{\rho_f}$	(10)

where λ_f and λ_m are the thermal conductivity of the fillers (EG) and the thermal conductivity of the matrix, respectively. As a single matrix with only one dispersed phase is considered by those theoretical models, the mixture of SBS and PEG1500 is taken as the matrix when the mass fraction of EG is less than 4 %, otherwise is taken as the filler [40]. λ_f and λ_m can be obtained by Equation (11), which was proposed by Woodside and Messmer [41].

Similarly, air is considered as a part of the dispersed phase due to its great influence on the thermal conductivity of EG, and λ_{EG} at 6 MPa is deduced for Equation (11). φ_{EG} is the mass fraction of EG and ϕ is the volume fraction of EG determined by the preparation process of the experimental samples using Equation (12). C_p and C_f are the coefficients determined by the properties of the matrix and the filler. The value of C_f is considered to be large when the conductive chains can be easily formed by fillers, while C_p is affected by the crystallinity and crystal size of the matrix. In this study, C_p and C_f are 1.665 and 0.850, respectively, derived from references [27,42].

$$\begin{cases} \lambda_m = (\lambda_{SBS}^{1/2}\phi_{SBS} + \lambda_{PEG}^{1/2}\phi_{PEG})^2 \\ \lambda_f = (\lambda_{EG}^{1/2}\phi_{EG} + \lambda_{air}^{1/2}\phi_{air})^2 \end{cases} \varphi_{EG} < 4\%$$
$$\begin{cases} \lambda_m = (\lambda_{EG}^{1/2}\phi_{EG} + \lambda_{air}^{1/2}\phi_{air})^2 \\ \lambda_f = (\lambda_{SBS}^{1/2}\phi_{SBS} + \lambda_{PEG}^{1/2}\phi_{PEG})^2 \end{cases} \varphi_{EG} \geq 4\% \quad (11)$$

$$\phi = \frac{m_{EG}/\rho_{EG}}{m_{SBS}/\rho_{SBS} + m_{PEG}/\rho_{PEG} + m_{EG}/\rho_{EG}} \quad (12)$$

where λ_{SBS} and λ_{PEG} are the thermal conductivities of SBS, PEG1500, which are measured in Section 3.2 to be 0.180 and 0.346 W/(m·K), respectively. λ_{EG} as the thermal conductivity of EG is 160.2 W/(m·K) [43], and λ_{air} as the thermal conductivity of air is 0.026 W/(m·K) [35]. ϕ_{SBS}, ϕ_{PEG} and ϕ_{air} are the volume fractions of SBS, PEG1500 and air, respectively. ρ_{SBS}, ρ_{PEG} and ρ_{EG} are measured to be 781, 1200 and 1353 kg/m^3 at the pressure of 6 MPa, respectively, ρ_{air} is 1.29 kg/m^3 and $\rho_{composite}$ are shown in Table 1. As the mass fractions of SBS in the samples increase from 0% to 50%, λ_m is considered to gradually decrease from 0.346 to 0.238 W/(m·K), and λ_f is revised to be 67.43 W/(m·K).

As shown in Figure 11, both the theoretical and experimental thermal conductivities of the composite PCMs with different mass ratios of SBS have the same trend, i.e., decreasing with the increase in the SBS content. It is notable that the variation rate of the experimental results is more apparent with different samples compared to those of Cheng-Vachon model while the prediction values are greatly overestimated by Ling model. In the present study, SBS obviously reduced thermal conductivity of the composite by forming a film on EG. However, the predicted value in the Ling model only depends on ρ_{EG} and $\lambda_{EG,}$ and the thermal conductivityof the matrix is completely ignored, inducing the large estimated values. As for Cheng-Vachon model, EG is fixedly regarded as the dispersed phase regardless of its volume fraction and leads to the low predicted thermal conductivities of the composites.

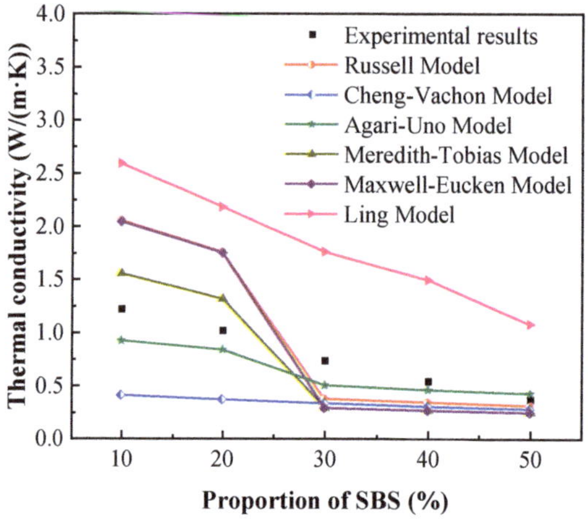

Figure 11. Theoretical and experimental thermal conductivities of composite flexible PCMs at 12.5 °C. The equations of Russell Model, Cheng-Vachon Model, Agari-Uno Model, Meredith-Tobias Model, Maxwell-Eucken Model, Ling Model are from [25–27,37–39], respectively.

The theoretical thermal conductivities of Russel model, Agari-Uno model, Meredith-Tobias model and Maxwell-Eucken model slightly deviate from the experimental results for PES50, among which the predicted value of the Agari-Uno model is the closest to the experimental result, overestimating the λ_{eff} by 15.8%. The deviation sharply increases for low concentration of SBS in samples, and those of theoretical thermal conductivities of PES10 are 68.4%, 24.2%, 27.4% and 67.6% of the experimental results, respectively. Agari-Uno model is considered to be a relatively suitable method to predict the thermal conductivities of such composites in the present study with the average deviation of 20.5%.

The air inside the composites greatly restricts the heat transfer at the interface between the matrix and nano-filler and increases the thermal resistance, which cannot be neglected in the theoretical prediction of thermal conductivities. For Rusell model, Meredith-Tobias model and Maxwell-Eucken model, the predicted values are not fully match with the experimental results and it is underestimated and overestimated when the concentration of SBS is high and low, respectively. The possible reason is that even the reduction of thermal conductivity caused by the theoretical volume fraction of air is considered in these models, it is difficult to avoid the involvement of air during the process of synthesizing composites as shown in Table 1. The neglected interfacial thermal resistance between particles and matrix leads to the theoretical predicted value being relatively large especially when EG is regarded as the matrix.

In Agari-Uno model, thermal resistance of interface is considered by C_f and C_p. It is determined by the matrix and filler, which affects the thermal conductivities of the composite PCMs by the effects of heat transfer path and it might be much more obvious when the mass fraction of EG is more than 20%. As the mass fraction of EG is less than 5%

in the present study, the prediction is sightly lower than the experimental results and the influence of air is considered by the model, leading a relatively accurate prediction.

4. Conclusions

In the present study, a steady-state experimental device was built that could measure the thermal conductivities of samples under different temperatures or phases. Then, the copolymer SBS was added into PEG1500 with 5 wt.% EG as an additive to form a flexible composite PCM. Extensive experiments were conducted to obtain the thermal conductivities of the composite PCMs under various conditions, and the following conclusions can be drawn:

(1) The accuracy of the experimental rig was verified by measuring the thermal conductivity of 304 stainless steel and deionized water. The results showed that the deviation between the thermal conductivity of 304 stainless steel and the literature was within 4%, and that of deionized water was within 8.3% at different temperatures.

(2) The thermal conductivity of SBS itself was 0.18 W/(m·K) and slightly increased with the increase in temperature, rather than the thermal conductivity of composite PCM, which decreased rapidly with the increase in the SBS content. After the integration with 10 wt.% SBS, the thermal conductivity of PES0 dropped sharply to 1.197 from 1.97 W/(m·K), and the decline ratio was 39.2%, which was up to 81.6% when the addition ratio of SBS was 50 wt.%. When the SBS content was constant, the thermal conductivity of all samples increased sightly at first and then decreased with the increase in temperature, and the peak thermal conductivity might appear near the melting phase-change temperature of the composite PCM.

(3) The deviations of the theoretical thermal conductivities predicted by the Russell model, Cheng-Vachon model, Agrai-Uno model, Meredith-Tobias model, Maxwell-Eucken model and Ling model were averagely 47.9%, 49.9%, 20.5%, 39.7%, 56.3% and 146.7% respectively. Agari-Uno model was considered to be the relatively suitable method to predict the thermal conductivities of samples in the present study. In general, the influence of air on the experimental measurement and theoretical prediction of thermal conductivity should be further investigated.

Author Contributions: Conceptualization, Z.F. and X.X.; methodology, Z.F. and X.X.; software, Z.F.; validation, Z.F.; formal analysis, Z.F.; investigation, X.X.; resources, X.X.; data curation, Z.F.; writing—original draft preparation, Z.F.; writing—review and editing, X.X.; visualization, Z.F.; supervision, X.X.; project administration, X.X.; funding acquisition, X.X. All authors have read and agreed to the published version of the manuscript.

Funding: This research was funded by Shanghai Pujiang Program, grant number 20PJ1400200; Yunnan Provincial Rural Energy Engineering Key Laboratory, grant number 2022KF001; Fundamental Research Funds for the Central Universities of China, grant number 2232021D-11; Shanghai Overseas High Level Talents Program.

Data Availability Statement: Not applicable.

Acknowledgments: The authors thank Junjie Zhou who helps the experimental tests.

Conflicts of Interest: The authors declare no conflict of interest.

References

1. Weng, J.; Yang, X.; Zhang, G.; Ouyang, D.; Chen, M.; Wang, J. Optimization of the detailed factors in a phase-change-material module for battery thermal management. *Int. J. Heat Mass Transf.* **2019**, *138*, 126–134. [CrossRef]
2. Jin, X.; Duan, X.; Jiang, W.; Wang, Y.; Zou, Y.; Lei, W.; Sun, L.; Ma, Z. Structural design of a composite board/heat pipe based on the coupled electro-chemical-thermal model in battery thermal management system. *Energy* **2021**, *216*, 119234. [CrossRef]
3. Yang, N.; Zhang, X.; Shang, B.; Li, G. Unbalanced discharging and aging due to temperature differences among the cells in a lithium-ion battery pack with parallel combination. *J. Power Sources* **2016**, *306*, 733–741. [CrossRef]
4. Chen, F.; Huang, R.; Wang, C.; Yu, X.; Liu, H.; Wu, Q.; Qian, K.; Bhagat, R. Air and PCM cooling for battery thermal management considering battery cycle life. *Appl. Therm. Eng.* **2020**, *173*, 115154. [CrossRef]
5. Zhang, W.; Qiu, J.; Yin, X.; Wang, D. A novel heat pipe assisted separation type battery thermal management system based on phase change material. *Appl. Therm. Eng.* **2020**, *165*, 114571. [CrossRef]

6. Hussain, A.; Abidi, I.H.; Tso, C.Y.; Chan, K.C.; Luo, Z.; Chao, C.Y.H. Thermal management of lithium ion batteries using graphene coated nickel foam saturated with phase change materials. *Int. J. Therm. Sci.* **2018**, *124*, 23–35. [CrossRef]
7. Zhang, N.; Yuan, Y.; Cao, X.; Du, Y.; Zhang, Z.; Gui, Y. Latent heat thermal energy storage systems with solid-liquid phase change materials: A review. *Adv. Eng. Mater.* **2018**, *20*, 1700753. [CrossRef]
8. Mettawee, E.B.S.; Assassa, G.M.R. Thermal conductivity enhancement in a latent heat storage system. *Sol. Energy* **2007**, *81*, 839–845. [CrossRef]
9. Bondareva, N.S.; Buonomo, B.; Manca, O.; Sheremet, M.A. Heat transfer inside cooling system based on phase change material with alumina nanoparticles. *Appl. Therm. Eng.* **2018**, *144*, 972–981. [CrossRef]
10. Zou, D.; Ma, X.; Liu, X.; Zheng, P.; Hu, Y. Thermal performance enhancement of composite phase change materials (PCM) using graphene and carbon nanotubes as additives for the potential application in lithium-ion power battery. *Int. J. Heat Mass Transf.* **2018**, *120*, 33–41. [CrossRef]
11. Ling, Z.; Li, S.; Cai, C.; Lin, S.; Fang, X.; Zhang, Z. Battery thermal management based on multiscale encapsulated inorganic phase change material of high stability. *Appl. Therm. Eng.* **2021**, *193*, 117002. [CrossRef]
12. Krishna, J.; Kishore, P.S.; Solomon, A.B. Heat pipe with nano enhanced-PCM for electronic cooling application. *Exp. Therm. Fluid Sci.* **2017**, *81*, 84–92. [CrossRef]
13. Mehrabi-Kermani, M.; Houshfar, E.; Ashjaee, M. A novel hybrid thermal management for Li-ion batteries using phase change materials embedded in copper foams combined with forced-air convection. *Int. J. Therm. Sci.* **2019**, *141*, 47–61. [CrossRef]
14. Lin, Y.; Kang, Q.; Wei, H.; Bao, H.; Jiang, P.; Mai, Y.W.; Huang, X. Spider web-inspired graphene skeleton-based high thermal conductivity phase change nanocomposites for battery thermal management. *Nano-Micro Lett.* **2021**, *13*, 316–329. [CrossRef] [PubMed]
15. Xiao, X.; Zhang, P.; Li, M. Effective thermal conductivity of open-cell metal foams impregnated with pure paraffin for latent heat storage. *Int. J. Therm. Sci.* **2014**, *81*, 94–105. [CrossRef]
16. Liu, C.; Yu, W.; Chen, C.; Xie, H.; Cao, B. Remarkably reduced thermal contact resistance of graphene/olefin block copolymer/paraffin form stable phase change thermal interface material. *Int. J. Heat Mass Transf.* **2020**, *163*, 120393. [CrossRef]
17. Ahadi, M.; Andisheh-Tadbir, M.; Tam, M.; Bahrami, M. An improved transient plane source method for measuring thermal conductivity of thin films: Deconvoluting thermal contact resistance. *Int. J. Heat Mass Transf.* **2016**, *96*, 371–380. [CrossRef]
18. Vitiello, D.; Nait-Ali, B.; Tessier-Doyen, N.; Tonnesen, T.; Laím, L.; Rebouillat, L.; Smith, D.S. Thermal conductivity of insulating refractory materials: Comparison of steady-state and transient measurement methods. *Open Ceram.* **2021**, *6*, 100118. [CrossRef]
19. Ruoho, M.; Valset, K.; Finstad, T.; Tittonen, I. Measurement of thin film thermal conductivity using the laser flash method. *Nanotechnology* **2015**, *26*, 195706. [CrossRef]
20. Xiao, X.; Zhang, P.; Li, M. Experimental and numerical study of heat transfer performance of nitrate/expanded graphite composite PCM for solar energy storage. *Energy Convers. Manag.* **2015**, *105*, 272–284. [CrossRef]
21. Jannot, Y.; Degiovanni, A.; Schick, V.; Meulemans, J. Apparent thermal conductivity measurement of anisotropic insulating materials at high temperature by the parallel hot-wire method. *Int. J. Therm. Sci.* **2021**, *160*, 106672. [CrossRef]
22. Yang, I.; Kim, D.; Lee, S. Construction and preliminary testing of a guarded hot plate apparatus for thermal conductivity measurements at high temperatures. *Int. J. Heat Mass Transf.* **2018**, *122*, 1343–1352. [CrossRef]
23. Wang, X.; Yu, H.; Li, L.; Zhao, M. Research on temperature dependent effective thermal conductivity of composite-phase change materials (PCMs) wall based on steady-state method in a thermal chamber. *Energy Build.* **2016**, *126*, 408–414. [CrossRef]
24. Feng, B.; Zhang, Y.H.; Tu, J.; Fan, L.W.; Yu, Z.T. Determination on the thermal conductivity and thermal contact resistance of thin composite phase change films as a thermal interfacial material. *Case Stud. Therm. Eng.* **2022**, *33*, 101979. [CrossRef]
25. Russell, H.W. Principles of heat flow in porous insulators. *J. Am. Ceram. Soc.* **1935**, *18*, 1–5. [CrossRef]
26. Cheng, S.C.; Vachon, R.I. The prediction of the thermal conductivity of two and three phase solid heterogeneous mixtures. *Int. J. Heat Mass Transf.* **1969**, *12*, 249–264. [CrossRef]
27. Agari, Y.; Ueda, A.; Nagai, S. Thermal-conductivity of a polymer composite. *J. Appl. Polym. Sci.* **1993**, *49*, 1625–1634. [CrossRef]
28. Wang, T.; Wang, S.; Wu, W. Experimental study on effective thermal conductivity of microcapsules based phase change composites. *Int. J. Heat Mass Transf.* **2017**, *109*, 930–937. [CrossRef]
29. Xu, J.Z.; Gao, B.Z.; Kang, F.Y. A reconstruction of Maxwell model for effective thermal conductivity of composite materials. *Appl. Therm. Eng.* **2016**, *102*, 972–979. [CrossRef]
30. Huang, Q.; Li, X.; Zhang, G.; Deng, J.; Wang, C. Thermal management of lithium-ion battery pack through the application of flexible form-stable composite phase change materials. *Appl. Therm. Eng.* **2021**, *183*, 116151. [CrossRef]
31. Zhang, J.; Li, X.; Zhang, G.; Wang, Y.; Guo, J.; Wang, Y.; Huang, Q.; Xiao, C.; Zhong, Z. Characterization and experimental investigation of aluminum nitride-based composite phase change materials for battery thermal management. *Energy Convers. Manag.* **2020**, *204*, 112319. [CrossRef]
32. Leng, Z.; Yuan, Y.; Cao, X.; Zhong, W.; Zeng, C. Heat pipe/phase change material coupled thermal management in Li-ion battery packs: Optimization and energy-saving assessment. *Appl. Therm. Eng.* **2022**, *208*, 118211. [CrossRef]
33. Wu, W.; Wu, W.; Wang, S. Thermal optimization of composite PCM based large-format lithium-ion battery modules under extreme operating conditions. *Energy Convers. Manag.* **2017**, *153*, 22–33. [CrossRef]
34. Huang, Q.; Li, X.; Zhang, G.; Wang, Y.; Deng, J.; Wang, C.; Chen, T. Pouch lithium battery with a passive thermal management system using form-stable and flexible composite phase change materials. *ACS Appl. Energy Mater.* **2021**, *4*, 1978–1992. [CrossRef]
35. Yang, S.; Tao, W. *Heat Transfer*, 4rd ed; Higher Education Press: Beijing, China, 2006.

36. Kou, Y.; Wang, S.; Luo, J.; Sun, K.; Zhang, J.; Tan, Z.; Shi, Q. Thermal analysis and heat capacity study of polyethylene glycol (PEG) phase change materials for thermal energy storage applications. *Chem. Thermodyn.* **2019**, *128*, 259–274. [CrossRef]
37. Meredith, R.E.; Tobias, C.W. *Advances in Electrochemistry and Electrochemical Engineering*; Wiley: New York, NY, USA, 1962; Volume 2, pp. 15–47.
38. Hashin, Z.; Shtrikman, S. A variational approach to the theory of the effective magnetic permeability of multiphase materials. *J. Appl. Phys.* **1962**, *33*, 3125–3131. [CrossRef]
39. Ling, Z.; Chen, J.; Xu, T.; Fang, X.; Gao, X.; Zhang, Z. Thermal conductivity of an organic phase change material/expanded graphite composite across the phase change temperature range and a novel thermal conductivity model. *Energy Convers. Manag.* **2015**, *102*, 202–208. [CrossRef]
40. Xia, L.; Zhang, P.; Wang, R. Preparation and thermal characterization of expanded graphite/paraffin composite phase change material. *Carbon* **2010**, *48*, 2538–2548. [CrossRef]
41. Woodside, W.; Messmer, J.H. Thermal conductivity of porous media. I. Unconsolidated Sands. *J. Appl. Phys.* **1961**, *32*, 1688–1699. [CrossRef]
42. Wu, S.; Yan, T.; Kuai, Z.; Pan, W. Preparation and thermal property analysis of a novel phase change heat storage material. *Renew. Energy* **2020**, *150*, 1057–1065. [CrossRef]
43. Xiao, X.; Zhang, P.; Li, M. Thermal characterization of nitrates and nitrates/expanded graphite mixture phase change materials for solar energy storage. *Energy Convers. Manag.* **2013**, *73*, 86–94. [CrossRef]

micromachines

MDPI

Article

Significance of Convection and Internal Heat Generation on the Thermal Distribution of a Porous Dovetail Fin with Radiative Heat Transfer by Spectral Collocation Method

G. Sowmya [1], Maha M. A. Lashin [2], M. Ijaz Khan [3,*], R. S. Varun Kumar [4], K. C. Jagadeesha [4,5], B. C. Prasannakumara [4], Kamel Guedri [6], Omar T Bafakeeh [7], El Sayed Mohamed Tag-ElDin [8] and Ahmed M. Galal [9,10]

1 Department of Mathematics, M S Ramaiah Institute of Technology, Bangalore 560054, India
2 Electrical Engineering Department, Princess Nourah bint Abdulrahman University, Riyadh 11564, Saudi Arabia
3 Department of Mathematics and Statistics, Riphah International University I-14, Islamabad 44000, Pakistan
4 Department of Studies in Mathematics, Davangere University, Tholhunse 577002, India
5 Department of Mathematics, I.D.S.G. Government First Grade College, Chikkamagaluru 577101, India
6 Mechanical Engineering Department, College of Engineering and Islamic Architecture, Umm Al-Qura University, P.O. Box 5555, Makkah 21955, Saudi Arabia
7 Department of Industrial Engineering, Jazan University, Jazan 82822, Saudi Arabia
8 Faculty of Engineering and Technology, Future University in Egypt, New Cairo 11835, Egypt
9 Mechanical Engineering Department, College of Engineering, Prince Sattam Bin Abdulaziz University, Wadi ad-Dawaser 11991, Saudi Arabia
10 Production Engineering and Mechanical Design Department, Faculty of Engineering, Mansoura University, Mansoura 35516, Egypt
* Correspondence: mikhan@math.qau.edu.pk

Abstract: A variety of methodologies have been used to explore heat transport enhancement, and the fin approach to inspect heat transfer characteristics is one such effective method. In a broad range of industrial applications, including heat exchangers and microchannel heat sinks, fins are often employed to improve heat transfer. Encouraged by this feature, the present research is concerned with the temperature distribution caused by convective and radiative mechanisms in an internal heat-generating porous longitudinal dovetail fin (DF). The Darcy formulation is considered for analyzing the velocity of the fluid passing through the fin, and the Rosseland approximation determines the radiation heat flux. The heat transfer problem of an inverted trapezoidal (dovetail) fin is governed by a second-order ordinary differential equation (ODE), and to simplify it to a dimensionless form, nondimensional terms are utilized. The generated ODE is numerically solved using the spectral collocation method (SCM) via a local linearization approach. The effect of different physical attributes on the dimensionless thermal field and heat flux is graphically illustrated. As a result, the temperature in the dovetail fin transmits in a decreasing manner for growing values of the porosity parameter. For elevated values of heat generation and the radiation-conduction parameter, the thermal profile of the fin displays increasing behavior, whereas an increment in the convection-conduction parameter downsizes the thermal dispersal. It is found that the SCM technique is very effective and more conveniently handles the nonlinear heat transfer equation. Furthermore, the temperature field results from the SCM-based solution are in very close accordance with the outcomes published in the literature.

Keywords: extended surface; porous extended surface; dovetail fin; spectral collocation method (SCM)

1. Introduction

Fins or extended surfaces are an efficient method to strengthen the dissipation of heat from a hot surface. They are utilized in several industrial applications, including processing plants, aerospace, and electronics. It is well recognized that the operations of certain industrial sectors, such as chemical, nuclear, and power plants, considered the increased production of an enormous amount of heat, necessitating a meticulous technique of heat dissipation. Even though diverse refrigerating strategies have been used to disperse heat from thermal systems, fins serve a crucial role in increasing the rate of heat transmission. It is especially possible to speed up heat transmission between a solid and an adjacent liquid by using an expanded surface. The analysis of fin design has attracted a lot of attention in past years, and much importance has been placed on the innovation of apposite techniques for solving nonlinear fin issues. Unsteady thermal dispersal in a fin with different cross-section profiles, namely, rectangular, concave, triangular, and convex, was discussed by Mosayebidorcheh et al. [1]. Moreover, they considered internal heat production and thermal conductivity as linear thermal components, whereas the convective heat coefficient is regarded as a nonlinear thermal function. Turkyilmazoglu [2] explored the heat transfer and thermal dispersal through a fin of exponential profile. Gouran et al. [3] debriefed the thermal variations in a fin of a rectangular profile by considering the conductive and convective heat flux with internal heat production. The features of stress caused due to the thermal load on an annular fin were explained by Kumar et al. [4] with the consideration of thermal dispersion driven by convection and heat generation. The impact of convection and a magnetic environment was considered by Sowmya et al. [5] to explicate the thermal features of a rectangular-profiled annular fin.

Heat transport mechanisms in porous media are important in a variety of applications, including chemical catalytic reactors, geothermal energy extraction, thermal insulation, heat exchanger design, and oil exploration. As a result, they have gained considerable priority in recent years of research [6–10]. Porous fins designed with a porous material are more effective and compact in heat exchange compared to all other fins in many industrial applications such as heat exchangers and solar collectors. Even though porous fins have limited thermal conductivity, they significantly strengthen fluid flow blending and enhance the contact between the material's surface area and the cooling liquid inside, and the use of porous materials has been noticed as an efficient and appropriate procedure of heat transmission enrichment. Consequently, the mechanism of heat transport from porous fins has evoked interest in many investigators. Das and Kundu [11] presented a significant approach for enhancing the rate of energy transmission through the porous fin by applying the combined magnetic and electrical effects. The collocation method with sine transformation was implemented by Nabati et al. [12] to explore the numerical result for the thermal equation of a permeable fin. Wang et al. [13] analyzed the upshot of inclination angle on the thermal diffusion of a radiative permeable fin with convection and internal heat generation. The nature of the thermal field of a longitudinal fin having various cross-section profiles was scrutinized by Hosseinzadeh et al. [14] with the consideration of wet surface conditions. Production of entropy in the permeable fin of an exponential profile was expounded by Din et al. [15] with radiation, internal heat production, and convective mechanisms. Recently, Kumar et al. [16] considered the porous concave fin with an inclined position to examine the unsteady thermal distribution in the extended surface. The rectangular fin has been explored extensively owing to its low cost and ease of fabrication. Fins with a parabolic structure, on the other hand, are usually recommended over rectangle profile fins in design features where weight is a consideration, such as in aircraft applications. However, due to their curved configuration, parabolic fins are more challenging to fabricate and thereby more expensive. Furthermore, if an effort is made to fabricate a triangular fin, the sharp edge causes safety concerns. A convenient and more generalizable effect is caused by a trapezoidal fin rather than by other shaped fins. As a result, in terms of light weight and safety, a trapezoidal fin is one of the best alternatives. Thus, fins with a constant slope profile, also known as trapezoidal profiles, are utilized in numerous applications, and several researchers considered trapezoidal-profiled fins in their thermal analysis [17–20]. Recently, Gireesha et al. [21] examined the temperature variation and heat transmission in an inclined permeable straight fin of dovetail and trapezoidal profile with radiation and convective heat transfer.

The spectral collocation method (SCM) solves nonlinear differential equations faster and more effectively. It is the methodological approach to approximation in which residuals (or errors) are strictly controlled. This methodology's prominence arises from its superior characteristics to traditional finite difference approaches, and they can perform quicker convergent approximations. The simultaneous verification of a solution's spectrum and verifying of resolution is enabled by the dual illustration in physical and transformed space. Particular symmetries in a solution can be manipulated using spectral methods. As an outcome, inconsistencies in phase and dissipation in the procedure are minor or nonexistent. The primary benefit of these methodologies is their consistency for a specific number of unknowns. Finite-difference and finite-element techniques produce only algebraic convergence rates. On the other hand, SCM provides an exponential rate of convergence for smooth problems in simple models. The most prevalent spectral representations are Galerkin, collocation, and tau techniques. Due to its computational simplification and good accuracy, the SCM has been shown to be an important strategy in the engineering and science fields [22–24]. In previous years, the SCM was employed to attain the solution for the heat transfer equation of fins, rods, moving fins, and fins with irregular cross-section areas [25–27].

Upon performing a detailed literature survey, it was observed that Gireesha et al. [21] and Kang [28] scrutinized temperature distribution in a reversed trapezoidal (dovetail) fin. Dovetail fins can be employed in double pipe heat exchangers, which transfer more heat with the same base area as a rectangle-structured fin. By using the dovetail fin, a better heat dispersal over the extended area furnished by the reversed trapezoidal can be acquired. However, this type of fin has received less attention. On the other side, the use of microchannel heat sinks with fin arrangements has received a great deal of attention from authors [29,30]. An increased surface area aided by improved blending is the core benefit of the fin configuration in the microchannel. In addition, considerable effort has been devoted to improving convective heat transfer and thus reducing thermal resistance by fully or partially implanting porous fins, blocks, or baffles into the microchannel. Therefore, the present investigation elucidates the thermal distribution and transfer of heat through a porous dovetail fin with convection and radiation mechanisms. Moreover, the novelty of this investigation is to scrutinize the effect of convection, internal heat generation, and radiation on thermal distribution through a porous dovetail fin. The Darcy formulation is considered for analyzing the fluid velocity passing via the fin, and the radiation heat flux is governed by the Rosseland approximation. The ODE portraying the heat transport of the porous fin is nondimensionalized and numerically simplified using SCM. Graphs and tables are developed to observe the characteristics of the thermal profile and temperature gradient for various values of dimensionless parameters.

2. Mathematical Formulation

The steady-state heat passage of a porous DF with internal heat generation is conferred in this model, as displayed in Figure 1. Heat is dissipated from the fin's surface to the ambient environment at temperature T_∞ via convection and radiation. The permeable medium is assumed to be saturated with a single-phase, homogeneous, and isotropic liquid. Heat exchange occurs in a permeable fin due to heat conduction and thermal radiation transfer heat from the adjacent side. Transfer of heat energy is due to natural convection that occurs when the fluid's mass flows through the porous structure.

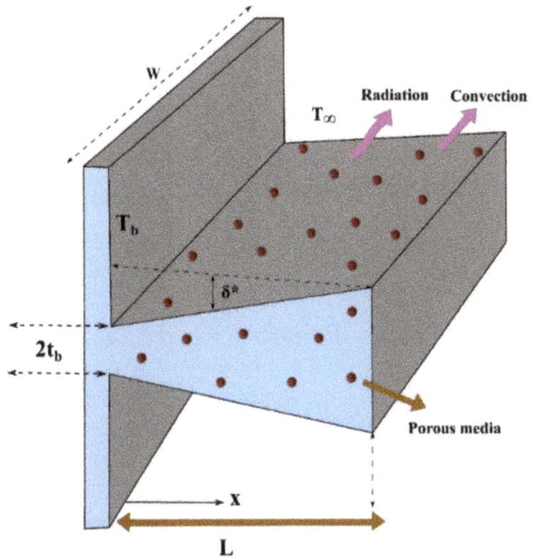

Figure 1. Physical model of a dovetail fin.

The mathematical model for the present fin problem is mentioned as follows (see Hatami and Ganji [31], Kundu and Yook [32]):

$$q_x - q_{x+dx} = (1-\phi)h^* w(T - T_\infty)dx + \dot{m}c_p(T - T_\infty) + w\varepsilon^*\sigma\left(T^4 - T_\infty^4\right)dx - w\,t(x)(1-\phi)q^*_{\rm int}(T) \tag{1}$$

At the fin's base, the heat flux vector of combined radiation and conduction is prescribed as

$$q_b = q_{cond} + q_{rad} \tag{2}$$

and the term for heat conductivity may be represented as follows using Fourier's rule of conduction (see Rostamiyan et al. [33]):

$$q_{cond} = -k_{eff}A_{cr}\frac{dT}{dx} \tag{3}$$

where, $k_{eff} = (1-\phi)k_s + \phi k_f$ is the effectual thermal conductivity and

$$A_{cr} = w\,t(x), \tag{4}$$

is the area of cross-section.

The thermal radiative flux term is represented using the Rosseland diffusion approximation as (see Bhanja et al. [34] and Alhejaili et al. [35])

$$q_{rad} = -\frac{4\sigma A_{cr}}{3\beta_R^*}\frac{dT^4}{dx} \tag{5}$$

Now, Equation (1) can be written as

$$\begin{array}{c} -\frac{dq_b}{dx} - (1-\phi)wh^*(T - T_\infty) + w\,t(x)(1-\phi)q^*_{\rm int}(T) - \dot{m}c_p(T - T_\infty) \\ -w\varepsilon^*\sigma\left(T^4 - T_\infty^4\right) = 0 \end{array} \tag{6}$$

and $\dot{m}$ is demarcated as

$$\dot{m} = \rho_f v_w(x) w\phi\, dx \tag{7}$$

where the passage velocity $v_w(x)$ is represented as follows by considering Darcy's model (see Moradi et al. [36] and Weera et al. [37]):

$$v_w(x) = \frac{gK\beta_f}{\nu_f}(T - T_\infty) \tag{8}$$

Utilizing Equations (3)–(5), Equation (6) is symbolized as follows:

$$\begin{aligned} &\frac{d}{dx}\left[k_{eff} w\, t(x)\frac{dT}{dx} + \frac{4\sigma w\, t(x)}{3\beta_R^*}\frac{dT^4}{dx}\right] - (1-\phi)wh^*(T - T_\infty) + w\, t(x)(1-\phi)q_{\text{int}}^*(T) \\ &-\dot{m}c_p(T - T_\infty) - w\varepsilon^*\sigma(T^4 - T_\infty^4) = 0 \end{aligned} \tag{9}$$

From the perspective of energy transfer, the porous fin may be referred to as the simple model when it is only operated in a natural convection environment. To formulate such a kind of heat transfer equation, the approach depends only on Darcy's model and energy balance. However, energy dissipation from the permeable fin is dependent not only on natural convection but also on solid surface convection and radiative heat exchange with the adjacent environment. Therefore, the scenario analyzed in this investigation is when a slight temperature variance occurs inside the material while the heat flows. Moreover, it has been revealed that the term T^4 can be demonstrated as a linear thermal function in such a particular circumstance (see Kiwan [38] and Madhura et al. [39]),

$$\text{i.e., } T^4 \cong 4T_\infty^3 T - 3T_\infty^4 \tag{10}$$

Using Equation (10) in Equation (9) yields

$$\begin{aligned} &\frac{d}{dx}\left[\left((1-\phi)k_s + \phi k_f\right)\frac{dT}{dx} + \phi\frac{16\sigma T_\infty^3}{3\beta_R^*}\frac{dT}{dx}\right] - (1-\phi)h^*(T)(T - T_\infty) \\ &+(1-\phi)q_{\text{int}}^*(T) - \phi\frac{\rho_f c_p g K\beta_f}{\nu_f}(T - T_\infty)^2 - \varepsilon^*\sigma(T^4 - T_\infty^4) = 0 \end{aligned} \tag{11}$$

The local semi-fin thickness is specified as

$$t(x) = t_b - \delta^*\left(\frac{x}{L}\right) \tag{12}$$

Putting Equation (12) into Equation (11) leads to

$$\begin{aligned} &\frac{d}{dx}\left[\left((1-\phi)k_s + \phi k_f\right)\left(t_b - \delta^*\left(\tfrac{x}{L}\right)\right)\frac{dT}{dx} + \phi\left(t_b - \delta^*\left(\tfrac{x}{L}\right)\right)\frac{16\sigma T_\infty^3}{3\beta_R^*}\frac{dT}{dx}\right] - (1-\phi)h^*(T)(T - T_\infty) - \phi\frac{\rho_f c_p g K\beta_f}{\nu_f}(T - T_\infty)^2 - \\ &\varepsilon^*\sigma(T^4 - T_\infty^4) + (1-\phi)\left(t_b - \delta^*\left(\tfrac{x}{L}\right)\right)q_{\text{int}}^*(T) = 0 \end{aligned} \tag{13}$$

$h^*(T)$ is generally assumed to be constant in the traditional analysis of fin problems. However, when the fin is being cooled by a boiling two-phase flow, where $h^*(T)$ is highly predicated on the local temperature surplus, such an assumption could lead to significant error. Thus, $h^*(T)$ is supposed to be a nonlinear component of the local temperature difference T causing the convection process and is given as follows (see Sowmya et al. [40]):

$$h^*(T) = h_b\left[\frac{T - T_\infty}{T_b - T_\infty}\right]^n \tag{14}$$

where n is a dimensionless constant that can range between −6.6 and 5, or, in the majority of practical uses, between −3 and 3.

In addition,

$$q_{\text{int}}^*(T) = q_\infty^*[1 + \lambda(T - T_\infty)] \tag{15}$$

The boundary conditions for DF are

$$\begin{aligned} T(0) &= T_b, \\ \frac{dT}{dx}_{x=L} &= 0 \end{aligned} \tag{16}$$

Nondimensionalization is executed utilizing the following nondimensional terms:

$$\Theta = \frac{T - T_\infty}{T_b - T_\infty}, \text{ and} X = \frac{x}{L} \tag{17}$$

and Equation (13) can be simplified as

$$\frac{d}{dX}\left[((1-\phi)+\phi k_r)(1-CX)\frac{d\Theta}{dX} + 4\,R_d^*\phi(1-CX)\frac{d\Theta}{dX}\right] - (1-\phi)Nc\,\Theta^{n+1} - \phi S_H\Theta^2 - Nr\left[(Nt+\Theta)^4 - Nt^4\right] + (1-\phi)(1-CX)Q_{\text{int}}(1+\gamma\Theta) = 0 \tag{18}$$

Nondimensional parameters in the above equation are listed in Table 1, and C is supposed to be negative $(C < 0)$, related to DF (Gireesha et al. [21]).

Correspondingly, Equation (16) is reduced to

$$\Theta(0) = 1, \quad \left.\frac{d\Theta}{dX}\right|_{X=1} = 0 \tag{19}$$

Now, Equation (18) is rearranged as follows to reduce nonlinear effects:

$$\frac{d}{dX}\left[((1-\phi)+\phi k_r)(1-CX)\frac{d\Theta}{dX} + 4\,R_d^*\phi(1-CX)\frac{d\Theta}{dX}\right] + \chi_1 Nc\,\Theta^m + \chi_2 Nr\,\Theta^4 = (1-\phi)Nc\,\Theta^{n+1} + \phi S_H\Theta^2 + Nr\left[(Nt+\Theta)^4 - Nt^4\right] - (1-\phi)(1-CX)Q_{\text{int}}(1+\gamma\Theta) + \chi_1 Nc\,\Theta^m + \chi_2 Nr\,\Theta^4 \tag{20}$$

Table 1. Nondimensional parameters present in the dimensionless fin equation.

Definition	Nondimensional Parameters
$Nc = \dfrac{h_b L^2}{k_s t_b}$	Convective–conductive parameter
$Nr = \dfrac{\sigma\,\varepsilon^* L^2}{k_s t_b}(T_b - T_\infty)^3$	Radiation number
$C = \frac{\delta^*}{t_b}$	Fin taper ratio
$\gamma = \lambda(T_b - T_\infty)$	Heat generation parameter
$S_H = \dfrac{\rho_f c_p g K \beta_f L^2}{\nu_f k_s t_b}(T_b - T_\infty)$	Porosity parameter
$k_r = \dfrac{k_f}{k_s}$	Relative thermal conductivity
$Q_{\text{int}} = \dfrac{q_\infty^*\,L^2}{(T_b - T_\infty)k_s}$	Heat generation number
$R_d^* = \dfrac{4\sigma}{3\beta_R^* k_s}(T_\infty)^3$	Radiation–conduction parameter
$Nt = \dfrac{T_\infty}{T_b - T_\infty}$	Temperature ratio parameter

3. Spectral Collocation Method Formulation

Over the past few decades, a significant attempt has been made to find exact solutions for mathematical models indicating the physical phenomenon. However, various models have been explored numerically and analytically by some investigators [41–43]. In the present work, SCM with Chebyshev–Gauss–Lobatto collocation points (Khorrami [44], Sun and Li [45]) are utilized for spatial discretization of Equation (18).

$$\xi_i = -\cos\left[\frac{\pi(j-1)}{\text{N}-1}\right], j = 1, 2 \ldots, \text{N} \tag{21}$$

The succeeding transformation maps $[X_{\min}, X_{\max}]$ into $[-1,1]$ to meet the Chebyshev polynomial requirement.

$$2X = [(X_{\max} - X_{\min})\xi + (X_{\max} + X_{\min})] \quad (22)$$

As per the SCM principle, Θ on collocation points can be simulated by Lagrange interpolation polynomials (see Khater et al. [22], Ma and Huang [23]).

$$\Theta(\xi) \approx \sum_{j=1}^{N} \Theta(\xi_j) g_j(\xi) \quad (23)$$

where

$$g_j(\xi) = \frac{w'_j / (\xi - \xi_j)}{\sum\limits_{l=1}^{N} w'_l / (\xi - \xi_j)} \quad (24)$$

and

$$w'_l = (-1)^{l-1} \Delta'_l, \quad \Delta'_l = \begin{cases} 1/2, \ l = 1, N \\ 1, \ \text{otherwise} \end{cases} \quad (25)$$

Putting Equation (21) into Equation (20) yields the matrix structure of a spectral discretized algebraic expression.

$$\sum_{l=1}^{N} E_{j,l} \Theta_l = \Gamma_j, \ j = 1, 2 \ldots N \quad (26)$$

where the matrices element expressions $E_{j,l}$ and Γ_j are demarcated as

$$E_{j,l} = \begin{cases} -(k_r\phi - \phi + 1)C\left(\frac{2}{X_{\mathrm{mix}} - X_{\min}}\right)D_{jl}^{(1)} + (k_r\phi - \phi + 1)\left(-C\left(\frac{X_{\max} - X_{\min}}{2}\right) + 1\right)\left(\frac{2}{X_{\max} - X_{\min}}\right)^2 D_{jl}^{(2)} - \\ 4R_d^*\phi C\left(\frac{2}{X_{\max} - X_{\min}}\right)D_{jl}^{(1)} + 4R_d^*\phi\left(-C\left(\frac{X_{\max} - X_{\min}}{2}\right) + 1\right)D_{jl}^{(2)} + \\ \chi_1 Nc\left(\Theta_j^*\right)^m + \chi_2 Nr\left(\Theta_j^*\right)^4 & , j = l \\ (k_r\phi - \phi + 1)\left(-C\left(\frac{X_{\max} - X_{\min}}{2}\right) + 1\right)\left(\frac{2}{X_{\max} - X_{\min}}\right)^2 D_{jl}^{(2)} - 4R_d^*\phi C\left(\frac{2}{X_{\max} - X_{\min}}\right)D_{jl}^{(1)} \\ +4R_d^*\phi\left(-C\left(\frac{X_{\max} - X_{\min}}{2}\right) + 1\right)D_{jl}^{(2)} - (k_r\phi - \phi + 1)C\left(\frac{2}{X_{\max} - X_{\min}}\right)D_{jl}^{(1)} & , j \neq l \end{cases} \quad (27)$$

and

$$\Gamma_j = (1 - \phi)Nc\left(\Theta_j^*\right)^{n+1} + \phi S_H\left(\Theta_j^*\right)^2 + Nr\left(\left(Nt + \Theta_j^*\right)^4 - Nt^4\right) - (1 - \phi)\left(1 - C\left(\frac{X_{\max} - X_{\min}}{2}\right)\right)Q_{\mathrm{int}}\left(1 + \gamma\left(\Theta_j^*\right)\right) + \chi_1 Nc\,\Theta^m + \chi_2 Nr\,\Theta^4 \quad (28)$$

Additionally, the spectral discretization of boundary conditions is provided in Ma et al. [26].

4. Results and Discussions

This segment explores the consequence of several parameters in the thermal analysis of a porous DF. The SCM technique is utilized to solve the nondimensional energy transfer issue. In addition, graphs and tables are used to examine the behavior of temperature profile $\Theta(X)$ and heat flux $-\Theta'(0)$ with the consequence of various thermal parameters, including convection–conduction parameter $(1 \leq Nc \leq 5)$, radiation–conduction parameter $(0.5 \leq R_d^* \leq 5)$, temperature ratio parameter $(0.1 \leq Nt \leq 0.5)$, porosity parameter $(0 \leq S_H \leq 20)$, radiation number $(1 \leq Nr \leq 10)$, coefficients of fin taper ratio $(-0.1 \leq C \leq -0.3)$, and heat generation number $(0 \leq Q_{\mathrm{int}} \leq 1)$. Table 2 portrays the variation in thermal profile and heat flux for varying values of porous fraction ϕ (0 and 0.1), thermal conductivity parameter k_r (0 and 0.3), Nc (2 and 5), R_d^* (0.5 and 0.7), temperature ratio parameter Nt (0.1 and 0.3), porosity parameter S_H (0 and 7), and radiation number Nr (2 and 4), while the coefficients of fin taper ratio $C = -0.3$, heat generation number $Q_{\text{int}} = 0.8$, power index $n = 0$, and fin length are taken as constant.

Table 2. Numerical values of $\Theta(X)$ (at $X = 0.5$) and $-\Theta'(0)$ for different parameters when $C = -0.3$, $n = 0$, $Q_{\text{int}} = 0.8$, and $\gamma = 0.1$.

ϕ	k_r	Nc	Nr	Nt	S_H	R_d^*	$\Theta(X)$	$-\Theta'(0)$
0							0.66972	0.32422
0.1							0.66394	0.33463
	0						0.66264	0.33536
	0.3						0.66652	0.33318
		2					0.66394	0.33463
		5					0.5071	0.47356
			2				0.59196	0.3729
			4				0.50587	0.40026
				0.1			0.72743	0.28435
				0.3			0.66394	0.33463
					0		0.68813	0.31494
					7		0.65494	0.34165
						0.5	0.67332	0.32714
						0.7	0.68322	0.32122

The values of thermal profile of the porous fin are noticed to be diminished as the ϕ, Nc, Nr, Nt, and S_H increases, but the heat flux augments for enhancement in these parameter values. Elevated scales of k_r and R_d^* lead to the enhancement of the temperature profile and cause the decrement of heat flux. For the various magnitude of selected parameters, Table 3 was constructed to compare the present results (PR) of heat flux to the benchmark solution of Gorla and Bakier [46] (GB).

Table 3. Validation of present work with existing literature by taking parameters values as $C = Nc = n = Q_{\text{int}} = \gamma = R_d^* = 0$, $\phi = 1$, and $k_r = 1$.

S_H	Nr	Nt	$-\Theta'(0)$	
			Gorila and Bakier [46]	Present Result
1	0.1	0.01	0.68610	0.68613
		0.1	0.70210	0.70210
		0.5	0.83890	0.83892
	10	0.01	2.1763	2.1765
		0.1	2.6413	2.6415
		0.5	5.4420	5.4423
10	0.1	0.01	2.5616	2.5616
		0.1	2.5660	2.5660
		0.5	2.6100	2.6100
	10	0.01	3.2791	3.2792
		0.1	3.6059	3.6060
		0.5	5.9690	5.9693

The comparison of the current analysis to the available result exhibits an excellent agreement with a maximum error of 0.00003. This analysis can be studied in detail by considering the graphical structure. Figure 2a indicates a comparison of the current numerical scheme with that of Gorla and Bakier [46] by taking $C = Nc = n = Q_{\text{int}} = \gamma = R_d^* = 0$, $\phi = 1$, and $k_r = 1$. In this figure, the value of $-\Theta'(0)$ is compared by setting the S_H and Nr values to 1 and 0.1, respectively, and $-\Theta'(0)$ also depends on the varying values of $Nt(0.01, 0.1, 0.5)$. From Figure 2a, an appropriate consistency between the data from this analysis and those published in Ref. [46] is observed, implying that the numerical technique for the simulation is accurate. Observations indicate an improvement in heat flux as a result of an increment in Nt, which is the ratio of ambient temperature to the difference among the base and ambient temperatures. The physical justification is that greater emission of heat into the environment is correlated with higher scales of Nt, which causes the thermal distribution to decrease and heat flux to augment. Figure 2b is provided to estimate the error between these comparisons using the correlation $Error = |\Theta_{PR} - \Theta_{GB}|$. According to the comparative outcomes, the maximum deviances are identified for $Nt = 0.01$ (see Figure 2b). Furthermore, the comparisons and error estimations show that the SCM has good accuracy due to their exponential convergence.

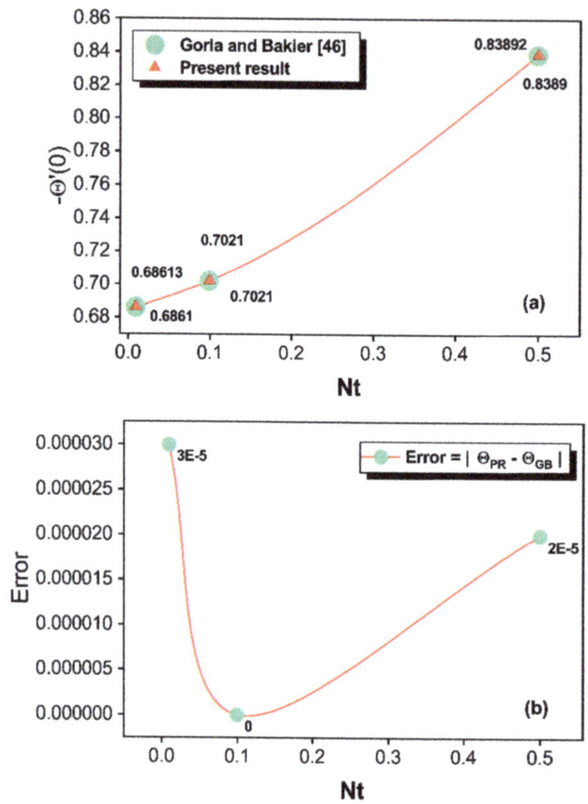

Figure 2. (**a**,**b**) Comparison of the results and error estimations.

The primary difference between solid and porous fins is the presence of an extensive number of pores involved with the permeable fin, which permits liquid to flow via these holes. Based on the current examination, Figure 3 is designed to demonstrate the thermal distribution and heat transfer rate in porous and solid dovetail fins as a component of fin length. At $X = 0$, the thermal profile for both solid and porous dovetail fins attain their maximum value, and the temperature profile shows decreasing characteristics as the X ranges from 0 to 1. In other words, the temperature is maximum at the fin base and gradually reduces from there to the tip. Finally, there will be less temperature distribution at its tip. The porous fin has a lower thermal distribution than that of the solid fin owing to the occurrence of a large number of pores that allow fluid to flow through the holes. Consequently, a porous fin offers a greater rate of heat transmission than is achieved with a solid fin.

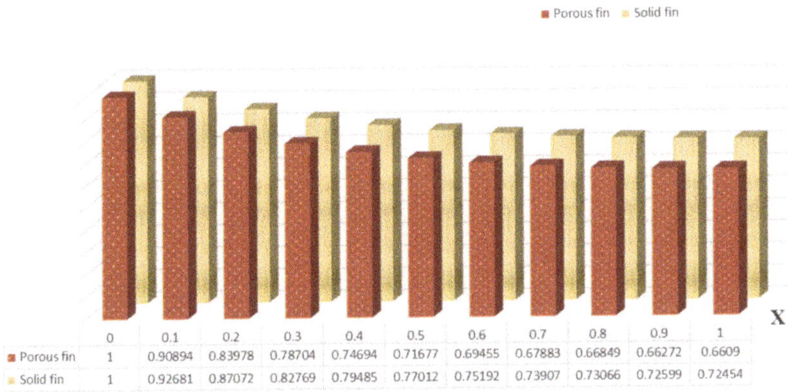

Figure 3. Thermal distribution along the longitudinal direction.

Figure 4 delineates the variation in thermal flux for increased values of Q_{int} and also elucidates the comparison between solid and permeable fins. From this figure, one can notice that the thermal gradient decreases with enrichment in Q_{int} for both kinds of fins. Moreover, the heat flux is more for permeable fin compared to solid fin.

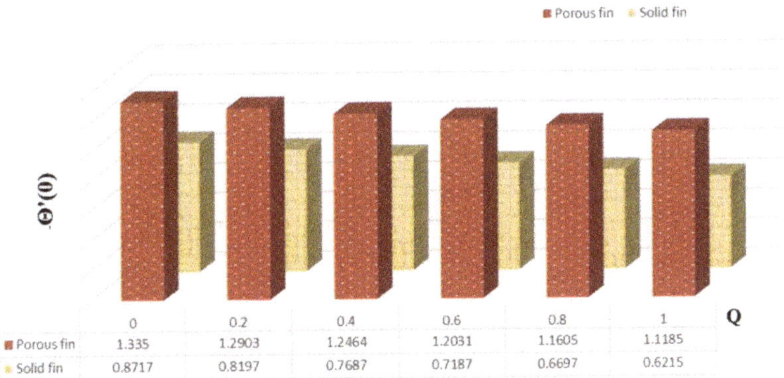

Figure 4. Heat flux of porous and solid fins.

The SCM is a numerical method for more rapidly and effectively solving nonlinear differential equations. It has advantages over traditional finite difference approaches, including the ability to perform faster convergent approximations. The major purpose of plotting Figure 5 is to determine the accuracy of the considered numerical method, SCM, by comparing with RKF-45. As a result of the comparison of the two methods, an excellent pact has been recognized in the figure.

The excellence of the proposed investigation can be spotted in Figure 6, which displays the nature of nondimensional temperature profile Θ as a function of Nc for solid and porous fins. The thermal profile of porous fin is reflected by solid thick lines in the figure, whereas dotted lines represent solid fin. For an increase in Nc, the thermal dispersion through both fins diminishes along the longitudinal direction. However, a solid fin has a higher thermal profile than that of a porous fin, as evidenced by the figure. As a result, heat transfer in dovetail porous fin is higher than that in solid fin.

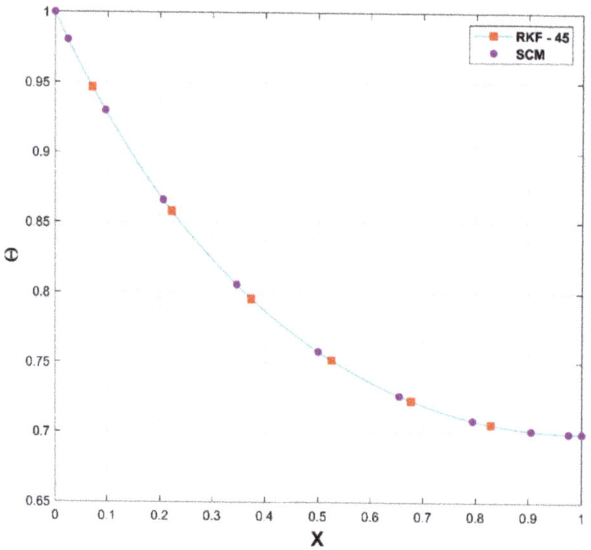

Figure 5. Comparison of the PR with RKF-45.

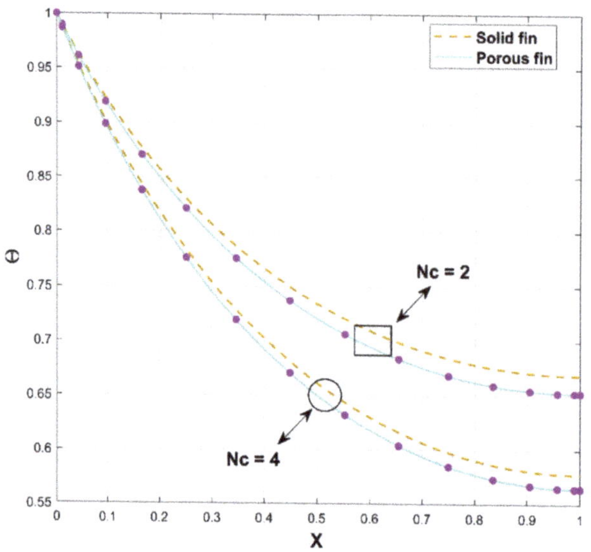

Figure 6. Thermal distribution through porous and solid fins.

The deviation of the thermal field for the different coefficients of Nr has been embodied in Figure 7. This figure reflects that the temperature field diminishes with corresponding augmented values of Nr. The enhanced radiation results in effective heat dissipation from the fin surface to the adjacent fluid, as implied by the steadily declining temperature in DF. As an upshot, heat energy transmission induced by radiation proliferates the rate of heat transfer from DF.

Figure 8 characterizes the behavioral patterns of Θ for various R_d^*. Higher values of R_d^* improve Θ, as can be seen in this illustration. The temperature in the fin increases as energy flux from heat dispersion and radiation is enhanced.

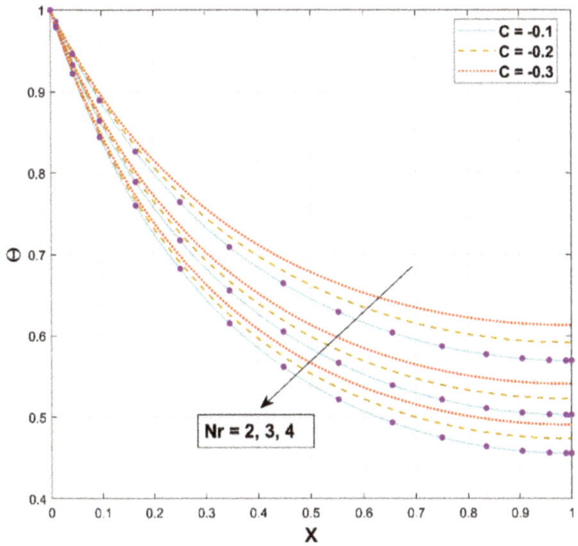

Figure 7. Upshot of Nr on Θ.

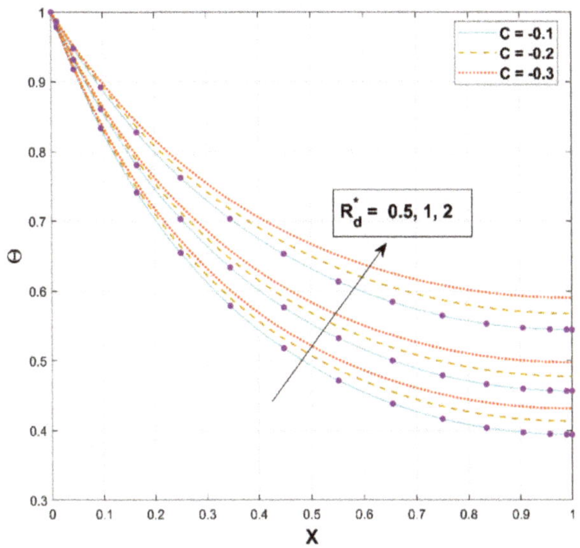

Figure 8. Upshot of R_d^* on Θ.

Surface convective heat transmission is involved in the rate of heat transmission and is indicated by the parameter Nc, which has the correlation between the convective and conductive processes. The influence of this parameter on the thermal field is indicated in Figure 9. According to this figure, improving this ratio lowers the temperature distribution. As the convection effect improves, more heat from the fin surface is transferred to the surroundings. The temperature consequently drops as the rate of heat transfer intensifies.

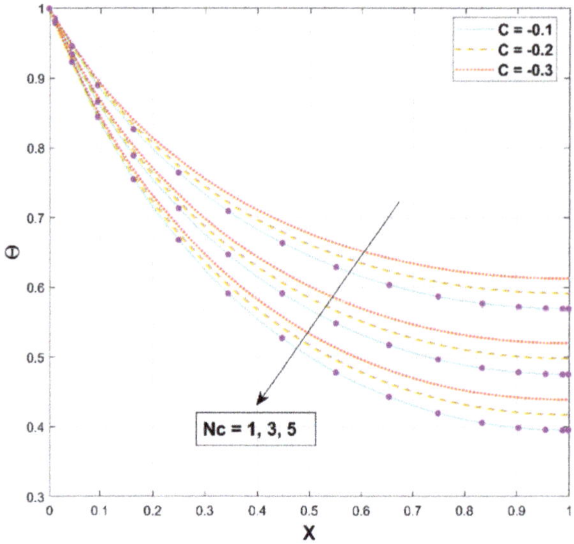

Figure 9. Upshot of Nc on Θ.

Figure 10 is illustrated to exhibit the impact of Q_{int} on the thermal distribution along the longitudinal axis. Θ shows the upsurge behavior for increased estimations of the heat generation parameter. An upsurge in Q_{int} accelerates the internal heat generation (temperature-dependent), as conveyed in Equation (15). Advanced internal heat generation results in elevated Θ since a greater quantity of heat must be dispersed by the fin to its surroundings.

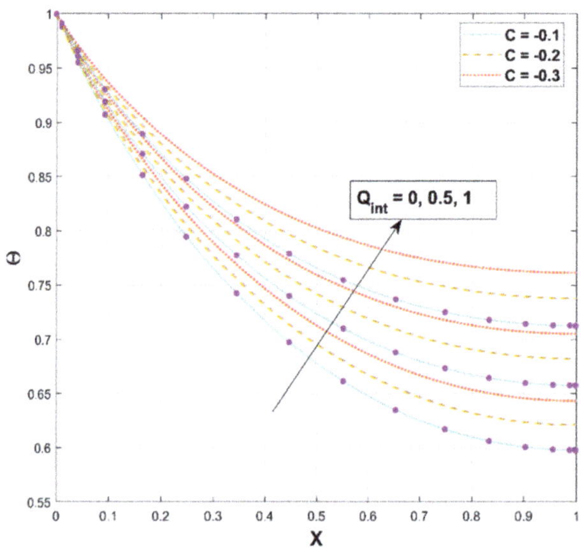

Figure 10. Upshot of Q_{int} on Θ.

Figure 11a,b represents the relationship between the thermal field and the porosity parameter S_H. When this parameter is increased, the temperature distribution is reduced, resulting in a higher heat transfer rate, as displayed in Figure 11a. When steady-state heat transmission via a fin with porous media is examined under the impacts of natural convection and radiation, it is discovered that porous materials significantly improve heat transfer. Since a permeable fin is composed of pores and solid, heat interaction from the solid part of the porous fin is caused by direct exposure to the environment. The decreasing trend in fin temperature triggered by an increment in the porosity factor is based on the fact that as the porosity parameter improves, so does the porousness of the permeable fin, and thus the capacity of the working liquid to penetrate through the fin pores augments. Additionally, as the buoyancy force increases, more heat is dissipated, which accelerates the fin's ability to transfer heat. In addition, a three-dimensional illustration is provided in Figure 11b for representing the nature of the thermal field for a specific value of the porous parameter. The energy dissipation rate lowers as the permeability improves, and the porous fin's efficient thermal conductivity significantly reduces as the solid material is removed. Radiation and convection, on the other hand, improve the rate of heat removal. This variable symbolizes the temperature ratio, and an increase in this factor causes an increment in heat flux.

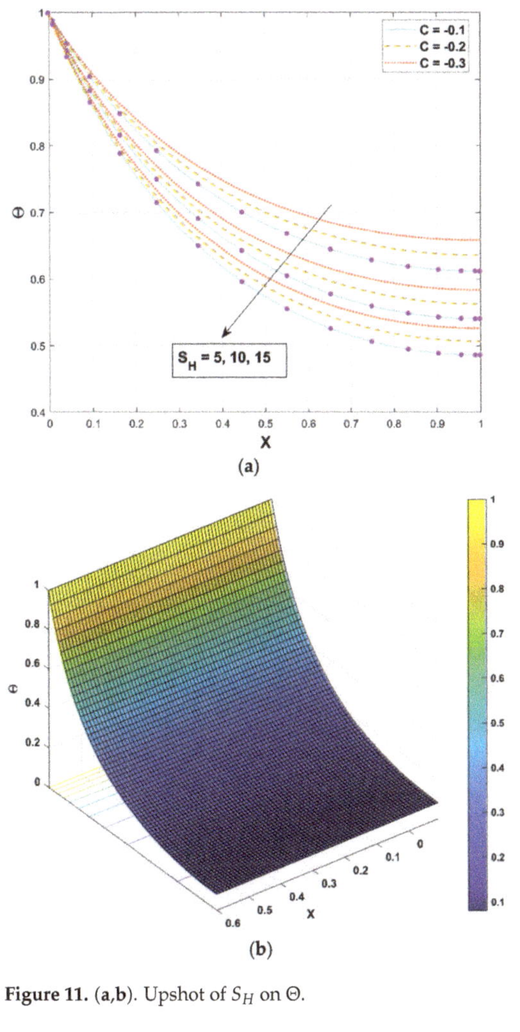

Figure 11. (**a**,**b**). Upshot of S_H on Θ.

Figure 12a,b illustrates the significant attributes of heat flux versus the convection–conduction parameter and the porosity parameter, respectively. According to the perceived notion, heat flux significantly augments when the convection–conduction parameter value is increased with respect to the porosity factor (Figure 12a), and an increment in thermal flux is caused by an increase in porosity factor (Figure 12b).

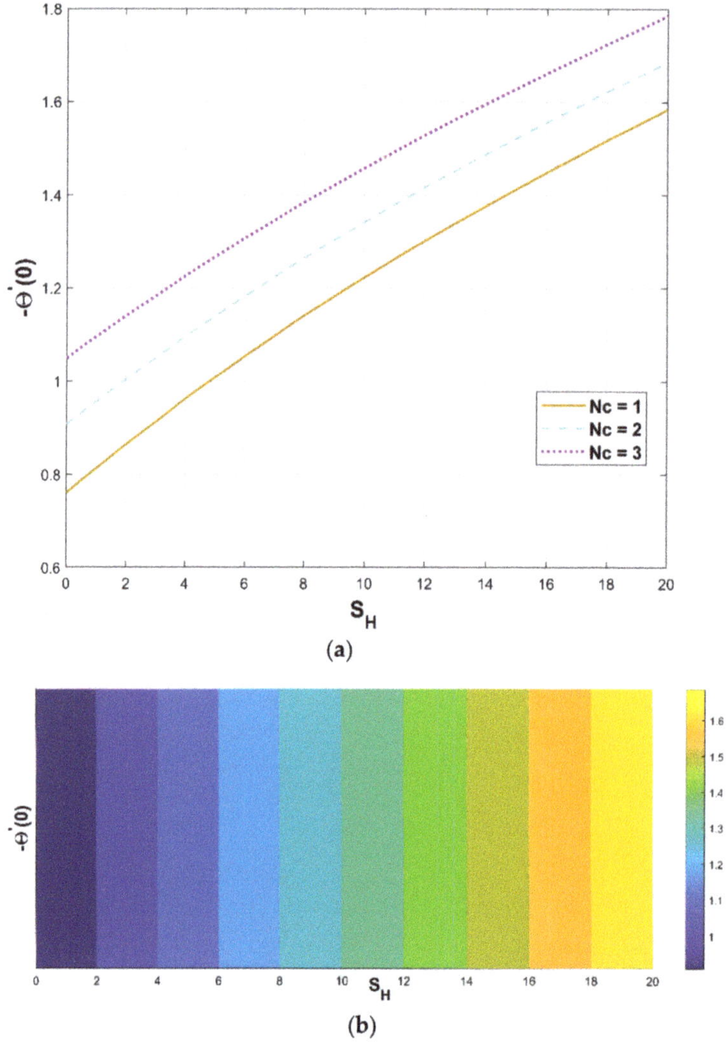

Figure 12. (**a**,**b**). Upshot of Nc and S_H on $-\Theta'(0)$.

Modifications in the quantity of Nr and porous factor have a consequence on the thermal flux, as revealed in Figure 13a,b. Figure 13a portrays the enriching behavior of heat flux as Nr increases against the porous factor. The heat flux decreases as a component of the porous factor, which is evidenced in Figure 13b.

For a varying radiation–conduction parameter, Figure 14a,b portray the performance of the heat generation parameter on heat flux. As per Figure 14a, elevating values of Q_{int} significantly decreases the temperature flux against R_d^*. Greater values of R_d^* implicate stronger radiative heat transfer and lead to an increase in the temperature profile, which causes less heat flux (Figure 14b).

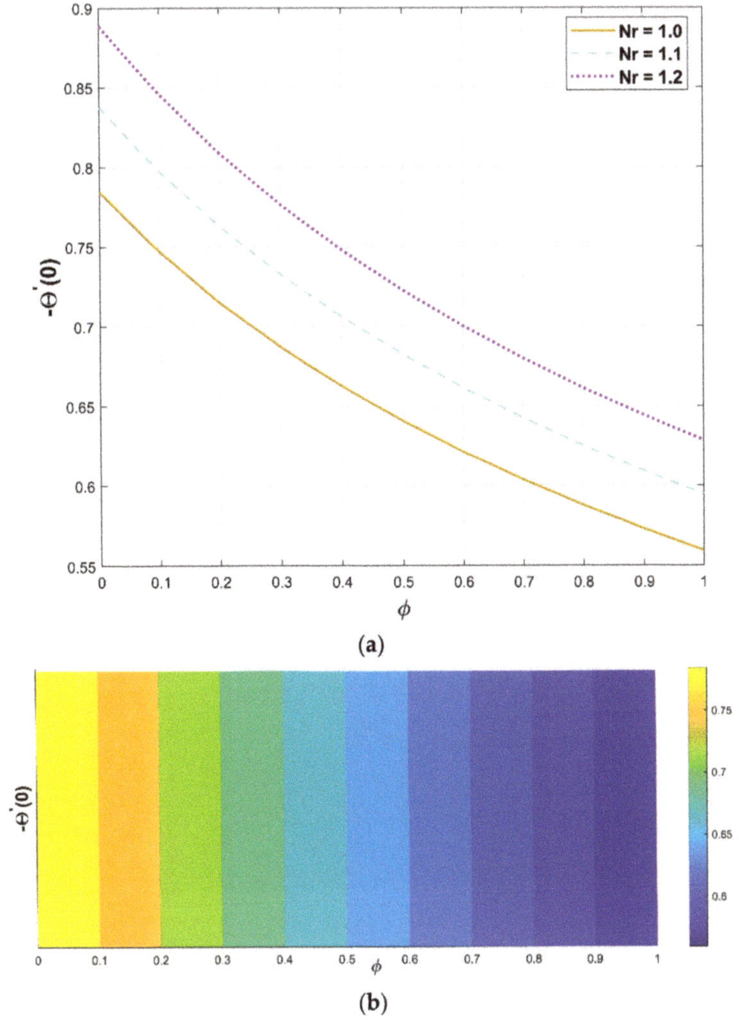

Figure 13. (**a**,**b**). Upshot of Nr and ϕ on $-\Theta'(0)$.

For lower variability in R_d^*, it is found that the heat flux enhances. The fundamental notion for using permeable fin is to enhance the efficacious surface area through which heat is convected to the nearby liquid. Even though the effectual thermal conductivity of the permeable fin weakens due to the discharge of solid substance, this decline is balanced by an expansion in surface area. As a consequence, increased porosities significantly augment the heat transfer rate. As predicted, improving R_d^* augments radiation heat exchange through the porous fin.

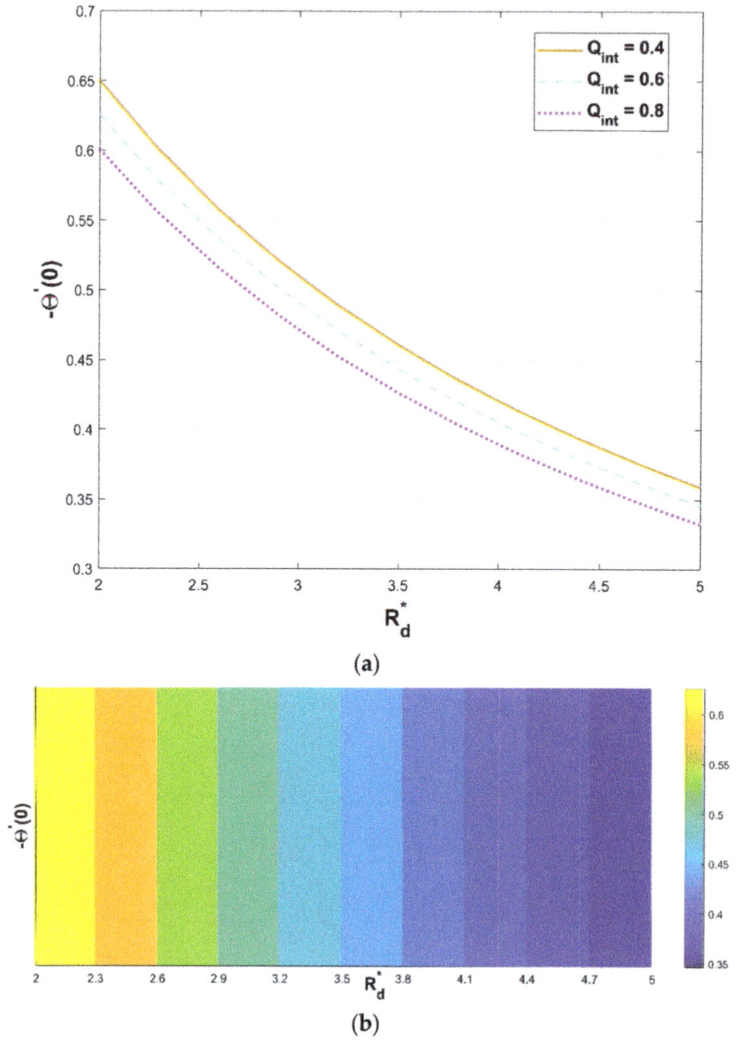

Figure 14. (**a**,**b**). Upshot of Q_{int} and R_d^* on $-\Theta'(0)$.

5. Conclusions

This analysis delves into the temperature behavior and heat transference of a permeable dovetail fin with internal heat generation. The nonlinear differential equation of a conducting–convecting–radiating dovetail fin with temperature-dependent thermal parameters is solved efficiently using SCM. The following are the major findings of the proposed study:

- SCM has an excellent agreement with the findings of available research, and to achieve precise results, only a few collocation points are required; thus, the technique is very effective in computation. This high level of accuracy suggests that the SCM is a viable alternative to other methods for solving fin problems with high nonlinearities.
- The thermal distribution through DF declines monotonically from the fin base to its tip, and the heat transfer rate in a porous fin is higher when compared to a nonporous fin.

- The temperature of the DF upsurges as the heat-generating number rises, which result in reduced heat flux.
- An increase in magnitude of the porosity parameter causes lower thermal dispensation and higher thermal flux in the fin.
- As the scale of the radiation number and convective–conductive parameter improves, the thermal distribution becomes lower and heat flux enhances through the dovetail fin.
- The increased magnitude of the temperature ratio parameter causes a decrement in the temperature distribution of the fin.
- The increment in the radiative–conductive variable leads to the upsurge of thermal dispersion and lessening of heat flux.

In the future, the numerical method can be extended to solve various fin models with more complex nonlinear differential equations. In addition, the current study can be extended to a transient thermal analysis of the dovetail fin under various physical mechanisms.

Author Contributions: Conceptualization: M.I.K. and B.C.P.; formal analysis: G.S. and R.S.V.K.; investigation: G.S. and K.C.J.; methodology: A.M.G. and B.C.P.; project administration: B.C.P.; resources: B.C.P. and K.G.; software: R.S.V.K. and M.M.A.L.; supervision: M.I.K. and B.C.P.; validation: K.G. and B.C.P.; visualization: G.S. and R.S.V.K.; mathematical modelling and Engligh language: O.T.B. and E.S.M.T.-E.; writing—original draft: K.C.J. and A.M.G.; writing—review and editing: M.I.K. and B.C.P. All authors have read and agreed to the published version of the manuscript.

Funding: The authors received no specific funding for this work.

Data Availability Statement: All relevant data are within the paper.

Acknowledgments: The authors express their gratitude to Princess Nourah bint Abdulrahman University Researchers Supporting Project (Grant No. PNURSP2022R152), Princess Nourah bint Abdulrahman University, Riyadh, Saudi Arabia.
The authors would like to thank the Deanship of Scientific Research at Umm Al-Qura University for supporting this work by Grant Code: 22UQU4331317DSR69.

Conflicts of Interest: The authors declare no conflict of interest.

Nomenclature

ν	Kinematic viscosity $(m^2\,s^{-1})$	γ	Heat generation parameter (dimensionless)
g	Acceleration due to gravity $(m\,s^{-2})$	t_b	Fin's thickness (m)
C	Fin taper ratio(dimensionless)	T	Temperature (K)
q	Heat transfer rate (W)	$t(x)$	Fin's semithickness (m)
w	Fin's width (m)	Θ	Nondimensional temperature
h^*	Convective heat transfer coefficient $\left(W\,m^{-2}\,K^{-1}\right)$	$v_w(x)$	Average velocity of the liquid passing through the fin $(m\,s^{-1})$
q^*_∞	Heat generation at ambient temperature $(W\,m^{-3})$	c_p	Specific heat $\left(J\,kg^{-1}\,K^{-1}\right)$
Nt	Temperature ratio parameter (dimensionless)	Nc	Convection–conduction parameter (dimensionless)
σ	Stefan–Boltzmann constant $\left(W\,m^{-2}\,K^{-4}\right)$	k	Thermal conductivity $\left(W\,m^{-1}\,K^{-1}\right)$
Nr	Radiation number (dimensionless)	q^*_{int}	Heat generation $(W\,m^{-3})$
ε^*	Emissivity	S_H	Porosity parameter (dimensionless)
L	Fin's length (m)	β^*_R	Rosseland extinction coefficient (m^{-1})
ρ	Density $(kg\,m^{-3})$	Q_{int}	Heat generation number (dimensionless)
x	Fin axial coordinate (m)	β	Thermal expansion coefficient $\left(K^{-1}\right)$
X	Length (dimensionless)	**Subscripts**	
ϕ	Porosity	∞	Ambient
$\dot{m}$	Mass flow rate $(kg\,s^{-1})$	b	Base
R^*_d	Radiation–conduction parameter (dimensionless)	f	Fluid
A_{cr}	Cross-sectional area (m^2)	s	Solid
K	Permeability (m^2)	eff	Effective properties

References

1. Mosayebidorcheh, S.; Farzinpoor, M.; Ganji, D. Transient thermal analysis of longitudinal fins with internal heat generation considering temperature-dependent properties and different fin profiles. *Energy Convers. Manag.* **2014**, *86*, 365–370. [CrossRef]
2. Turkyilmazoglu, M. Thermal performance of optimum exponential fin profiles subjected to a temperature jump. *Int. J. Numer. Methods Heat Fluid Flow* **2021**, *32*, 1002–1011. [CrossRef]
3. Gouran, S.; Ghasemi, S.; Mohsenian, S. Effect of internal heat source and non-independent thermal properties on a convective–radiative longitudinal fin. *Alex. Eng. J.* **2022**, *61*, 8545–8554. [CrossRef]
4. Varun Kumar, R.; Sowmya, G.; Essa, F.A.; Prasannakumara, B.; Alsehli, M.; Saleh, B. Thermal analysis of an annular fin under multi-boiling heat transfer coefficient using differential transform method with Pade approximant (DTM-Pade). *Proc. Inst. Mech. Eng. Part E J. Process Mech. Eng.* **2022**, 09544089221076255. [CrossRef]
5. Sowmya, G.; Kumar, R.S.V.; Alsulami, M.D.; Prasannakumara, B.C. Thermal stress and temperature distribution of an annular fin with variable temperature-dependent thermal properties and magnetic field using DTM-Pade approximant. *Waves Random Complex Media* **2022**, 1–29. [CrossRef]
6. Zhou, S.-S.; Bilal, M.; Khan, M.; Muhammad, T. Numerical analysis of thermal radiative maxwell nanofluid flow over-stretching porous rotating disk. *Micromachines* **2021**, *12*, 540. [CrossRef]
7. Jbeili, M.; Zhang, J. Effects of microscopic properties on macroscopic thermal conductivity for convective heat transfer in porous materials. *Micromachines* **2021**, *12*, 1369. [CrossRef]
8. Elsafy, K.M.; Saghir, M.Z. Forced convection in wavy microchannels porous media using TiO_2 and Al_2O_3–Cu nanoparticles in water base fluids: Numerical results. *Micromachines* **2021**, *12*, 654. [CrossRef]
9. Khan, U.; Zaib, A.; Ishak, A.; Waini, I.; Raizah, Z.; Galal, A.M. Analytical approach for a heat transfer process through nanofluid over an irregular porous radially moving sheet by employing KKL correlation with magnetic and radiation effects: Applications to thermal system. *Micromachines* **2022**, *13*, 1109. [CrossRef]
10. Mourad, A.; Abderrahmane, A.; Younis, O.; Marzouki, R.; Alazzam, A. Numerical simulations of magnetohydrodynamics natural convection and entropy production in a porous annulus bounded by wavy cylinder and Koch snowflake loaded with Cu–Water nanofluid. *Micromachines* **2022**, *13*, 182. [CrossRef]
11. Das, R.; Kundu, B. Prediction of heat-generation and electromagnetic parameters from temperature response in porous fins. *J. Thermophys. Heat Transf.* **2021**, *35*, 761–769. [CrossRef]
12. Nabati, M.; Salehi, G.H.; Taherifar, S. Numerical solution for a porous fin thermal performance problem by application of Sinc collocation method. *Math. Methods Appl. Sci.* **2021**. [CrossRef]
13. Wang, F.; Kumar, R.V.; Sowmya, G.; El-Zahar, E.R.; Prasannakumara, B.; Khan, M.I.; Khan, S.U.; Malik, M.; Xia, W.-F. LSM and DTM-Pade approximation for the combined impacts of convective and radiative heat transfer on an inclined porous longitudinal fin. *Case Stud. Therm. Eng.* **2022**, *35*, 101846. [CrossRef]
14. Hosseinzadeh, S.; Hasibi, A.; Ganji, D. Thermal analysis of moving porous fin wetted by hybrid nanofluid with trapezoidal, concave parabolic and convex cross sections. *Case Stud. Therm. Eng.* **2022**, *30*, 101757. [CrossRef]
15. Din, Z.U.; Ali, A.; De la Sen, M.; Zaman, G. Entropy generation from convective–radiative moving exponential porous fins with variable thermal conductivity and internal heat generations. *Sci. Rep.* **2022**, *12*, 1791. [CrossRef]
16. Varun Kumar, R.S.; Sowmya, G.; Jayaprakash, M.C.; Prasannakumara, B.C.; Khan, M.I.; Guedri, K.; Kumam, P.; Sitthithakerngkiet, K.; Galal, A.M. Assessment of thermal distribution through an inclined radiative-convective porous fin of concave profile using generalized residual power series method (GRPSM). *Sci. Rep.* **2022**, *12*, 13275. [CrossRef]
17. Khani, F.; Aziz, A. Thermal analysis of a longitudinal trapezoidal fin with temperature-dependent thermal conductivity and heat transfer coefficient. *Commun. Nonlinear Sci. Numer. Simul.* **2010**, *15*, 590–601. [CrossRef]
18. Das, R. Estimation of feasible materials and thermal conditions in a trapezoidal fin using genetic algorithm. *Proc. Inst. Mech. Eng. Part G J. Aerosp. Eng.* **2015**, *230*, 2356–2368. [CrossRef]
19. Jayaprakash, M.; Alzahrani, H.A.; Sowmya, G.; Kumar, R.V.; Malik, M.; Alsaiari, A.; Prasannakumara, B. Thermal distribution through a moving longitudinal trapezoidal fin with variable temperature-dependent thermal properties using DTM-Pade approximant. *Case Stud. Therm. Eng.* **2021**, *28*, 101697. [CrossRef]
20. Varun Kumar, R.S.; Sarris, I.E.; Sowmya, G.; Madhukesh, J.K.; Prasannakumara, B.C. Effect of electromagnetic field on the thermal performance of longitudinal trapezoidal porous fin using DTM–Pade approximant. *Heat Transf.* **2022**, *51*, 3313–3333. [CrossRef]
21. Gireesha, B.J.; Keerthi, M.; Eshwarappa, K. Heat transfer analysis of longitudinal fins of trapezoidal and dovetail profile on an inclined surface. *Phys. Scr.* **2021**, *96*, 125209. [CrossRef]
22. Khater, A.; Temsah, R.; Hassan, M. A Chebyshev spectral collocation method for solving Burgers'-type equations. *J. Comput. Appl. Math.* **2008**, *222*, 333–350. [CrossRef]
23. Ma, X.; Huang, C. Spectral collocation method for linear fractional integro-differential equations. *Appl. Math. Model.* **2014**, *38*, 1434–1448. [CrossRef]
24. Yang, C.; Jian, Y.; Xie, Z.; Li, F. Electromagnetohydrodynamic electroosmotic flow and entropy generation of third-grade fluids in a parallel microchannel. *Micromachines* **2020**, *11*, 418. [CrossRef] [PubMed]
25. Sun, Y.-S.; Ma, J.; Li, B.-W. Spectral collocation method for convective–radiative transfer of a moving rod with variable thermal conductivity. *Int. J. Therm. Sci.* **2015**, *90*, 187–196. [CrossRef]

26. Ma, J.; Sun, Y.; Li, B.; Chen, H. Spectral collocation method for radiative–conductive porous fin with temperature dependent properties. *Energy Convers. Manag.* **2016**, *111*, 279–288. [CrossRef]
27. Ma, J.; Sun, Y.; Li, B. Spectral collocation method for transient thermal analysis of coupled conductive, convective and radiative heat transfer in the moving plate with temperature dependent properties and heat generation. *Int. J. Heat Mass Transf.* **2017**, *114*, 469–482. [CrossRef]
28. Kang, H.S. Optimization of a reversed trapezoidal fin using a 2-D analytic method. *J. Mech. Sci. Technol.* **2008**, *22*, 556–564. [CrossRef]
29. Ahmed, N.; Mughal, M.P.; Shoaib, W.; Raza, S.F.; Alahmari, A.M. WEDM of copper for the fabrication of large surface-area micro-channels: A prerequisite for the high heat-transfer rate. *Micromachines* **2020**, *11*, 173. [CrossRef]
30. Xie, G.-F.; Zhao, L.; Dong, Y.-Y.; Li, Y.-G.; Zhang, S.-L.; Yang, C. Hydraulic and thermal performance of microchannel heat sink inserted with pin fins. *Micromachines* **2021**, *12*, 245. [CrossRef]
31. Hatami, M.; Ganji, D.D. Thermal behavior of longitudinal convective–radiative porous fins with different section shapes and ceramic materials (SiC and Si3N4). *Ceram. Int.* **2014**, *40*, 6765–6775. [CrossRef]
32. Kundu, B.; Yook, S.J. An accurate approach for thermal analysis of porous longitudinal, spine and radial fins with all nonlinearity effects–analytical and unified assessment. *Appl. Math. Comput.* **2021**, *402*, 126124. [CrossRef]
33. Rostamiyan, Y.; Ganji, D.D.; Petroudi, R.; Nejad, K. Analytical investigation of nonlinear model arising in heat transfer through the porous fin. *Therm. Sci.* **2014**, *18*, 409–417. [CrossRef]
34. Bhanja, D.; Kundu, B.; Aziz, A. Enhancement of heat transfer from a continuously moving porous fin exposed in convective–radiative environment. *Energy Convers. Manag.* **2014**, *88*, 842–853. [CrossRef]
35. Alhejaili, W.; Kumar, R.V.; El-Zahar, E.R.; Sowmya, G.; Prasannakumara, B.; Khan, M.I.; Yogeesha, K.; Qayyum, S. Analytical solution for temperature equation of a fin problem with variable temperature-dependent thermal properties: Application of LSM and DTM-Pade approximant. *Chem. Phys. Lett.* **2022**, *793*, 139409. [CrossRef]
36. Moradi, A.; Hayat, T.; Alsaedi, A. Convection-radiation thermal analysis of triangular porous fins with temperature-dependent thermal conductivity by DTM. *Energy Convers. Manag.* **2014**, *77*, 70–77. [CrossRef]
37. Weera, W.; Kumar, R.V.; Sowmya, G.; Khan, U.; Prasannakumara, B.; Mahmoud, E.E.; Yahia, I.S. Convective-radiative thermal investigation of a porous dovetail fin using spectral collocation method. *Ain Shams Eng. J.* **2022**, 101811. [CrossRef]
38. Kiwan, S. Effect of radiative losses on the heat transfer from porous fins. *Int. J. Therm. Sci.* **2007**, *46*, 1046–1055. [CrossRef]
39. Madhura, K.R.; Babitha; Kalpana, G.; Makinde, O.D. Thermal performance of straight porous fin with variable thermal conductivity under magnetic field and radiation effects. *Heat Transf.* **2020**, *49*, 5002–5019. [CrossRef]
40. Sowmya, G.; Sarris, I.E.; Vishalakshi, C.S.; Kumar, R.S.V.; Prasannakumara, B.C. Analysis of transient thermal distribution in a convective–radiative moving rod using two-dimensional differential transform method with multivariate pade approximant. *Symmetry* **2021**, *13*, 1793. [CrossRef]
41. Bimbato, A.M.; Alcântara Pereira, L.A.; Hirata, M.H. Study of surface roughness effect on a bluff body—the formation of asymmetric separation bubbles. *Energies* **2020**, *13*, 6094. [CrossRef]
42. Rakhsha, M.; Kees, C.E.; Negrut, D. Lagrangian vs. Eulerian: An analysis of two solution methods for free-surface flows and fluid solid interaction problems. *Fluids* **2021**, *6*, 460. [CrossRef]
43. Kumar, R.S.V.; Kumar, R.N.; Sowmya, G.; Prasannakumara, B.C.; Sarris, I.E. Exploration of temperature distribution through a longitudinal rectangular fin with linear and exponential temperature-dependent thermal conductivity using DTM-Pade approximant. *Symmetry* **2022**, *14*, 690. [CrossRef]
44. Khorrami, M.R. A Chebyshev spectral collocation method using a staggered grid for the stability of cylindrical flows. *Int. J. Numer. Methods Fluids* **1991**, *12*, 825–833. [CrossRef]
45. Sun, Y.-S.; Li, B.-W. Chebyshev collocation spectral approach for combined radiation and conduction heat transfer in one-dimensional semitransparent medium with graded index. *Int. J. Heat Mass Transf.* **2010**, *53*, 1491–1497. [CrossRef]
46. Gorla, R.S.R.; Bakier, A. Thermal analysis of natural convection and radiation in porous fins. *Int. Commun. Heat Mass Transf.* **2011**, *38*, 638–645. [CrossRef]

micromachines

MDPI

Article

Study of New Mini-Channel Trans-Critical CO_2 Heat Pump Gas Cooler

Jiawei Jiang [1,2], Shiqiang Liang [2,3,4,*], Can Ji [5], Longyan Wang [1] and Chaohong Guo [2,3,4]

1 Research Center of Fluid Machinery Engineering and Technology, Jiangsu University, Zhenjiang 212013, China; 2212011015@stmail.ujs.edu.cn (J.J.); longyan.wang@connect.qut.edu.au (L.W.)
2 Institute of Engineering Thermophysics, Chinese Academy of Sciences, Beijing 100190, China; guochaohong@iet.cn
3 School of Engineering Science, University of Chinese Academy of Sciences, Beijing 100049, China
4 Dalian National Laboratory for Clean Energy, Dalian 116000, China
5 Energy Research Institute, Qilu University of Technology (Shandong Academy of Sciences), Jinan 250014, China; jic@sderi.cn
* Correspondence: liangsq@mail.etp.ac.cn; Tel.: +86-136-9319-6177

Abstract: A gas cooler is one of the important parts of a carbon dioxide (CO_2) heat pump water heater, and it must meet the needs of not only pressurization but also heat transfer. It is important to study gas coolers. In this paper, a heat exchanger with a spiral channel is studied. ANSYS CFX software was used to analyze the flow and heat transfer characteristics of the heat exchanger (single-plate model). The influences of the cooling pressure of CO_2, the mass flux of CO_2, the mass flux of water and the channel radius of CO_2 are discussed. In this paper, the results show that the cooling pressure of CO_2, the mass flux of CO_2 and the channel radius of CO_2 all have a large influence on the local heat transfer coefficient: with an increase in the cooling pressure of CO_2, the peak value of the heat transfer coefficient of CO_2 decreases and the average heat transfer coefficient decreases; with an increase in the mass flux of CO_2, the peak value of the heat transfer coefficient of CO_2 increases and the average heat transfer coefficient increases; and with a decrease in the channel radius of CO_2, the peak value of the heat transfer coefficient of CO_2 increases. The water mass flux has only a slight effect on heat transfer, and the lower cooling pressure of CO_2 corresponds to a higher peak heat transfer coefficient, which can reach 27.5 $kW \cdot m^{-2} \cdot K^{-1}$ at 9 MPa.

Keywords: trans-critical carbon dioxide heat pump gas cooler; spiral heat exchanger; numerical simulation

1. Introduction

CO_2 is an excellent candidate due to its nontoxicity, incombustibility, safety, low cost and environmentally benign nature (ODP = 0, GWP = 1). It is widely implemented in refrigeration and heat pump systems, air conditioning and various industrial uses. Lorentzen first proposed the application of trans-critical carbon dioxide cycles to automotive heat pumps in 1993, and this was the beginning of the widespread application of carbon dioxide heat pumps in various fields [1]. The gas cooler is one of the important parts of a CO_2 heat pump. According to the different cooling media used, gas coolers are mainly divided into air-cooled and water-cooled. The air-cooled type is mainly used for automotive heat pumps, divided into tube-fin and microchannel coolers; the water-cooled type is mainly used for trans-critical CO_2 heat pump water heaters, mostly using double pipes. Chang analyzed tube-fin gas coolers with different circuits through an experimental study, as well as the effect of geometric factors on heat transfer, and found that three circuits were optimal [2]. Chai simulated a tube-fin gas cooler with distributed modeling and used the ε-NTU method to research the effect of design factors on the cooling performance of the whole gas cooler [3]. Wang studied a herringbone wavy-fin gas cooler and proposed a new heat transfer correlation with an average error of 8.82% [4]. Ge first constructed a mathematical model by dividing the tube-fin gas cooler into four regions and verified it with experimental results, which matched well; then, a finite difference model was developed to simulate the performance of the gas cooler; finally, the model was used to simulate and analyze the gas cooler with different circuits [5–7]. Zhang analyzed the heat transfer performance of an existing tube-fin gas cooler model with the CFD method [8,9]. Zilio and Simone designed three different shell and tube gas coolers and performed experimental and simulation analyses; they found that the influence of different tube types of inlet water flow rates on the heat production rate was reduced at 10 MPa, and the heat production rate was relatively high for smooth and internally grooved tubes [10]. Fronk designed a cross-flow plate type microchannel gas cooler and researched the heat transfer and pressure drop characteristics of the gas cooler with both five water flow channels and with seven water flow channels, at different inlet flow rates and inlet temperatures, and he summarized the heat transfer correlations based on the experimental results and also simulated the performance of a microchannel heat exchanger with 12 water flow channels. The study showed that the microchannel gas cooler was superior to the double pipe gas cooler [11,12]. Starace proposed a method to investigate and find the overall performance of a heat exchanger starting from CFD simulations at the micro-scale; the study showed that the multi-scale approach leads to a better accuracy level than the full-scale one [13]. Fiorentino carried out some numerical simulations of the evaporative condenser heat and mass transfer processes at the tube scale, then studied two different types of flow by varying the water-to-air mass flow ratio. He found that a decrease of 25% in the water-to-air mass flow ratio led to the film separating into droplets. A test rig to investigate the evaporative condenser at small scale had been designed and built. An increase of 37.5% in the air flow rate (at a constant rate of sprayed water) led to a maximum reduction in the heat transfer rate of 50% [14,15]. Bilal used the fouling model to investigate the risk-based thermal performance of these evaporative heat exchangers and carried out sensitivity analysis, finding that the condensing temperature is the most sensitive parameter; for any increase in the inlet's relative humidity, the normalized sensitivity of the surface area increased [16,17]. Scattina developed a new methodology that allows scaling-up from the CFD analysis of a small element to the prediction of the performance of an entire compact heat exchanger [18].

Current research on trans-critical CO_2 heat pump gas coolers is focused on the double pipe and tube-fin types, with no structural innovation. Gas coolers must be able to withstand high pressure, with high reliability, low costs, compactness, efficiency and other requirements. A double pipe gas cooler is not compact enough, not easy to scale and is difficult to clean, so it is not good enough to meet the above needs. The concept of a spiral heat exchanger was first proposed in the late nineteenth century and was reinvented in Sweden during the 1930s. Up to now, much research relating to the characteristics of spiral heat exchangers has been presented by researchers. According to Bes, spiral heat exchangers have high thermal efficiency and compactness, and they have only one cross-sectional channel for each fluid, in which the flow direction is constantly changing, thus eliminating the flow stagnation zone channel and making it difficult for fouling to remain, and it rarely scales [19]. If the heat exchanger is fouled, this can be soaked in white vinegar, dissolved with anhydrous ethanol, and finally blown with dry compressed air since the CO_2 channel must be kept dry. In this paper, a new spiral plate mini-channel gas cooler with asymmetric channels on both sides is proposed for CO_2 heat pump gas coolers with relatively large water and CO_2 flow rates.

ANSYS CFX software (ANSYS, Canonsburg, PA, USA) was used to simulate the flow and heat transfer characteristics of supercritical carbon dioxide in the new mini-channel gas cooler, and to analyze the influence of different structural parameters and working conditions on heat transfer, so as to provide more directions and new ideas for research on trans-critical carbon dioxide heat pump gas coolers.

2. Physical Model

The geometric details of the new mini-channel heat exchanger are shown in Figure 1. The mini-channel heat exchanger consists of two different plates alternating in sequence. Figure 1a shows one of them with a semi-circular channel of CO_2 above and a semi-elliptical channel of CO_2 below; the other plate has a semi-elliptical channel above and a semi-circular channel below, and the two plates are alternately stacked in sequence to form the whole heat exchanger. The overall schematic is shown in Figure 1b, and the important dimensions of the gas cooler are listed in Table 1. The hydraulic diameters of the CO_2 channels and the water channels were selected according to plate thicknesses available on the market, pressure-bearing capacity and mass flow rate. CO_2 and water are counter-flowing, with CO_2 flowing from inside to outside and water flowing from outside to inside, as seen in Figure 1a. Two different channels are distributed on both sides of the plate, which can prevent the mixing of fluids. Each plate is stacked and combined into a whole, connected together by diffusion welding, and can withstand high pressure. Diffusion welding is a kind of pressure welding and refers to the process of surfaces being brought into mutual contact under the action of high temperature pressure. With the connected surfaces close to each other, and local plastic deformation, after a certain period of time the atoms of the bonding layer diffuse and intersperse with each other, forming a reliable connection of the whole.

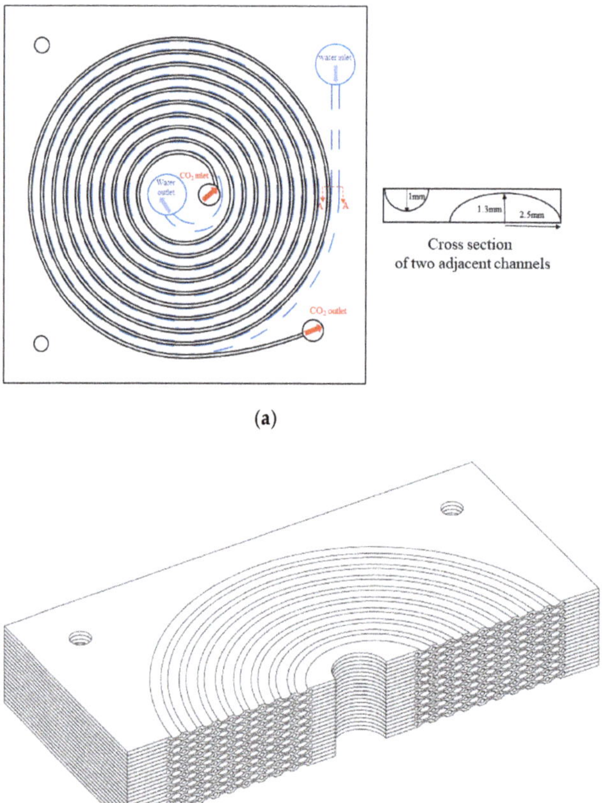

Figure 1. (**a**) A type of structural plate; (**b**) Overall schematic.

Table 1. Important dimensions of the gas cooler.

Overall Dimensions	
Gas cooler length	200 mm
Gas cooler width	200 mm
Gas cooler height	270 mm
Gas cooler single layer plate thickness	1.5 mm
Carbon Dioxide Side	
Carbon dioxide channel length	3.2 m
Carbon dioxide channel radius	1 mm
Water Side	
Elliptical channel length	2.8 m
Elliptical channel short semi-axis	1.3 mm
Elliptical channel long semi-axis	2.5 mm

3. Structural Strength Analysis

As the CO_2 side of the gas cooler must be able to withstand high pressures of at least 13 MPa, the strength analysis of the designed heat exchanger was first performed to determine whether its pressure bearing capacity meets the requirements. The ANSYS Static Structural test was used for the analysis, and the model was simplified to a straight section of 1 cm in length, as shown in Figure 2. The bottom and right side of the models were set as symmetric boundary conditions, and the front face was set as frictionless support. After setting up the conditions, the stress and strain solutions were carried out. The stress and strain are displayed in Figure 3. Figure 3 shows the results with the boundary conditions of 20 MPa applied on the CO_2 side and 0.5 MPa applied on the water side. It was found that the elliptical channel can bear 20 MPa pressure, which was far beyond the pressurization.

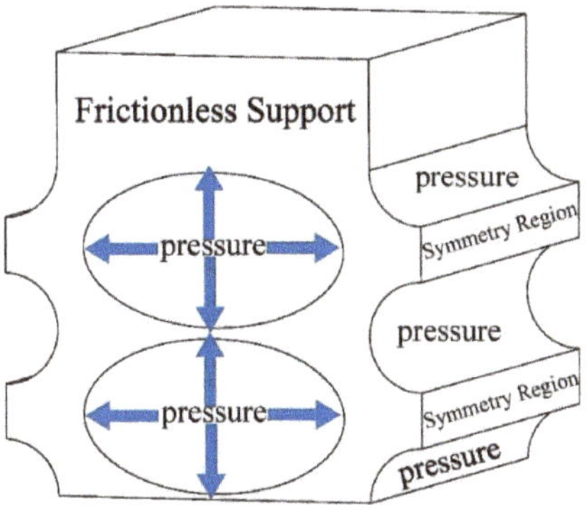

Figure 2. Strength analysis model and boundary condition setting.

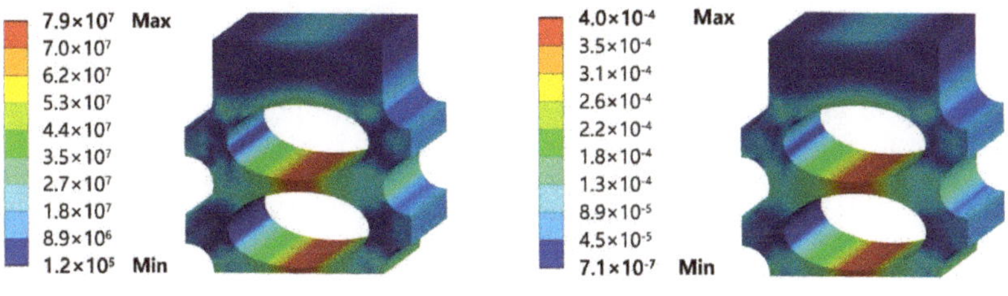

(a) (b)

Figure 3. (**a**) Stress; (**b**) Strain.

4. Numerical Model

4.1. Data Reduction

The relevant governing equations for the turbulence model are shown in Appendix A. The local bulk temperature of the supercritical CO_2 and the cooling water, the inner wall temperature of the channel and the heat flux on the wall along the channel were obtained from the numerical results. The heat transfer coefficient of the supercritical CO_2 flowing in the channel was calculated as:

$$h = \frac{q_x}{T_{b,CO_2} - T_{W,x}} \tag{1}$$

where q_x and $T_{W,x}$ are the circumferentially averaged heat flux and temperature on the inner wall along the channel, respectively, which were calculated as:

$$q_x = \frac{\int_L qdL}{L} \tag{2}$$

$$T_{W,x} = \frac{\int_L TdL}{L} \tag{3}$$

L is the length of the semicircle arc on the inner wall where the axial position is equal to x. x represents the axial location along the channel. Unlike the traditional definition of the bulk temperature of fluids with constant properties, the bulk temperature of the supercritical CO_2 was determined by the average enthalpy, which was calculated as:

$$H_b = \frac{\int \rho uHdA}{\int \rho udA} \tag{4}$$

The bulk temperature of the supercritical CO_2 was defined as:

$$T_{b,CO_2} = f(H_b, P) \tag{5}$$

The mass flux was defined as:

$$G = \frac{m}{A} \tag{6}$$

4.2. Reliability Verification

Because of the novel structure proposed in this paper, there were no relevant experiments for CO_2 heat exchangers that could be verified. In Paisarn's study, numerical and experimental results of the heat transfer and flow characteristics of a horizontal spiral-coil tube (Figure 4) were investigated. The RNG $k - \varepsilon$ two-equation turbulence model was applied to simulate turbulent flow and heat transfer characteristics. Reasonable agreement was obtained by comparison between the results from the experiment and those obtained from the model [20]. The channel of CO_2 in this paper was a circular cross-section spiral channel, and a similar spiral tube was selected for turbulence model verification. Experimental results for the conjugate heat transfer between supercritical CO_2 and water by Zhang were used to validate the numerical method [21]. The geometric model of the spiral tube is shown in Figure 5. The specific dimensions of the spiral tube used in the simulation were: inner diameter 9 mm, outer diameter 12 mm, coil diameter 283 mm, pitch 32 mm. Supercritical CO_2 flowed from the bottom to the top inside the spiral tube and was heated by the wall. The pressure was 8.02 MPa, the inlet mass flux was 97.92 $kg/(m^2{\cdot}s)$, the inlet temperature was 288.15 K and the heat flux was 10.4 kW/m^2. Figure 5 shows the variation in the bulk temperature with x/d at different cross sections simulated by the two turbulence models compared with the experimental data of Zhang. The horizontal coordinate x/d in the figure indicates the ratio of the distance from this cross-section to the inlet cross-section to the pipe diameter. It can be seen from Figure 6 that the results obtained from the RNG $k - \varepsilon$ model simulations are in better agreement with the experimental data, indicating that the RNG $k - \varepsilon$ model has higher accuracy in calculating the rotational flow.

The large deviation in the two turbulence models in the high temperature prediction may be attributed to the constant heat flow conditions that cause heat to accumulate at the outlet. The RNG $k - \varepsilon$ model was used in the subsequent simulations.

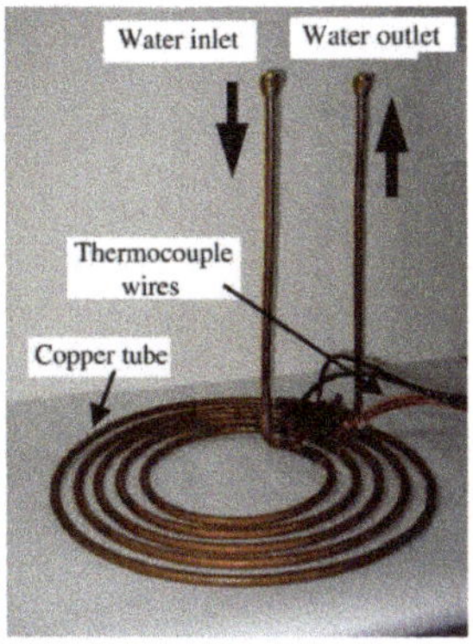

Figure 4. The physical model of spiral-coil tube.

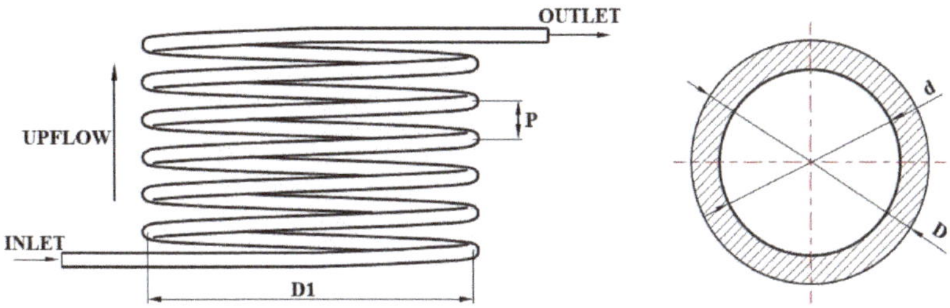

Figure 5. Schematic of spiral tube geometry model.

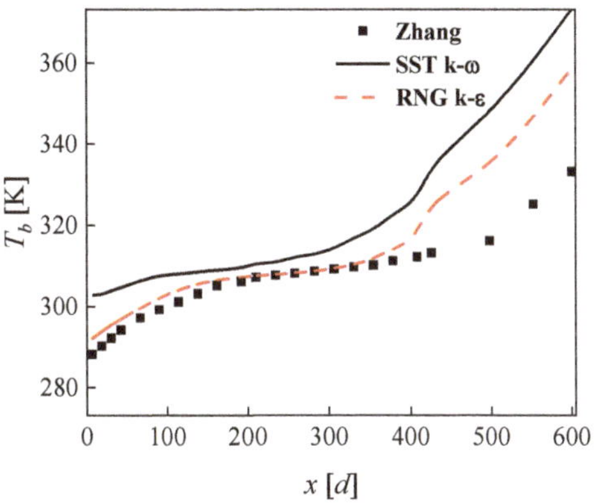

Figure 6. Comparison of simulation and experimental values of the average temperature of the inner cross-section of the spiral tube.

4.3. Grid Independence

Due to the large computational resources consumed by the overall analysis, a single plate of the heat exchanger was used for the analysis in this paper, as shown in Figure 1a. The spiral channel of CO_2 on one side of the plate had a 1 mm radius semicircle cross-section, a spiral start radius of 10 mm, an end radius of 90 mm and a pitch of 8 mm. On the other side, for the spiral channel of water, the long semi-axis of the channel cross-section was 2.5 mm, the short semi-axis was 1.3 mm, the spiral start radius was 14 mm, the end radius was 86 mm and the pitch was 8 mm. The veneer was a 200 mm × 200 mm square.

Workbench meshing was used to divide the fluid domain and the solid domain into meshes, and the cross-sections of two adjacent channels of heat transfer units in a single plate are displayed in Figure 7. To eliminate the impact of the entrance and exit regions, some extensions were made to the channel. In order to obtain more accurate calculation results, and at the same time avoid excessive calculation, it was necessary to compare the calculation results under different grid numbers, and that grid division is shown in Figure 8. The first layer thickness in the CO_2 region was set to 0.0000003 mm to make sure that the dimensionless wall distance y+ < 1. The mass flow rates of CO_2 and water were 0.0033 $kg \cdot s^{-1}$ and 0.01 $kg \cdot s^{-1}$, respectively. The inlet temperatures of CO_2 and water were 373.15 K and 293.15 K, respectively. The pressure of CO_2 and water were 9 MPa and 0.5 MPa, respectively. Setting symmetric boundary conditions on the surface of the two fluid domains, the residuals reached 10^{-6} and the calculation converged. In this paper, five sets of different grid dimensions with total grid numbers of 3.0 million, 3.7 million, 4.3 million, 5.2 million and 6.0 million were selected to carry out subsequent numerical simulations by calculating the pressure drop of CO_2, as displayed in Figure 9, and finally, a 5.2 million grid was selected.

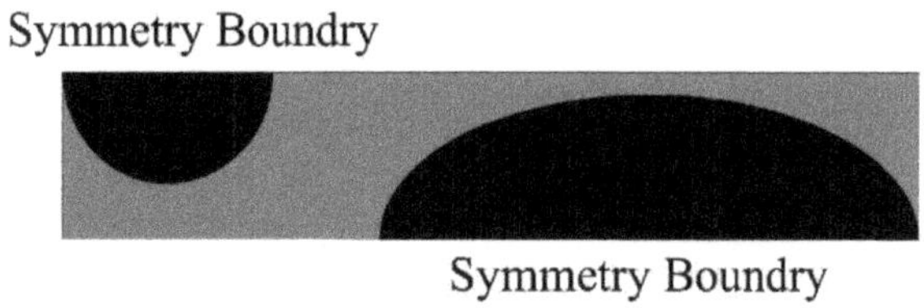

Figure 7. Heat exchanger unit boundary conditions.

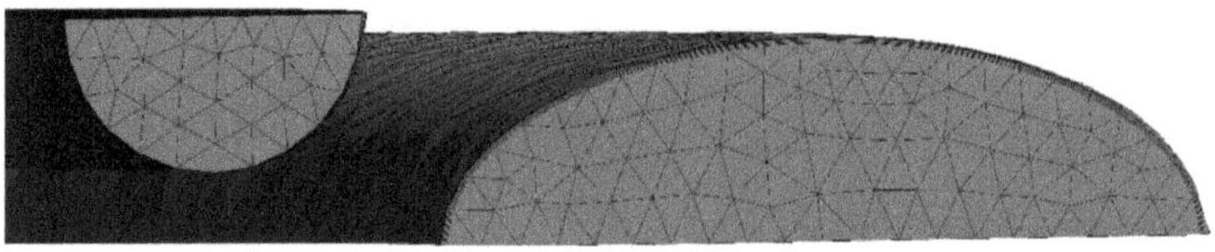

Figure 8. Grid partition.

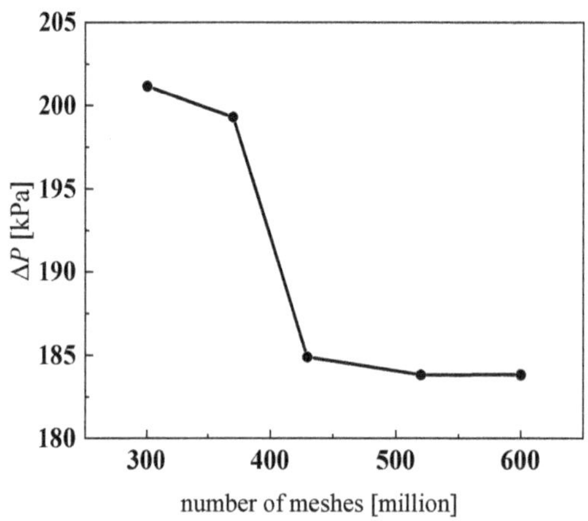

Figure 9. Grid independence.

5. Results and Discussion

5.1. Effect of Cooling Pressure of CO_2 on Heat Transfer

In this section, the main goal is to research the effects of different cooling pressures on heat transfer. The mass flow rates of CO_2 and water were 0.0033 kg·s^{-1} and 0.01 kg·s^{-1}, respectively. The inlet temperatures of CO_2 and water were 373.15 K and 293.15 K, respectively. The pressure of CO_2 and water were 10 MPa and 0.5 MPa, respectively. The CO_2 heat transfer coefficient under cooling pressure is shown in Figure 10. The specific calculation conditions are listed in Table 2.

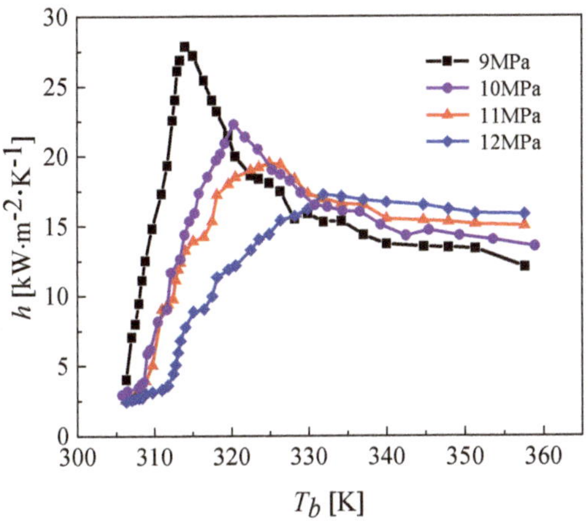

Figure 10. Effect on the heat transfer coefficient of CO_2 at different cooling pressures.

Table 2. Calculation conditions.

Case	G_{CO2} (kg·m^{-2}·s^{-1})	P_{water} (MPa)	T_{water} (K)	P_{CO2} (MPa)	T_{CO2} (K)
1	1047	0.5	293.15	9	373.15
2	1047	0.5	293.15	10	373.15
3	1047	0.5	293.15	11	373.15
4	1047	0.5	293.15	12	373.15

It can be seen from Figure 10 that the peak value of the local heat transfer coefficient of CO_2 decreases as the cooling pressure increases, and the critical temperature corresponding to the peak value is different. The fluid temperatures corresponding to the peak heat transfer coefficients at cooling pressures of 9, 10, 11 and 12 MPa were 313.15 K, 320.15 K, 324.15 K and 334.15 K, respectively. The closer to the pseudocritical point, the higher the constant pressure specific heat capacity and thermal conductivity of CO_2, hence the higher the corresponding local heat transfer coefficient. The reason the heat transfer coefficients are relatively close at different pressures away from the pseudocritical point is that the specific heat capacity of carbon dioxide changes slightly at different pressures and the larger thermodynamic changes are mainly concentrated near the pseudocritical point. The peak heat transfer coefficient of CO_2 reaches 27.5 kW·m^{-2}·K^{-1} at a cooling pressure of 9 MPa. Figure 11 shows that the average CO_2 heat transfer coefficient decreases gradually as the cooling pressure increases. Figure 12 illustrates the pressure distribution of CO_2.

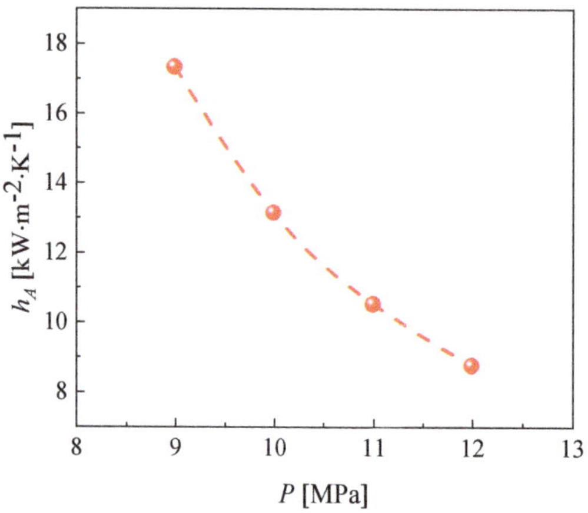

Figure 11. Variation in the average heat transfer coefficient of CO_2 at different cooling pressures.

5.2. Effect of Inlet Mass Flux of CO_2 on Heat Transfer

In this section, the effects of different mass fluxes of CO_2 on the heat transfer characteristics are discussed. The conditions were the same as in the previous section except for the mass flux of CO_2. The mass flow rate of water was 0.01 kg·s^{-1}. The inlet temperatures of CO_2 and water were 373.15 K and 293.15 K, respectively. The pressure of CO_2 and water were 10 MPa and 0.5 MPa, respectively. The specific calculation conditions are given in Table 3.

Table 3. Calculation conditions.

Case	G_{CO2} (kg·m^{-2}·s^{-1})	P_{water} (MPa)	T_{water} (K)	P_{CO2} (MPa)	T_{CO2} (K)
1	523	0.5	293.15	10	373.15
2	654	0.5	293.15	10	373.15
3	785	0.5	293.15	10	373.15
4	916	0.5	293.15	10	373.15
5	1047	0.5	293.15	10	373.15

The variation in the convective heat transfer coefficient with different mass fluxes in the CO_2 channel is given in Figure 13. As the mass flux increases, the convective heat transfer coefficient of CO_2 also increases. The reason for this is that as the mass flux increases, the intensity of the turbulence in the channel increases and the fluid continuously scours the wall, affecting the formation of the boundary layer. The changing pattern of the average heat transfer coefficient at different mass fluxes is described in Figure 14. Figure 15 shows the variation in temperature of CO_2 at different mass fluxes. It can be seen from the figure that the heat transfer in the outer ring is becoming worse and the channel length can be considered to be shortened in the subsequent optimization. Figures 16 and 17 give the variation in the pressure drop in the channel of CO_2 with the mass flux. With the increase in the mass flux, the pressure drop in the channel increases continuously, and the variation in the pressure drop in the channel at the lower mass flux is slightly smaller than that at the larger mass flux. The difference in the pressure drop tends to increase gradually with the increasing of the mass flux. According to the changing of the pressure drop and the heat transfer coefficient with the mass flux, it is necessary to balance the two conditions to choose the most suitable working condition. In the conventional type of heat exchangers, the average heat transfer coefficient of double pipe and shell and tube heat exchangers is mostly below 5 kW·m^{-2}·K^{-1}, while the average heat transfer coefficient of PCHE and microtube heat exchangers can reach 10 kW·m^{-2}·K^{-1}; the pressure drop is mostly below 100 kpa, so the disadvantage of spiral heat exchangers is the large pressure drop. According to the ngo study, the volumetric thermal capacity of PCHE can reach 25,000 kw/m^3 with a hydraulic diameter of 1 mm and the volumetric thermal capacity of a spiral heat exchanger is 19,000 kw/m^3 [22].

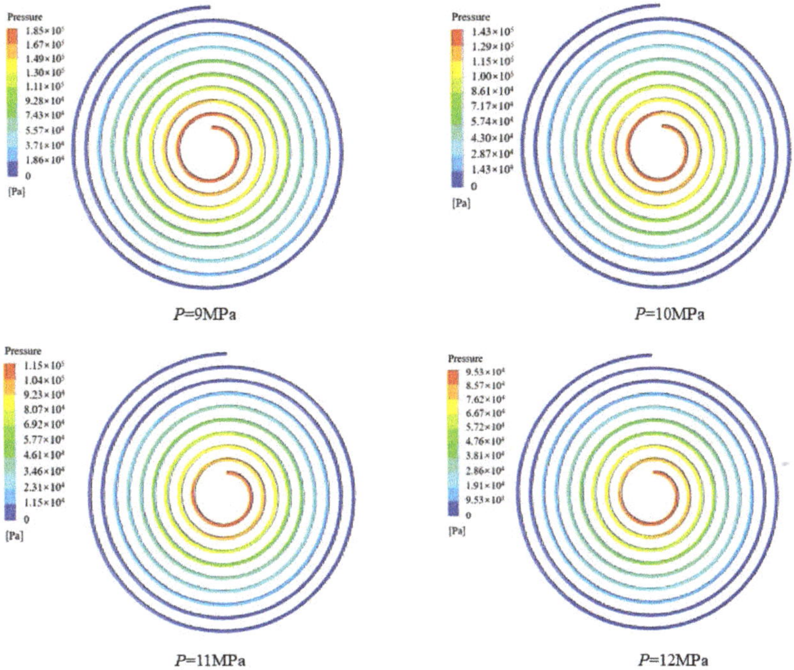

Figure 12. Variation in the pressure contour of CO_2 at different cooling pressures.

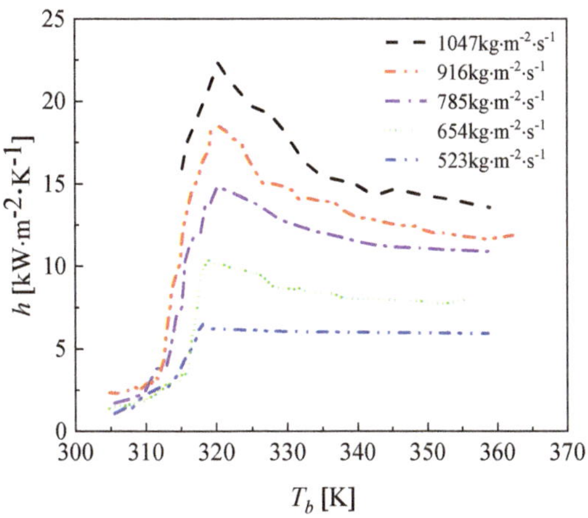

Figure 13. Variation in the heat transfer coefficient of CO_2 at different mass fluxes.

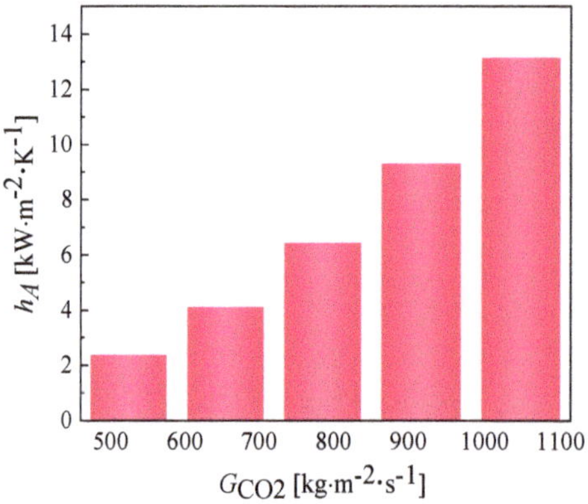

Figure 14. Variation in the average heat transfer coefficient of CO_2 at different mass fluxes.

Figure 18 shows the effect of different mass fluxes in the water channel on the heat transfer coefficient of CO_2. From Figure 15, it can be seen that increasing the inlet mass flux on the water side slightly increases the heat transfer coefficient of CO_2. Due to the heat transfer between the water and the outer surface of the metal wall increasing with the intensity of the water turbulence as the mass flux of the water increases, the heat transfer coefficient between carbon dioxide and the metal wall increases accordingly. The water flow rate affects the outer wall surface temperature of CO_2, and enhanced heat transfer can be achieved by changing the outer wall surface temperature of the CO_2. This also corroborates the study of Jiang that proposed that, to enhance heat transfer, the heat transfer of water is an influential factor that needs to be considered comprehensively [23].

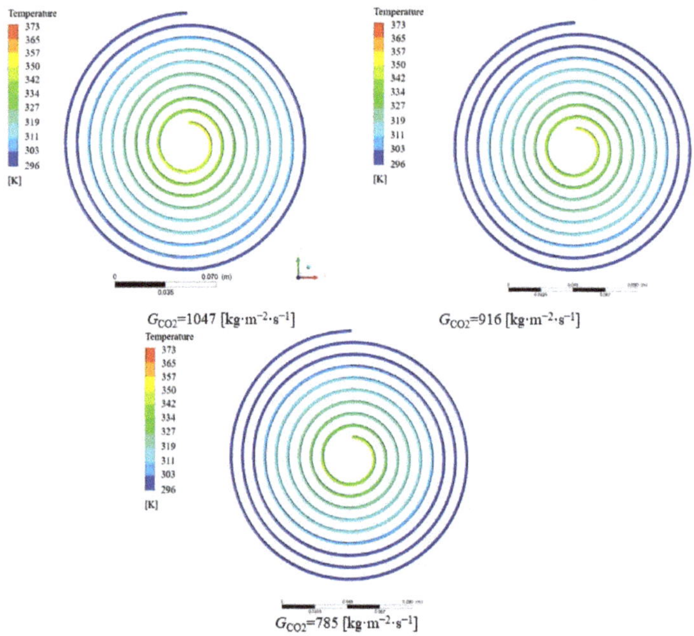

Figure 15. Variation in the temperature of CO_2 at different mass fluxes.

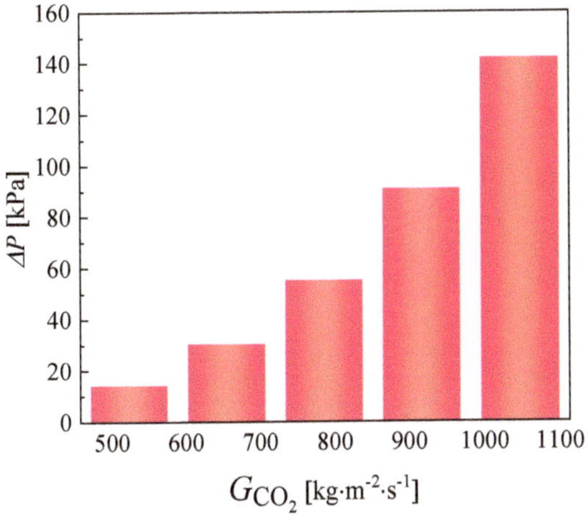

Figure 16. Variation in the pressure drop of CO_2 at different mass fluxes.

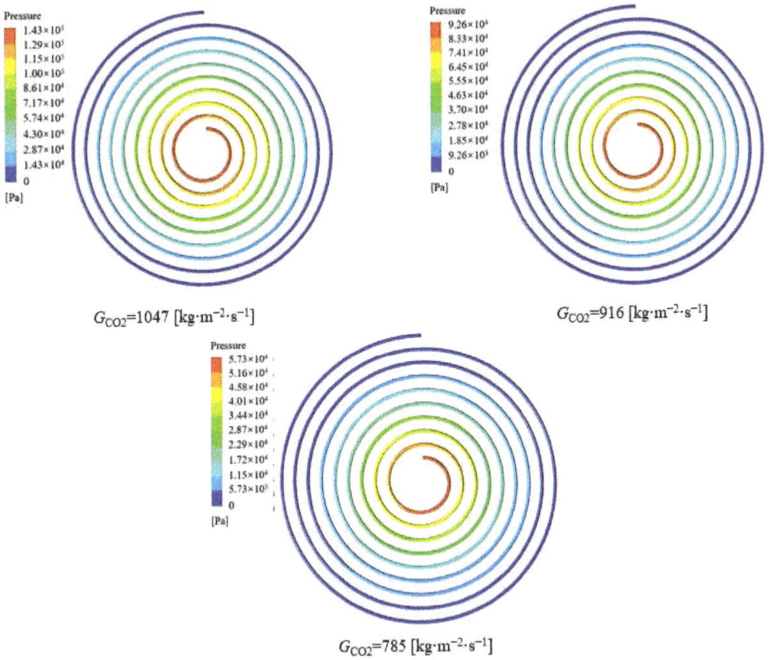

Figure 17. Variation in the pressure contour of CO_2 at different mass fluxes.

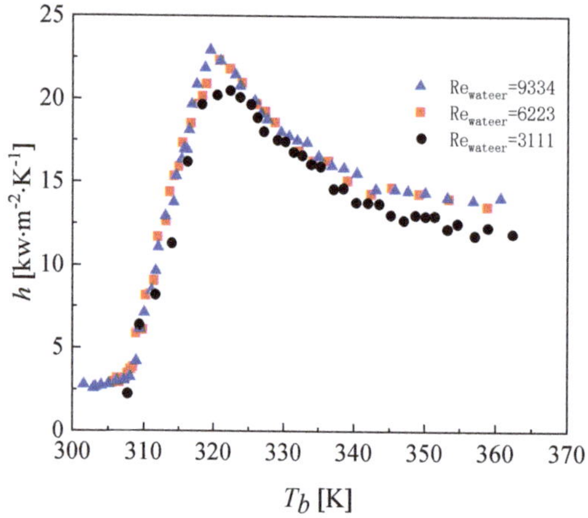

Figure 18. Variation in the heat transfer coefficient of CO_2 at different water mass fluxes.

5.3. *Effect of Channel Size of CO_2 on Heat Transfer*

In this section, the effects of different channel radiuses of CO_2 on the heat transfer characteristics are investigated. The mass flow rates of CO_2 and water were 0.002466 kg·s^{-1} and 0.01 kg·s^{-1}, respectively. The inlet temperatures of CO_2 and water were 373.15 K and 293.15 K, respectively. The pressure of CO_2 and water were 10 MPa and 0.5 MPa, respectively. The effects on the heat transfer coefficient of CO_2 at different channel radiuses are shown in Figure 19.

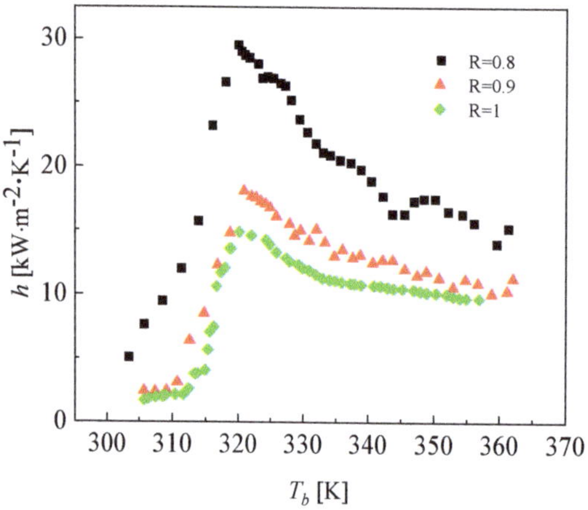

Figure 19. Effect of channel radius on the heat transfer coefficient of CO_2.

Figure 19 illustrates the effect of the channel radius on the heat transfer coefficient of CO_2. It can be seen that the heat transfer coefficient of CO_2 increases as the channel radius of CO_2 decreases. The reason is that reducing the channel radius of CO_2 corresponds to reducing the cross-section, which is equivalent to increasing the mass flux under the same conditions and increasing the turbulence intensity. Furthermore, the increase in the heat transfer coefficient was larger when the channel radius of CO_2 was 0.8 mm, and this was accompanied by a dramatic increase in the pressure drop. Under this condition, the heat transfer coefficients of CO_2 with all three different channel radiuses reached the peak at the pseudocritical point. Figures 20 and 21 display the effect of the channel radius on the pressure drop of CO_2. As the channel radius decreases, the pressure drop increases.

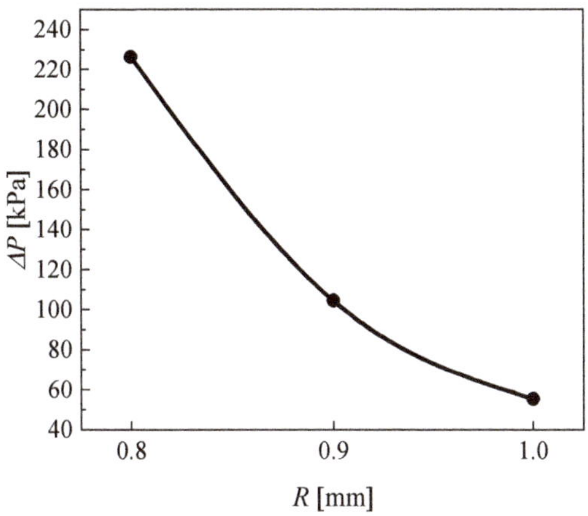

Figure 20. Effect of channel radius on the pressure drop of CO_2.

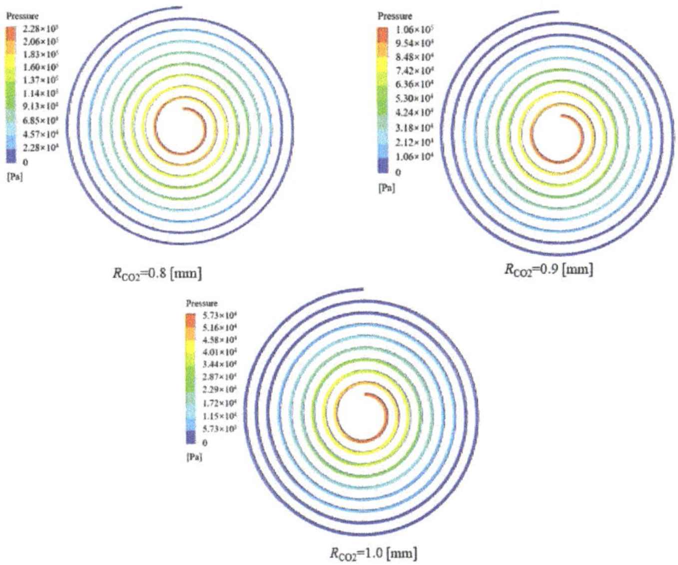

Figure 21. The pressure contour of CO_2 at different channel radiuses.

5.4. Field Synergy Principle of Heat Exchange

A further analysis of the heat transfer characteristics of CO_2 used the field synergy principle. This starts from the two-dimensional boundary layer convective heat transfer differential equation. Guo deduced the field synergy principle by considering convection as a special endothermic source [24]. Generally, the most frequently used field synergy principle analysis mainly uses the synergy angle as a primary indicator of synergy degree to analyze the mechanism of convective heat transfer enhancements. The synergy angle can be calculated with the following equation:

$$\beta = \cos^{-1}\left(\frac{|\vec{U} \cdot \nabla T|}{|\vec{U}| \cdot |\nabla T|}\right) \tag{7}$$

Figure 22 shows the long-range synergy angle distribution within the channel of CO_2. The smaller synergy angle is concentrated in the center of the channel, mainly because the temperature gradient at the center of the channel is along the axial direction and the flow direction consistency is higher. Heat transfer from the channel wall is dominated by thermal conduction near the wall boundary layer, so the temperature gradient is mainly along the wall normal direction, with a large angle to the flow direction. The location of the better synergy within the channel of CO_2 is at the peak heat transfer coefficient ($x/L = 0.468$).

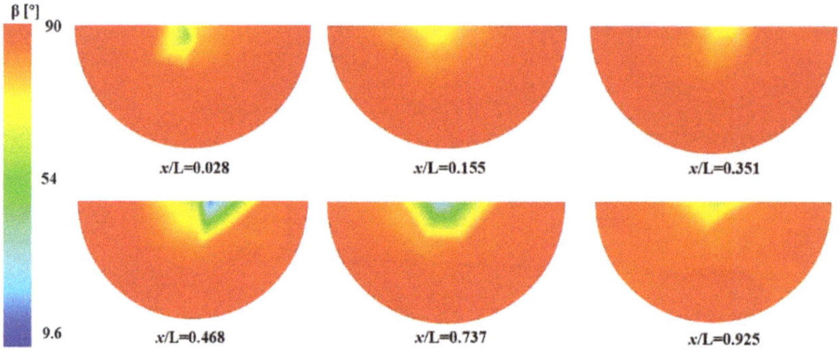

Figure 22. The variation in the synergy angle in different axial positions.

6. Conclusions

In this study, the heat transfer between CO_2 and water in a spiral heat exchanger is numerically analyzed. The main results of this study are as follows:

1. Spiral plate heat exchanger pressurization is good; when the radius of the circular channel is 1 mm, the elliptical channel short axis is 2.3 mm and the long axis is 5 mm, it can withstand the high pressure of 20 MPa.
2. As the cooling pressure decreases, the average heat transfer coefficient of CO_2 increases, the heat transfer capacity is enhanced and the peak heat transfer increases. Mainly owing to the fact that the lower the pressure, the higher the specific heat will be at constant pressure, the average heat transfer coefficient of CO_2 is also higher.
3. The higher the mass flux of CO_2, the higher the corresponding peak local convective heat transfer coefficient, which occurs near the pseudocritical temperature corresponding to the cooling pressure.
4. The effects of changing the mass fluxes of water and CO_2 on the heat transfer coefficient of CO_2 are different. The increase in the mass flux of CO_2 and the decrease in the channel radius have a more obvious effect on the heat transfer coefficient, while the increase in the mass flux of water affects the temperature of the wall surface of CO_2.
5. The locations with smaller synergy angles within the channel of CO_2 are distributed in the center of the channel, and the locations with better synergy occur at the location of the peak heat transfer coefficient of CO_2.

Author Contributions: J.J. conceived this review and wrote the manuscript; C.G. and C.J. designed section structures and produced figures; L.W. reviewed; S.L. reviewed, proposed the idea, edited and was responsible for funding acquisition. All authors have read and agreed to the published version of the manuscript.

Funding: This work was supported by the general fund project of the National Natural Science Foundation of China (No. 52176090); the special project for clean energy leading science and technology, Chinese Academy of Sciences (Grant No. XDA21010201); and the Key Research & Development Program of Shandong Province (Grant No. 2019GSF109019).

Conflicts of Interest: The authors declare no conflict of interest.

Appendix A. Governing Equations

In this section, we present governing equations of design and analysis.

By summarizing and analyzing the existing numerical simulation studies of supercritical pressure fluid heat transfer, it can be found that the RNG $k-\varepsilon$ model and the SST $k-\omega$ model are more suitable turbulence models. Their continuity equation, momentum equation and energy equation are consistent, as shown in Equations (A1)–(A3):

$$\frac{\partial(\rho U_j)}{\partial x_j} = 0 \tag{A1}$$

$$\frac{\partial}{\partial x_j}(\rho U_i U_j) = -\frac{\partial p\prime}{\partial x_i} + \frac{\partial}{x_j}[(\mu+\mu_t)(\frac{\partial U_i}{\partial x_j} + \frac{\partial U_j}{\partial x_i})] + S_M \tag{A2}$$

$$\frac{\partial}{\partial x_j}(\rho U_j c_p T) = \frac{\partial}{\partial x_j}(\lambda \frac{\partial T}{\partial x_j}) \tag{A3}$$

The equation k and equation ε of the RNG $k-\varepsilon$ model are shown in Equations (A4) and (A5):

$$\frac{\partial(\rho U_j k)}{\partial x_j} = \frac{\partial}{\partial x_j}[(\mu + \frac{\mu_t}{\sigma_k})\frac{\partial k}{\partial x_j}] + P_k - \rho\varepsilon + P_{kb} \tag{A4}$$

$$\frac{\partial(\rho U_j \varepsilon)}{\partial x_j} = \frac{\partial}{\partial x_j}[(\mu + \frac{\mu_t}{\sigma_{\varepsilon RNG}})\frac{\partial \varepsilon}{\partial x_j}] + \frac{\varepsilon}{k}(C_{\varepsilon 1RNG}P_k + C_{\varepsilon 1RNG}P_{\varepsilon b} - C_{\varepsilon 2RNG}\rho\varepsilon) \tag{A5}$$

The equation k and equation ω of the SST $k-\omega$ model are shown in Equations (A6) and (A7):

$$\frac{\partial(\rho U_j k)}{\partial x_j} = \frac{\partial}{\partial x_j}[(\mu + \frac{\mu_t}{\sigma_k})\frac{\partial k}{\partial x_j}] + P_k - \beta\prime\rho k\omega + P_{kb} \tag{A6}$$

$$\frac{\partial(\rho U_j \omega)}{\partial x_j} = \frac{\partial}{\partial x_j}[(\mu + \frac{\mu_t}{\sigma_\omega})\frac{\partial \omega}{\partial x_j}] + \alpha\frac{\omega}{k}P_k - \beta\rho\omega^2 + P_{\omega b} \tag{A7}$$

References

1. Lorentzen, G.; Pettersen, J. A new, efficient and environmentally benign system for car air-conditioning. *Int. J. Refrig.* **1993**, *16*, 4–12. [CrossRef]
2. Chang, Y.S.; Kim, M.S. Modeling and Performance Simulation of a Gas Cooler for a CO_2 Heat Pump System. *HVAC&R Res.* **2007**, *13*, 445–456. [CrossRef]
3. Chai, L.; Tsamos, K.M.; Tassou, S.A. Modelling and Evaluation of the Thermohydraulic Performance of Finned-Tube Supercritical Carbon Dioxide Gas Coolers. *Energies* **2020**, *13*, 1031. [CrossRef]
4. Wang, C.C.; Hwang, Y.M.; Lin, Y.T. Empirical correlations for heat transfer and flow friction characteristics of herringbone wavy fin-and-tube heat exchangers. *Int. J. Refrig.* **2002**, *25*, 673–680. [CrossRef]
5. Ge, Y.; Cropper, R. Performance evaluations of air-cooled condensers using pure and mixture refrigerants by four-section lumped modelling methods. *Appl. Therm. Eng.* **2005**, *25*, 1549–1564. [CrossRef]
6. Ge, Y.; Cropper, R. Performance simulation of refrigerated display cabinets operating with refrigerants R22 and R404A. *Appl. Energy* **2008**, *85*, 694–707. [CrossRef]
7. Ge, Y.T.; Cropper, R.T. Simulation and performance evaluation of finned-tube CO_2 gas coolers for refrigeration systems. *Appl. Therm. Eng.* **2009**, *29*, 957–965. [CrossRef]
8. Zhang, X.; Ge, Y.; Sun, J.; Li, L.; Tassou, S.A. CFD Modelling of Finned-tube CO_2 Gas Cooler for Refrigeration Systems. *Energy Procedia* **2019**, *161*, 275–282. [CrossRef]

9. Zhang, X.; Ge, Y.; Sun, J. Performance analysis of finned-tube CO_2 gas cooler with advanced 1D-3D CFD modelling development and simulation. *Appl. Therm. Eng.* **2020**, *176*, 115421. [CrossRef]
10. Zilio, C.; Mancin, S. Shell and tube carbon dioxide gas coolers-Experimental results and modelling. *Int. J. Refrig.* **2015**, *56*, 224–234. [CrossRef]
11. Fronk, B.M.; Garimella, S. Water-coupled carbon dioxide microchannel gas cooler for heat pump water heaters: Part I-Experiments. *Int. J. Refrig.* **2011**, *34*, 7–16. [CrossRef]
12. Fronk, B.M.; Garimella, S. Water-coupled carbon dioxide microchannel gas cooler for heat pump water heaters: Part II-Model development and validation. *Int. J. Refrig.* **2011**, *34*, 17–28. [CrossRef]
13. Starace, G.; Fiorentino, M.; Longo, M.; Carluccio, E. A hybrid method for the cross flow compact heat exchangers design. *Appl. Therm. Eng.* **2017**, *111*, 1129–1142. [CrossRef]
14. Fiorentino, M.; Starace, G. Numerical and Experimental Performance Analysis of Evaporative Condensers. *Energy Procedia* **2016**, *101*, 26–33. [CrossRef]
15. Fiorentino, M.; Starace, G. Sensitivity analysis of evaporative condensers performance using an experimental approach. *Energy Procedia* **2017**, *126*, 345–352. [CrossRef]
16. Qureshi, B.A.; Zubair, S.M. A comprehensive design and rating study of evaporative coolers and condensers. Part I. Performance evaluation. *Int. J. Refrig.* **2006**, *29*, 645–658. [CrossRef]
17. Qureshi, B.A.; Zubair, S.M. A comprehensive design and rating study of evaporative coolers and condensers. Part II. Sensitivity analysis. *Int. J. Refrig.* **2006**, *29*, 659–668. [CrossRef]
18. Ciuffini, A.; Scattina, A.; Carena, F.; Roberti, M.; Rivalta, G.T.; Chiavazzo, E.; Fasano, M.; Asinari, P. Multiscale Computational Fluid Dynamics Methodology for Predicting Thermal Performance of Compact Heat Exchangers. *J. Heat Transf.* **2016**, *138*, 071801. [CrossRef]
19. Bes, T.; Roetzel, W. Thermal theory of the spiral heat exchanger. *Int. J. Heat Mass Transf.* **1993**, *36*, 765–773. [CrossRef]
20. Naphon, P. Study on the heat transfer and flow characteristics in a spiral-coil tube. *Int. Commun. Heat Mass Transf.* **2011**, *38*, 69–74. [CrossRef]
21. Zhang, W.; Wang, S.; Li, C.; Xu, J. Mixed convective heat transfer of CO_2 at supercritical pressures flowing upward through a vertical helically coiled tube. *Appl. Therm. Eng.* **2015**, *88*, 61–70. [CrossRef]
22. Ngo, T.L.; Kato, Y.; Nikitin, K.; Tsuzuki, N. New printed circuit heat exchanger with S-shaped fins for hot water supplier. *Exp. Therm. Fluid Sci.* **2006**, *30*, 811–819. [CrossRef]
23. Jiang, P.X.; Zhao, C.R.; Shi, R.F.; Chen, Y.; Ambrosini, W. Experimental and numerical study of convection heat transfer of CO_2 at super-critical pressures during cooling in small vertical tube. *Int. J. Heat Mass Transf.* **2009**, *52*, 4748–4756. [CrossRef]
24. Guo, Z. Mechanism and control of convective heat transfer—Coordination of velocity and heat flow fields. *Chin. Sci. Bull.* **2000**, *19*, 2118–2122.

micromachines

MDPI

Article

Considerations for the Maximum Heat Load and Its Influence on Temperature Variation of the Evaporator in Flat MHPs in Transient Regimes

Ioan Mihai [1,*], Cornel Suciu [1,2,*] and Claudiu Marian Picus [1]

[1] Faculty of Mechanical Engineering, Automotive and Robotics, Stefan cel Mare University, 720229 Suceava, Romania; claudiu.picus@usm.ro

[2] Faculty of Electrical Engineering and Computer Science, Stefan cel Mare University, 720229 Suceava, Romania

* Correspondence: mihai.i@usm.ro (I.M.); suciu@usm.ro (C.S.); Tel.: +40-735173288 (I.M.); +40-742007820 (C.S.)

Abstract: The present paper describes a series of considerations for the occurrence of capillary boundaries in flat micro heat pipes (flat MHPs) and the conditions required for their stable operation in relation to the working circumstances and to the type of liquids inside the pipes. Particularities of heat transfer in a flat MHP are analyzed for situations of either excessive or deficient working liquid. Depending on the physical properties of the working liquids (acetone, methanol and distilled water), the maximum rate of heat flow that can be applied to a flat MHP is determined analytically. The calculus is made with the assumption that constant vaporization of the liquid is ensured in the flat MHP's evaporator, with no overheating. The considered analytical models allow for the evaluation of the liquid film thickness and the mass flow corresponding to the vaporization region. The temperature difference between the inner and outer walls of a flat MHP is found in the case of a transient regime and a variable thermal flow is applied in the evaporation region. The interior of flat MHPs was modeled in MATLAB using an FTCS (Forward-Time Central-Space) method, which is a finite difference method used for numerically solving the heat equation.

Keywords: microchannels; flat MHP's; heat load; evaporator; capillary radius

1. Introduction

Power electronic equipment is present in most modern specialties. The current extent of miniaturization requires cooling solutions capable of ensuring increased efficiency, reduced external power consumption, minimum dimensions and low-as-possible production costs [1,2]. Heat pipes and flat MHPs are among the systems able to ensure efficient cooling of electronic components, the latest ones are currently employed in many applications.

The first historical record of thermal pipes is in 1836 when Jacob Perkins used the principle of gravitational thermo-siphoning [3]. In 1929, F.W. Gay according to [3,4] used, for the first time, Perkins tubes to construct an air-to-air heat exchanger. It can be stated that modern heat pipes first appeared in 1944 when R.S. Gaugler (patent U.S. No. 2350348) [3,5] employed an internal structure that facilitated the motion of condensed liquid within a thermal tube. In 1962, Trefethen according to [6], proposed the use of thermal tubes in cosmic space equipment, and in 1964, Groover, et al., according to [7], were the first to introduce the expression "heat pipe".

Micro Heat Pipes were first mentioned relatively late, in 1984, by T.P. Cotter [8]. These pipes are characterized by the fact that the mean curvature of the working liquid meniscus at the liquid-vapor interface is comparable in size to the inverse of the hydraulic radius of the flowing channel. Such devices were later introduced for semiconductor technology due to their extremely small dimensions [9].

Constructively, flat MHPs are divided into three regions namely, vaporization, adiabatic and condensation. The cyclic thermodynamic transformations of fluids inside a flat

MHP are vaporization, due to absorbing heat from the power electronic component, vapor transport, through the adiabatic region, followed by condensation with the formation of liquid droplets. This step includes heat loss in the environment. The condensed liquid is then returned to the vaporization region by way of a "pumping" mechanism obtained via the capillarity effect in one or more internal capillary layers of flat MHPs.

Most cooling systems based on flat MHPs are passive devices, as they use the bi-phasic transformations of the working liquids. Even though flat MHPs are viable cooling solutions they present functional limits determined by their constructive shape, working liquid and the ability to transport important variable heat flows. Research carried out so far by other authors has focused on the cooling of flat MHPs, while Sprinceană and Mihai [2] propose an external evaporator cooling. The proposed cooling process takes place in addition to the cooling of the condenser with a fan and can be done by adding a cold liquid or using Peltier cells. The additional cooling of the evaporator leads in its turn, to an increase in the amount of liquid in the trapezoidal microchannels and thus of the capillary radius. Since the proposed idea is novel in the authors' opinion, the present paper aims to determine by analytical calculations the maximum heat load in the case of additional evaporator cooling. In addition, since during operation, there is a significant change in the various rates of heat flow, another objective of the paper is to study the temperature variation in the evaporator wall under the considered conditions. For this purpose, the temperature difference between the inside and outside of the evaporator wall from the operation onset until the regime stabilization was calculated for a variation of the heat flow between 10 W and 40 W. The calculations were performed for the case of transient operating regimes in flat MHPs. The evaporator temperature change was also determined by numerical modeling of the heat equation using the FTCS finite difference method. To verify the idea of additional evaporator cooling, Sprinceană and Mihai [2] performed a series of experiments using various liquids inside flat MHPs such as distilled water, methanol and acetone. The results show that, regardless of the used liquid, external evaporator cooling strongly increases the maximum cooling capacity of the flat MHPs. A comparison of the experimental results obtained by the authors with analytical or numerical results including those of other researchers was made.

2. Factors That Influence the Maximum Heat Transport Capacity for a Flat MHP

Figure 1 presents a flat MHP [2], where the operating regions can be observed, as well as some of the geometrical parameters.

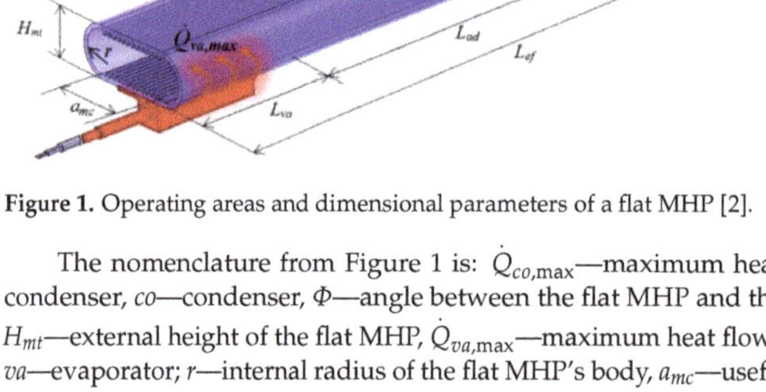

Figure 1. Operating areas and dimensional parameters of a flat MHP [2].

The nomenclature from Figure 1 is: $\dot{Q}_{co,\max}$—maximum heat flow released at the condenser, *co*—condenser, Φ—angle between the flat MHP and the horizontal direction, H_{mt}—external height of the flat MHP, $\dot{Q}_{va,\max}$—maximum heat flow applied at evaporator, *va*—evaporator; *r*—internal radius of the flat MHP's body, a_{mc}—useful width of rectangular microchannels, L_{va}, L_{ad}, L_{co}—lengths of condensation, adiabatic and vaporization regions, respectively, L_{ef}—effective length of the flat MHP's working regions.

The internal capillary layers of a flat MHP ensure the motion of the condensed liquid phase due to their structure made of microchannels, micro-spheres made of synthesized copper and woven microfibers, etc. A series of hypotheses [3,10–14], were made in order to analyze the heat transport capacity of a flat MHP. It is assumed that flat MHP works when placed horizontally, and the wall is heated on one side by the thermal source (the power electronic component) via a conduction process.

Heat transfer between the inner capillary layer and the working liquid is accomplished only in a radial direction (the process in the axial direction is neglected). The working liquid and its vapor move only in an axial direction. The condenser of flat MHPs is not blocked by the liquid obtained through vapor condensation while the capillary layer is assumed isotropic and saturated with liquid. The change of the liquid's meniscus in the microchannels of the capillary layer is determined by the capillary pressure.

The vapor and working liquid flow rates are considered constant and their flow laminar near the inner wall over the entire length of the flat MHP, and turbulent in the core of the vapor flow where the axial Reynolds number is greater than Re = 12,000 [15]. From a directional point of view, heat transfer from a flat MHP can either happen from exterior to interior (vaporization region) by receiving heat from the warm source in the vaporization region, inside the micro heat pipe (adiabatic area) when vapor moves from the vaporization region to the condensation region or from inside to the outside (condensation region). Heat transfer through the flat MHP's microchannels can be conductive and convective, with and without changing of phase, depending on the work area.

In the vaporization region, generated vapor moves at high speed, "often several hundred mph and approaching the speed of sound" [16], from this region to the condensation region. In a stationary regime, at the liquid-vapor interface, the principle of momentum conservation must be obeyed at any point from both transverse and longitudinal cross-sections. The Laplace–Young equation can be applied to find the radius of the liquid meniscus formed at the liquid-vapor interface. The continuity equations for the liquid and vapor flows through trapezoidal microchannels of a flat MHP through the central section from the condenser to the evaporator and in reverse are presented in [11,17].

The flat MHP's channels have been considered to have a trapezoidal cross-section. Starting from the relations given by [18,19] a code was written in Mathcad [2] and a series of parameters such as the equivalent hydraulic diameter, the Poiseuille number, the curvature radius of the liquid's meniscus versus the angle of the meniscus with the wall of the trapezoidal microchannel (considered circular by hypothesis) were found. The same code allowed for finding the pressure variation and the loss of pressure for vapor in the evaporator region using relations from the models developed in [12,19–21].

The variation of capillary pressure on the axial direction of the trapezoidal microchannel was calculated by determining the values of the Reynolds number for the working fluids flow. The minimum condition required for undiminished heat transport capacity through a flat MHP [22–24] states that the interior maximum capillary pressure must be no less than the pressures of vapor and working liquid.

Improper functioning of a flat MHP is often associated with a phenomenon called capillary limit which is directly related to the volume of working liquid existent within the capillary layers. The capillary limit phenomenon can be correlated to the pressure and temperature regimes inside the flat MHP. This phenomenon consists of impeding the motion of the freshly condensed liquid in the condenser towards the vaporization region via mechanisms that will be later described. The capillary limit phenomenon may produce the "drying" of the vaporization region with negative effects upon the quantity of vapor produced and implicitly upon the maximum heat transport capacity.

Case of excessive working liquid: When the quantity of liquid is in excess inside the flat MHP, the capillary limit may be attained. This is due to the fact that too much working liquid vaporizes and the volume of the generated vapor increases. As a result, the pressure inside the flat MHP rises.

Friction between the liquid and the walls of the microchannels generate adhesion forces in the boundary layer. The above-mentioned increase in internal pressure leads to augmented adhesion forces and implicitly to an alteration of the capillary limit (the pumping pressure through the capillary layer). As a consequence, the excess liquid formed in the condenser by vapor condensation cannot return to the evaporator by capillary effect at the same flow rate as in the regular operating regime. The flow rate decrease at the evaporator determines the diminishing or even cease of heat transfer, which means that the capillary limit has been reached.

Case of insufficient working liquid: If a heat-generating thermal flux is applied to the evaporator zone and the working liquid is scarce inside the flat MHP, the whole quantity of liquid envisaged for cooling will vaporize in a short period of time. Reducing the quantity of vapor obtained in the evaporation zone will diminish the quantity of condensed fluid produced in the condenser of the flat MHP. This situation also leads to a failure to meet the condition of the capillary. The evaporator region will dry out and the flat MHP will cease to function properly.

For a correct estimation of the maximum heat transport capacity of a flat MHP the following parameters were taken into account: nature of the flow, Poiseuille, Reynolds and Nusselt numbers, liquid and vapor flow velocity in trapezoidal microchannels and sintered layers.

3. Evaluation of Maximum Heat Load through a Flat MHP

Flat MHPs are thermodynamic systems able to transport heat even for small temperature differences between the evaporator and condenser [25]. They are used to transfer and transform the rate of heat flow [26]. This fact is a feature related to the bi-phase transformations and internal pressure. The inner structure of a flat MHP varies from one case to another but some of the elements are found in all constructive types. Further, the transport capacity is linked to the effective heat flux.

3.1. Analytical Determination of the Maximum Heat Flux Passing through the Evaporator of a Flat MHP

According to [3,27], the Young–Laplace conditions concerning the total variation of capillary pressure must be fulfilled so that the condensed liquid from the condenser returns to the evaporator through the internal capillary structure of a flat MHP:

$$\Delta P_{cp,t} \geq \Delta P_l + \Delta P_v + \Delta P_g, \quad (1)$$

where $\Delta P_{cp,t}$—total drop of capillary pumping pressure, ΔP_l—liquid pressure drop in microchannels, ΔP_v—vapor pressure drop along the flat MHP, ΔP_g—pressure drop due to gravity (only for the flat MHP working tilted with respect to the horizontal direction).

The geometric parameters of a flat MHP over the cross-section of the trapezoidal microchannels filled with working liquid and details for the common liquid-wick region [28], are shown in Figure 2.

The notations from Figure 2 are presented subsequently when used in equations. For nominal loads, Boughey B.W. [29] associated the capillary limit to the evaluation of thermal limit for which proper operation of flat MHP is ensured. As inside the flat MHP different working liquids are used, the surface tension of the liquid σ_l, is considered and the following inequality is written:

$$\Delta P_{cp,t} \geq \frac{2\sigma_l}{r_{cl}}. \quad (2)$$

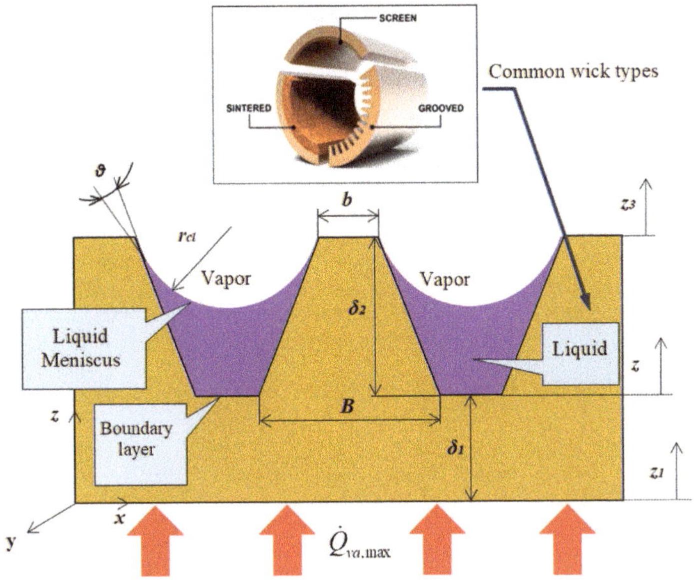

Figure 2. Cross-section through a flat MHP with capillary layer made of trapezoidal microchannels.

The pressure drop of the liquid flowing inside the flat MHP [29], is:

$$\Delta P_l = \frac{\mu_l \dot{m}_l L_e}{\rho_l \left(\sum_{i=1}^{N_{mc}} A_{mc}, i \right) k_l}, \tag{3}$$

where μ_l—dynamic viscosity of the liquid, $\dot{m}_l$—mass flow of the liquid, L_e—length of the evaporator (see Figure 1), ρ_l—liquid density, $\sum_{i=1}^{N_{mc}} A_{mc}, i$—the sum of areas for N_{mc} number of microchannels where the liquid flows, k_l—the effective thermal conductivity of the liquid-wick region.

The pressure drop of vapor passing inside the flat MHP from the evaporator to the condenser is given by:

$$\Delta P_v = \frac{\mu_v \dot{m}_v L_{va}}{(a_{mc} H_{mt} + \pi r^2) \rho_v k_v}, \tag{4}$$

where μ_v—dynamic viscosity of vapor, $\dot{m}_v$—mass flow of vapor, L_{va}—length of the evaporator, a_{mc}—width of the region with microchannels (see Figure 1), H_{mt}—interior height of the flat MHP, r—interior radius of the flat MHP, ρ_v—vapor density, k_v—effective thermal conductivity of the vapor-wick combination.

The liquid pressure drop due to gravitational acceleration is:

$$\Delta P_g = \rho_l L_{ef} g \sin \Phi, \tag{5}$$

where ρ_l—density of the liquid from microchannels, L_{ef}—effective length of the flat MHP, g—the gravitational acceleration, Φ—angle made by the flat MHP with the horizontal direction (see Figure 1). The pressure drop of the liquid ΔP_g due to gravity can be zero, positive or negative, depending on the values of sinΦ. The total area of microchannels is considered in calculus. When the inequality from relation (1) is fulfilled, it can be said that the capillary limit is influenced by the flow of the liquid from the condenser toward the evaporator and by the pressure drops of both liquid and its vapor.

When the heat flux is applied only on one side of the vaporization region of a flat MHP [20,29,30], the produced capillary limit differs from when it is applied on both sides. Theoretically, in the case of heat flux being applied on both sides of the evaporator the capillary limit is superior to the case of one side application. For the situation of a flat MHP placed horizontally where the heat flux is applied on one side, only the microchannels from that side are considered for computation.

Functionally, the flat MHP is in a state of thermal equilibrium and thus the mass flow of liquid vapor and that of liquid from the evaporator are equal in the flow cross-section. The equation of energy balance is obtained taking into account [29,31,32] $\dot{Q}_{va,\max}$—the maxim heat flux ($\dot{Q} = \frac{\partial Q}{\partial t}$ quantity of heat, Q, transmitted over time (t)) that can be transported from the evaporator zone (Figure 1), $\dot{m}_l$—mass flow rate of the liquid, $\dot{m}_v$—mass flow of vapor through the flow cross-section in the evaporator's elements and h_{lg}—latent heat at constant pressure:

$$\dot{m}_l = -\dot{m}_v = \frac{\dot{Q}_{va,\max}}{h_{lg}}. \quad (6)$$

Neglecting the pressure loss due to vaporization and condensation, in 1995, Faghri [29], deduced the equation for the capillary limit ($Q_{cpl,\max}$), given by:

$$Q_{cpl.\max} = \frac{\Delta P_{cp,t} - \Delta P_g}{f_l - f_v}, \quad (7)$$

where f_l and f_v are the coefficients of friction losses for liquid and vapor, respectively. For the case of heat flux applied on one side of the flat MHP, the total maximum drop of capillary pumping pressure is:

$$\Delta P_{cp,\max} = \frac{12\mu_v \dot{m}_v \overline{L}_{ef}}{\rho_v (aH_{mt} + \pi r^2)\delta_v^2} + \frac{12\mu_l \dot{m}_l \overline{L}_{ef}}{\rho_l \sum_{i=1}^{N_{mc}} A_{mc} \Psi}, \quad (8)$$

where δ_v is the mean thickness of vapor core in the cross-section of the flat MHP and ψ represents the permeability of the capillary layer. The mean length of the flat MHP is assumed $\overline{L}_{ef} = \frac{L_{va}+L_{co}}{2}$ since the liquid and the vapor pass over a shorter track than the actual one. The balance equation can be written as:

$$\frac{2\sigma_l}{r_{cl}} = \overline{L}_{ef} \frac{\dot{Q}_{va,\max}}{h_{lg}} \left(\frac{12\mu_v}{\rho_v (aH_{mt} + \pi r^2)\delta_v^2} + \frac{\mu_l}{\rho_l \sum_{i=1}^{N_{mc}} A_{mc} \psi} \right), \quad (9)$$

where r_{cl} is the liquid meniscus radius in the channels of the flat MHP and h_{lg} is the latent heat of vaporization for the working liquid. The maximum rate of heat flux to be transported from the evaporator region to the condenser is obtained as:

$$\dot{Q}_{va,\max} = \frac{2\sigma_l h_{lg}}{r_{cl} \overline{L}_{ef} \left[\frac{12\mu_v}{\rho_v (aH_{mt}+\pi r^2)\delta_v^2} + \frac{\mu_l}{\rho_l \sum_{i=1}^{N_{mc}} A_{mc} \psi} \right]}. \quad (10)$$

3.2. Evaluation of Maximum Heat Flux of the Evaporator of a Flat MHP in Relation to the Capillary Radius and the Working Liquids

The calculations were made using a Mathcad environment [33] and various calculus codes were developed by Mihai and Sprinceana [2]. For the trapezoidal microchannels of the flat MHP the next geometrical parameters were considered (Figure 2): length of trapezoidal solid small base b = 0.200 × 10^{-3} m, length of the large base B = 0.270 × 10^{-3} m and the height δ_1 = 0.275 × 10^{-3} m.

The capillary radius was found via calculations for different values of the angle θ and it was established that it can take values in the range r_{cpl} = 3.48 × 10^{-4}÷9.00 × 10^{-4} m. The heat exchange takes place through a total surface of the trapezoidal microchannels A_{mc}= 9.454 × 10^{-4} m^2 for a length of the evaporator L_{va} = 30 × 10^{-3} m, of the adiabatic zone L_{ad} = 80 × 10^{-3} m and L_{co} = 40 × 10^{-3} m of the condenser, respectively. The total length of the flat MHP is $L_{ef} = L_{va} + L_{adb} + L_{co}$. The thermo-physical properties of the working liquids typically used in flat MHP [34–37], are presented in Table 1.

Table 1. Thermo-physical properties of working liquids used in flat MHPs.

Working Liquid	Liquid Density	Thermal Conductivity	Surface Tension	Dynamic Viscosity	Boiling Point
Notations	ρ_{lic}	λ_{lic}	σ_{lic}	μ_{lic}	T_{sat}
Unit	[kg/m^3]	[W/m°C]	[N/m]	[Pa·s]	[°C]
Distilled water	996	0.632	0.0695	7.99 × 10^{-4}	100
Methanol	792	0.203	0.0262	5.21 × 10^{-4}	65
Acetone	791	0.181	0.0237	3.23 × 10^{-4}	56

Starting from the parameters illustrated in Table 1, the maximum heat flux $\dot{Q}_{va,\max}$ [2], was calculated as a function of the capillary radius of the liquid meniscus from the trapezoidal microchannel, and its evolution is represented in Figure 3. In order to analyze the maximum heat flux rate that can be transported in the evaporator region of a flat MHP when the capillary limit is reached, the latent heat of vaporization h_{lg} dependent on the working fluid was considered in the calculus.

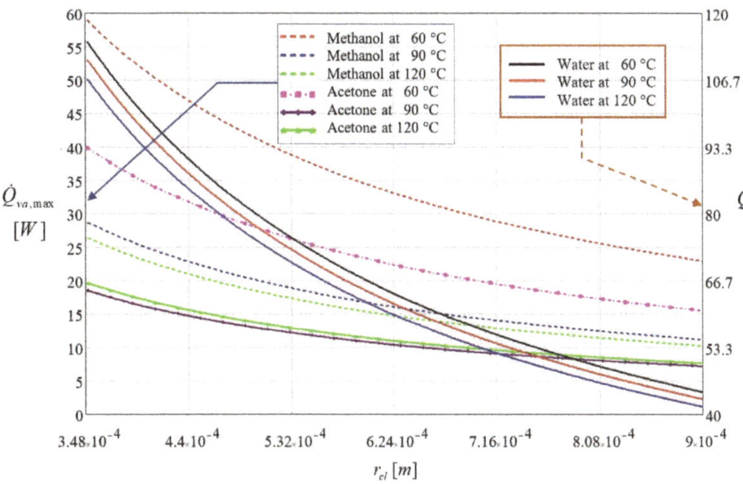

Figure 3. Maximum heat flux rate as a function of capillary radius and temperature for various fluids.

Examining the results illustrated in Figure 3 it can be observed that distilled water has the greatest heat transport capacity from the evaporator to the condenser. The maximum heat flux transported from the vaporization region of the flat MHP increases as the capillary radius r_{cpl} diminishes.

For the distilled water, calculus yielded a maximum heat flux $\dot{Q}_{va,\max} = 114.37W$ at T = 60 °C corresponding to a capillary radius r_{cl} = 3.48 × 10^{-4} m. When the vaporization region is overheated at a temperature T = 120 °C, the calculus is conducted to a heat flux value of $\dot{Q}_{va,\max} = 106.62W$ for the same value of capillary radius.

Methanol and acetone show completely different behavior. Considering the same capillary radius as for the distilled water, the maximum heat flux corresponding to a temperature of 60 °C is 60.24 W for methanol and 21.27 W for acetone. Increasing the temperature to 120 °C, the resulting heat flux was 44.43 W for methanol and 19.12 W for acetone. Methanol and acetone are liquids with vaporization temperatures in the vicinity of 60 °C and for these cases, the variation of the maximum rate of heat flow is seriously influenced by the latent heat of vaporization.

The plots shown in Figure 3 reveal that the heat flux applied in the vaporization region influences the quantity of produced vapors. It depends on the nature of the working fluid and on the value of the vaporization temperature. For too high rates of vaporization, the quantity of produced vapors exceeds the flow rate of the condensed vapors produced in the condenser of a flat MHP. In this case, a dry region appears inside the evaporator. For the three analyzed liquids, the maximum rate of heat flow to be transported by a flat MHP reduces with the increase in the capillary radius and increase in temperature.

4. Temperature Variation during Transient Regime for a Flat MHP

The constructive parameters of the flat MHP presented in Figures 1 and 2 are used for calculations. It is assumed that the flat MHP has a base made of copper and a liquid-wick structure accomplished by longitudinal microchannels with a trapezoidal cross-section.

4.1. Temperature Variation during Transient Regime through the Wall and Liquid-Wick Structure for the Evaporator of the Flat MHP

The equation of the thermal conduction for the non-stationary regime, unidirectional in the wall of the flat MHP that has not included the wick structure [1,13,38,39], is:

$$\rho_{Cu}c_{p,Cu}\frac{\partial T_{p,va}}{\partial t} = k_{Cu}\frac{\partial^2 T_{p,va}}{\partial z_1^2},\ for\ 0 \leq \delta_{tot} \leq \delta_1,\ \ t \geq 0,\ \ \delta_{tot} = \delta_1 + \delta_2. \tag{11}$$

where ρ_{Cu}—copper density, $c_{p,Cu}$—specific heat of copper, $T_{p,va}$—temperature of the evaporator, k_{Cu}—thermal conductivity of copper, t—time, z—radial direction (Figure 2).

For the microchannels composing the internal capillary layer (*mc* index) the Equation (11) can be written as:

$$(\rho c_p)_{eq}\frac{\partial T_{mc}}{\partial t} = k_{eq}\frac{\partial^2 T_{mc}}{\partial z_2^2},\ for\ \delta_1 \leq \delta_{tot} \leq (\delta_1 + \delta_2),\ \ t \geq 0, \tag{12}$$

where the index "*eq*" has the meaning of "equivalent term". The value of the temperature on the exterior side of the wall of the flat MHP (in the central zone of the evaporator) is considered equal to the external temperature of the heat source associated with an electronic component. The heat transfer in the vaporization region of the flat MHP is non-stationary, unidirectional along z axis (Figure 2); it is conductive in the region of δ thickness, mixed regime in the δ_1 region (conductive in the solid sector of the trapezoidal microchannels and convective in the occupying liquid) and convective along z_3 direction that corresponds to the motion of the vapors inside the flat MHP. For a non-stationary regime and considering the temperature generated at the wall by the heat source, the equations for the temperature variation in the wall of the evaporator and of the microchannel of the flat MHP can be written as:

$$\Theta_{p,va}(t, z_1) = T_{p,va}(t, z_1) - T_i, \tag{13}$$

$$\Theta_{mc}(t, z_2) = T_{mc}(t, z_2) - T_i, \tag{14}$$

where $\Theta_{p,va}$ (t,z_1), Θ_{mc} (t,z_2)—the difference in temperature between the wall of the evaporator and the one from inside the microchannel, t—time, z_1—the thickness of the wall, z_2 thickness of the liquid-wick structure which in this particular case corresponds to the microchannels filled with working liquid (Figure 2), $T_{p,va}$ (t,z_1) temperature of the evaporator's wall, identical to the one of the warm source, $T_{p,va}$ (t,z_2)—temperature at the tip of the microchannel's nervure, T_i—initial temperature.

At the moment $t = 0$, depending on the considered thickness of the wall ($\delta = 0$) up to the base of the trapezoidal microchannels of the flat MHP, the Equations (13) and (14) become:

$$t = 0 \left| \begin{array}{l} \Theta_{p,va}(t,z_1) = 0 \ \Rightarrow T_{p,va}(t,z_1) - T_i = 0 \\ \Theta_{mc}(t,z_2) = 0 \Rightarrow T_{mc}(t,z_2) - T_i = 0 \end{array} \right. . \quad (15)$$

It is assumed that the general equation for the distribution of temperature through the wall of the flat MHP in the evaporation zone has the form of a second-degree polynomial with respect to z:

$$\Theta_{p,va}(t,z_1) = a(t) + b(t)z_1 + c(t){z_1}^2. \quad (16)$$

When a thermal flow (generated by a source of heat) is applied on the external surface of the wall of the flat MHP the wall of the evaporator will warm up through conductive thermal transfer. After traversing the wall of the flat MHP the rate of heat flow will worm up the working liquid from the trapezoidal microchannels ensuring the capillarity. By hypothesis, at the moment $t = 0$ there is no variation of temperature at the inner and outer sides of the wall of the condenser of the flat MHP. It is considered the instantaneous distance for the displacement of the lines of the heat field through the wall of the flat MHP, δ_T, and $\dot{q}_{va}$ the density of thermal flow (internal source of generating heat) in the evaporation zone. The boundary conditions for the wall of the evaporator are stated:

$$\left\{ \begin{array}{ll} z_1 = 0 & \dot{q}_{p,va} = -k_{Cu}\frac{\partial\Theta_{p,va}(t,z_1)}{\partial z_1} \\ z_1 = \delta_T(t)|_{\max} & \frac{\partial\Theta_{p,va}(t,z_1)}{\partial z_1} = 0 \end{array} \right. . \quad (17)$$

Considering the temperature variation for transient regime, Equation (12) becomes:

$$\rho_{Cu}c_{p,Cu}\frac{\partial\Theta_{p,va}}{\partial t} = k_{Cu}\frac{\partial^2\Theta_{p,va}(t,\delta)}{\partial {z_1}^2}. \quad (18)$$

Using the boundary conditions from Equation (17) and the Equation (12), it becomes:

$$a(t) = \frac{\dot{q}_{p,va}\delta_T(t)}{2k_{Cu}}, b(t) = -\frac{\dot{q}_{p,va}}{k_{Cu}}, c(t) = \frac{\dot{q}_{p,va}}{2k_{Cu}\delta_T(t)}. \quad (19)$$

The temperature variation in the wall of the evaporator with respect to time and to the thickness of the wall of the flat MHP, when the rate of heat flow passes through it is:

$$\Theta_{p,va}(t,z_1) = \frac{\dot{q}_{p,va}\delta_T(t)}{2k_{Cu}}\left(1 - \frac{z_1}{\delta_T(t)}\right)^2. \quad (20)$$

Since for plane-parallel walls both the density of the rate of heat flow $\dot{q}_{p,va}$ and the rate of heat flow $\dot{Q}_{va}$ are conserved, the single unknown from Equation (20) is $\delta_T(t)$. The rate of heat flow that passes the solid region (without microchannels) of the flat MHP between z_1 and z_2 is:

$$\dot{Q}_{va} = \dot{q}_{p,va}\,A_{p,va}. \quad (21)$$

The notations from Equation (21) are: $\dot{Q}_{va}$—the thermal flow that moves inside the evaporator, $A_{p,va}$—the surface of the evaporator's wall that is crossed by the thermal flux. It must be considered that heat exchange is a transient process. As the material of the wall of the flat MHP accumulates heat, the coefficient of thermal diffusivity must be considered.

The equation of the thermal conduction is integrated in order to calculate the evolution in time of the distance of the thermal field pattern $\delta_T(t)$, with respect to the wall:

$$\int_0^{\delta_T(t)} \frac{\partial \Theta_{p,va}(t,z_1)}{\partial t} dz_1 = \int_0^{\delta_T(t)} a_{p,va} \frac{\partial^2 \Theta_{p,va}(t,z_1)}{\partial {z_1}^2} dz_1, \quad (22)$$

where:

$$a_{p,va} = \frac{k_{Cu}}{\rho_{Cu} c_{p,Cu}}, \quad (23)$$

is the coefficient of thermal diffusivity of the evaporator's wall. The difference in temperature $\Theta_{p,va}$ (t,z_1) from Equation (20) is replaced into the relation (22) and after solving, it results:

$$d\delta_T^2(t) = 6a_{p,va}dt. \quad (24)$$

By integrating the Equation (24), the distance traveled by the lines of the thermal field through the wall of the evaporator as a function of the material thermal diffusivity coefficient and of time is obtained:

$$\delta_T(t) = \sqrt{6a_{p,va}t}. \quad (25)$$

The surface of the evaporator where the thermal flow is applied was found $A_{va} = 9.88 \times 10^{-4}$ m^2 for a thickness of the evaporator's wall, δ_1. The variation of temperature difference as a function of time (480 s) in the wall of the evaporator of the flat MHP (from the side of the wall in contact with the source of heat to the distance δ_1 (Figure 2) is represented in Figure 4. The outer side of the flat MHP is subjected to a variable thermal flow in the range of 10 ÷ 40 W. The calculus was made until the stabilization of the thermal regime.

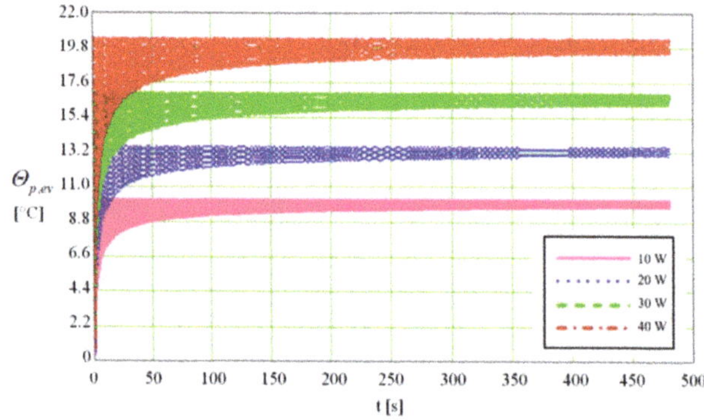

Figure 4. Difference in temperature versus time in the wall of the evaporator of flat MHP for different values of the thermal flow (10 ÷ 40 W).

From the plot presented in Figure 4, it is observed that the difference in temperature in the wall of the flat MHP is higher as the rate of heat flow increases. The thermal balance is reached when after a certain time the difference in temperature between the two faces of the wall of the flat MHP becomes constant. From the calculus it results that the period necessary to attain the thermal balance differs as a function of the values of the input rate of heat flow. For an input rate of heat flow of 10 W the temperature stabilizes after 148 s while for the maximum value of 40 W the time required for attaining the steady temperature increases to 396 s.

Using the same code, for the same time duration of 480 s (on logarithmic scale), the difference in temperature between the outer and inner faces of the wall of the evaporator was obtained for rate of heat flow values of 10, 20, 30 and 40 W and for wall thickness in the range 0÷2.50 × 10^{-3} m, were traced as shown in Figure 5.

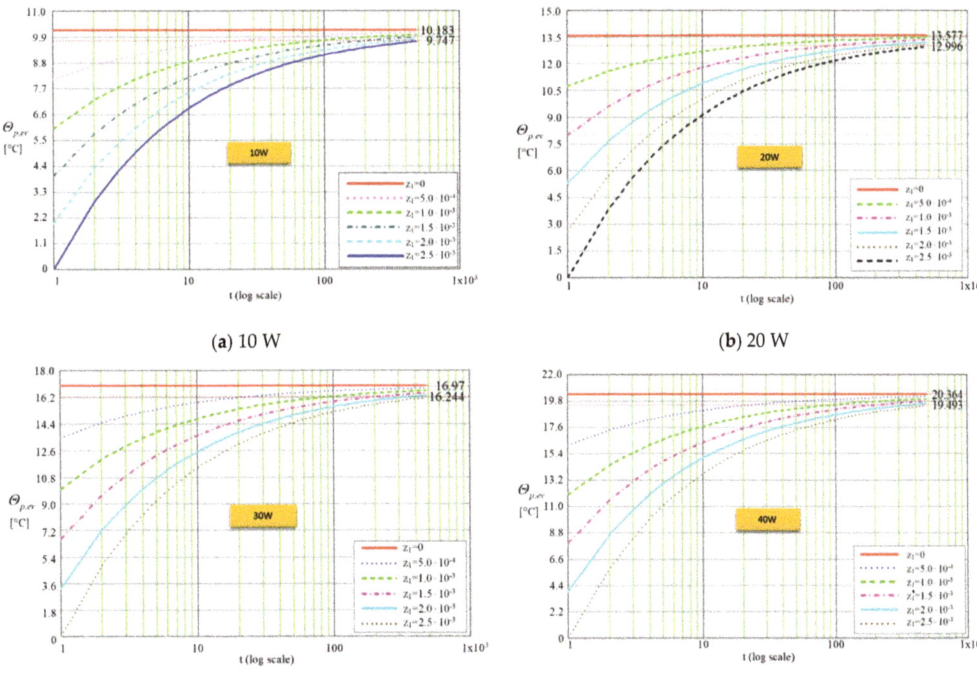

(**a**) 10 W

(**b**) 20 W

(**c**) 30 W

(**d**) 40 W

Figure 5. Difference in temperature versus time through the flat MHP wall for various rates of heat flow.

From the plots presented in Figure 5a–d it is observed that the entire thickness of the wall of the evaporator is passed by the lines of the thermal field during a period of time until *480 s*. It can be noticed that after the crossing of the thickness δ_1, (Figure 2), the temperature difference reaches a maximum of 10.18 °C for 10 W and 20.36 °C at 40 W, respectively. After passing the copper layer of thickness δ_1, the rate of heat flow will pass through the liquid-wick structure of thickness δ_2 (Figure 2). Taking into account that z_2 represents the heat propagation direction, the temperature variation in the liquid-wick structure can take the form:

$$T_{mc}(t,\delta_2) = a_1(t) + b_1(t)z_2 + c_1(t){z_2}^2. \tag{26}$$

The boundary conditions for the conductive heat transfer for unidirectional transient regime through the fins of the microchannels are:

$$\begin{cases} z_2 = 0 & \dot{q}_{mc} = -k_{eq}\frac{\partial T_{mc}(t,z_2)}{\partial z_2} \\ z_2 = \delta_{2,T}(t)\big|_{\max} & \frac{\partial T_{mc}(t,z_2)}{\partial z_2} = 0 \end{cases}, \tag{27}$$

where $\delta_{2,T}(t)$—is the distance passed by the rate of heat flow through the trapezoidal microchannels (with the maximum value δ_2), k_{eq}—is the equivalent coefficient of thermal conduction for the liquid-wick structure. It is considered that the wall has a homogenous and isotropic structure. Analogous to the precedent case, the energy balance for the evolution of rate of heat flow through the liquid-wick structure is written as:

$$T_{mc}(t,z_2) = T_{p,va}(t)\left[1-\frac{z_2}{\delta_{2,T}(t)}\right]^2 \; ; \; \dot{q}_{mc} = \frac{2k_{eq}T_{mc}(t,z_2)}{\delta_{2,T}(t)}. \tag{28}$$

The Fourier equation, for passing through the liquid-wick structure of the flat MHP with δ_2 thickness, for unidirectional (along z_2 direction) transient regime is:

$$\frac{\partial T_{mc}}{\partial t} = \frac{k_{eq}}{\rho_{eq}c_{p,eq}}\frac{\partial^2 T_{mc}}{\partial z_2{}^2} + \frac{\dot{q}_{va}}{\rho_{eq}c_{p,eq}}. \tag{29}$$

The energy balance for the capillary layer made of trapezoidal microchannels is:

$$\int_0^{\delta_{2,T}(t)} \frac{\partial T_{mc}(t,z_2)}{\partial t}dz_2 = \frac{\delta_{2,T}(t)}{2k_{eq}T_{mc}(t,z_2)}\int_0^{\delta_{2,T}(t)} a_{mc}\frac{\partial^2 T_{mc}(t,z_2)}{\partial z_2{}^2}dz_2, \tag{30}$$

where a_{mc} is the coefficient of thermal diffusivity of the microchannel.

$$T_{mc}(t,z_2) = \frac{\dot{q}_{va}\left(6\frac{k_{eq}}{\rho c_{p,eq}}t\right)^{1/2}}{2k_{eq}}\left[1-\frac{z_2}{\delta_{2,T}(t)}\right]^2. \tag{31}$$

In order to find the equivalent thermal conductivity of the evaporation region, there are considered the materials of the wall and of the capillary layer of the flat MHP together with the working liquid. These all together will have an equivalent thermal conductivity. According to Figure 6, it is considered that the trapezoidal microchannel is totally or partially filled with liquid.

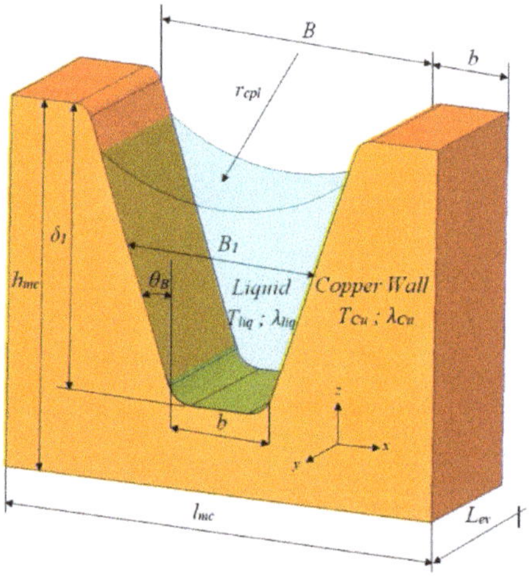

Figure 6. Geometry of the cross-section through a trapezoidal microchannel of the flat MHP [2].

In the evaporation zone, the microchannel is not totally filled with liquid. This fact is explained by the permanent vaporization of the working liquid along the entire path between the condenser and evaporator. Conversely, in the condenser's region, due to permanent condensation of the vapors, the liquid may exceed the height of the microchannels. For the working fluid in the evaporator microchannels, a longitudinal section of a right-angled trapezium is formed between the condenser and the evaporator.

The trapezoidal microchannels are built into the copper wall, on the inside of the flat MHP. As the thickness of the liquid layer and the velocity of the liquid through the microchannels of the evaporator decrease, the convective phenomenon may be neglected. The hypotheses considered for calculus are that the heat transfer through the liquid-wick structure is conductive and under the same conditions for each fin of the microchannel, while for the convective heat transfer through the liquid from the microchannels, the dimensions of the microchannels are considered. For the vaporization case at the liquid-vapors interface [40], the coefficient of convective thermal transfer is estimated based on the kinetic theory of gases:

$$h = \frac{2\kappa_a}{2-\kappa_a}\frac{\rho_v h_{lg}^2}{T_v}\left(2\pi\frac{R}{M}T_v\right)^{-1/2}\left(1-\frac{P_v}{2\rho_v h_{lg}}\right), \tag{32}$$

where: κ_a—thermal accommodation coefficient, ρ_v—density of vapors, h_{lg}—latent heat of vaporization of the vapors, T_v vaporization temperature, M—molecular mass, P_v—partial vapor pressure of the vapors.

For distilled water and methanol [41], the thermal accommodation coefficient can be considered equal to unit, $\kappa_{dW} = \kappa_M = 1$ and for acetone is $\kappa_{ac} = 0.32$. The mass accommodation coefficient for vapor is quoted for $\kappa_{vW} = 0.04$, for acetone $\kappa_{vac} = 0.026$ and for methanol $\kappa_{vM} = 0.056$. For the vaporization in the region of liquid meniscus [42], the mass flow rate of the vaporized liquid $\dot{m}_v$ can be determined depending on the temperature of the liquid film, its thickness and the pressure and temperature at the vapor-liquid interface:

$$\dot{m}_v = a\left(T_{\delta_{fl,va}} - T_v\right) + b\left(P_l - P_v\right), \tag{33}$$

where $T_{\delta_{fl,va}}$ is the temperature of the film of liquid from the evaporator, T_v—the temperature of the liquid vapors, P_l, P_v—pressure of the liquid and of the liquid vapors, respectively. According to [42], the parameters a and b are:

$$a = \frac{2\kappa_a}{2-\kappa_a}\left(\frac{M}{2\pi R T_{\delta_{fl,va}}}\right)^{1/2}\left(\frac{P_v M h_{lg}}{R T_v T_{\delta_{fl,va}}}\right), \tag{34}$$

$$b = \frac{2\kappa_a}{2-\kappa_a}\left(\frac{M}{2\pi R T_{\delta_{fl,va}}}\right)^{1/2}\left(\frac{V_{m,l} P_v}{R T_{\delta_{fl,va}}}\right), \tag{35}$$

where $V_{m,l}$—the molar volume of the liquid. The coefficient of accommodation for distilled water [41] can be considered equal to unit. The mass flow rate of the evaporated liquid in the evaporator is found applying the equation of unidirectional conduction:

$$\dot{m}_v = \frac{k_{eq}\left(T_{p,va} - T_{\delta_{fl,va}}\right)}{\delta_{fl,va} h_{lg}}. \tag{36}$$

The temperature of the liquid from the evaporator can be calculated using the relation:

$$T_{\delta_{fl,va}} = \frac{\delta_{fl,va} h_{lg}\left[a T_v - b(P_l - P_v)\right] + k_{eq} T_{p,va}}{k_{eq} + a\,\delta_{fl,va} h_{lg}}. \tag{37}$$

The equivalent thermal conductivity of the liquid-wick structure in the vaporization zone [43] is calculated by the relation:

$$k_{eq,va} = \frac{\dot{Q}_{va} L_{va}}{A_{va}\left(T_{p,va} - T_{\delta_{fl,va}}\right)}. \tag{38}$$

The variation of the equivalent thermal conductivity of the liquid-wick structure in the vaporization region as function of the thermal flow for different temperatures is presented in Figure 7, for the distilled water as working fluid.

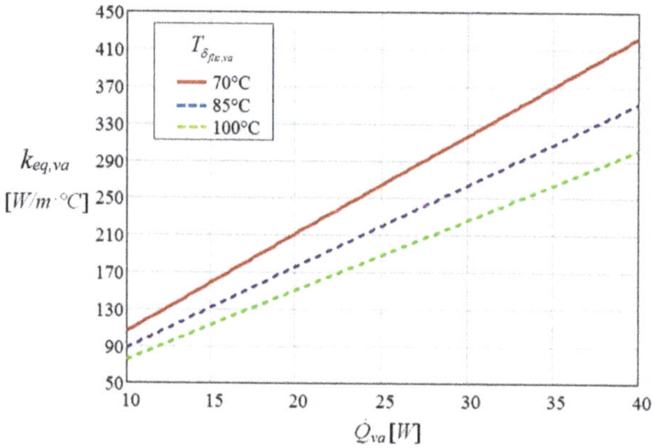

Figure 7. The equivalent thermal conductivity of the liquid-wick structure in the vaporization region as function of the thermal flow and temperature.

The equivalent thermal conductivity of the vaporization region changes directly proportional to the applied thermal flow in the region of the evaporator and inversely proportional to the variation of temperature of the film of liquid from the microchannels of the flat MHP. From Figure 7 it is observed that for the distilled water as working liquid the equivalent thermal conductivity of the vaporization region $k_{eq,va}$ presents the highest increase at 70 °C and for a thermal flow of 40 W.

A source code was developed in Mathcad [2], assuming that the equivalent thermal conductivity depends on the thermal flow, the interior structure of the flat MHP, material, etc., and presents the maximum value in the region where the warm source is applied. The computations were made for the same values of the input heat ranging from 10 ÷ 40 W. The code allows for the determination of the temperature through the liquid-wick structure of a flat MHP in a transient regime.

For the wall of the evaporator (Figure 2) it was adopted $\delta_1 + \delta_2 = 5.50 \times 10^{-3}$ m. After the calculus was made, the plots from Figure 8a–d were traced for a time interval of 180 s (logarithmic scale), representing the variation of temperature from the base of the liquid-wick structure that is in contact with the warm source till the interior of the flat MHP. On each figure, the software shows on the upper right side the maximum values that can be reached in the structure of the flat MHP. The thickness values considered in calculus are given in the legend of each figure. The variation of temperature with time through the liquid-wick structure of flat MHP is calculated for four values of the rate of heat flow, 10, 20, 30 and 40 W for different values of the wall thickness. Theoretically, three cases can occur: the first corresponds to input heats of 10 W and 20 W when the temperature reaches a maximum of 65.410 °C and 78.348 °C, respectively, values for which the electronic components are heated but not necessarily dangerous for the adequate functioning. In this case, the flat MHP can cool the electronic device without problems, since the liquid inside it attains the vaporization temperature corresponding to the partial vacuum from inside. The second case corresponds to an input heat of 30 W when the maximum temperature reached is 96.866 °C. In this situation, the flat MHP must employ working liquids with higher vaporization temperatures capable to ensure superior heat transport capacity.

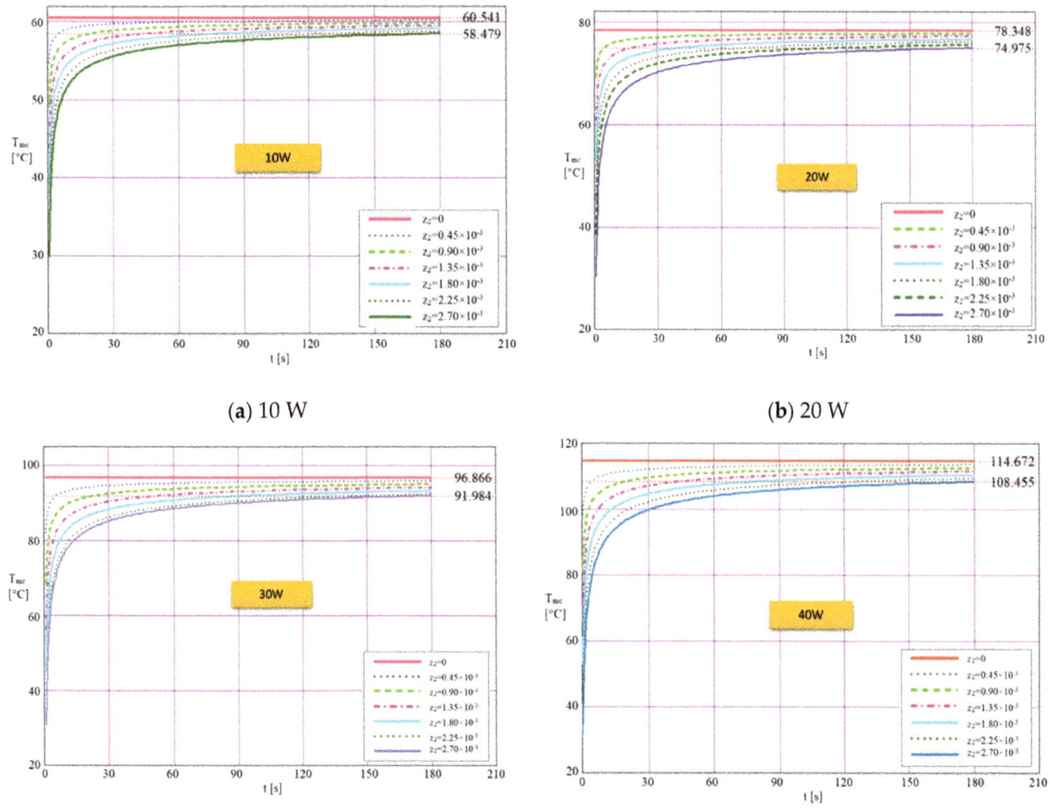

(**a**) 10 W (**b**) 20 W

(**c**) 30 W (**d**) 40 W

Figure 8. Temperature variation in time through the wall of flat MHP for different heat flow rates.

This second class of flat MHP operates at a limit if the maximum temperature of the warm source is reached. Though, it can be affirmed that a stable running regime exists. The third case refers to the input heat of 40 W when the maximum temperature reached is 114.672 °C and in this situation, the flat MHP operates under overload. For these values of input heat, the flat MHP must be designed with adequate working liquids, materials and capillary structures. The mathematical model employed demonstrates that the variation of temperature depends on the value of the input heat (that imposes the maximum temperature in the evaporation zone of the flat MHP), the physical characteristics of the material, the summated thickness of the wall and liquid-wick structure and time. The manner the rate of heat flow influences the temperature decrease is presented in Table 2.

After the transient operation, the working regime of the flat MHP becomes steady if fluctuations of the input heat are not occurring. The influence of the input heat upon the values of the temperature inside the flat MHP in the evaporation region is fundamental for their functioning both for nominal loads and overloads.

Table 2. The temperature drop in the walls of the flat MHP depending on the variation of the thermal flow.

Parameters	U.M.	Values			
Input rate of heat flow	[W]	10	20	30	40
Maximum temperature outside evaporator of the flat MHP	[°C]	60.541	78.348	96.866	114.672
Internal temperature inside the evaporator of the flat MHP	[°C]	58.479	74.975	91.984	108.455
Temperature difference	[°C]	2.062	3.373	4.882	6.217

4.2. Modeling of Heat Transfer inside a Flat MHP

The model of heat transfer was accomplished considering both the heat and the mass transfer. The displacement of the vapors within the flat MHP from the evaporator to the condenser is subjected to a convective heat transfer. Considering $\vec{v}$—the vector of displacement of the field of heat, the equation of thermal balance for the heat transfer is:

$$\frac{\partial T_{p,va}}{\partial t}+\frac{\partial}{\partial z}\left(\vec{v}T_v\right)=a_{p,va}\frac{\partial^2 T_{p.va}}{\partial z^2}+\alpha_{co}(T_v-T_i)L_{ef}(y), \tag{39}$$

where $T_{p,va}$—is the temperature of the wall of the evaporator, T_v—temperature of vapors, T_i—the temperature inside the structure of the flat MHP; $a_{p,va}$—coefficient of thermal diffusivity of the evaporator's wall, α_{co}—the coefficient of convective transfer, L_{ef}—the effective length of the flat MHP (see Figure 1).

A source code was developed in MATLAB [2,44], using FTCS—a finite difference method, for numerically solving the heat transfer equation through the flat MHP, by considering i points in the space and k points in time:

$$\frac{\left(T_i^{k+1}-T_i^k\right)_{p,va}}{\Delta t}+v\frac{\left(T_{i+1}^{k+1}-T_{i-1}^{k+1}\right)_v}{2\Delta z}=a_{p,va}\frac{\left(T_{i+1}^{k+1}-2T_i^{k+1}+T_{i-1}^{k+1}\right)_{p,va}}{\Delta z^2}+\alpha_{co}\left(T_v-T_i^k\right)L(y_i). \tag{40}$$

It is assumed that v is constant and $T_i^k = T_{ini}$ and Equation (40) becomes:

$$\frac{\left(T_i^{k+1}-T_i^k\right)_{p,va}}{\Delta t}+v\frac{\left(T_{i}^{k+1}-T_{i-1}^{k+1}\right)_v}{2\Delta z}=a_{p,va}\frac{\left(T_{i+1}^{k+1}-2T_i^{k+1}+T_{i-1}^{k+1}\right)_{p.va}}{\Delta z^2}+\alpha_{co}\left(T_v-T_i^k\right)L_{ef}(y_i), \tag{41}$$

$$\begin{aligned}&-\left(\tfrac{a_{p.va}\Delta t}{\Delta z^2}+\tfrac{v\Delta t}{\Delta z}\right)T_{i-1}^{k+1}+\left(2\tfrac{a_{p.va}\Delta t}{\Delta z^2}+\tfrac{v\Delta t}{\Delta z}+1\right)T_i^{k+1}-\left(\tfrac{a_{p.va}\Delta t}{\Delta z^2}\right)T_{i+1}^{k+1}=\\&=T_i^k+\alpha_{co}\left(T_{vap}-T_i^k\right)L_{ef}(y_i).\end{aligned} \tag{42}$$

The initialization parameters are set in the MATLAB code and for the convective transfer under the transient regime during the period of time t the isothermal curves are obtained.

In order to compare the results of the model with the analytical and experimental ones, the input heat values were set in the range of 10 ÷ 40 W.

The isothermal curves that resulted from the MATLAB simulation are represented in Figure 9 and highlight the transient evolution of the temperature for different rates of heat flows when the working liquid is distilled water. The curves are obtained for a certain number of iterations specifies in the source code. It is observed that according to Figure 9a, when the flat MHP operates at 10 W the temperature reached in the evaporation zone is only 48.7 °C compared to the maximum value attained within the structure, 58.479 °C. For the cases from Figure 9b,c, the curves present similar shapes, only the values differ in time.

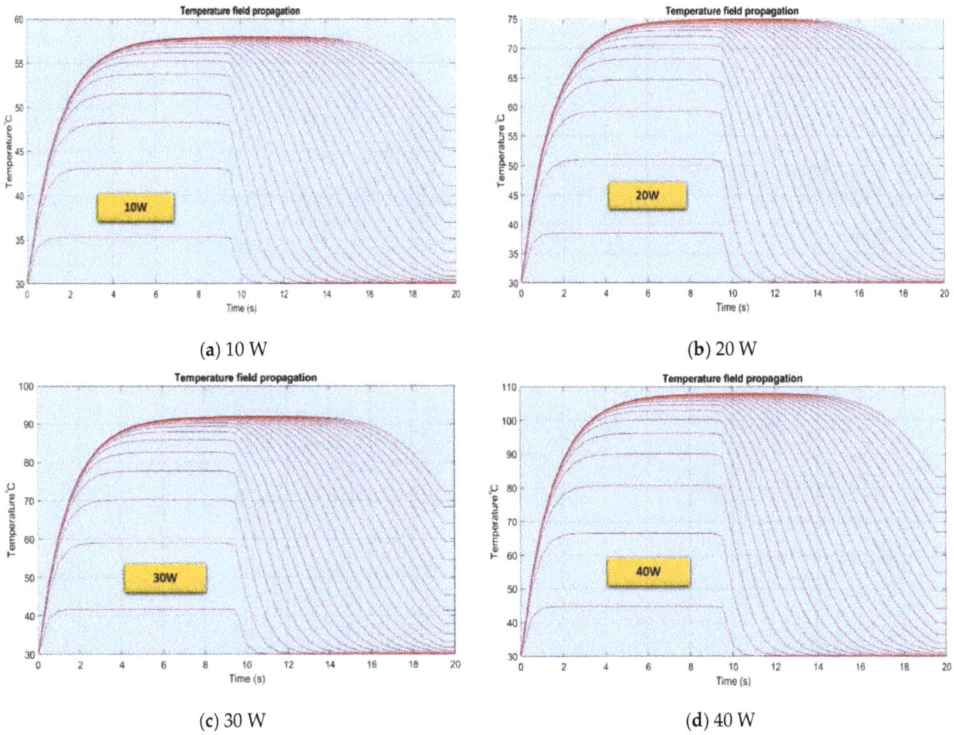

(**a**) 10 W (**b**) 20 W

(**c**) 30 W (**d**) 40 W

Figure 9. Modeling temperature variation versus time within a flat MHP.

For the maximum rate of heat flow of 40 W in the case presented in Figure 9d, the temperature reached within the evaporation zone is 83.270 °C and for the internal structure, the temperature is 108.455 °C. It is noticed that the differences in temperature augment significantly as the input heat increases. The shape of the curves shows that initially, the working liquid warms during an interval of 6 ÷ 8 s until the maximum value is reached, and subsequently, the vaporization process begins.

The vaporization process is accompanied by a mass transfer process since the vapors migrate towards the condenser. This fact determines the decrease in the temperature inside the flat MHP followed by stabilization at the values indicated in the figures.

4.3. Result Validation

In order to validate the results obtained analytically and numerically, a comparison with own experimental results as well as with other results from the literature [45–47], was made. The experimental data was obtained using the set-up shown in Figure 10.

The front panel of the experimental set-up 1 allows different heating regimes to be preset using the keypad 2 while switch 3 allows the set-up to be switched on and off. The parameters required to control the heating system can be monitored in real-time by means of display 4. Various heating regimes are obtained by means of a heating element 12, made by depositing a 4 Ω resistance graphite layer on a metal support with a, supplied from a current source. The set-up is fitted with a cooling control system 6 which can be switched on and off by means of switch 5. This controls a fan 14, located in the condensing area. The fan airflow can be varied by changing its operating speed. The electronic thermometer 7 equipped with eight channels, acquires data from several temperature sensors 11 and allows monitoring of the temperature along the entire length of the MHP. Switch 8 allows temperature monitoring through the flat (interchangeable) micro-thermal tubes 13. A data logger 9, linked to the electronic thermometer stores the temperatures read by the thermocouples and ensures the interface to a computer 10 via a USB port. Using data acquisition software, temperature variation graphs are obtained for eight points.

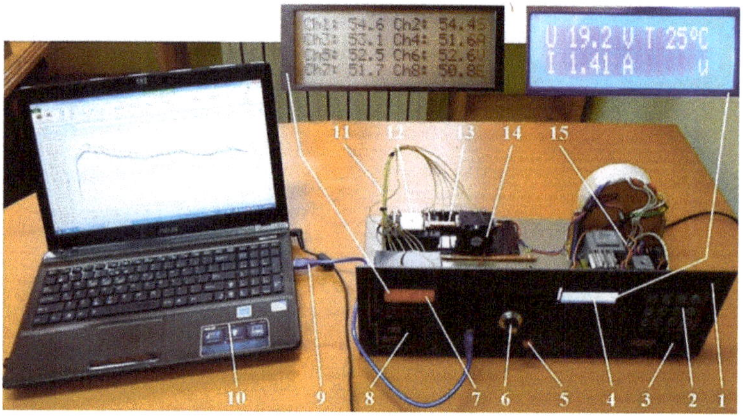

Figure 10. Experimental set-up for the study of heat transfer through flat MHPs [2].

The calculus models used allow for determining the temperature difference between the hot surface and the inside of the heat pipe, in the evaporator region of the flat MHP. The used equipment allows precise evaluation of the external temperature evolution in the transient regime for the evaporator wall. In Table 3, the data obtained by the authors for the analytical experimental and FTCS modeling cases are compared with results obtained by other authors.

Table 3. Comparison between present study results and the literature.

Parameters	U.M.	Values				
Heat flow	[W]	10	10	10	6	20
Thermal regime stabilization time	[s]	175	175	175	197	278
MHP surface temperature	[°C]	59.2	60.54	58.75	66.9	64.5
MHP inner wall temperature	[°C]	57.1	58.48	56.60	60.1	58.8
Temperature difference through the MHP wall	[°C]	2.1	2.06	2.15	6.8	5.7
Data source		Experimental	Analytic	Numeric	[45]	[46]

For the distilled water, the difference in temperature between the two faces of the liquid-wick structure is 2.06 °C for 10 W and 6.21 °C for 40 W. One can state that difference in temperature increases with the thickness of the flat MHP, so thinner walls are recommendable. But an optimum thickness must be ensured since the wall of the flat MHP must resist the pressure differences from within.

Results shown in Table 3 illustrate that the experimental data obtained by the authors are in good agreement with those determined by calculus, which means that the method of determining the temperature field within the MHP's wall can be applied regardless of the used liquid and the heat source power. Differences found for results obtained numerically as well as for the results of other authors are due to the imposed initial conditions, used materials, different powers of the heat sources, etc. However, the curves obtained by the authors are similarly shaped to those found in the literature.

With regards to the maximum cooling capacity of MHPs, in Table 4, the values obtained by the authors are compared with those from other similar research.

Table 4. MHP's maximum heat transport capacity.

	MHP's Heat Transport Capacity	Operating Temperature	Capillary Radius	Wall/Working Fluid
U.M.	W	°C	m	
Various functional limits (present study)	106.62	120	3.48×10^{-4}	Cooper/distilled water
	44.43	120	3.48×10^{-4}	Cooper/methanol
	19.12	120	3.48×10^{-4}	Cooper/acetone
[47]—horizontal	88.70	150	6.25×10^{-3}	Cooper/distilled water
[47]—horizontal	83.50	120	6.25×10^{-3}	Cooper/distilled water
[47]—vertical	46.50	130	6.25×10^{-3}	Cooper/distilled water
[47]—vertical	44.60	120	6.25×10^{-3}	Cooper/distilled water

By analyzing the data in Table 4 and from the practice of operating MHPs, it was deduced that there are different factors influencing the maximum heat transport capacity. It can be observed that depending on the capillary radius, used materials, working temperature and position of MHPs different functional limits can occur with a strong impact on the cooling of components.

5. Conclusions

Thermal tubes ensure the cooling of power electronic components by transporting the heat from the vaporization zone to the condensation zone. The investigations and results presented herein allow the formulation of the following conclusions:

1. The method, acknowledged to provide efficient vaporization and condensation, consists in creating a partial vacuum at pre-set values. The effectiveness of the cooling depends on the liquid used and on its quantity. When the working liquid is too abundant or insufficient, malfunctions of the flat thermal microtubes may occur.
2. Assessment of the maximum heat load and its influence on temperature variation of the evaporator in flat MHPs in a transient regime is useful in the field of cooling components that need fast dissipation of important heat. The present study allows determining the correlation between some parameters on the microchannel capillary radius, for three fluids and different temperatures that can be applied to different thermal control systems. Such systems can be used in various fields, such as the cooling of biological objects and drying technology.
3. In the present paper the maximum rate of heat flow that can be applied in the vaporization region of a flat MHP was found analytically as a function of the capillary radius and for different working liquids such as acetone, methanol or distilled water. Finding the maximum rate of heat flow is important for practical applications since beyond this value the functional blocking of the flat MHP, by overheating, occurs.
4. Calculations performed to determine the maximum heat transport capacity in MHPs take into account the physical characteristics of the working fluids, the geometrical parameters of the microchannels and the structure providing capillarity. The maximum heat transport capacity of MHPs is conditioned by the applied heat flux. At too high a heat flux applied in the vaporization zone, MHPs cannot be functionally stabilized if the working fluid flow rate is too low or in excess. Excess liquid flow can cause such intense vapor flow that it prevents the liquid from flowing out of the microchannels through the vaporization and promotes boundary layer detachment.
5. The capillarity limit influences the functioning of MHPs, as for low temperatures it is restricted by the reduced pressure of vapors and the high viscosity of the liquid. When temperatures increase near the critical point, latent vaporization heat, surface tension as well as maximum power, decrease drastically.
6. The decrease in the input heat takes place with the temperature increase and it is explained by the diminution of the thickness of condensation film in the vaporization

region, due to the faster vaporization rate. Decreasing the rate of heat flow with increasing temperature also depends on the working fluid. The explanation for the pronounced decrease in maximum input heat for acetone and methanol is found in their much higher volatility.

7. The calculus codes developed in Mathcad permit predicting the difference in temperature between the two faces of the evaporator (exterior and interior). The hypothesis adopted is that the external temperature to the flat MHP corresponds to the warm source. The source codes consider the values of the input heat applied to the flat MHP, the material, structure and thickness of the solid walls, the working liquid, specific heat, coefficient of thermal diffusivity and the equivalent coefficient of thermal conduction, etc.
8. The selection of the working liquid must be made to ensure the cooling of the electronic device even for overloading. As known [1] the functional difference concerning a temperature disparity of over 10 °C with respect to the constructive maximum temperature may conduct to the physical collapse of the electronic components.

Author Contributions: Conceptualization, I.M.; methodology, I.M.; software, I.M.; validation, I.M., C.S. and C.M.P.; formal analysis, I.M.; investigation, I.M., C.S. and C.M.P.; resources, I.M. and C.S.; data curation, I.M.; writing—original draft preparation, I.M.; writing—review and editing, C.S.; visualization, I.M., C.S. and C.M.P.; supervision, I.M.; project administration, I.M.; funding acquisition, I.M. All authors have read and agreed to the published version of the manuscript.

Funding: This research received no external funding.

Institutional Review Board Statement: Not applicable.

Informed Consent Statement: Not applicable.

Data Availability Statement: Some or all data, models, or code generated or used during the study are available from the corresponding author by request.

Conflicts of Interest: The authors declare no conflict of interest.

References

1. Mihai, I.; Pîrghie, C.; Zegrean, V. Research Regarding Heat Exchange through Nanometric Polysynthetic Thermal Compound to Cooler–CPU Interface. In *Heat Transfer Engineering*; Taylor & Francis: Abingdon, UK, 2010; Volume 31, pp. 90–97. ISSN1 1521-0537. ISSN2 0145-7632.
2. Sprînceană, S. Heat Transfer Intensification Methods in Micro Heat Pipes. Ph.D. Thesis, Stefan cel Mare University of Suceava, Suceava, Romania, 2017; pp. 1–308.
3. Reay, D.; Kew, P.A.; McGlen, R.J. *Heat Pipes Theory, Design and Applications*, 6th ed.; Butterworth-Heinemann: Oxford, UK, 2014; 208p.
4. Hua, L. Heat Exchanger Development for Waste Water Heat Recovery. Master's Thesis, University of Canterbury, Department of Mechanical Engineering, Christchurch, New Zealand, 2005; pp. 1–116.
5. Nemec, P.; Čajal, A.; Malchol, M. Testing Thermal Properties of the Cooling Device with Heat Pipes. In *EPJ Web of Conferences*; EDP Sciences: Les Ulis, France, 2013; pp. 1066–1070. [CrossRef]
6. Faghri, A. Heat pipes: Review, opportunities and challenges. *Front. Heat Pipes* **2014**, *5*, 1–48. [CrossRef]
7. UNILAB SRL. Heat Pipes—Part I: From Aerospace to Air Conditioning Applications. Available online: https://www.unilab.eu/articles/technical-articles/thermodynamic-engineering-articles/heat-pipes-pt-1/ (accessed on 31 July 2020).
8. Cotter, T.P. Principles and Prospects of the Micro Heat Pipe. In Proceedings of the 5th International Heat Pipe Conference, Tsukuba, Japan, 14–18 May 1984; pp. 328–335.
9. Ma, H. Heat Pipes—Chapter 9. In *Mechanical Engineers' Handbook: Energy and Power*, 3rd ed.; Kutz, M., Ed.; John Wiley & Sons, Inc.: Hoboken, NJ, USA, 2006; Volume 4, pp. 335–360.
10. Choondal, B.S. Modeling of the flow and heat transfer in micro heat pipes. In Proceedings of the ASME 2004 2nd International Conference on Microchannels and Minichannels, Rochester, NY, USA, 17–19 June 2004.
11. Jeong, S.; Young, S.P. Analysis of Thermal Performance in a Micro Flat Heat Pipe with Axially Trapezoidal Groove. *Tamkang J. Sci. Eng.* **2003**, *6*, 201–206.
12. Sprinceana, S.; Mihai, I. The modeling of the speed, of the friction coefficient and the capillary pressure in the heat convective transfer through porous media. In Proceedings of the 18th International Conference, Suceava, Romania, 8–9 May 2015; pp. 371–377.

13. Zhu, N.; Vafai, K. Analytical modeling of the startup characteristics of asymmetrical flat-plate and diskshaped heat pipes. *J. Hear Mass Transfer.* **1998**, *41*, 2619–2631. [CrossRef]
14. Wang, Y.; Vafai, K. Transient characterization of at plate heat pipes during startup and shutdown operations. *Int. J. Heat Mass Transf.* **2000**, *43*, 2641–2655. [CrossRef]
15. Jang, J.M.; Faghri, A.; Chang, W.S. *Analysis of the Transient Compressible Vapor Flow in Heat Pipe*; NASA Contract Report 185119; Wright State University: Dayton, OH, USA, 1989; pp. 1–10.
16. William Anderson, G.W.; Tarau, C. TFAWS Heat Pipes. Presentation. In Proceedings of the Thermal & Fluids Analysis Workshop (TFAWS 2019), Hampton, VA, USA, 26–30 August 2019; p. 194.
17. Ranjan, R.; Patel, A.; Garimella, S.V.; Murthy, J. *Wicking and Thermal Characteristics of Micropillared Structures for Use in Passive Heat Spreaders*; CTRC Research Publications: West Lafayette, IN, USA, 2012; pp. 1–35. [CrossRef]
18. Oshman, C.; Li, Q.; Liew, L.A.; Yang, R. Thermal performance of a flat polymer heat pipe heat spreader under high acceleration. *J. Micromech. Microeng.* **2012**, *22*, 45018. [CrossRef]
19. Thomas, K.S.; Likins, R.C.; Yerkes, L.K. Fully-developed laminar flow in trapezoidal grooves with shear stress at the liquid-vapor interface. *Int. J. Heat Mass Transf.* **2001**, *44*, 3397–3412. [CrossRef]
20. Odabasi, G. Modelling of Multidimensional Heat Transfer in a Rectangular Grooved Heat Pipe. Master's Thesis, Middle East Technical University, Ankara, Turkey, 2014; pp. 1–156.
21. Sprinceana, S.; Mihai, I.; Beniuga, M.; Suciu, C. Heat transfer intensification by increasing vapor flow rate in flat heat pipes. In *Advanced Topics in Optoelectronics, Microelectronics, and Nanotechnologies VII*; SPIE: Bellingham, WA, USA, 2015; Volume 9258, pp. 415–422. [CrossRef]
22. Mohammed, N.; Gerner, F.M.; Henderson, H.T. Steady-state model of the loop heat pipe (LHP) with coherent porous silicon (CPS) wick in the evaporator. In Proceedings of the Semiconductor Thermal Measurement and Management Symposium, San Jose, CA, USA, 11–13 March 2003; pp. 88–96. [CrossRef]
23. Kimura, Y.; Nakamura, Y.; Sotani, J.; Katsuta, M. Steady and transient heat transfer characteristics of flat micro hea tpipe. *Furukawa Rev.* **2005**, *27*, 3–8.
24. Peterson, G.P.; Babin, B.R. *Analitical and Experimental Investigation of Miniature Heat Pipes—Phase II*; Report NoWRDC-TR-89-2067; Mechanical Engineering Department Texas, A&M University College Station: College Station, TX, USA, 1989; pp. 1–65.
25. Lips, S.; Lefèvre, F.; Bonjour, J. Physical mechanisms involved in grooved flat heat pipes: Experimental and numerical analyses. *Int. J. Therm. Sci.* **2011**, *50*, 1243–1252. [CrossRef]
26. Kakaç, S.; Vasiliev, L.L.; Bayazitoglu, Y.; Yener, Y. *Microscale Heat Transfer—Fundamentals and Applications: Proceedings of the NATO Advanced Study Institute on Microscale Heat Transfer—Fundamentals and Applications in Biological and Microelectromechanical Systems, Cesme-Izmir, Turkey, 18–30 July, 2004*; Springer Science & Business Media: Berlin/Heidelberg, Germany, 2006; p. 509.
27. Camargo, H.V.R.; Bazzo, E. Thermal behavior of microgrooved capillary for different thermal working fluids. In Proceedings of the 11th Brazilian Congress of Thermal Sciences and Engineering, Curitiba, Brazil, 5–8 December 2006; pp. 1–12.
28. Celsia, T.M. Heat Pipe & Vapor Chamber Technology Overview. Available online: https://celsiainc.com/heat-pipe-and-vapor-chamber-technology-overview/ (accessed on 9 April 2019).
29. Boughey, B.W. *Design, Construction, and Analysis of a Flat Heat Pipe*; U.S. Naval Academy, Report; USNA: Annapolis, MD, USA, 1999; pp. 1–102.
30. Faghri, A.; Khrustalev, D. *Micro-Miniature Heat Pipe Analysis*; Departament of Mechanical and Materials Engineering, Wright State University Dayton: Dayton, OH, USA, 1994.
31. Ma, H.B.; Hanlon, M.A.; Chen, C.L. An investigation of oscillating motions in a miniature pulsating heat pipe. *Microfluid. Nanofluid.* **2006**, *2*, 171–179. [CrossRef]
32. Singh, M.; Kondaraju, S.; Bahga, S.S. Enhancement of thermal performance of micro heat pipes using wettability gradients. *Int. J. Heat Mass Transf.* **2017**, *104*, 400–408. [CrossRef]
33. Maxfield, B. *Engineering with Mathcad: Using Mathcad to Create and Organize Your Engineering Calculations*; Elsevier: Amsterdam, The Netherlands, 2006.
34. EngineeringToolbox. Thermophysical Properties Methanol. Available online: http://www.engineeringtoolbox.com/material-properties-t_24.html (accessed on 25 January 2020).
35. Acetone Solvent Properties, CAS 67-64-1. Available online: https://macro.lsu.edu/HowTo/solvents/acetone.htm (accessed on 24 February 2020).
36. Methanol Institute, Methanol Technical Data Sheet. Available online: http://www.methanol.org/Technical-Information/Resources/Technical (accessed on 30 January 2020).
37. Dortmund Data Bank, Thermal Conductivity of Methanol. Available online: http://www.ddbst.com/en/EED/PCP/TCN_C110.php (accessed on 15 February 2020).
38. Bejan, A.; Kraus, A.D. *Heat Transfer Handbook*; John Wiley & Sons Inc.: Hoboken, NJ, USA, 2003; pp. 161–571.
39. Vafai, K.; Zhu, N.; Wang, W. Analysis of Asymmetric Disk Shaped and Flat-Plate Heat Pipes. *J. Heat Transf.* **1995**, *117*, 209–218. [CrossRef]
40. Revellin, R.; Rullière, R.; Lefèvre, F.; Bonjur, J. Experimental validation of an analytical model for predicting the thermal and hydrodynamic capabilities of flat micro heat pipes. *Appl. Therm. Eng.* **2008**, *29*, 1114–1122. [CrossRef]

41. Davidovits, P.; Worsnop, D.R.; Jayne, J.T.; Kolb, C.E.; Winkler, P.; Vrtala, A. Mass accommodation coefficient of water vapor on liquid water. *Geophys. Res. Lett.* **2004**, *31*, 122111. [CrossRef]
42. Do, K.H.; Kim, S.J.; Garimella, S.V. A Mathematical Model for Analyzing the Thermal Characteristics of a Flat Micro Heat Pipe with a Grooved Wick. *Int. J. Heat Mass Transf.* **2008**, *51*, 4637–4650. [CrossRef]
43. Iverson, B.D.; Davis, W.; Garimella, S.V.; North, M.T.; Kang, S.S. Heat and Mass Transport in Heat Pipe Wick Structures. *J. Thermophys. Heat Transf.* **2007**, *21*, 392–404. [CrossRef]
44. Sprinceana, S.; Mihai, I. Analysis of the convective heat transfer at flat thermal micro-pipe through Crank-Nicolson's method. In Proceedings of the 18th International Conference, Suceava, Romania, 8–9 May 2015; pp. 244–250.
45. Liu, X.; Chen, Y. Transient thermal performance analysis of micro heat pipes. *Appl. Therm. Eng.* **2013**, *58*, 585–593. [CrossRef]
46. Singh, R.; Akbarzadeh, A.; Mochizuki, M.; Saito, Y. Thermal performance of miniature loop heat pipe operating under different heating modes. In Proceedings of the Thermal and Thermomechanical Proceedings 10th Intersociety Conference on Phenomena in Electronics Systems, San Diego, CA, USA, 30 May–2 June 2006; pp. 557–562.
47. Heat Pipe Limits, Advanced Cooling Technologies, Online Software Calculating Transport Capacity. Available online: https://www.1-act.com/resources/heat-pipe-performance/ (accessed on 30 May 2020).

micromachines

Article

Numerical Investigation of Special Heat Transfer Phenomenon in Wire-Wrapped Fuel Rod of SFR

Xuefeng Tan [1], Bing Wang [2], Yun Guo [1] and Miao Hu [3,*]

1 School of Nuclear Science and Technology, University of Science and Technology of China, Hefei 230026, China; dkdmy@mail.ustc.edu.cn (X.T.); guoyun79@ustc.edu.cn (Y.G.)
2 China Ship Development and Design Center, Wuhan 430064, China; wb14he@mail.ustc.edu.cn
3 Department of Chemical and Materials Engineering, Hefei University, Hefei 230601, China
* Correspondence: miaohu1979@163.com; Tel.: +86-551-62158395

Abstract: Sodium-cooled reactors (SFR) have always been recognized as one of the most promising candidates for the fourth-generation nuclear systems as announced by the Generation-IV International Forum. In the design of SFR, helical wire-wrapped rod is applied to stabilize the structure of the rod bundle and enhance coolant mixing. Although there has been considerable research on SFR in computational fluid dynamics (CFD), the phenomenon of heat transfer has rarely been paid attention to. This article discovered that there exists reversed heat flux from coolant to wrapped wire, which is contrary to our usual understanding. This phenomenon has not been reported in previous CFD calculations. Hence, a solid heat conduction model is proposed to prove this phenomenon and analyze the heat transfer process. The simulation results show that the wrapping wire embedding depth, the shape of the calculation domain and the physical properties of all components have great influence on the magnitude of the reversed heat flux. The present findings will have strong influence on the temperature field and maximum value of the fuel rod as well as profound reference value for future flow calculation, especially in grid generation and treatment of the junction between the winding wire and fuel rod.

Keywords: wire-wrapped fuel bundle; heat transfer; SFR; microchannel

1. Introduction

It has been widely acknowledged that fast reactors have a higher neutron utilization compared to thermal reactors. So far, historical experiences of design, construction, testing, operation, inspection and repair of past or existing demonstration and/or prototype SFRs, such as PFR (UKAEA), BN-350 (Kazakhstan), BN-600 (Russia) and Monju (Japan), have established the technological basis of the SFR [1]. Wire-wrapped pins have been presented as a solution to enhance heat transfer in the sub-channel between the fuel pins due to the small core geometry and higher heat flux compared with PWR; the general idea is that a helical wire wrapping forces the fluid to rotate around the fuel pins [2]. The flow and heat transfer are a long-standing research topic owing to the special constructure. Much related numerical and experimental research has been done to help understand the effects of wire-wrapping on the fuel pin.

Novendstern developed a semi-empirical model to predict pressure drop in a wire-wrapped fuel pin bundle, which is able to predict the value of the pressure drop within 14% over a wide range of geometries in the turbulent flow regime [3]. Gajapathy et al. analyzed the sodium flow and temperature distributions in heat generating fuel pin bundles with helical spacer wires by using a commercial CFD code, in which 7, 19 and 37 fuel pin bundles were taken into accountant. It was found that normalized outlet velocities were nearly equal to unity, and this is in great agreement with published hydraulic experimental measurements [4]. Hamman and Berry simulated three-dimensional flow distributions of sodium in the sub-channels of a 19-pin fuel bundle for large-scale problems, in which the results have a difference of 9–15% in comparison with some correlations [5]. Volkov et al. found that using high-Reynolds k-ε turbulence model to simulate the flow in the fuel pin bundle could reduce the dimension of the computational mesh. The results are in good agreement with the empirical dependences and international calculations [6]. Talebi et al. conducted a parameter research of the Reynolds number, inlet temperature and heat flux on the fluid flow, and heat transfer inside the sub-channel with a CFD code that solves the algorithm [7]. The results show that the wire-wrapped spacers created a uniform fluid velocity and temperature in the cross-section of the channel.

Tao Wu et al. conducted numerical research on microchannel heat sinks under high temperature conditions for diverse cross-section shapes, different liquid metal and various inlet velocities [8]. It was found that lithium and circle are the most appropriate choices for working fluid and microchannel cross-section shape. In addition, the inlet velocities has a significant influence on the pressure drop and heat transfer. Ying et al. analyzed different dimples geometries of heat exchanger channel numerically and experimentally to find that hemispherical dimple channels have the highest heat transfer coefficient and present better overall thermal performance [9].

Bovati et al. used Reynolds-Averaged Navier-Stokes (RANS) models to predict the axial and transverse pressure drops for a range of Reynolds numbers from 1270 to 100,000 in the numerical investigation of the 61-pin wire-wrapped bundle. They demonstrated that RANS is a suitable approach in predicting velocity and pressure fields in wire-wrapped rod bundles, with a relatively low computational effort [10].

Jeong and Song used an SST turbulence model to conduct a numerical investigation of a JAEA 7-pin fuel assembly experiment with local blockage for SFR. They compared the results with the experimental data and found small-scale vortex structures significantly enhance the convective heat transfer while large-scale vortex structures supply thermal energy near the heated cladding wall surface [11].

From all the above summaries, the conclusion can be drawn that, previous researches focused more on the fluid field because of the complex geometry, and so some details in heat transfer and temperature were ignored. At the same time, the grids or meshes near the wrapping wire are always sacrificed or compromised for the flow field calculations for CFD analysis. The most common method is to embed the winding wire into the fuel rod. Obviously, it will have a strong influence on the local heat transfer. This article discovered a phenomenon that reversed heat fluxes exist in the junction between the wrapping wire and the fuel rod which do not appear to have been reported. Hence, a solid model of heat transfer is proposed to prove this phenomenon and explore the factors which may have influence on it.

2. Reversed Heat Fluxes in Fluid Flow

As Merzari et al. [12] reported in 2012, two equation turbulence models are sufficient to predict turbulent properties governed by the tight lattice rod bundle and wire-wrapped spacer, and to estimate turbulent heat flux in liquid sodium. Employing a temperature-dependent property, the system of time-averaged governing equations for mass, momentum, and energy are given by:

$$\frac{\partial \rho U_j}{\partial x_j} = 0 \tag{1}$$

$$\frac{\partial(\rho U_i U_j)}{\partial x_j} = -\frac{\partial P}{\partial x_i} + \frac{\partial}{\partial x_j}\left(\mu_{eff}\frac{\partial U_i}{\partial x_j}\right) - \rho g_i \tag{2}$$

$$\frac{\partial(\rho U_j T)}{\partial x_j} = \frac{\partial}{\partial x_j}\left(\left(\frac{\mu}{Pr} + \frac{\mu_t}{Pr_t}\right)\frac{\partial T}{\partial x_j}\right) \tag{3}$$

where U, T and P are velocity, temperature and pressure. Pr and Pr_t are Prandtl and turbulent Prandtl number, respectively. In this article, the $k - \varepsilon$ model was selected as turbulence model, and more details about the wall treatment can be found in user guide of

Star CCM+ [13]. Kays proposed a model to evaluate the value of turbulent Prandtl number as follows:

$$\mathrm{Pr_t} = 0.85 + \frac{0.7}{\mathrm{Pe_t}} \tag{4}$$

The initial turbulent intensity I is set to 5% which is a default value of Star CCM+. Based on turbulent intensity, turbulent kinetic energy is evaluated by $\mathrm{k} = \frac{3}{2}(\mathrm{I}\overline{\mathrm{U}})^2$, in which $\overline{\mathrm{U}}$ is the averaged coolant velocity and hydraulic diameter of rod bundle, and is chosen as the characteristic length.

2.1. Geometry Model of Fluid Flow

Figure 1 shows the total geometry model including fluid domain, wire-wrapped winding and fuel bundle. Figure 2 shows the wire-wrapped fuel bundle. Figure 3 shows the planform of the flow region and fuel assembly. The geometries of rod and wrapping wire are the same as reference [14]. The geometry parameters are listed in Table 1. The wires are embedded into the claddings with 3.375 mm from the center of the rod. For the fuel rod bundle, double helical pitch length was selected in calculation. In order to decrease the influence of boundary conditions for inlet and outlet, 50 mm and 80 mm flow lengths are added at inlet and outlet, separately, as shown in Figure 1. Hence, the inlet velocity condition and pressure outlet condition can be imposed on both ends. The walls of wire, cladding and duct are defined as no slip boundary conditions. Volume heat resource (1.3974×10^9 W/m^3) is used only in the rod bundle (Table 2).

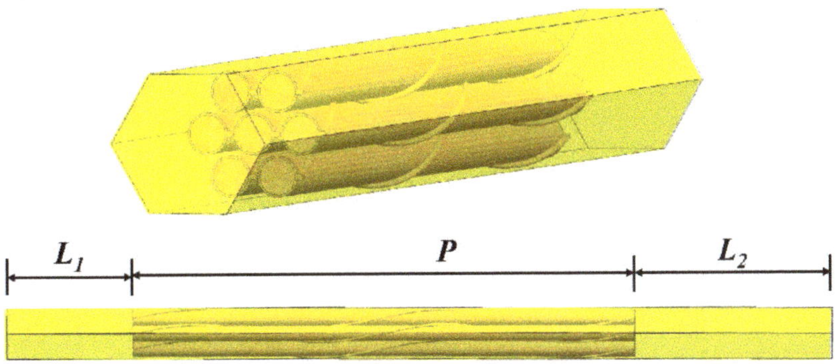

Figure 1. The calculation domain of 7-pin fuel bundle with inlet and outlet sections.

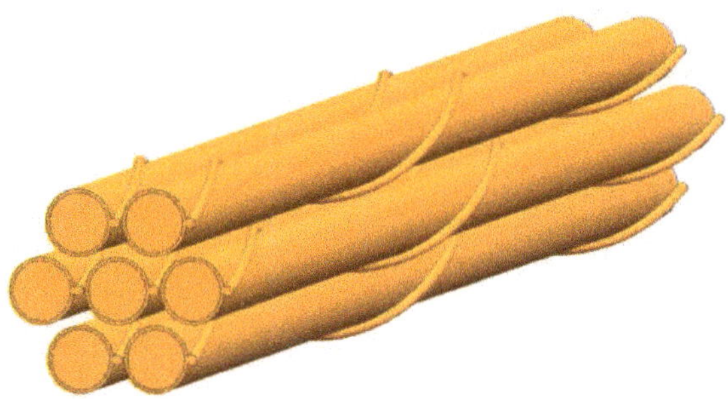

Figure 2. The wire-wrapped fuel bundle.

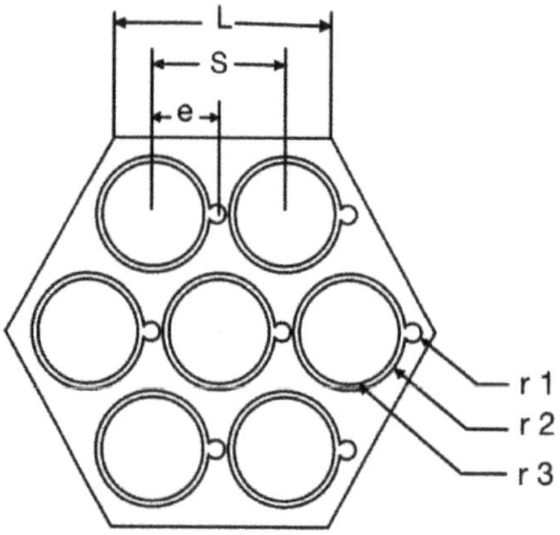

Figure 3. The planform of the flow region and fuel assembly.

Table 1. Geometric parameters.

Parameter	Value
Helical pitch	100 mm
Duct side length (L)	11.58 mm
Pin radius (r1)	2.58 mm
Cladding radius (r2)	3 mm
Wire radius (r3)	0.95 mm
Pin pitch (S)	7 mm
Distance from rod to wire (e)	3.375 mm
Pitch number	2
Number of pins	7

Table 2. Boundary conditions.

Boundary	Conditions
Inlet	Mass flow inlet
outlet	Pressure outlet
Inlet temperature	633.15 K
Duct walls	Adiabatic
All the walls	No slip
Mass flow (Q)	1.0 kg/s
Heat source	Volume source in rod bundles
Volume source	1.3974×10^9 W/m^3
Reynolds number	60,285

Fluid was regarded as a Newtonian and incompressible liquid; its physical properties change with temperature. The fitting polynomial of the properties are as follows [15]:

$$\rho = 1016.82 - 0.239T \tag{5}$$

$$C_p = 1629.1 - 0.83267T + 0.00046208T^2 \tag{6}$$

$$\lambda = 105.1843 - 0.049T \tag{7}$$

$$\mu = 0.00385 - 1.66448 \times 10^{-5}T + 3.00411 \times 10^{-8}T^2 - 2.48661 \times 10^{-11}T^3 + 7.8085 \times 10^{-15}T^4 \tag{8}$$

The applicable temperature range of the above fitting formulas was 373.15 K to 1073.15 K. In the simulation of flow, the reference temperature was selected as 645 K. The materials of cladding and fuel rod bundle were stainless steel Ti316 and UO_2, separately. Their physical properties are listed in Table 3.

Table 3. Physical properties.

	UO_2	Ti316
Density	9560 kg/m^3	7995 kg/m^3
Thermal conductivity	4.4709 W/m·K	21.8 W/m·K
Specific heat	316.6071 J/kg·K	560 J/kg·K

2.2. Meshing

The mesh was created by Star CCM+ [13]. The prism layer mesh was employed to generate the grids in the near-wall region. Meanwhile, a polyhedral mesh generator and surface reconstruction were applied to get high-quality grids [13]. Figures 4 and 5 show the mesh model and local details, Y^+ is kept around 1 at these areas and other near-wall region to meet the wall function requirements. It can be found that near the wall, not only the fluid region but also the cladding region the meshes are refined. In traditional calculation, the meshes for thermal conductivity are coarse. Here near the heat transfer wall, the mesh densities in the fluid and solid domains are comparable; this idea comes from the critical heat flux numerical simulation [16]. Due to the extreme inhomogeneity of wrapping wire and fuel rod circumferential heat transfer [14], more meshes are used to resolve these local temperature fields. The meshes applied in this article are the same as the densest mesh case in [14] by Wang et al., which has been verified for grid independence. The total mesh number is 8,357,610, with 261,048 in the fuel rod bundle, 2,637,278 in the wrapping wire and 5,459,284 in the fluid region. Using these parameters, the phenomenon worth mentioning below were discovered.

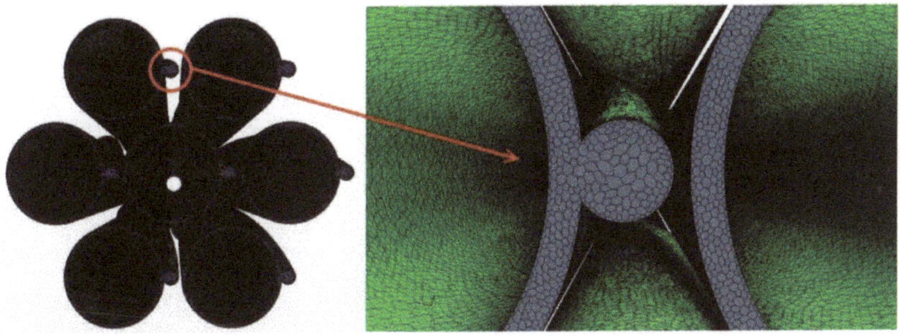

Figure 4. The meshes between wire and rod.

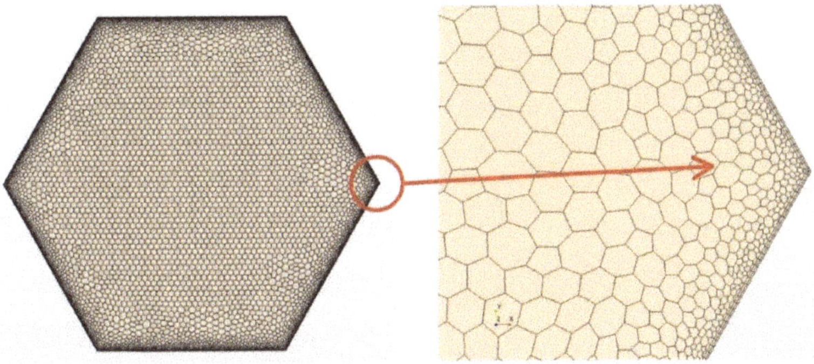

Figure 5. The meshes of inlet section and the corner.

Volume change is a very important value which can measure the quality of the mesh unit in Star CCM+. If the value is 1.0, it means that the mesh has a larger or equal volume than its neighbors, and if the value is 0.01 or less, it is considered a poor mesh quality [13]. As shown in Figure 6, 92.530% of the grids are in the interval 0.1 to 1, and 7.470% of the grids are in the interval 0.01 to 0.1. Therefore, under the evaluation criteria of Star CCM+, the mesh quality is guaranteed.

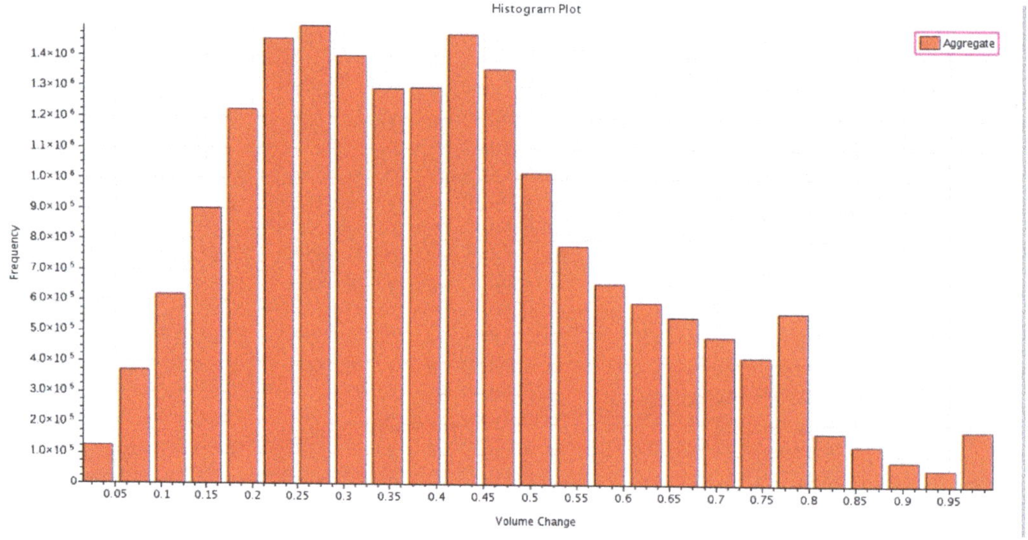

Figure 6. Volume change of mesh unit.

2.3. Reversed Heat Fluxes in Fluid Flow

In this study, the fluid case was calculated by commercial CFD codes of Star CCM+ on a 32-core computer. The convergence criterion was set to 10^{-5} and it took around 10 h to reach the criteria by 32-core parallel calculation. As per usual understanding, the heat flux will pass through the cladding and wrapping wire into fluid. However, in the result of the present steady simulation, we found that heat fluxes exist from fluid back to wire in the junction area between wire and cladding, as Figure 6 shows. This phenomenon exists in all the 7 pins for the whole wire, but the maximum value of heat flux in each wire is different, probably owing to flow rotation. For easy observation, Figure 7 only shows the phenomenon of one pin; the picture on the left shows the overall heat fluxes in the wrapped wire and the picture below in the right hand shows reversed heat fluxes. The average value of the reversed heat fluxes is about 8×10^5 W/m^2 which is pretty close to the normal fluxes in the picture in the upper right of Figure 7, and cannot be a numerical error. Figure 8 shows the temperature field of the vertical section plane (the middle part of the bundle) of flow direction. Near the junction there are some areas where the temperatures of fluid are higher than those in the wire or cladding region. This may explain how the reversed heat fluxes are generated.

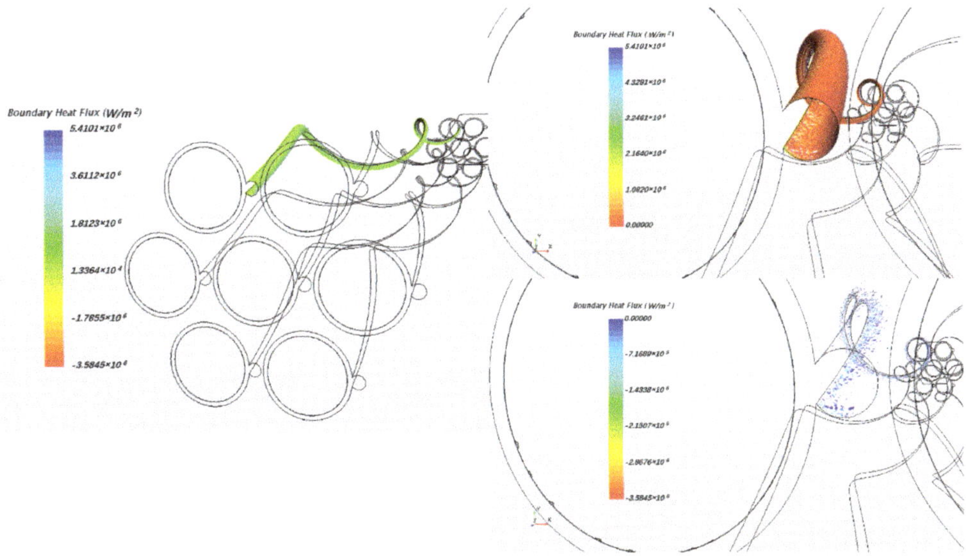

Figure 7. The heat flux on one wire surface with the positive and reversed heat fluxes.

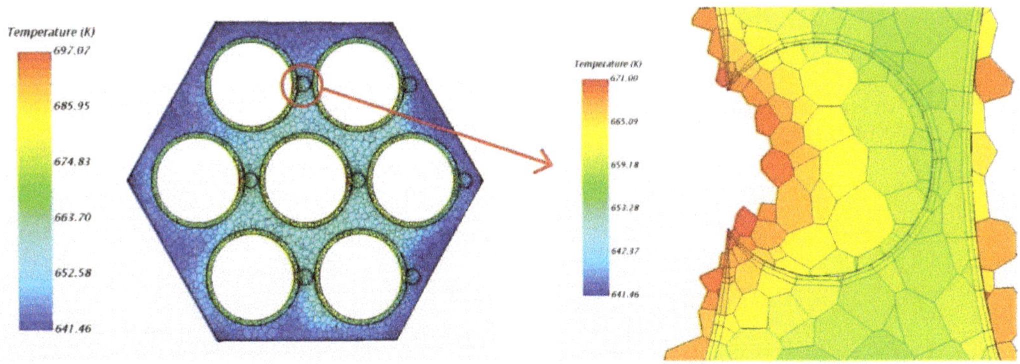

Figure 8. The temperature field of cross section and local magnification around wrapping wire.

Figure 9 shows the direction of the angle and temperature spots, which will be used in the next analysis. Circumferential temperature distributions of wire and sodium can be seen in Figure 10. As the figure shows, the maximum temperature is located in sodium compared with the wrapping wire. While the angle is around 50 degrees, the temperature in sodium is higher than that in the wire. Within a 20 degrees angle difference, the temperature difference can even reach almost 6 K, which is a very impressive value, and that is why the reversed heat fluxes can reach such a high level. Hence, the heat transfer from sodium to wrapping wire becomes a necessity. Furthermore, in flow condition, the flow near the junction is close to the wall so the velocity is very low, which can lead to insufficient heat convection. Heat conduction probably plays an important role here.

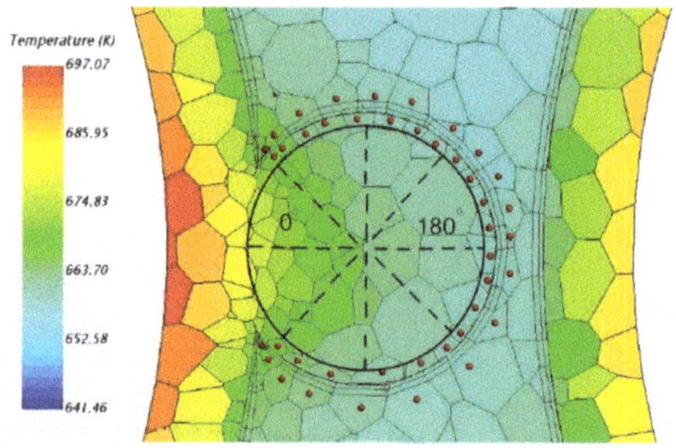

Figure 9. The direction of the angle and temperature spots.

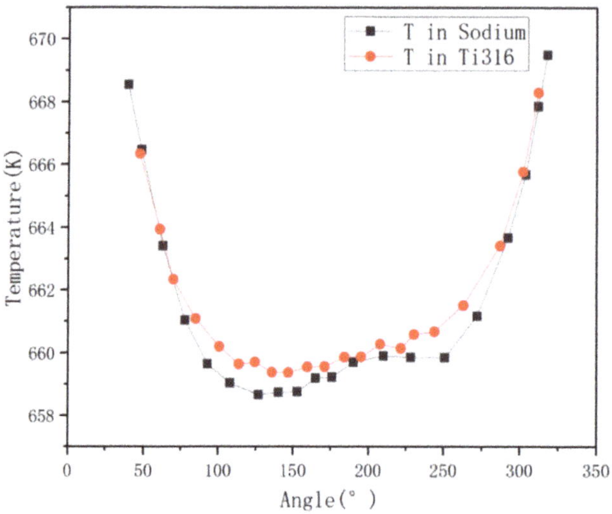

Figure 10. Temperature curves of sodium and wrapping wire around the wrapping wire.

3. Solid Model

There are many factors that will have influence on heat exchange in the simulation of fluid and as the above shows, the phenomenon is not very obvious in fluid calculation. The velocity of flow in the junction is much smaller than in the main flow so the heat conduction plays a major part of heat transfer in the junction. After considering these, a solid heat conduction model is proposed, in which only the fluid region is turned into a solid one, to prove the existence of this phenomenon. In the solid model, the properties of sodium are kept the same as fluid. There are four potential paths that the heat flux may pass through the wire and cladding to the sodium as Figure 11 shows. Figure 12 shows the thermal resistance of the four paths. If we fix attention on the junction, it can be found that there might be two directions for the flux to pass as path 2 and path 3 show. It is pretty hard to tell in which path the thermal resistance is smaller, but we believe that the heat flux through path 3 produces this special phenomenon.

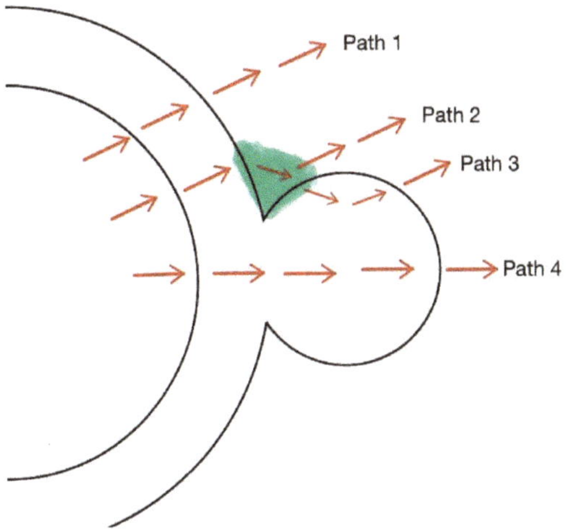

Figure 11. Potential paths of heat fluxes.

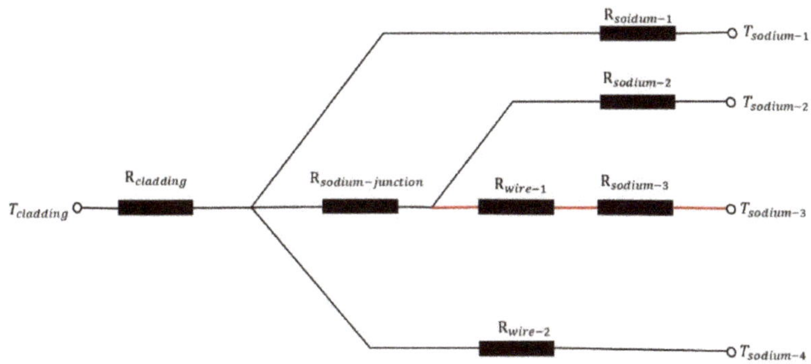

Figure 12. Thermal resistance for each path.

Furthermore, since the thermal resistance from flow to wrapping wire in fluid is obviously smaller than that in solid, the tendency of reversed heat fluxes in fluid is stronger than it in the solid. Therefore, if reversed heat fluxes still exist in the solid model, they must also exist in the fluid model.

3.1. Geometry Model of the Solid

In order to reduce the amount of calculation, a one-pin model was selected as simplification. Geometric dimensions of rod, winding and wire are the same as fluid model shown in Table 1. The external side length ("a") is 5 mm. Figure 13 shows the shape and materials of solid model and Table 4 shows the boundary conditions. The physical properties of rod, wire cladding and external solid region(sodium) are listed in Table 5. Thus, we can now proceed to use the solid model for investigation of some interesting factors related to the phenomenon described.

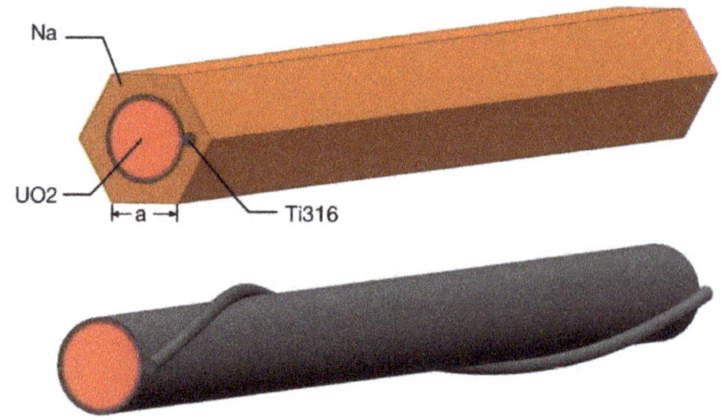

Figure 13. Shape and material of the solid model.

Table 4. Boundary conditions.

Boundary	Conditions
Duct walls	Temperature
Wall temperature	633.15 K
Heat source	Volume source in rod
Volume source	1.3974×10^9 W/m^3

Table 5. Physical properties.

	Na	UO_2	Ti316
Density	862.665 kg/m^3	9560 kg/m^3	7995 kg/m^3
Thermal conductivity	73.5793 W/m·K	4.4709 W/m·K	21.8 W/m·K
Specific heat	1284.2647 J/kg·K	316.6071 J/kg·K	560 J/kg·K

3.2. Meshing

Surface reconstruction and repair technology and polyhedral mesh generator of star CCM+ are applied to generate high-quality grids. Curve control near the junction is used to generate denser grids so that the shape of the junction is closer to reality and the accuracy of simulation can be higher. Furthermore, it can get more accurate values such as temperature, fluxes, and so on. Figure 14 shows the grids and details of typical mesh near the helical wire spacer. Table 6 shows the detailed information about the mesh.

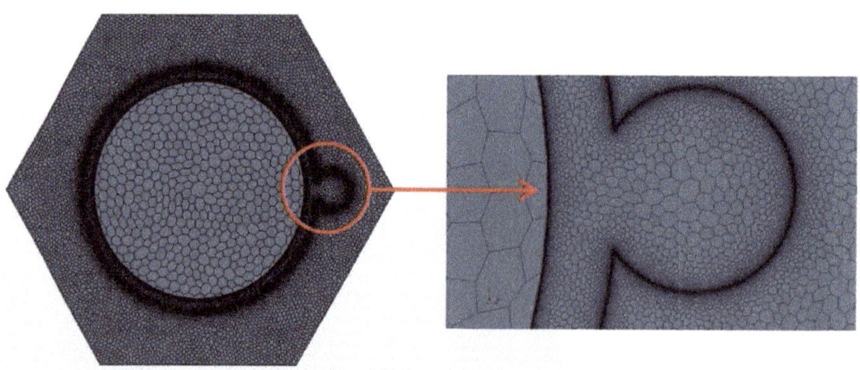

Figure 14. The mesh of solid model and local detail view profile.

Table 6. Details of solid model grids.

Region	Base Size	Number of Grids
Na	0.1 mm	4,614,753
Wires and cladding	0.1 mm	10,115,338
Rod	0.26 mm	106,596
Total		14,836,687

3.3. *Sensitivity Analysis of Mesh*

Mesh resolution can have a great influence on the simulation accuracy. Thus, to evaluate the dependency of simulation accuracy on the mesh resolution, three different mesh types were considered in the validation calculation. Base size is an important value which can decide the scale and number of the grids generated by Star CCM+. With a larger base size, less mesh is expected. In the three cases, mesh base sizes in region of Na and wire cladding are respectively set to 0.1 mm, 0.12 mm and 0.16 mm, while base size of rod is still 0.26 mm. Mesh numbers for each case are shown in Table 7. For the solid cases, the convergence criterion is that the scaled residual falls below 10^{-13} and it took about 40 min to get the results while using 28 threads for parallel computing. The time of calculation was only 1/15 that of the fluid calculation. Hence, the solid model can greatly decrease the costs and save time.

Table 7. Mesh number.

	Base Size	Rod	Wire Cladding	Na	Total
Case 1	0.1 mm	106,596	10,115,338	4,614,753	14,836,687
Case 2	0.12 mm	106,596	7,126,721	3,560,360	10,793,677
Case 3	0.16 mm	106,596	4,086,034	2,038,983	6,231,613

To prove the calculation results are independent of mesh resolution, 15 points with radially equidistant interval of 2 mm were selected in the section perpendicular to the pitch direction, at the middle length of fuel rod. Figure 15 shows temperatures at those points. From the three temperature lines in Figure 15, it is easy to see that the temperature distributions of case 1 and case 2 are close. Therefore, base size in case 1 was selected in subsequent calculations.

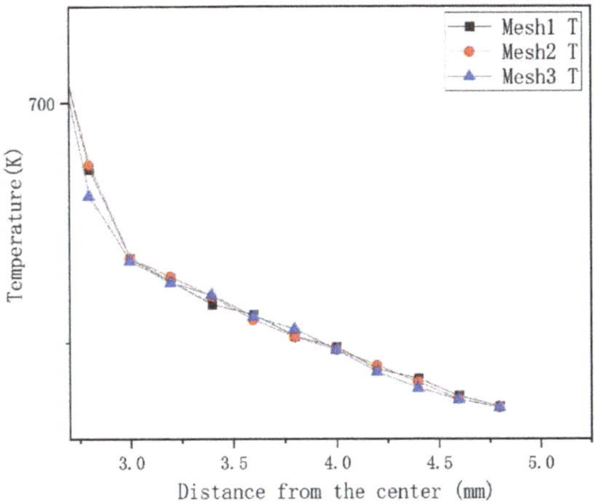

Figure 15. Radial temperature profiles for different mesh.

3.4. Simulation Results of Solid Model

Figure 16 shows the phenomenon of reversed heat flux for solid model. Here positive heat flux values means that the direction of the flux is from solid sodium to wire and so the second picture in Figure 16 indicates the location and magnitude of the reversed heat flux. The maximum value exceeds 1.8 MW/m^3 which is about 40% of the normal fluxes, very notable. Obviously, it is more pronounced than in the flow condition. The local temperature field is shown in Figure 17.

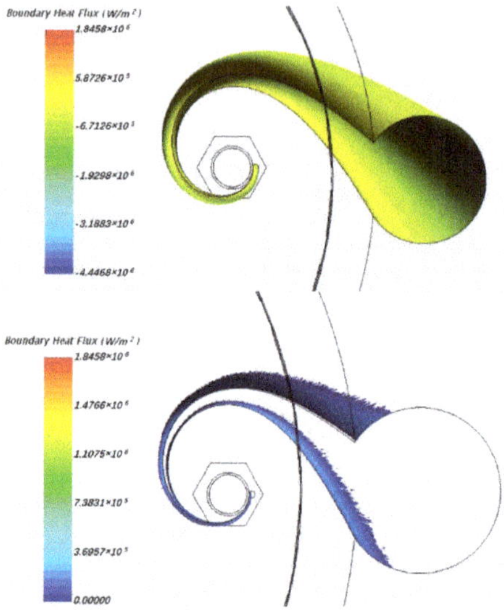

Figure 16. The schematic of heat flux along the wrapping wire surface under solid model.

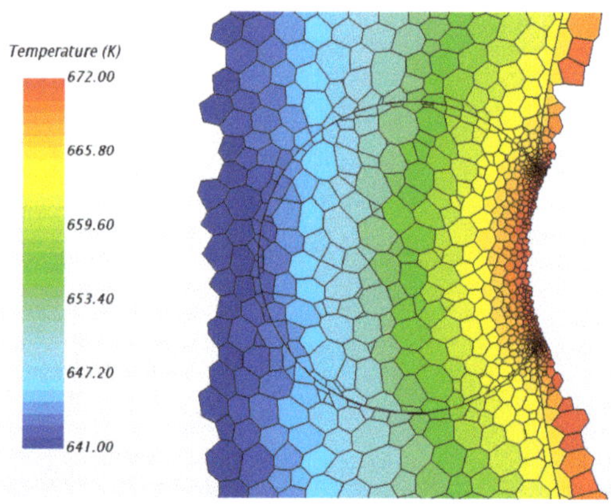

Figure 17. Local temperature field of wrapping wire.

In Figure 17, it can be found that, near the junction, there is a decrease of temperature from sodium to wrapping wire. Although the temperature difference is not obvious, the fluxes still cannot be ignored. It is about the order of 2×10^5 in main area of reversed fluxes while the normal fluxes in wrapping wire is about 4×10^5 W/m^2. Meanwhile, the structure of the wrapping wire also has a significant effect on the heat flux distribution of the surface of cladding. Figure 18 shows that the fluxes near the contact area between wrapping wire and cladding is obviously higher than other places, meaning that it is unreasonable to set boundary conditions for surface heat fluxes and to ignore heat transfer in fuel rod and wire cladding as a simplification, which are often applied in previous numerical simulations of fluid. Because the thermal resistance passing through the wrapping wire is large, the heat passing through the wrapping wire originally will go through the rod surface near the junction where the heat resistance is smaller, and so the heat flux on the rod surface next to the wrapping wire increases. In fluid calculation, similar results can also be found. In addition to the similar cause of heat flux redistribution caused by thermal resistance, the following situation may also be responsible. Due to the presence of wrapping wire, although the flow rate will be reduced due to narrow size, the secondary flow will impact this position directly, so the heat transfer will be increased [14]. Thus, more heat is concentrated in this narrow space, causing the sodium temperature to rise even higher than that of the wire. Thus, the phenomenon of reverse heat transfer occurs. The influence of some obvious factors on this phenomenon will be discussed in detail, hoping to provide guidance for CFD calculation.

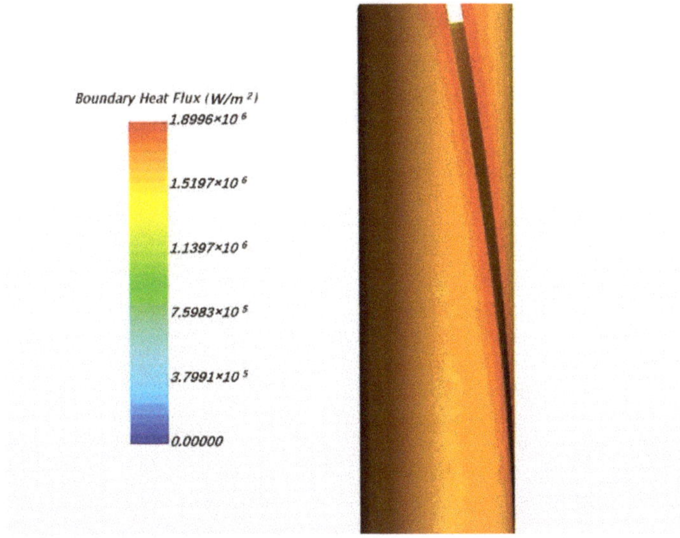

Figure 18. The surface heat flux of the rod.

4. Factors Influencing the Phenomenon

4.1. Embedding Depth of Wrapping Wire

The contact between wrapping wire and cladding has a significant effect on the calculation results because the geometry at the junction will change with it. The embedding depth is often described by the distance from wire to rod ("e") as defined in Table 1. Three different embedding depths are shown in Table 8. Besides the embedded contacts, there is also another contact method in which the wrapped wire is separated from the cladding with a short distance 0.005 mm, as in case 4.

Table 8. Embedding depths.

No.	Distance from Wrapping Wire to Rod "e"
Case 1	3.375 mm
Case 2	3.425 mm
Case 3	3.475 mm
Case 4	3.480 mm

From case 1 to case 3, the embedding depth gradually decreases which means the distance "e" increases and the sharp corner at the junction gets more distinct. In this way the situation is approaching engineering practice. Figure 19 shows the details of four embedding depths which decrease from case 1 to case 4.

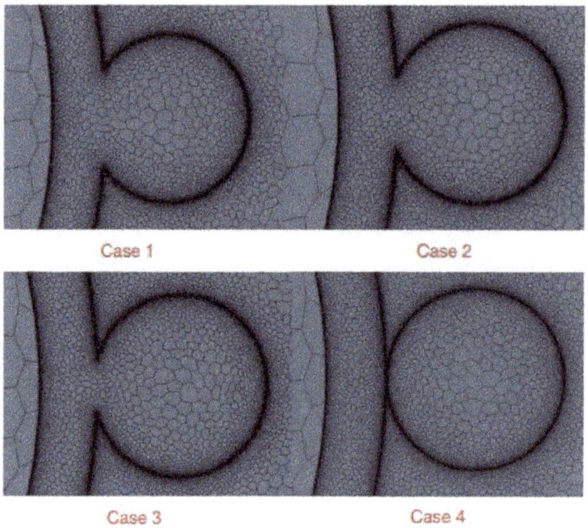

Figure 19. Four embedding depths.

As shown in Figure 20, with the embedding depth decreasing, the magnitude of reversed heat fluxes increases. The reversed heat flux region in case 3 and case 4 are almost twice as large as it in case 1. Furthermore, areas where the phenomenon exists are also gradually expanding; in case 4, the area of reversed heat fluxes is about half of the area of wrapping wire. Therefore, the embedding depth can have a great influence on this phenomenon because it can change the thermal resistance at the junction while geometry varies. Hence, in flow analysis, the commonly used wrapping wire embedding method may weaken the inverse heat transfer phenomenon. This may be why the phenomenon has not been found and studied before.

4.2. Calculation Domain

Since the solid model is a simplification of the flow model, the size of calculation domain may have an effect on the results. Therefore, three cases in which the value of side length "a" in Figure 13 varies were calculated to find out those influences. The side lengths for the three cases were 5 mm, 6 mm and 7 mm, respectively.

Considering that for smaller embedding depth, the phenomenon is more obvious and easily observed, the embedding depth "e" is set to 3.45 mm, and in Figure 21 the three figures from left to right are case 1, case 2 and case 3. As for the maximum value of reversed heat fluxes in the three cases, in case 1, it is 8.0×10^5 W/m^2 and in case 2 it is 7.5×10^5 W/m^2, while in case 3 it is 5.2×10^5 W/m^2. The maximum value is near the junction in all cases, and the phenomenon is obviously stronger in case 1 than in the other two cases. Besides the maximum value, the size of the area for reversed heat fluxes also changes with the side length; the area ratio for these three cases is around 11:9:8. Therefore, the conclusion is that the reversed fluxes will get weaker if the areas where the fluxes exist get smaller and the calculation domain increases.

4.3. Physical Properties

In SFR the coolant is sodium and material of cladding and wrapping wire is Ti316 alloy. Because thermal conductivities vary in different materials, which directly affects the thermal resistance, physical properties have a significant impact on such phenomenon. To figure out how the physical properties affect the phenomenon, four cases were calculated for comparison and are shown in Table 9. As an original comparison, the embedding depth is set to 3.45 mm and the side length "a" is 5 mm in case 0. All the following cases are derived from case 0.

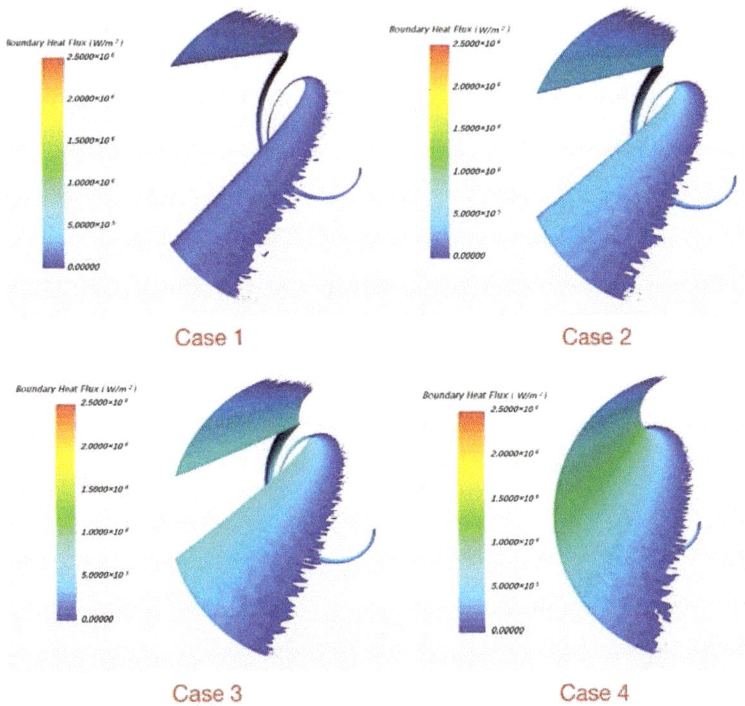

Figure 20. The reversed heat flux varies with embedding depth.

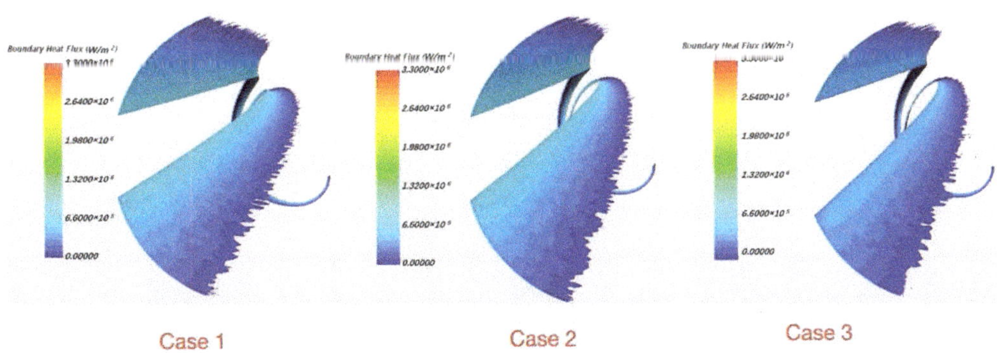

Figure 21. The reversed heat flux varies with calculation domain.

In Case 1, conductivity of sodium doubles as before which means its thermal resistance becomes half of the previous value. Thus, heat transfer is better in the coolant, and heat fluxes only transferred by coolant are more than going back through the wrapping wire. In this way, the phenomenon of reversed fluxes become weaker. Under the same logic, the phenomenon gets stronger with the decrease of thermal conductivity of sodium as seen in Figure 22a. In the same way, increasing the thermal conductivity of the material of the wrapping wire can strengthen the phenomenon as Figure 22b shows. The result of the case for original comparison is case 0 in Figure 22a,b. Hence, under flow condition, due to convection it looks like the thermal conductivity of sodium is increased. Then the reversed heat flux phenomenon is reduced. However, because of the narrow junction, the convection heat transfer is very weak; thus, heat conduction is actually the most important factor. Moreover, we boldly increased the thermal conductivity to find out that when the thermal conductivity is eight times its actual value, the phenomenon will disappear, as shown in Figure 23. It is believed that with such a large thermal conductivity, almost all the heat fluxes are passed by path 1 in Figure 11. Under such condition, there is almost no heat flux transferred by the wrapping wire. However, in engineering practice, a coolant with such a large thermal conductivity does not exist. Thus, this phenomenon has a large possibility to exist in practice.

Table 9. Thermal conductivities of sodium and Ti316 for five cases.

	Na	Ti316
Case 0	λ_a	λ_b
Case 1	$2\lambda_a$	λ_b
Case 2	$0.5\lambda_a$	λ_b
Case 3	λ_a	$2\lambda_b$
Case 4	λ_a	$5\lambda_b$

Note: λ_a and λ_b are conductivities of sodium and Ti316 alloy.

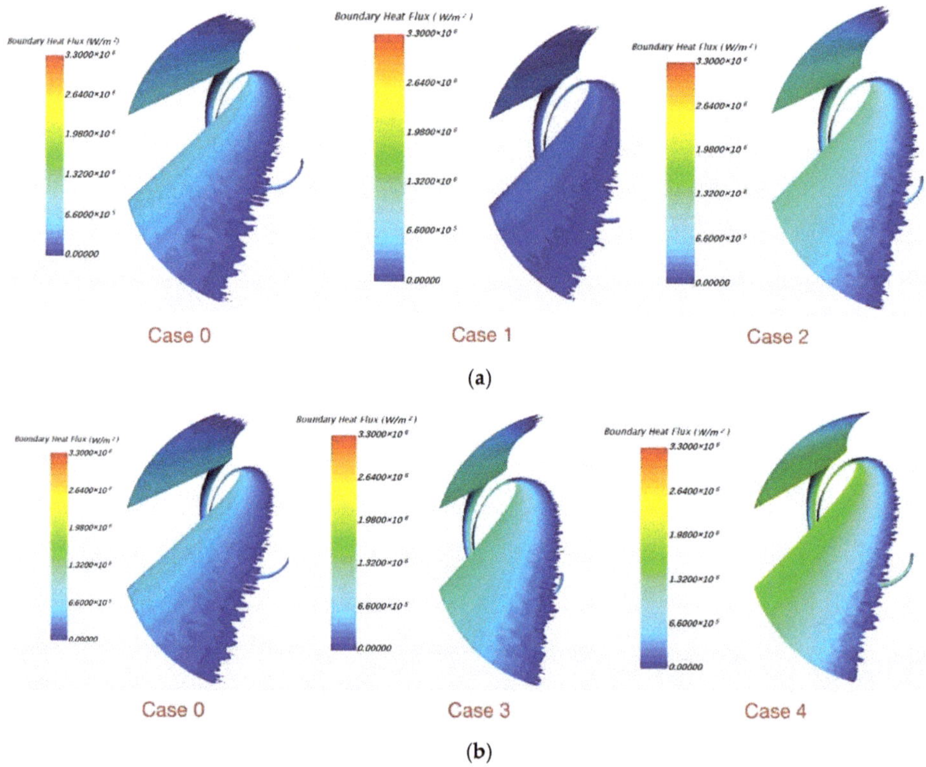

Figure 22. Physical property variation: (**a**) Thermal conductivity changes of sodium; (**b**) thermal conductivity changes of Ti316 alloy.

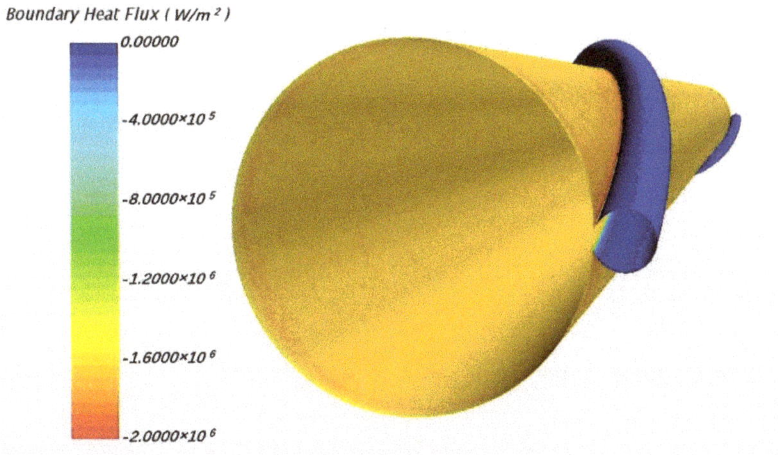

Figure 23. Heat fluxes at eight times thermal conductivity of sodium.

5. Conclusions

As described in this article, a reversed heat flux phenomenon near the junction between wrapping wire and cladding was discovered by accident in the simulation of a sodium-cooled 7-pin wire-wrapped rod bundle model. The same phenomenon in an imaginary solid thermal conductivity model confirms that it does exist. Some factors have been considered and researched. The main conclusions are listed as follows:

With an increase in embedding depth, the phenomenon of reversed heat flux is enhanced and the average value of the reversed heat flux gets larger. The traditional embedding method in CFD may not get this phenomenon because it may obtain an incorrect local temperature field.

Thermal conductivities of coolant and wire cladding both have great influence on the reversed heat flux by the thermal resistance on the path of reversed flux. Increasing the thermal conductivity of the wrapping wire or decreasing the thermal conductivity of the coolant can significantly enhance this phenomenon. When the thermal conductivity of sodium is eight times its actual value, the phenomenon will disappear.

Finally, in previous simulation research of flow, regarding the walls of wrapping wire as adiabatic and choosing homogeneous surface heat fluxes boundary condition are often adopted to reduce computation quantity. However, according to present results, it is obvious that such simplifications are unreasonable to some extent. Moreover, though high temperature near the junction between the wrapping wire and cladding has already been noticed, it is often explained by low coolant velocity near the wall, but the results in this article may provide another explanation. Furthermore, the work in this article can provide reference values for future flow calculation, especially in grid generation and treatment of the junction between the wrapping wire and the fuel rod.

Author Contributions: Conceptualization, M.H. and Y.G.; methodology, X.T. and Y.G.; software, X.T. and B.W.; validation, formal analysis, investigation, resources, data curation, X.T.; writing—original draft preparation, X.T.; writing—review and editing, Y.G.; visualization, B.W.; supervision, project administration, Y.G. All authors have read and agreed to the published version of the manuscript.

Funding: This research received no external funding.

Data Availability Statement: The data presented in this study are available on request from the correspond author. The data are not publicly available due to privacy.

Conflicts of Interest: The authors declare no conflict of interest.

Nomenclature

Q_v	Calorific value per unit volume (W/m^3)
C_p	Specific heat capacity (J/kg/K)
Λ	Thermal conductivity of fuel rod (W/m/K)
T	Temperature (K)
P	Density of fuel rod (kg/m^3)
T	Time (s)
R	Radius of any point in fuel rod (mm)
R_f	Radius of fuel rod (mm)
T_s	Temperature at the edge of fuel rod (K)
M	Molecular viscosity (kg/m/s)
U_i, U_j	Cartesian velocity components (m/s)
x_i, x_j	Cartesian coordinates (m)
Pr	Prandtl number
Pr_t	Turbulent Prandtl number
Pe_t	Turbulent Peclet number
P	Pressure (Pa)

References

1. Kazumi, A.; Philippe, D.; Yang, H.; Jean Paul, G.; Yeong-il, K.; Yury, A.; Robert, H.; Nariak, U. A summary of sodium-cooled fast reactor development. *Prog. Nucl. Energy* **2014**, *77*, 247–265.
2. Park, S.D.; Seo, H.; Jeong, Y.S.; Jerng, D.W.; Bang, I.C. Thermal-hydraulic analysis of a 7-pin sodium fast reactor fuel bundle with a new pattern of helical wire wrap spacer. *Ann. Nucl. Energy* **2018**, *114*, 198–205. [CrossRef]
3. Novendstern, E.H. Turbulent flow pressure drop model for fuel rod assemblies utilizing a helical wire-wrap spacer system. *Nucl. Eng. Des.* **1972**, *22*, 28–42. [CrossRef]
4. Gajapathy, R.; Velusamy, K.; Selvaraj, P.; Chellapandi, P.; Chetal, S.C. CFD investigation of helical wire-wrapped 7-pin fuel bundle and the challenges in modeling full scale 217 pin bundle. *Nucl. Eng. Des.* **2007**, *237*, 2332–2342. [CrossRef]
5. Hamman, K.D.; Berry, R.A. A CFD simulation process for fast reactor fuel assemblies. *Nucl. Eng. Des.* **2010**, *240*, 2304–2312. [CrossRef]
6. Volkov, V.Y.; Belova, O.V.; Krutikov, A.A.; Skibin, A.P. A coolant flow simulation in fast reactor wire-wrapped assembly. *Therm. Eng. (Engl. Transl. Teploenerg.)* **2013**, *60*, 429–433. [CrossRef]
7. Talebi, M.; Abdolahi, R.; Nadooshan, A.A. Investigation of effects of helical wire-wrapped spacers on thermo hydraulic properties of a fluid flow along to heated fuel bundle. *Int. J. Appl. Innov. Eng. Manag.* **2013**, *2*, 388–398.
8. Wu, T.; Wang, L.; Tang, Y.; Yin, C.; Li, X. Flow and Heat Transfer Performances of Liquid Metal Based Microchannel Heat Sinks under High Temperature Conditions. *Micromachines* **2022**, *13*, 95. [CrossRef] [PubMed]
9. Ying, P.; He, Y.; Tang, H.; Ren, Y. Numerical and Experimental Investigation of Flow and Heat Transfer in Heat Exchanger Channels with Different Dimples Geometries. *Machines* **2021**, *9*, 72. [CrossRef]
10. Bovati, O.; Yildiz, M.A.; Hassan, Y.; Vaghetto, R. RANS simulations for transition and turbulent flow regimes in wire-wrapped rod bundles. *Int. J. Heat Fluid Flow* **2021**, *90*, 108838. [CrossRef]
11. Jeong, J.-H.; Song, M.-S. CFD investigation of a JAEA 7-pin fuel assembly experiment with local blockage for SFR. *Nucl. Eng. Technol.* **2021**, *53*, 3207–3216. [CrossRef]
12. Merzari, E.; Pointer, W.D.; Smith, J.G.; Tentner, A.; Fischer, P. Numerical simulation of the flow in wire-wrapped pin bundles: Effect of pin-wire contact modeling. *Nucl. Eng. Des.* **2012**, *253*, 374–386. [CrossRef]
13. STAR-CCM + User Guide. Version 13.06.011. 2018. Available online: https://docs.sw.siemens.com/documentation/external/PL20200805113346338/en-US/userManual/userguide/html/index.html?param=WXD84authLoc=https://thesteveportal.plm.automation.siemens.com/AuthoriseRedirect (accessed on 17 May 2022).
14. Wang, D.L.; Peng, C.H.; Guo, Y. Thermal-hydraulic analysis of a 7-pin sodium-cooled fast reactor wire-wrapped fuel bundle. *Int. J. Heat Mass Transf.* **2020**, *160*, 120178. [CrossRef]
15. Chen, J.; Zhang, D.; Song, P.; Wang, X.; Wang, S.; Liang, Y.; Qiu, S.; Zhang, Y.; Wang, M.; Su, G.H. CFD investigation on thermal-hydraulic behaviors of a wire-wrapped fuel subassembly for sodium-cooled fast reactor. *Ann. Nucl. Energy* **2018**, *113*, 256–269. [CrossRef]
16. Dan, L. *Evaluation and Optimization of CHF Calculation Model for Curved Surface*; University of Science and Technology of China: Hefei, China, 2019.

micromachines

MDPI

Article

Particularities of R134a Refrigerant Temperature Variations in a Transient Convective Regime during Vaporization in Rectangular Microchannels

Ioan Mihai [1,*], Cornel Suciu [1,2,*] and Claudiu Marian Picus [1]

[1] Faculty of Mechanical Engineering, Automotive and Robotics, Stefan cel Mare University, 720229 Suceava, Romania; claudiu.picus@usm.ro

[2] Faculty of Electrical Engineering and Computer Science, Stefan cel Mare University, 720229 Suceava, Romania

* Correspondence: mihai.i@usm.ro (I.M.); suciu@usm.ro or camil.suciu@student.usv.ro (C.S.); Tel.: +40-735173288 (I.M.); +40-742007820 (C.S.)

Abstract: An analysis of the R134a (tetrafluoroetane) coolant's non-stationary behavior in rectangular microchannels was conducted with the help of a newly proposed miniature refrigerating machine of our own design and construction. The experimental device incorporated, on the same plate, a condenser, a lamination tube and a vaporizer, all of which integrated rectangular microchannels. The size of the rectangular microchannels was determined by laser profilometry. R-134a coolant vapors were pressurized using a small ASPEN rotary compressor. Using the variable soft spheres (VSS) model, the mean free path, Knudsen and Reynolds numbers, as well as the dimensionless velocity profile can be assessed analytically. In order to determine the average dimensionless temperature drop in the vaporizer's rectangular microchannels, in non-stationary regime, an analytical solution for incompressible flow with slip at the walls, fully developed flow and laminar regime was used, by aid of an integral transform approach. In the experimental study, the transitional distribution of temperature was tracked while modifying the R134a flow through the rectangular microchannels. Coolant flow was then maintained at a constant, while the amount of heat absorbed by the vaporizer was varied using multiple electric resistors. A comparative analysis of the analytical and experimental values was conducted.

Keywords: microchannels; R134a refrigerant; vaporization; heat exchanger

1. Introduction

The increasingly advanced miniaturization of electronic components and their exposure to increasingly higher voltages has resulted in the development of high-performance cooling solutions. In practice, the aim is to decrease the size of these cooling systems, whilst at the same time try to conserve their efficiency. Although not limited to these fields, the following applications can be mentioned: microelectronics (first introduced by Tuckerman and Pease) [1–5]; micro-electro-mechanical systems (MEMS) [6–8]; micro-heat pipe spreaders [4]; micro heat exchangers, condensers, evaporators and boilers [4,9]; miniature refrigerators [10]; biomedical devices [8,11] cited by [12,13]; fuel cell technology [14–16], etc.

The above referenced equipment contains mini, micro or nano channels that can configure mini or micro heat exchangers. The microchannels can be circular, rectangular or have a different shape [3,13,17].

The need to use flow sections that are increasingly smaller for the heat exchange convective effect is based on the assertions of Kandlikar, Aubert et al. [3,8,13]. Since 1997, Phillips, R.J. et al. [18] indicated that for surfaces reaching a maximum of 120 °C, microchannel heat sinks can dissipate heat loads as high as 1000 W/cm^2. Such values must be attained in high performance cooling systems.

The present paper aims to analyze how the R134a coolant reacts in the rectangular microchannels of the vaporizer after an isenthalpic Joule-Thomson lamination process. Research in the field indicates that applying classic calculus methods to determine heat exchange in micro and nano channels is no longer a viable option. This can be explained by the fact that the size of the microchannel(s) is so small that it only gets close in scale to a couple of molecule layers. Hence, with regards to such extremely small scales, in 2003 Kandlikar and Grande [3] asserted that a rarefaction effect of common gases at atmospheric pressure is generated. Experimental research conducted so far shows that the heat transfer coefficients are either larger [19] or smaller [20] than the predicted values. It was also determined that deviations from the predicted values are registered depending on the flow conditions. In order for water to flow through rectangular microchannels, Peng et al. [21], demonstrated that the Reynolds number can have values ranging between 200–700 whilst Xu et al. [22] aimed for channels with similar dimensions, with Reynolds number values ranging between 20–4000. Liu, Harms, Popescu et al. [23–25] showed that the Reynolds number values vary significantly depending on the microchannel dimensions, near limit conditions, type of coolant used and the adopted mathematical model. These discrepancies can have direct effects on the design of microchannels in cooling systems.

The aim of this study is to examine if the values predicted through mathematical modelling for rectangular microchannels are close to the experimental values when subjected to a non-stationary heat exchange. Temperature distribution was analyzed when the R134a coolant flowed through rectangular microchannels at different flow rates, and also when the vaporizer was heated. In order to determine how the R134a coolant performs in rectangular microchannels, an experimental setup was designed and constructed as further shown.

2. Experimental Setup

Figure 1 illustrates an image of the experimental setup that allows monitoring of the temperature distribution of the R134a coolant flow through the components of the installation.

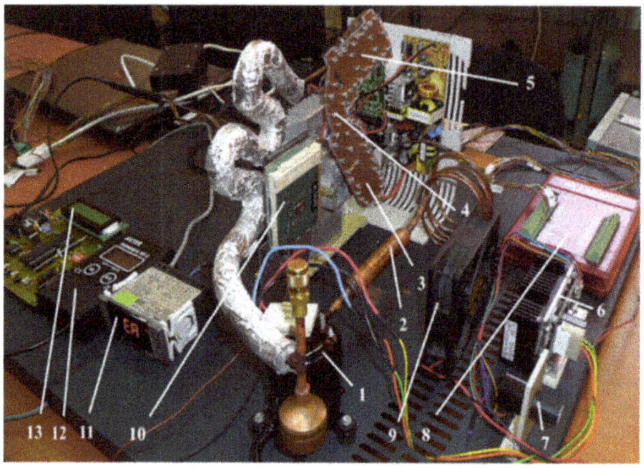

Figure 1. Image of the experimental setup: 1—ASPEN rotary compressor; 2—filter; 3—condenser; 4—lamination tube; 5—evaporator; 6—driver; 7—rotary switch used to adjust the speed; 8—LabJack U12 interface; 9—cooler; 10—CPU with a heat source located on the back side, not visible in the image; 11, 12, 13—thermometers.

The test bench consists of an ASPEN Miniature Rotary Compressor 24 V, referenced as "1", model 14-24-1101, preloaded with 21 cm^3 of RL68 POE oil. The three-phase, direct current electric motor (up to 9.5 A) powered at 24 V, is equipped with a driver that ensures operation within 1800–7000 rev/min. When using the R134a coolant, at maximum revolutions, a heat dissipation power of 360 W is generated. From a design point of view, depending on the power supply parameters provided by the driver, the speed of the compressor can be varied accordingly. The mini cooling system's compressor can provide different pressure variations whose values can be found in the technical documentation [26,27]. From compressor 1, the coolant flows through filter 2, and then through condenser 3 which is incorporated in the cooling system. Next, from the condenser, the R134a coolant flows through the lamination micro tube 4 and then through the vaporizer 5 which is incorporated in a cooling-lamination-heating plate. The elements positioned on the cooling plate have rectangular microchannels. The speed of compressor 1 can be varied through driver 6, using the rotary switch 7.

The LabJack interface, indicated by 8, allows determination of the temperature at various predefined measuring points in the system: at the compressor's intake and exit points, as well as at the condenser's and vaporizer's entrance, middle and exit points. On the stand there is also a cooler 9 that serves the purpose of cooling the condenser. The vaporizer of the installation is in direct contact (through a thermal paste) with the CPU heat generating element, indicated by 10. The CPU was taken from a computer mainboard and was fitted, on its back side, with three DC powered electronic resistors, capable of reaching up to 110 °C. By raising the DC voltage of the power supply, any temperature ranging between the ambient value and maximum value can be obtained. It is required to know in advance the temperature to voltage ratio. Three other thermometers are also installed on the test bench. The one labeled 11 has the purpose of monitoring the temperature of the compressor and also of triggering a halt command for the compressor in case its temperature exceeds the maximum allowed temperature of 135 °C, per its manufacturing specifications. The thermometer labeled 12 is of a high precision type, serving the purpose of calibrating and measuring the local temperature at any point in the system, using a probe. Lastly, the thermometer labeled 13 allows for measurements of the ambient temperature. The installation is connected through a valve to a coolant tank. This allows monitoring of the volumes of coolant that are used by the mini-installation and to supplement the coolant when it registers losses.

If the drive motor speed is varied, different pressure values for the coolant can be obtained as it flows through the rectangular microchannels, which, in turn, determine flow-speed variations in the mini-refrigerating unit and, consequently, flow rate variations.

3. Characteristics of Rectangular Microchannels

The cooling-lamination-heating plate was crafted out of copper with a 2.5 mm thickness and was designed for cooling the CPU (central processing unit). A technical problem that arose during the design stage was the high thermal conduction coefficient of this metal. Even if the channels are placed on the same board (which sustains an isobaric-isothermal condensation process, a lamination process Joule-Thomson, followed by an isobar-isothermal vaporization process), the conduction effect was avoided by the addition of channels and by creating thermal bridges on the back of the plates. A detailed image of the plate can be seen in Figure 2.

Tube 1 from Figure 2, connected with vaporizer E, removes the coolant that was vaporized by the heat generated from the CPU located in central area 2. In this area the vaporizer tends to balance the heat flow emanated by the CPU with the cold flow coming from the rectangular microchannels. The lowest temperature experimentally obtained at the vaporizer level was −22.3 °C. The coolant flows in the vaporizer through rectangular microchannels 3. Shaft 4 was designed in order to avoid conductive heat transfer between the vaporizer and the condenser. Due to the scale of the rectangular microchannels, and in order to avoid micro impurities or accidental vapor bubbles, a micro filter, 5, was mounted between the lamination valve 6 and microchannels 3. The coolant that results from the compressor's output passes through tube 7 to micro filters 8 and 9 (intake, output) that are a part of condenser C. The coolant is directed to area 12 through the rectangular microchannels 10, while over area 12 a cold air stream passes as it comes from the mini-installation cooler. Because, at the delivery end of the condenser it is imperative that the coolant is fully converted to liquid, and also because to the low coolant flow through the installation, only a maximum quantity of 17 g of coolant was inserted, three cylindrical tanks of different volumes were fitted, indicated by 11. During the experiments, the coolant was observed in these tanks in its liquid stage. Throughout the heat exchange plate-outline, 13 threaded holes were drilled that allowed mounting of an acrylic transparent plate PMMA XT with a 2.5 mm thickness. This plate has a softening temperature of 115 °C, a specific heat capacity of 1.47 J/kg, a linear coefficient of expansion of 7×10^5 K^{-1}, heat conductivity of 0.18 W/mK, a maximum long-term temperature of 90 °C, a maximum short-term temperature of 110 °C and a refraction index of 1492 nD.

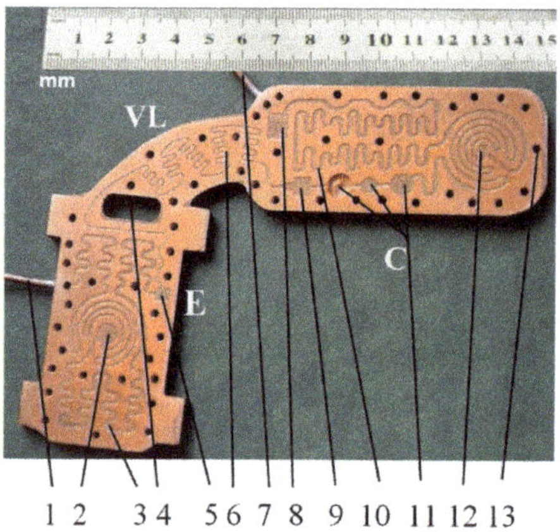

Figure 2. Condenser, lamination tube and the vaporizer of the mini refrigerating unit: 1—vaporizer drainage tube; 2—central area of the vaporizer; 3, 10—rectangular microchannels of the vaporizer, condenser, 4—shaft; 5, 8—mini filters; 6—rectangular microchannels in the lamination area; 7—compressor intake tube; 9—convergent nozzle; 11—mini liquid coolant tanks; 12—condenser cooling area; 13—mounting holes.

The manufacturing of the rectangular microchannels was carried out using a laboratory method that consisted of multiple stages based on the photochemical imprinting of the material, followed by acid submersion. In order to compute the calculus and to determine the size of the mini-refrigerating unit it is important to know the scale of the rectangular microchannels and the roughness degree that resulted from their processing. The measured values can be used to determine any pressure decrease, the volume of the area through which the coolant flows, and the flow speed, as well as to study the influence of the roughness degree over the flow rate. For this purpose, the surface of the rectangular microchannels was mapped by the aid of laser profilometry. Typical measurement results obtained from the scanning process with regards to the main component parts of the mini refrigerating unit are further presented. Figure 3 shows a 3D representation of the rectangular microchannels that are identical for both condenser and vaporizer, and Figure 4 shows their 2D profile and dimensions.

For each scanned surface, values for the height and width of the rectangular microchannels were obtained in μm, as well as the inclination of the walls, compared to the vertical axis, and the surface roughness. Figure 4 illustrates an example of a transverse profile, highlighting the depth and width of the microchannels. The 2D profile shown in Figure 4 corresponds to the arrow shown in Figure 3. Based on the obtained results, no significant variations were found between the widths of different channels, with the difference amounting to maximum 0.8 μm in width and, respectively, 0.9 μm in depth. The shape of the channels at the base of the microchannels is shown to be different from a rectangular shape due to the limitations of the scanning equipment and image vertical scaling being similar, however, to the ones described in similar research.

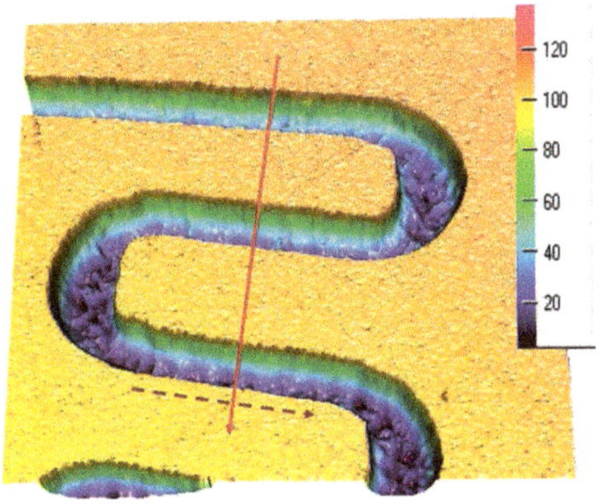

Figure 3. Rectangular microchannels from the condenser and the vaporizer, vertical scale in μm.

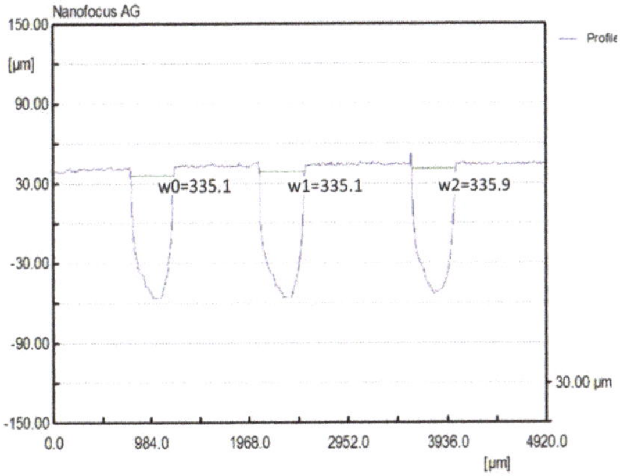

Figure 4. Transverse 2D profile and dimensions of the rectangular microchannels, obtained by laser profilometry.

Figure 5 illustrates how the curvature radii of the rectangular microchannels were measured and in Table 1 a sample of such measurement results is presented.

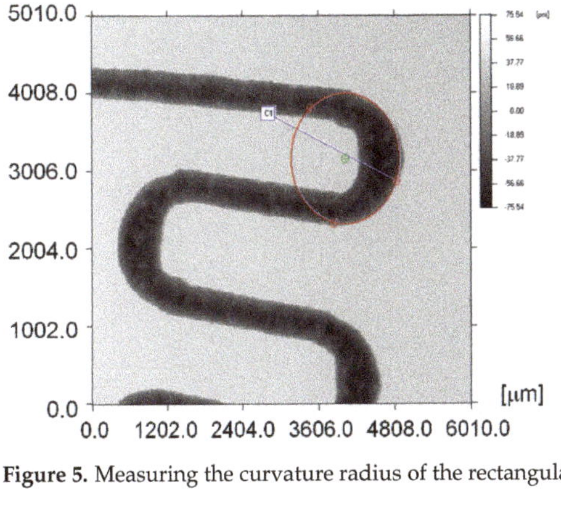

Figure 5. Measuring the curvature radius of the rectangular microchannels.

Table 1. Rectangular microchannels measured curvature radii.

Circle	X [μm]	Y [μm]	r [μm]	l [μm]
C1	4052	3166	852.68	5143
C2	3814	3161	466.58	4953

In Table 1, X and Y represent the coordinates of the circle center by reporting to the initial reference point set for measuring. r is the circle radius and l is the length of the circle center position-vector versus the initial reference point. Values for circle C1 correspond to the exterior of the rectangular microchannel curvature and the values for circle C2 correspond to the interior of the rectangular microchannel.

An example of the surface microtopography can be observed in Figure 6, including the additional R134a coolant tanks, as well as the convergent nozzle that precedes the lamination channel. Figure 7 shows a transverse profile and dimensions for the R1 tank.

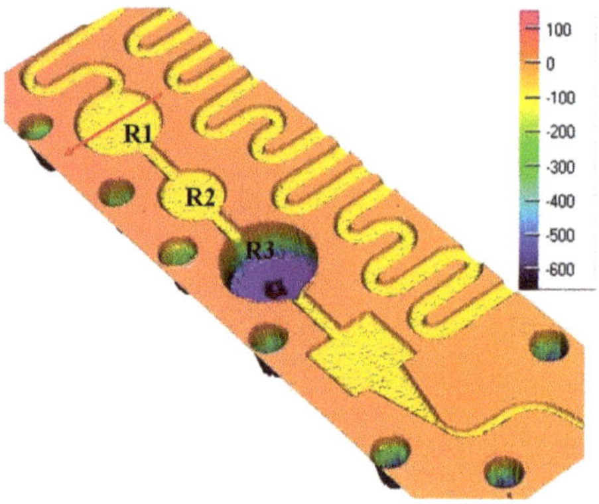

Figure 6. Additional coolant tank area and convergent nozzle, vertical scale in μm.

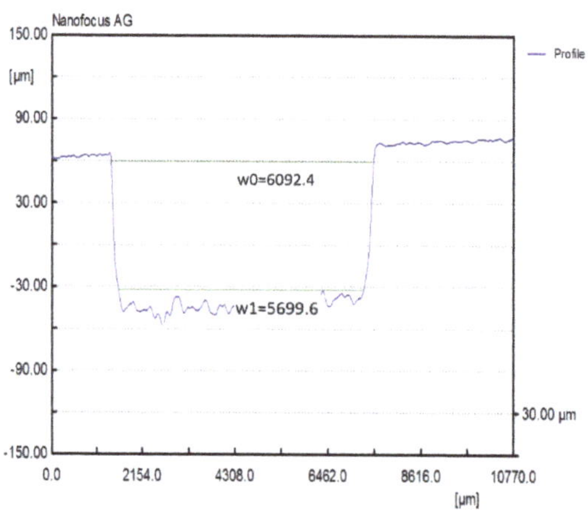

Figure 7. 2D Transverse profile and dimensions of R1 coolant tank.

The measurements computed in the laboratory, together with the ones presented in Figure 7, indicate the radius of the tanks $r_{R1,R3}$ = 3046.2 μm, r_{R2} = 2311.4 μm and an average depth of $h_{R1,R2}$ = 98.7 μm, while h_{R3} = 703 μm, based on other measurements.

Figure 8a shows a scanned image of the convergent nozzle that ensures the coolant flow from tanks R1–R3 to the lamination microchannel (scale is in μm), and Figure 8b presents all the areas where the measurements mentioned in Table 2 were carried out.

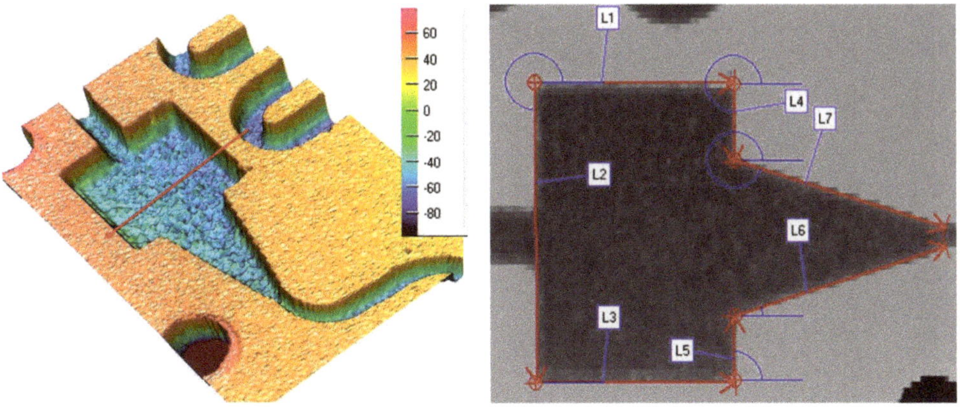

(a) (b)

Figure 8. Convergent lamination nozzle area: (**a**) 3D representation, (**b**) measured regions.

The convergent nozzle region dimensions shown in Table 2 are relevant when computing the parameters at the intake end of the lamination tube that ensures the Joule Thomson effect is achieved.

Figure 9 illustrates the direction used to generate the profile shown in Figure 10 from which the rectangular microchannels used to laminate the R134a coolant were measured. It is immediately obvious that these microchannels are much smaller in width compared with the ones from the condenser and the vaporizer. Due to technological reasons, the height of the rectangular microchannels is the same at the condenser, the lamination tube and the vaporizer, throughout the entire flow path of the coolant.

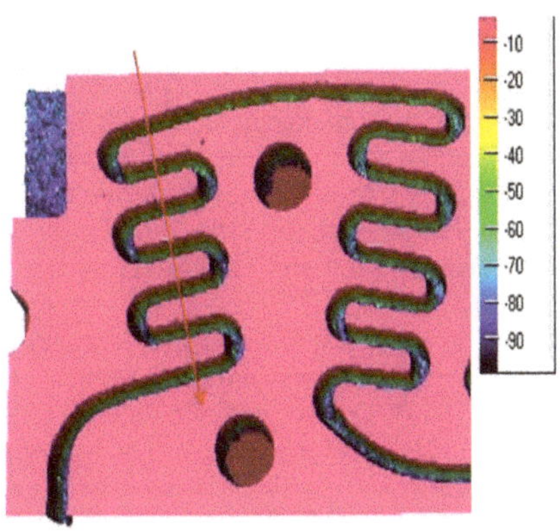

Figure 9. Rectangular microchannels used for laminating the HFC-R134a coolant, scale in μm.

Table 2. Dimensions of the convergent nozzle.

Line	L [µm]	dz [µm]	Angle [°]
L1	4554	−4.66	0.000
L2	5163	3.24	270.290
L3	4528	−0.76	0.000
L4	1345	−79.25	272.230
L5	1091	−90.33	90.000
L6	4965	18.42	14.081
L7	4976	19.39	346.645

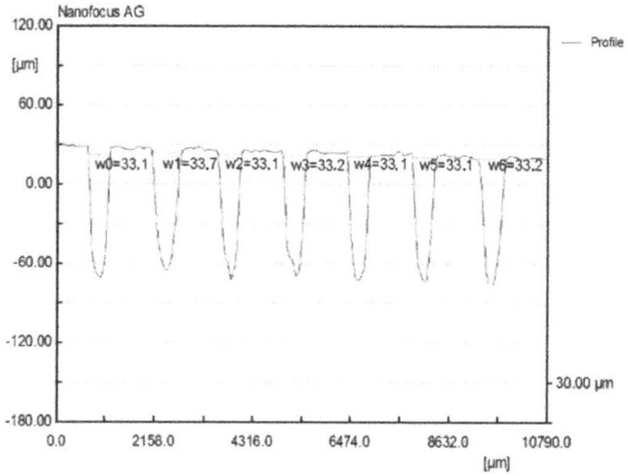

Figure 10. 2D profile and dimensions of rectangular microchannels used for coolant lamination.

4. Temperature Variation of R134a Refrigerant during Vaporization in Rectangular Microchannels in a Transient Convective Regime

The aim of this research was to determine how the temperature variation of R134a coolant varies in rectangular microchannels. When coolant enters the vaporizer in a liquid state from the lamination capillary tube, at a certain well-determined distance, a liquid cone forms in the microchannel, inside which coolant vapors appear. This is followed by a portion of the microchannel in which the mixture is biphasic. The working fluid from the lamination capillary tube is discharged into the rectangular microchannels (with a much larger cross-section) which leads to negative temperatures. During the passage through the rectangular microchannels of the vaporizer, the coolant starts to boil, so that is in a vapor state at the outlet. It is necessary for the coolant to be in this state at the outlet of the vaporizer for it to be sucked in by the rotary compressor. In this process, transient convection heat transfer takes place.

Figure 11 shows a sequence of rectangular micro-channels belonging to a mini heat exchanger and subjected to a heat flow in the lower part in the case of the vaporizer.

For the calculations, it was considered that the R134a coolant had constant physical properties for a transient-state heat transfer, forced laminar flow regime. It was also considered that the flow was incompressible. It can be appreciated that the natural convection of heat was negligible. To simplify the calculations, it was assumed that the inlet temperature distribution was uniform and the temperature at the channel wall was prescribed and uniform.

For heat transfer, rectangular microchannels with mean width b = 335.36 µm, mean height h = 98.7 µm, mean distance between channels s = 378.9 µm and total length L1 = 8010 µm were considered.

The experimental setup, described above, allows the speed of the ASPEN rotary compressor model 14-24-1101 [26,27] to be changed between 1800–7000 rpm by means of the motor control driver at voltages between 1.13–4.94 V. It is necessary to consider in the calculations that changing the speed of the compressor also changes the flow rate of coolant through the rectangular microchannels. Applying the calculation model in [13], for the mini heat exchanger with efficiency $\eta_f = 0.688$, through which the R134a refrigerant flows the pressure drop, Reynolds number, fluid temperature at the inlet and outlet of the rectangular microchannels can be determined.

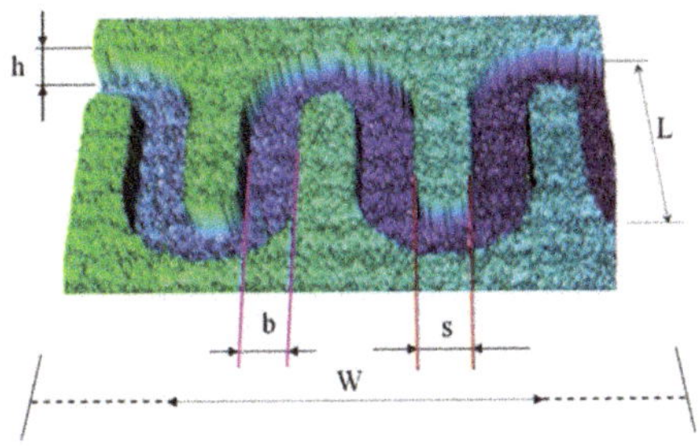

Figure 11. Micro-channel parameters considered for the analytical model.

For the R134a refrigerant, the literature [28] recommends the values given in Table 3.

Table 3. Properties of Coolant R134a in gaseous and liquid state at atmospheric pressure.

Gaseous Phase			Liquid Phase		
Property	**Value**	**UM**	**Property**	**Value**	**UM**
Gas density (1.013×10^5 Pa)			Liquid density		
(at boiling point)	5.28	kg/m^3	(1.013×10^5 Pa and 25 °C)	246.6	kg/m^3
(1.013×10^5 Pa and 15 °C)	4.25	kg/m^3			
Compressibility factor (Z)			Boiling point		
(1.013 bar and 15 °C)	1	-	(1.013×10^5 Pa)	1206	K
Specific volume			Latent heat of vaporization	215.9	kJ/kg
(1.013×10^5 Pa and 15 °C)	0.235	m^3/kg	(1.013 bar at boiling point)		
Heat capacity at constant pressure (Cp)			Vapor pressure		kPa
(1.013×10^5 Pa and 25 °C)	87.54	J/kgK	(at 5 °C)	350	
			(at 15 °C)	490	
			(at 20 °C)	570	
			(at 50 °C)	1320	
			Heat capacity at constant pressure (Cp)		
			(1.013×10^5 Pa and 25 °C)	1441	J/kg·K

As the pressure in the installation changes depending on the speed of the ASPEN rotary compressor, there may be different values specific to the refrigerant properties. An example shown in Table 4 contains the values for the gaseous state of R134a refrigerant, for a pressure $p_{R134a} = 2 \times 10^5$ Pa, and a Temperature $T_{R134a} = 22$ °C, as given in the literature [29].

For the gaseous state of R134a coolant, the specific constant $\Re_{R134a}$ can be determined with Equation (1):

$$\Re_{R134a} = \frac{p_{R134a}\frac{1}{\rho_{R134a}}}{Z_{cR134a}T_{R134a}}, \quad (1)$$

where: p_{R134a}—coolant pressure in rectangular microchannels [Pa]; ρ_{R134a}—R134a density [kg/m^3]; Z_{cR134a}—coefficient of compressibility (dimensionless); T_{R134a}—coolant temperature in rectangular microchannels [K].

For the calculation of the mean free path of coolant molecules R134a through rectangular microchannels, there are, according to Stéphane Colin [13], several models developed. For the adopted Variable Soft Spheres model, denoted VSS, the coefficient in the mean free path expression, k_{mfp} (dimensionless), is expressed with the relation:

$$k_{mpf} = \frac{4\alpha_{VSS}(7-2\omega)(5-2\omega)}{5(\alpha_{VSS}+1)(\alpha_{VSS}+2)\sqrt{2\pi}}, \tag{2}$$

Table 4. Gaseous state R134a refrigerant properties.

Gaseous State R134a Coolant		
Property	**Value**	**UM**
Density	9.0613	kg/m^3
Specific enthalpy	419.88	kJ/kg
Specific entropy	1.8547	kJ/kg K
Specific isobar heat capacity	0.87506	kJ/kg K
Isobar coefficient of thermal expansion	4.077	10^{-3} (1/K)
Heat conductance	13.5875	10^{-3} (W/mK)
Dynamic viscosity	11.885	10^{-6} (Pa s)
Kinematic viscosity	2.2199	10^{-6} (m^2/s)
Thermal diffusivity	29.655	10^{-7} (m^2/s)
Prandtl-number	0.765	-
Compressibility coefficient	0.95505	-

where α_{VSS} is a characteristic coefficient for the VSS molecular model, for which a value of $\alpha_{VSS} = 0.722$ (dimensionless) was adopted. The temperature exponent of the viscosity coefficient, ω (dimensionless) is:

$$\omega = \frac{\eta+3}{2(\eta-1)}. \tag{3}$$

For the exponent in the inverse power law model, denoted by η, a value of $\eta = 13$ (dimensionless) was adopted, according to the values given by Lennard-Jones [7].

The mean free path λ, was determined as:

$$\lambda = \frac{k_{mpf}\mu_{R134a}}{\rho_{R134a}\sqrt{\Re_{134a}T_{R134a}}}. \tag{4}$$

The dynamic viscosity of the gaseous state R134a coolant was noted with μ_{R134a}, and its value was taken from Table 4.

If the width of the rectangular microchannel $b_{ch} = 335.36$ μm and its height is $h_{ch} = 98.7$ μm the equivalent hydraulic diameter D_h can be calculated with the relation:

$$D_h = \frac{4(b_{ch}h_{ch})}{(b_{ch}+h_{ch})}. \tag{5}$$

Calculations showed that $\lambda = 8.063 \times 10^{-3}$ μm and $D_h = 305.03$ μm.

For the VSS model, the Knudsen and Reynolds numbers are determined with the relations:

$$Kn_{VSS} = \frac{\lambda}{D_h}, \mathrm{Re} = \frac{U(Y)D_h}{\nu_{R134a}}. \tag{6}$$

The value obtained for $K_{nVSS} = 2.643 \times 10^{-5}$ indicates a continuum flow, according to the classification made by Stéphane Colin [13], with classical no-slip boundary conditions. It can be considered that the flow is accurately modelled by the compressible Navier-Stokes equations.

The developed velocity profile, U(Y), found by (Mikhailov and Cotta, 2004) [7] on the flow direction y, will thus be:

$$U(Y) = \frac{6Kn_{VSS}\beta_v + 3\frac{(1-Y^2)}{2}}{1+6Kn_{VSS}\beta_v}, \; \mathrm{Y} = \frac{y}{L_1}. \tag{7}$$

The dimensionless coefficient βv is obtained from:

$$\beta_v = \frac{(2-\alpha_m)}{\alpha_m}. \quad (8)$$

In Equation (8), the dimensionless α_m coefficient represents the Eigenvalues for the velocity distribution in a rectangular microchannel.

If it is considered that the total length of the vaporizer plate in which the rectangular microchannels are embedded is L_1, from the calculations we obtained the representation in Figure 12 for the developed velocity profile is U(Y) and the Reynolds number, Re(Y). The figure shows how the velocity of Freon R134a along the rectangular microchannels in the vaporizer changes with respect to the dimensionless size Y.

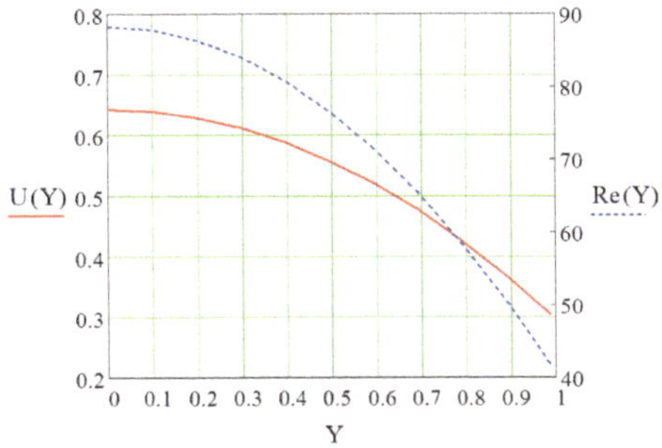

Figure 12. Velocity distribution U(Y) and Reynolds number Re(Y) for refrigerant circulation in rectangular microchannels as a function of path travelled (dimensionless values).

Analyzing in Figure 12 the dimensionless flow velocity of the R134a coolant through the rectangular microchannels, there is a rather large drop of 53.81%, from the inlet value U(Y) = 0.695 to U(Y) = 0.374, which corresponds to the flow through the entire vaporizer duct. The shape of the velocity curve is similar to those presented by Stéphane Colin and Aubert [8,13]. The input/output value ratio determined in the present work of value Ui/Uo = 2.178, differs slightly from that determined by Aubert W*i/W*o = 2.25, respectively, Colin of value U*i/U*o = 2.66. According to the representation in Figure 12, the Reynolds number reaches a maximum of 88.258 and a minimum of 41.723, values that show, according to the classification made by Stéphane Colin [13] that the flow regime is laminar.

For dimensionless velocities of coolant in microchannels, considering ψ(Y) as the Eigen functions of the following Sturm-Liouville problem [7], we can write in simplified form, Equation (9):

$$\frac{d^2\Psi_i(Y)}{dY^2} + \mu_i^2 U(Y)\Psi_i(Y) = 0, \quad (9)$$

where Y will have values between 0 and 1 and the original eigenvalues are obtained from:

$$\mu_i = \sqrt{\frac{2}{3}(1+6Kn_{VSS})\nu_i}, \; i = 1,\, 2,\, 3\ldots.$$

The results are obtained by solving Equation (9) in terms of the confluent hyper geometric function, also known as the Kummer function $_1F_1$ [a; b; z]:

$$_1F_1(a;b;z) = \sum_{i=0}^{\infty} \frac{(a)_i z^i}{(b)_i i!}, \quad (10)$$

where:

$$a_{(i)} = \frac{1-\nu_i(1+Kn_{VSS}\beta_v)}{4}, \quad (11)$$

$b_{(i)} = 1/2$ for the considered case, and z will be:

$$z = \nu_i^2,$$

This leads to:

$$\Psi_i(Y) = \sum_{i=0}^{\infty} \frac{(a)_i z^i}{(b)_i i!} e^{\nu_i \frac{Y^2}{2}}. \tag{12}$$

The deduced Equation (12) satisfies Equation (9) and the conditions:

$$\left.\frac{d\Psi_i(Y)}{dY}\right|_{Y=0} = 0,\ Kn\beta_v\beta \left.\frac{d\Psi_i(Y)}{dY}\right|_{Y=1} = -\frac{1}{2}\Psi_i(1). \tag{13}$$

The β coefficient is determined with the relation:

$$\beta = \frac{\beta_t}{\beta_v}, \beta_t = \frac{\frac{2-\alpha_i}{\alpha_i}\frac{2\chi}{\chi+1}}{\text{Pr}}, \tag{14}$$

where α_t is the thermal accommodation coefficient, χ the adiabatic exponent and Pr is the Prandtl number.

In this case the Eigen condition is set:

$$\left\{2Kn\beta_v\beta_1 F_1\left[\frac{5}{4} - T_1, \frac{3}{2}, \nu_i\right]\nu_i(1-4T_i) +_1 F_1\left[\frac{1}{4} - T_1, \frac{1}{2}, \nu_i\right](1-2Kn\beta_v\beta\nu_i)\right\}e^{\frac{\nu_i}{2}} = 0, \tag{15}$$

where: $T_1 = \frac{\nu_i(1+4Kn\beta_v)}{4}$.

The equation is a function of two parameters $Kn\beta_v$ and β. From Equation (15), the eigenvalues, ν_i, and the original eigenvalues, μ_i can be determined.

The normalization integral coefficient of the function is:

$$N_i = \int_0^1 U(Y)(\Psi_i(Y))^2 dY. \tag{16}$$

The normalized form of the Eigen function is:

$$\tilde{\Psi}_i(Y) = \frac{\Psi_i(Y)}{\sqrt{N_i}}. \tag{17}$$

The transform-inverse pair is:

$$\theta_{im}(Z,\tau) = \int_0^1 U(Y)\tilde{\Psi}_i(Y)\theta(Y,Z,\tau)dY, \tau = \frac{\alpha t}{L_1^2}, Z = \frac{1}{Pe}\frac{z}{L_1}, \tag{18}$$

where τ is the dimensionless time, equal to the product of α ($\alpha = k/\rho c_p$ where k is the thermal conductivity divided by the product of density and specific heat at constant pressure) and a characteristic time, divided by the quadratic distance from the middle of the body through which the heat passes to the surface, Pe is the Peclet number. The inverse of the function is:

$$\theta(Z,Y,\tau) = \sum \tilde{\Psi}_i(Y)\theta_{im}(Z,\tau), \tag{19}$$

The value of the dimensionless mean temperature drops as a function of τ is:

$$\theta_{av}(Z,\tau) = \frac{\int_0^1 \left[\frac{6Kn_{VSS}\beta_v + 3\frac{1-Y^2}{2}}{1+6Kn_{VSS}\beta_v} \sum_{i=1}^{\infty} \tilde{\Psi}_i(Y)\theta_{im}(Z,\tau)\right] dY}{\int_0^1 \frac{6Kn_{VSS}\beta_v + 3\frac{1-Y^2}{2}}{1+6Kn_{VSS}\beta_v} dY}, \tag{20}$$

Figure 13a,b, illustrate the evolution of the dimensionless mean temperature obtained by implementing the above described equations in a Mathcad code [30]. The presence and absence of a heat source at the vaporizer are considered.

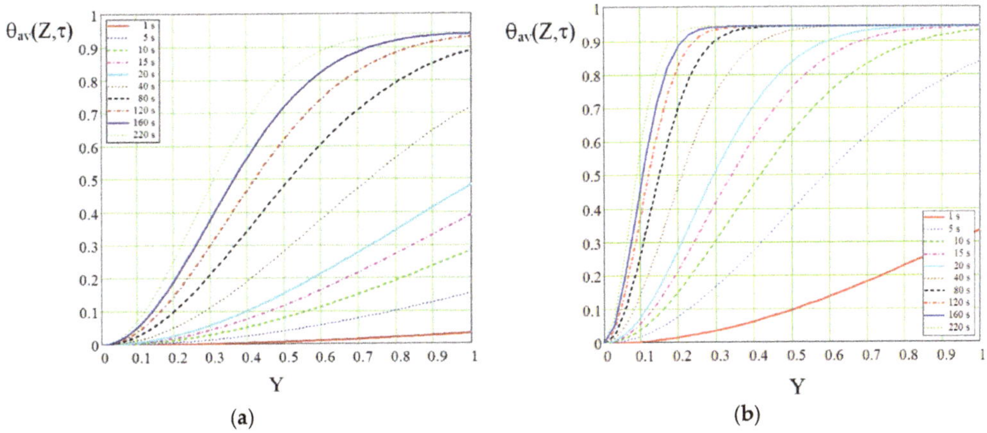

Figure 13. Evolution of the mean dimensionless temperature over time: (**a**) without heat source at the vaporizer, (**b**) with heat source at the vaporizer.

In order to determine the temperature drop along rectangular microchannels, taking into account velocity and time, the following equation is used:

$$T_{R134a} = T_{in} - \theta_{av}(Z,\tau)(T_{in} - T_{ou}), \tag{21}$$

where T_{in}, T_{ou} are the temperatures of R134a coolant at the inlet and outlet of the vaporizer's rectangular microchannels. Figure 14 shows the temperature evolution in the rectangular microchannels belonging to the vaporizer as a function of time and the dimensionless value Y.

If it is assumed that the vaporizer is not heated by the hot source, according to the calculations and Figure 14a, it is found that the temperature decreases in the center of the vaporizer from an initial value of +21.8 °C to a minimum value of −12.3 °C. If the non-stationary behavior is analyzed in the same figure, it can be observed that after 8 s the coolant temperature drops to −9.8 °C. This confirms that the Joule-Thompson effect followed by expansion in the vaporizer takes place rapidly in the rectangular microchannels, but a strong rarefaction effect also occurs. After 60 s of flowing through the rectangular microchannels, the cooling regime stabilizes, and the curves have the same shape and close values.

For the case of heating the center of the vaporizer from the calculations and Figure 14b, the refrigerating agent which initially has a temperature of +34.1 °C cools down in the center of the vaporizer to a minimum value of +5.6 °C. Concerning the non-stationary behavior, the calculations show that after 5 s, the coolant temperature in the rectangular microchannels decreases to +9 °C.

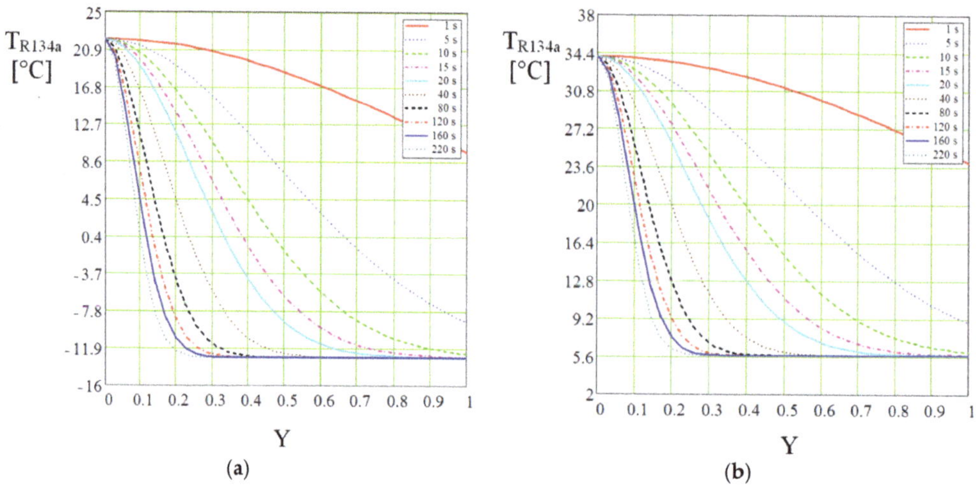

Figure 14. Temperature change in non-stationary mode for R134a coolant along the rectangular microchannels of the vaporizer at compressor speed of 4400 rpm: (**a**) without hot source, (**b**) with hot source at T = 34 °C.

5. Experimental Determinations of R134a Coolant Temperature Evolution during Mechanical Compression in Rectangular Microchannels

In order to be able to make a comparative analysis of the data obtained from the calculations with experimental results, it is considered appropriate to present below the values determined on the mini refrigeration unit with rectangular microchannels mechanically compressing the R134a coolant.

Temperature was measured experimentally at various points of interest in the installation using a LabJack U12 digital-analogue interface to which eight LM15 temperature sensors were connected. The LabVIEW environment was used to read the data. The temperature sensors provide values for: CI—compressor inlet; CD—compressor discharge; IC—condenser inlet; CC—condenser central area; OC—condenser outlet; IV—vaporizer inlet; CV—vaporizer central area; and OV—vaporizer outlet.

The temperature distribution of R134a coolant in rectangular microchannels at different points of the plant was obtained for the following situations:

- the rotational speed of the ASPEN rotary compressor is changed, in which case according to the technical documentation [26] and the research carried out [27] different flow rates are obtained in the microchannels of the mini refrigeration plant;
- the same compressor speed is maintained, but the heat generated by a centrally located hot source behind the evaporator is modified.

Changing the speed of the ASPEN rotary compressor is easily obtained in practice, using a potentiometer attached to the control driver.

The temperature change in the central area of the evaporator was achieved with the help of three electrical resistors located inside an AMD CPU. These were DC powered at different voltages using a voltage source. The electrical resistors simulate the heart of the CPU and can reach a maximum of 110 °C, their temperature being measured with an electronic thermometer.

Experimental results for the two cases are presented in the following sections.

5.1. R134a Coolant Temperature Evolution at Variable Flow Rate through Rectangular Microchannels

Since the experimental determinations were carried out under real operating conditions, it is interesting to see how the temperature evolves in the rectangular microchannels and in the various elements of the mini refrigeration unit. From the many possible speeds ranging from 1800–7000 rpm, it was considered that for the interpretation of the evolution of the parameters in the mini system, the results obtained for the minimum, the average and the maximum speed were sufficient. In Figure 15a–c, the results of the experimental determinations can be visualized as time-dependent plots.

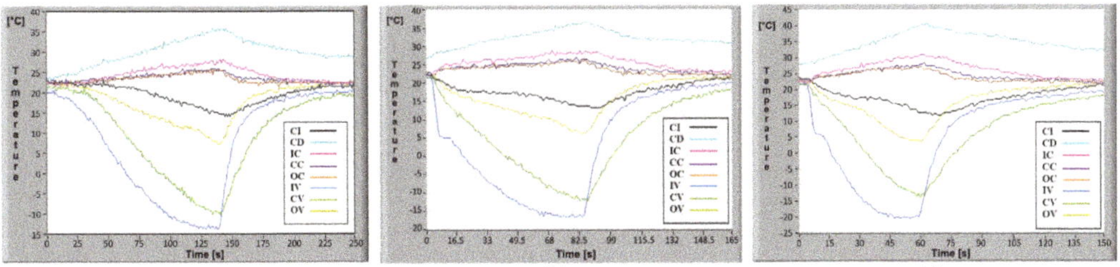

(**a**) (**b**) (**c**)

Figure 15. Temperature evolution in rectangular microchannels of the mini refrigeration unit for ASPEN rotary compressor different speeds: (**a**) 1800 rpm, (**b**) 4400 rpm, (**c**) 7000 rpm.

The obtained results allow analysis of how the system temperature evolves as a function of increasing coolant flow in the rectangular microchannels. It should be noted that since the ASPEN rotary compressor and the drive motor are encapsulated, the speeds are assessed based on technical data provided by the supplier [26,27]. The analysis of the parameters obtained is the basis for determining the optimum operating speed.

5.2. R134a Coolant Temperature Evolution at Constant Flow Rate through Rectangular Microchannels and Central Heating of the Vaporizer

As stated, the mini refrigeration system was tested in the case of central heating of the vaporizer by means of the processor denoted by 10 in Figure 1, which had three DC powered electrical resistors applied at various voltages. Basically, three heating presets were used and the obtained measurement results are shown in Figure 16a–c.

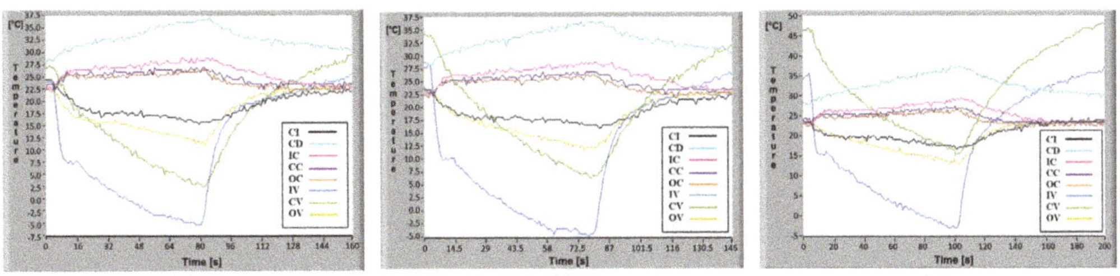

(**a**) (**b**) (**c**)

Figure 16. Temperature evolution in rectangular microchannels of the mini refrigeration unit for electrical resistors in the central area of the evaporator heated to (**a**) 26 °C, (**b**) 34 °C, and (**c**) 47 °C.

It was experimentally found that significant differences occur between the temperature of the electrical resistors (50°, 70°, 90 °C) and the values reached at the CPU-vaporizer interface (26°, 34°, 47 °C). This can be explained by the existence of inner layers of the CPU, surface roughness, etc.

The experimental results shown in Figures 15a–c and 16a–c illustrate the temperature variation in a transient regime in any point of the mini refrigeration unit with rectangular microchannels and highlight the temperature decrease speed at the vaporizer level, depending on the compressor speed.

6. Discussion

The present paper describes temperature distribution in the rectangular microchannels of a mini refrigeration unit using mechanical compression of R134a coolant, depending on operating conditions. It was experimentally found that a minimum temperature of −22.3 °C is reached in the vaporizer. The dimensions of the vaporizer are only 47.2 mm × 18.7 mm.

To perform the calculations it is necessary to know in detail both the dimensions of the rectangular microchannels and the values of the roughness of the flow channels. In usual flow channels the roughness influences the flow parameters only to a small extent, but in the case of micro and nano channels, the thickness of the liquid film or vapor jet can have values sensitively close to those of the surface asperities. Rectangular microchannel sizes differ depending on the operating area. Rectangular microchannels, as shown in Figures 3–5, have values that are close to each other at the condenser and evaporator, with mean widths of 335.36 μm, mean depths of 98.7 μm and an average radius of channel curvature of 659.63 μm. For the same mean depth, in the laminating region, according to Figures 9 and 10, the rectangular microchannel width is only 33.21 μm. This narrowing of the flow section is mandatory to obtain the Joule-Thompson effect. Since microchannels are used, it was considered useful in the design to use three tanks for storing the freon in the liquefied state. Tanks R1 and R3 in Figure 6 have a radius of 3046.2 μm while the radius of reservoir R2 is 2311.4 μm. At an average depth of 98.7 μm for R1 and R2, respectively, 703 μm for R3, a liquid freon reservoir of 25.027 mm^3 is obtained, which ensures the avoidance of its gaseous penetration into the rolling tube.

Starting from the dimensionless mean values of the coolant temperature at transient flow through the rectangular microchannels determined by calculus for the vaporizer, the temperature change as a function of time can be analyzed.

Calculations were performed for two cases, as further discussed.

The average experimental values determined for the mini refrigerating unit's vaporizer are shown in Table 5.

Table 5. Experimental parameters in the vaporizer region.

Heat Source	Initial Vaporizer Temperature	Temperature at Vaporizer Inlet (IV)	Temperature in Vaporizer Central Region (CV)	Temperature at Vaporizer Outlet (OV)	Time	Rotary Compressor Speed
Unit	°C	°C	°C	°C	s	rpm
NO	23	−13.3	−8.9	+7.5	120	1800
	23	−16.1	−12.3	+6.8	69	4400
	23	−22.3	−14.4	+4.2	48	7000
YES	26	−5.0	+3.2	+11.3	71.2	4400
	34	−3.2	+7.1	+12.5	72.5	4400
	47	−2.5	+17.0	+13.7	93.4	4400

Experimentally, it was analyzed how the R134a coolant behaves while traveling through the rectangular microchannels in three situations: changing the compressor speed (implicitly the flow rate), and maintaining the compressor speed at 4400 rpm without and with heating of the vaporizer central region.

For the case of changing the flow rate when the ASPEN rotary compressor speed reaches 1800, 4400 and 7000 rpm without heating the vaporizer, as shown in Figure 15a–c, the minimum temperatures reached at the vaporizer vary in quite a wide range, from −13.3 °C to −22.3 °C. The time in which the mentioned negative values are reached ranges from 120 s, corresponding to the minimum speed, up to 48 s, which corresponds to a compressor speed of 7000 rpm. It is immediately clear that the duration of the cooling process is 2.5 times shorter at maximum speed than at minimum speed. Experimentally, however, it was found that although the minimum temperatures are reached much faster at maximum speed, the compressor also heats up much faster. Analysis of the behavior of a cooling agent in rectangular microchannels with a changing flow rate is necessary because of the strong rarefaction effect due to extremely small flow cross-sections. In operation, this strongly affects the temperature and noise of the compressor which changes significantly with speed. Based on the minimum temperature value required to be achieved with the mini refrigeration unit, the noise produced and the compressor's maximum bearable heating temperature, it was found that operation is optimal at 4400 rpm.

In the case where the vaporizer does not receive additional heat, at 4400 rpm the corresponding experimental results shown in Figure 15b (CV curve), indicate the temperature variation in the central region of the vaporizer independent of time. After 8 s there is a sudden drop in temperature from 21.8 °C to +12 °C, not reaching the −9.8 °C determined theoretically. This can be explained by the inertia of the temperature sensor and the fact that it was not placed inside the rectangular microchannel, but close to it. However, after 69 s when the system is switched off (the stabilized minimum temperature regime has been reached) the temperature value in the experimental case is −11.7 °C compared to the theoretical case when it is −12.3 °C. The temperature difference of 0.5 °C between the results of the mathematical calculations and the experimental values can be explained by losses in the mini refrigeration unit and measurement errors.

If the center of the vaporizer is heated, the experimental determinations shown in Figure 16b (CV curve) indicate a sudden drop in temperature after 8 s of operation from 34.1 °C to 27.1 °C. It is also noted that the theoretical value of 7.8 °C is not reached. After the 72.5 s, when the unit stops at the minimum stabilized temperature regime, the temperature value in the theoretical case is 5.6 °C and in the experimental case 6.3 °C. As in the previous case there is a small difference in temperature for the two cases of 0.7 °C but also in this case it can be appreciated that there is a very good correlation between the results of the mathematical calculations and the experiment.

The mini-installation with rectangular microchannels mechanically compressing R134a refrigerant using an ASPEN rotary compressor, although using only 17 g of fluid, provides a maximum refrigerating capacity of 1230 Btuh. In the refrigerating mini-installation's vaporizer, of only 47.2 mm × 18.7 mm, a temperature of −22.3 °C was obtained experimentally compared to the theoretical value of −24 °C. It was found that although the flow cross-section of the rectangular microchannels has a strong rarefaction effect it is possible to functionally obtain mini refrigeration units for cooling in various fields of application.

7. Conclusions

The mini refrigeration unit experimentally built and presented in this paper offers the novelty of implementing microchannels in all its regions. The dimensions of the rectangular microchannels shown in Figures 6–9 were obtained by laser profilometry. The experimental measurements show how the rectangular microchannels were dimensioned for each region of the mini refrigerating unit so that a maximum specific cooling capacity was obtained. Another novelty element introduced by the presented experimental setup was that all the functional regions of the refrigerating unit were placed on a single plate with rectangular microchannels, as shown in Figure 2, which allows for a very compact unit.

Table 6 presents some comparative specifications of the above presented mini refrigerating unit with microchannels, compared to similar units found in the literature [31–33].

Table 6. Mini refrigeration unit.

Model	Application	Power Source [V]	Cooling Capacity [W]	Specific Cooling Capacity [W/kg]	Condenser PCS	Dimension L × W × H [mm]
Proposed mini refrigerating unit with microchannels	PCC/MF/PR/ LSC/HBC	DC24	360	328.18	Microchannel	170 × 120 × 115
Purswave Qx36W, [31]	PCC/MF/PR/ LSC/HBC	DC24	365	122.48	Fin	470 × 180 × 160
HACF405DC12, [32]	MBP/LBP	DC12	405	135.00	Fin	250 × 160 × 170
HAC550DC24, [32]	MBP	DC24	550	157.14	Fin	250 × 160 × 260
HAF115DC24-L, [32]	LBP	DC24	115	32.85	Fin	320 × 110 × 200
HAF115DC24-V, [32]	LBP	DC24	115	32.85	Fin	190 × 160 × 200
FSCD019Z12, [33]	MC/MAC/MF/MCH	DC12	365	202.77	Microchannel	180 × 170 × 128
FSCD0325Z24, [33]	TCW/MWC/EV	DC24	550	183.33	Microchannel	250 × 160 × 170

where: PC, CPU, sever cooling—PCC; medical facilities—MF; portable refrigerator—PR; limited space cooling—LSC; human body cooling—HBC; mini-cooler—MC; medium back pressure—MBC; low back pressure—LBC; mini air conditioner—MAC; mini freezer—MF; mini chiller—MCH; tiny cold-water machine—TCW; micro water chiller—MWC; electric vehicle air conditioning—EV.

The convenient placement of eight temperature sensors and real time signal acquisition allowed plotting of the real temperature evolution in key points of the mini refrigerating unit, as shown in Figures 15 and 16, which in turn allows study of the temperature-variation laws over time.

The mini refrigerating unit described in the present paper offers several advantages, starting with an important specific cooling capacity of 328.18 W/kg, due to the implementation of rectangular microchannels in the condenser, vaporizer and lamination regions. The data illustrated in Table 6 show that the use of rectangular microchannels that are only 33.21 μm wide (in average) significantly increases the specific cooling capacity (by 61.85% in the studied case) by reporting to other mini cooling units. This is achieved while the unit is very compact as its dimensions are 59.89% smaller than the highest performing similar unit. The obtained performances are also remarkable for the fact that the described mini refrigerating unit only uses 17 g of R134a coolant.

The described mini refrigerating unit equipped with rectangular microchannels was tested in a laboratory over the course of a year without intervening on its constructive elements. This shows that this set-up can be used to obtain very reliable cooling equipment that uses microchannels and only small amounts of cooling agent.

8. Patents

System with rectangular micro- and nano-channels for freon processor cooling.

- Patent Number(s): RO129915-A2
- Inventor(s): MIHAI I, OLARIU E
- Patent Assignee Name(s) and Code(s): UNIV SUCEAVA STEFAN CEL MARE (UYSU-Non-standard)
- Derwent Primary Accession Number: 2015-10385E

Author Contributions: Conceptualization, I.M.; methodology, I.M.; software, I.M.; validation, I.M., C.S. and C.M.P.; formal analysis, I.M.; investigation, I.M., C.S. and C.M.P.; resources, I.M. and C.S; data curation, I.M.; writing—original draft preparation, I.M.; writing—review and editing, C.S.; visualization, I.M., C.S. and C.M.P.; supervision, I.M.; project administration, I.M.; funding acquisition, I.M. All authors have read and agreed to the published version of the manuscript.

Funding: This research received no external funding.

Institutional Review Board Statement: Not applicable.

Informed Consent Statement: Not applicable.

Data Availability Statement: Some or all data, models, or code generated or used during the study are available from the corresponding author by request.

Conflicts of Interest: The authors declare no conflict of interest.

References

1. Tuckerman, D.B.; Pease, R.F.W. High Performance Heat Sinking for VLSI. *IEEE Electron. Device Lett.* **1981**, *2*, 126–129. [CrossRef]
2. Kim, D.-K.; Kim, S.J. Closed-form correlations for thermal optimization of microchannels. *Int. J. Heat Mass Transf.* **2007**, *50*, 5318–5322. [CrossRef]
3. Kandlikar, S.; Grande, W. Evolution of microchannel flow passages-thermohydraulic performance and fabrication technology. *Heat Transf. Eng.* **2003**, *24*, 3–17. [CrossRef]
4. Chen, Y.-T.; Kang, S.-W.; Tuh, W.-C.; Hsiao, T.-H. Experimental Investigation of Fluid Flow and Heat Transfer in Microchannels. *Tamkang J. Sci. Eng.* **2004**, *7*, 11–16.
5. Jung, J.-Y.; Kwak, H.-Y. Fluid flow and heat transfer in microchannels with rectangular cross section. *Heat Mass Transf.* **2008**, *44*, 1041–1049. [CrossRef]
6. Ho, C.M.; Tai, Y.C. Micro-Electro-Mechanical-Systems (MEMS) and Fluid Flows. *Annu. Rev. Fluid Mech.* **1998**, *30*, 579–612. [CrossRef]
7. Castelloes, F.; Cardoso, C.; Couto, P.; Cotta, R. Transient analysis of slip flow and heat transfer in microchannels. In In Proceedings of the 10th Brazilian Congress of Thermal Sciences and Engineering—ENCIT, Rio de Janeiro, Brazil, 29 November–3 December 2004; Paper CIT04-0640. 2004; pp. 1–10.
8. Aubert, C. *Ecoulements Compressibles de Gaz dans les Microcanaux: Effets de Raréfaction, Effets Instationnaires, These*; l'Université Paul Sabatier: Toulouse, France, 1999; pp. 1–250.
9. ALShqirate, A.; Hammad, M.; Tarawneh, M. Cooling of Superheated Refrigerants Flowing Inside Mini and Micro Tubes, Study of Heat Transfer and Pressure Drop, CO_2 Case Study. *Jordan J. Mech. Ind. Eng.* **2012**, *6*, 199–203.
10. Ratchananusorn, W. Flow Phenomena, Heat and Mass Transfer in Microchannel Reactors. Master's Thesis, Lappeenranta University of Technology, Lahti, Finland, 2007; pp. 1–96.
11. Effenhauser, C.S.; Manz, A.; Widmer, H.M. Glass Chips for High-Speed Capillary Electrophoresis Separations with Submicrometer Plate Heights. *Anal. Chem.* **1993**, *65*, 2637–2642. [CrossRef]
12. Shahsavari, S.; Tamayol, A.; Kjeang, E.; Bahrami, M. Convective Heat Transfer in Microchannels of Noncircular Cross Sections: An Analytical Approach. *J. Heat Transf.* **2012**, *134*, 091701. [CrossRef]
13. Kandlikar, G.S.; Garimella, S.; Li, D.; Colin, S.; King, M.R. *Heat Transfer and Fluid Flow in Minichannels and Microchannel*; Elsevier: Amsterdam, The Netherlands, 2006; pp. 2–27, 89–133, 201–297, 340–386. ISBN 0-0804-4527-6.
14. Lee, J.; Kjeang, E. A Perspective on Microfluidic Biofuel Cells. *Biomicrofluidics* **2010**, *4*, 041301. [CrossRef] [PubMed]
15. Kjeang, E.; Michel, R.; Harrington, D.A.; Djilali, N.; Sinton, D. A Microfluidic Fuel Cell with Flow-Through Porous Electrodes. *J. Am. Chem. Soc.* **2008**, *130*, 4000–4006. [CrossRef] [PubMed]
16. Jang, H.S.; Park, D.S. Microfabrication of Microchannels for Fuel Cell Plates. *Sensors* **2010**, *10*, 167–175. [CrossRef] [PubMed]
17. Brandner, J.; Bohn, L.; Henning, T.; Schygulla, U.; Schubert, K. Microstructure Heat Exchanger Applications in Laboratory and Industry. *Heat Transf. Eng.* **2007**, *28*, 761–771. [CrossRef]
18. Phillips, R.J.; Glicksman, L.R.; Larson, R. Forced-Convection, Liquid Cooled, Microchannel Heat Sinks for High Power-Density Microelectronics. In Proceedings of the International Symposium Cooling Technology for Electronic Equipment, Honolulu, HI, USA, 17–21 March 1987; pp. 227–248.
19. Choi, S.B.; Barron, R.F.; Warrington, R.O. Fluid flow and heat transfer in microtubes. *Micromech. Sens. Actuat. Syst. ASME* **1991**, *32*, 123–134.
20. Silvério, V.; Moreira, A. Friction losses and heat transfer in laminar microchannel single-phase liquid flow. In Proceedings of the Sixth International ASME Conference on Nanochannels, Microchannels and Minichannels, ICNMM2008, Darmstadt, Germany, 23–25 June 2008; pp. 1–9.
21. Peng, X.F.; Peterson, G.P.; Wang, B.X. Frictional flow characteristics of water flowing through microchannels. *Exp. Heat Transf.* **1994**, *7*, 249–264. [CrossRef]
22. Xu, B.; Ooi, K.T.; Wong, N.T.; Choi, W.K. Experimental investigation of flow friction for liquid flow in microchannels. *Int. Commun. Heat Mass Transf.* **2000**, *27*, 1165–1176. [CrossRef]
23. Liu, D.; Garimella, S.V. Investigation of liquid flow in microchannels. *AIAA J. Thermophys. Heat Transf.* **2004**, *18*, 65–72. [CrossRef]
24. Harms, T.M.; Kazmierczak, M.J.; Gerner, F.M. Developing convective heat transfer in deep rectangular microchannels. *Int. J. Heat Fluid Flow* **1999**, *20*, 149–157. [CrossRef]
25. Popescu, A.; Welty, J.R.; Pfund, D.; Rector, D. Thermal measurements in rectangular microchannels. In Proceedings of the ASME 2002 International Mechanical Engineering Congress and Exposition, New Orleans, LA, USA, 17–22 November 2002.
26. ASPEN COMPRESSOR, LLC. ASPEN Miniature Rotary Compressor. 2012, pp. 1–2. Available online: www.aspencompressor.com (accessed on 5 March 2022).
27. ASPEN COMPRESSOR, LLC. ASPEN Recommended Compressor Installation Practices. 2012, pp. 1–11. Available online: https://docplayer.net/21586235-Recommended-compressor-installation-practices.html (accessed on 12 March 2022).

28. Gas Encyclopedia, H2FC-CF3, 1,1,1,2-Tetrafluoroethane (R134A), CAS Number: 811-97-2, UN3159. Available online: http://encyclopedia.airliquide.com/encyclopedia.asp? (accessed on 15 April 2022).
29. Available online: http://www.peacesoftware.de/einigewerte/calc_r134a.php5 (accessed on 25 April 2022).
30. Mathcad 14, Licensed to: Stefan cel Mare University, Partial Product Code JE140709XX2311-XXD9-7VXX
31. ** Purswave Qx36W Refrigeration Units Water Cooling Units Mini Refrigeration System 220V/50Hz R134A. Available online: https://purswave.en.made-in-china.com/product/AyGxEefHVPkb/China-Purswave-Qx36W-Refrigeration-Units-Water-Cooling-Units-Mini-Refrigeration-System-220V-50Hz-R134A.html (accessed on 26 April 2022).
32. ** HVAC Pioneer Company, 12v 24v Mini Refrigerator Cooling Unit with Miniature Refrigeration Compressor. Available online: https://www.coowor.com/p/20190731153903140G6GKAJ/12v-24v-mini-refrigerator-cooling-unit-with-miniature-refrigeration-compressor.htm (accessed on 28 April 2022).
33. ** FS THERMO Mini Condensing, DC 12v 24v 48v Smallest Condensing Unit with Miniature Refrigeration Compressor for Mini Freezer. Available online: https://www.alibaba.com/product-detail/dc12v-24v-48v-smallest-condensing-unit_62283432657.html (accessed on 2 May 2022).

micromachines

MDPI

Article

Particle Deposition in the Vicinity of Multiple Film Cooling Holes

Yubo Peng, Guoqiang Xu, Xiang Luo *, Jian He and Dongdong Liu *

National Key Laboratory of Science and Technology on Aero-Engine Aero-Thermodynamics, The Collaborative Innovation Center for Advanced Aero-Engine of China, Beihang University, Beijing 100191, China; dream070141@buaa.edu.cn (Y.P.); 04822@buaa.edu.cn (G.X.); clhejian@126.com (J.H.)

* Correspondence: xiang.luo@buaa.edu.cn (X.L.); ddliu@buaa.edu.cn (D.L.)

Abstract: Particle deposition on film cooling surface is an engineering issue that degrades the thermal protection of turbine blade. Here, we present a combined experimental and numerical investigation on the particle deposition in the vicinity of multiple film cooling holes to reveal the effect of interactions between cooling outflows on particle deposition. The numerical simulation of film cooling with a group of three rows of straight film cooling holes is conducted and validated by experimental data with blowing ratios ranging from 0 to 0.08. Wax particles with size range from 5 to 40 μm are added in the heated mainstream to simulate the particle deposition in the experiment. The simulation results show the decrease of particle deposition with blowing ratio and various deposition characteristics in different regions of the surface. The flow fields from numerical results are analyzed in detail to illustrate deposition mechanism of the particles in different regions under the interactions of cooling outflows. The cooling air from the holes in the first row reduces the particle concentration near the wall but causes particle deposition in or between the tail regions by the generated flow disturbance. The cooling air from the latter hole separates the diluted flow in the upstream from the wall, and creates a tail region without particle deposition. This revealed particle deposition characteristics under the effect of outflows interaction can benefit the understanding of particle deposition in engineering applications, where multi-row of cooling holes are utilized.

Keywords: film cooling; heat transfer; particle deposition

1. Introduction

For thermal protection of the turbine blade, cooling air from secondary air system is supplied through the elaborately arranged film cooling holes on the blade to form a thin cooling air film for reduction of heat transfer from the hot mainstream flow. This so-called film cooling technology has been widely used for decades in turbomachinery for its high reliability in turbine blade cooling. Numerous studies have been carried out to reveal the mechanism of film cooling, optimize the design of hole, and improve cooling efficiency [1–6].

One of the challenges facing in film cooling is the deposition of particles, which are mainly originated from the pollutants, e.g., sand, debris, and ash, in inlet air or fuel, on the turbine blade [7–9]. These deposits can easily change the surface morphology and block the cooling holes, resulting in the deviation of the film cooling from its original design and thus degrading of the film cooling efficiency and endangering the safety of the turbine blade. To obtain further insights into the characteristics of particle deposition on different areas of the turbine blade, the particle deposition was divided into the external surface deposition and internal wall deposition.

For external particle deposition, Demling et al. [10] presented the effects of deposition accumulation in the upstream and downstream of the film hole on the film cooling efficiency. Ai et al. [11] focused on the deposition of pollutants in coal derived syngas near the film cooling hole. It was found that the deposition mainly occurred in the upstream of the film hole and the downstream area between the two film holes. However, there was less deposition coverage in the downstream of the hole. Somawardhana et al. [12] investigated the effect of geometric parameters on the cooling efficiency and found that the deposition shape had little effect on the cooling efficiency, but the deposition height had a great impact on the cooling efficiency. For the internal hole blockage, Pan et al. [13] studied the effect of particle blockage in the monoclinic hole on film cooling, and analyzed the influence of different positions of the blockage in the hole on film cooling. The results showed that the blockage can severely affect the film cooling when it was on the upstream surface of the hole's inner wall. Cardwell et al. [14] studied the effect of different particle blockages on the mass flow of cooling air under a specific pressure ratio. Wolff et al. [9] explored the impact of particle size on the deposition on the film cooling. The results showed that the particle with smaller size was easier to deposit, but the large particle was likely to destroy the deposition. Barker et al. [15] experimentally studied the influence of hole diameter, hole spacing, and hole inclination on hole blockage, and found that the hole inclination was the most important parameter in affecting the blockage inside the hole. Moreover, some researchers [16] studied the effects of both the external deposition and internal blockage on film cooling.

Many studies also focused on the dynamics of the deposition process in film cooling by changing the flow parameters or hole geometry. Rozati et al. [17] simulated the deposition of particles at the leading edge of film cooling blades, and the results indicated that the increase of blowing ratio reduced the deposition. Ai et al. [18] studied the effects of blowing ratio on deposition growth and blade surface temperature. It was found that the increase of blowing ratio led to the decrease of blade surface temperature and the growth of deposition. Lawson et al. [19] experimentally investigated the effect of particle deposition on film cooling of single row inclined holes. The results illustrated that with the increase of film cooling momentum ratio, the reduction of film cooling efficiency caused by particle deposition had a peak. Mensch et al. [20] investigated the particle deposition on the blade end wall with vapor film cooling. The experiment result revealed that the particle deposition had both positive and negative effects on the wall heat transfer. On one hand, the deposition was helpful to the heat insulation. On the other hand, the deposition changed the wall roughness and increased the wall heat transfer. For the geometry of the film cooling hole, Wang et al. [21] studied the effects of different film hole shapes on particle deposition. Wang et al. [22] numerically simulated the film cooling of single row laidback fan-shaped cooling holes to study the interaction between particles and cooling wall.

Researches on the particle deposition were mostly limited to the configuration of a single hole or a single row of holes, neglecting the interactions between the outflows from holes in different rows, which may play an important role in practical applications, as multiple rows of film cooling holes were utilized. This multiple row configuration aims not only for approaching the reality, but also for simulating the transpiration cooling, the extreme form of film cooling, by using the straight holes [23,24]. In this study, we utilize a multi-row configuration, three rows of staggered straight holes to both simulate the transpiration cooling in engineering application and simplify the geometry factor, to investigate the effect of outflow interactions on particle deposition characteristics on flat surface. Commercial software Fluent and its own discrete phase model are applied to numerically simulate the particle deposition process. Experiment of particle deposition using wax particle is conducted to compare with the simulation data. The calculation reveals how the interactions of the outflows from different cooling holes affect the particle deposition characteristics on a flat surface.

2. Materials and Methods

2.1. Experimental Apparatus

A low speed open loop wind tunnel is utilized for the experiment of the particle deposition. The schematic of the experimental setup is shown in Figure 1. An air compressor supplies the air flow, including the flow in the mainstream and the cooling air. These two flows are controlled by two gas control valves separately. One FCI-ST98 thermal flowmeter with the uncertainty of ±0.5% is assembled in each of these two flow channels to measure the flow rate. After the flowmeter, the air in the mainstream is heated by an electric heater to temperature T_g of 102 °C. After heating, melted wax with melting temperature at 44 °C supplied from a pressurized container, is sprayed into the mainstream channel via a spray nozzle, which gives a particle size range between 5 and 40 µm. The particles fully mix with the hot air and flow into the rectangular channel of test section, which is 30 mm wide and 45 mm high. For the cooling air, it is led into a coolant chamber for pressure stabilization, in which its temperature T_c and pressure P_c are measured. A stainless plate with 204 staggered arranged film cooling holes is assembled between the coolant chamber and mainstream channel, allowing the cooling air to flow through the cooling hole into the mainstream. The duration of the particle deposition experiment is 30 s. In the outlet of the tunnel, a filter is placed to collect the added wax particles. In the experiment, the weight of the wax container and film cooling plate are measured before and after the experiment to calculate the weight of used wax and deposited wax on the surface, respectively. Two calibrated K type thermocouples are fixed in the flow channels to monitor the temperature of the cooling air in the cooling chamber and hot air in the mainstream.

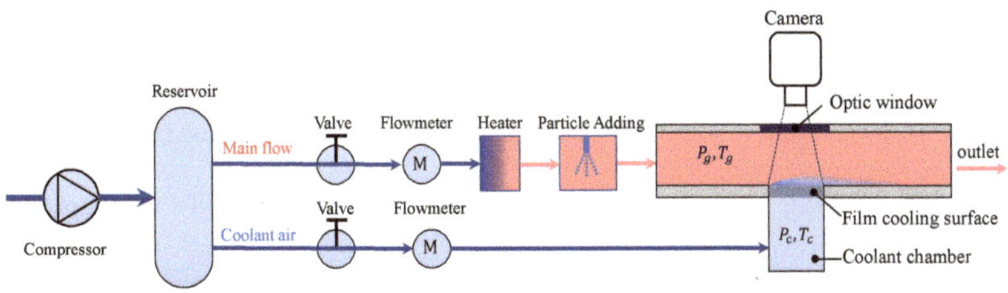

Figure 1. Schematic of the experiment apparatus for particle deposition experiment.

The blowing ratio F, which measures the mass flow ratio between the cooling air and hot air in the mainstream [25], can be defined as follows

$$F = \frac{\rho_c u_c}{\rho_g u_g}, \tag{1}$$

where ρ_c and ρ_g are the densities of the coolant and main flow, respectively, and u_c and u_g respectively, are the velocities of the coolant in the cooling chamber and main flow. Based on the parameters in engineering applications, we vary the blowing ratio F from 0 to 0.08 for both experiment and simulation. The cooling efficiency η [6] is defined as follows

$$\eta = \frac{T_{aw} - T_w}{T_{aw} - T_c}, \tag{2}$$

where T_{aw} is the adiabatic wall temperature, T_w the surface temperature, and T_c the cooling air temperature.

2.2. Particle Deposition Simulation

In numerical simulation, we choose a representative unit, including seven straight film cooling holes, one in the center and the other six surrounding holes arranged in the corners of a hexagon. As illustrated in Figure 2a, the simulation model consists of a coolant chamber and a mainstream channel with a width of 9 mm along the x axis, a height of 15 mm along the z axis, and a length of 49 mm along y axis. Seven straight film cooling holes connect the coolant chamber and mainstream channel. The diameter of the cooling hole is 1 mm, and the distances between the holes in the flow direction y and the transverse direction x are both 3 mm, same as the experiment.

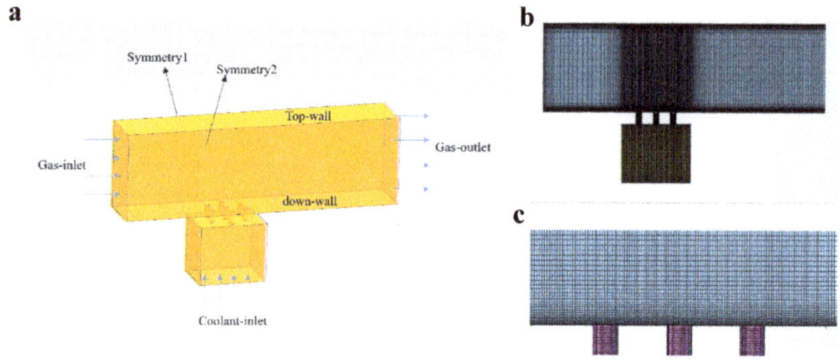

Figure 2. (**a**) Schematic of the simulation model and the boundary conditions. The detail of the generated mesh from the side view (**b**) and in the near wall region (**c**).

The commercial software ANSYS ICEM is used to generate Hexahedral mesh in the fluid domain. O-grid mesh is applied in the region of film cooling holes for high mesh quality. The detail of the mesh is shown in Figure 2b,c. Mesh refinement is conducted near the wall to keep the Yplus smaller than 1.2. The total quantity of elements in the mesh is 3.55 million. Fluent is chosen for numerical simulation in this study for its convenience in modeling the particle deposition. We use the idea gas as the working fluid due to the low flowing velocity and small range of temperature in this study. For the boundary condition, the inlet velocity in the mainstream is kept at 20 m/s and the temperature is set at 475 K. All the walls are set as non-slippery and adiabatic surfaces. The mass flow rate of the cooling gas with inlet temperature of 300 K is adjusted according to the blowing ratio F of the case. Since this calculation intends to simulate the particle deposition in transpiration cooling, therefore the outflow angle between the film cooling hole and y-axis is set at 90°, and the blowing ratio F is varied between 0 and 0.08—the same as in transpiration cooling.

To calculate the particle deposition and track the particle in the mainstream channel, discrete phase model (DPM) is implemented upon the flow field and heat transfer simulation. The trajectory of the particle is calculated using the Lagrangian method, which tracks the motion of each particle in the flow field. The motion of the particle is mainly determined by the flow field, but is also affected by its intrinsic random motion. To capture the motion of the particles and avoid the difficulties in coupled simulation caused by the interaction between the particles and flow field, several assumptions are made: (1) each particle is considered as a non-spinning hard sphere with uniform density; (2) the interaction between particles are neglected due to the low mass fraction ($\sim 10^{-6}$ kg/m^3) of the particles in this study; (3) the added particles do not affect the flow field; (4) the rebound of the particle caused by the surface roughness is neglected. Based on these assumptions, the particle is considered to be trapped on the wall when it contacts with the surface. If the particle is not colliding with the surface, its motion will be calculated until it leaves the controlled volume. Therefore, we are able to simulate the process of particle deposition in the mainstream channel. It has to be noted that the simulation is a steady state simulation by setting the mass flow rate of the added particle at 3×10^{-8} kg/s for duration of 1 s.

In this steady state simulation, the initial velocity of the added particle is set at 0 m/s, since the initial velocity is found to barely affect the deposition result. Based on the experiment, the range of the particle size in calculation is between 5 and 40 μm, following the Rosin–Rammler distribution and having an average diameter of 10 μm. The physical properties of the added wax particle are shown in Table 1.

Table 1. Physical properties of the added wax particle.

Particle Property	Value
Size range	5–40 μm
Density	$900\,\mathrm{kg/m^3}$
Specific heat	$984\,\mathrm{J/(kg{\cdot}K)}$
Thermal conductivity	$0.5\,\mathrm{W/(m{\cdot}K)}$

2.3. Simulation Validation

To choose a suitable turbulent model in the calculation of film cooling, we calculate the film cooling of a single hole based on the reference [26] using four turbulent models. The particle deposition is not considered in this simulation.

The geometry of the model is shown in Figure 3a. According to the settings in reference [26], we build a similar model with the mainstream channel having a length of 294 mm, a width of 18 mm, and a height of 90 mm. The diameter of the cooling hole is 6 mm and its inclination angle is 35°. The inlet velocity in the mainstream is set at 20 m/s, and the gas temperature at 298 K. The gas temperature in the cooling flow is kept at 188 K, and the blowing ratio is set at 0.08.

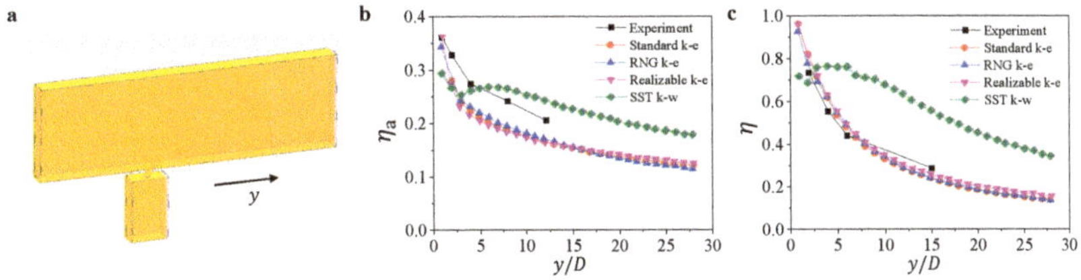

Figure 3. (**a**) Schematic of the simulation for model validation. The variation of (**b**) the averaged cooling efficiency η_a in transverse direction and (**c**) the cooling efficiency η in the center line along the flow direction for simulation result and experiment data from reference [26].

We present the simulation results, including the averaged cooling efficiency η_a in transverse direction and the cooling efficiency in the center line along the flow direction, comparing it with the experiment data from reference [26] in Figure 3b,c. Except for the result calculated using the turbulent model SST $k-\omega$, other simulation results have similar trends with the experimental data. Since the result from the turbulent model, RNG $k-\varepsilon$, is the best when comparing with the experimental data, we choose this turbulent model in the particle deposition simulation in this study.

2.4. The Initial Velocity of Added Particles

In this section, we explore the effect of the initial velocity V_p of the added particle on the particle deposition on the surface. In the experiment, the particles are added into the mainstream via a spraying nozzle, which is difficult to simulate in the calculation. Therefore, we assume the particles are uniformly added into the mainstream with same initial velocity V_p.

We calculate three cases with same blowing ratio $F = 0.08$ but change the initial velocity of the added particle from 0 to 20 m/s. The simulation results for different V_p are presented in Figure 4. In general, the contours of deposited particle thickness h between three cases are qualitatively similar (Figure 4a–c). We also extract the deposition thickness along the x axis on the surface at $y = 1.5\,\mathrm{mm}$, the particle mass concentration c along x-axis direction at ($z = 1\,\mathrm{mm}$, $y = 1.5\,\mathrm{mm}$) and ($z = 5\,\mathrm{mm}$, $y = 1.5\,\mathrm{mm}$) (Figure 4d–e). Although the thickness and concentration in some areas change, which can be attributed to the random motion of the particles, the data from three cases generally show insignificant differences, suggesting negligible effect of the particle's initial velocity on the particle deposition simulation. Therefore, we use the result of $V_p = 0$ for analysis of particle deposition in this study.

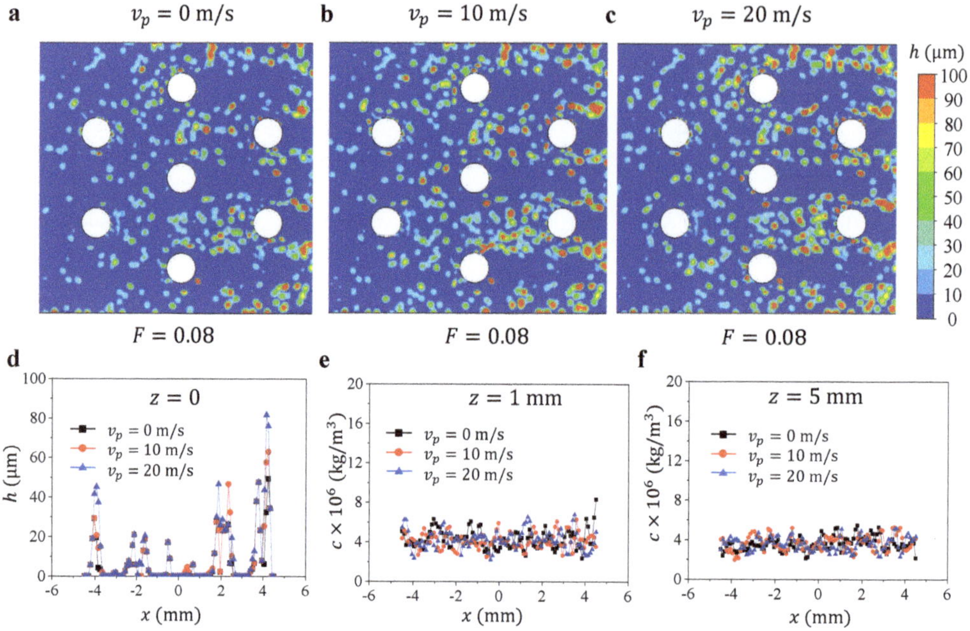

Figure 4. Contours of particle deposition thickness h at $F = 0.08$ with different V_p: (**a**) $V_p = 0\,\text{m/s}$, (**b**) $V_p = 10\,\text{m/s}$, and (**c**) $V_p = 20\,\text{m/s}$. (**d**) The deposition thickness h along x direction on the surface at $y = 1.5\,\text{mm}$. The particle mass concentration c along x direction at (**e**) ($z = 1\,\text{mm}$, $y = 1.5\,\text{mm}$) and (**f**) ($z = 5\,\text{mm}$, $y = 1.5\,\text{mm}$).

3. Results and Discussion

3.1. Particle Capture Efficiency

In the experiment, the particle deposition on the surface is a time-dependent process. The particles added in the mainstream would keep accumulating on the surface with time, resulting in an ever-increasing deposition thickness. However, we use steady state simulation for modeling the particle deposition. To validate our modeling result on particle deposition, we compare the particle capture efficiency ε, the ratio between the mass of the deposited particles on the surface and that of the added particles in the mainstream, from experiment and simulation, as shown in Figure 5.

The result in Figure 5 shows the monotonic decrease of the particle capture efficiency ε with blowing ratio F from both the experiment result and simulation data. The increase in F means more cooling air is supplied through the film cooling holes on the flat surface to prevent the particle deposition, leading to less deposition of the particles on the surface. The simulation result has a similar trend with the experimental data, although the simulation underestimates the particle deposition. The underestimation of ε from the simulation may be caused by the surface roughness on the wall or the effect of added particle on the mainstream flow, which are both not considered in the simulation. However, the comparison validates our simulation result of particle deposition on the surface.

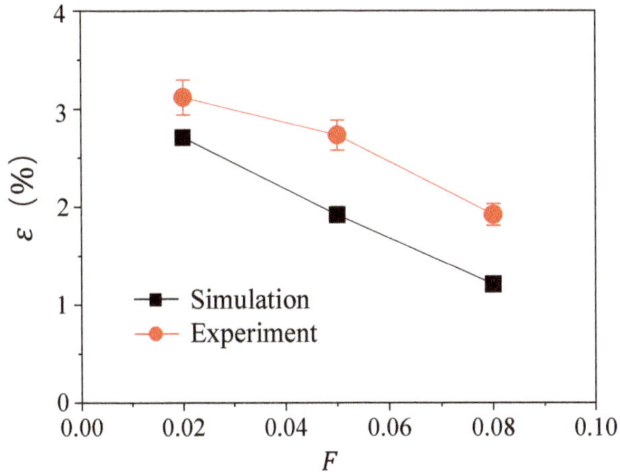

Figure 5. The variation of particle capture efficiency ε with blowing ratio *F* for experiment result and simulation data.

3.2. The Effect of Blowing Ratio on Particle Deposition

For the added particle, due to its higher density than that of surrounding gas, it intends to fall upon the surface under the gravitational force. Although the particle's small size and high gas velocity in the mainstream induce the added particles to follow the gas and move through the channel, the low velocity field in the boundary layer near the wall makes the particle deposition possible. When the blowing ratio is zero, i.e., no cooling air is supplied through the film cooling holes in the wall, the particles collide with the surface and deposit on the wall under the gravity (Figure 6a). With the increase of the flow distance and correspondingly the duration of falling particles, more particles are likely to contact the surface, inducing the increase of h in the flow direction. The irregular pattern of the deposited particles that appeared on the surface (Figure 6a) may be caused by the random motion of the particles. We measure the averaged deposition thickness h_a in several 1 mm × 1 mm areas along the flow direction y-axis with the center points lying on the lines s1 and s2 (Figure 6b), and plot the h_a with the y coordinate y_c of the center point of the chosen areas in Figure 6c. The data show the increase of h_a with y_c in general. Here, we define a thickness h_0, the averaged deposition thickness ($h_0 = 128$ μm) in the square region where $-4.5\,\text{mm} \leq y \leq 4.5\,\text{mm}$ and $-4.5\,\text{mm} \leq x \leq 4.5\,\text{mm}$ for $F = 0$.

With the supply of the cooling air on the surface ($F = 0.02$), the thickness h of the deposited particle substantially decreases (Figure 6a). The injected cooling air not only gives an upward momentum to the particles near the surface but also separates the particles from the surface, both preventing the particles from depositing on the surface. With the increase of blowing ratio F, the deposition thickness h generally decreases on the surface. For the convenience of the description, we denote the holes from "H1" to "H7", and their tail regions, i.e., the regions behind the holes along the flow direction from "HT1" to "HT7", as shown in Figure 6b. In particular, we observe some deposition patterns of the particles in the tail regions of the film cooling holes. For the tail regions in HT1 and HT2, the deposited thickness of particles are almost the same as the region in between these two regions when $F = 0.02$ and $F = 0.08$. For $F = 0.05$, h in HT1 and HT2 becomes larger than the surrounding region. However, the thickness h in the tail regions behind other film cooling holes, from H3 to H7, are significantly smaller than the surrounding regions, particularly when $F = 0.08$. These phenomena are also found in the experiment, as shown in Figure 6d.

To obtain a quantitative result of the particle deposition on the surface when $F > 0$, we measure the h_a in several 1 mm × 1 mm areas along s1 and s2, avoiding the film cooling holes, and normalize h_a using h_0 (Figure 7). The variation of h_a/h_0 with y_c on s1 and s2 are generally different. For h_a/h_0 on s1 (Figure 7a), we observe the increase of h_a/h_0 with y_c before the film cooling hole H4 (where $y_c = 0$), followed by the drop of h_a/h_0 around H4 and then the increase of h_a/h_0 with y_c. The drop of h_a/h_0 is attributed to the cooling flow from H4, which reduces the particle deposition, while the increase of h_a/h_0 after H4 may be induced by the reduced effect of the film cooling resulting from the detachment of cooling air from the surface. For h_a/h_0 on s2 (Figure 7b), the variation of h_a/h_0 with y_c over two cooling holes, H2 and H7, are significantly different. The dimensionless thickness h_a/h_0 keeps increasing after the first cooling hole H2 (where $y_c = -3$ mm), and the drop of h_a/h_0 is only observed after the second hole H7 (where $y_c = 3$ mm). The difference in the deposition characteristics after H4 and H2 could be resulted from the different incoming flow conditions of these two holes. For the flow before H2 on s2, it is undisturbed and can be considered as uniform; while for the flow before H4 on s1, the incoming flow field is already been affected by cooling outflow from H1 and H2. Therefore, the interaction between outflows from different film cooling holes results in different particle deposition characteristics, compared with that generated by one single film cooling outflow.

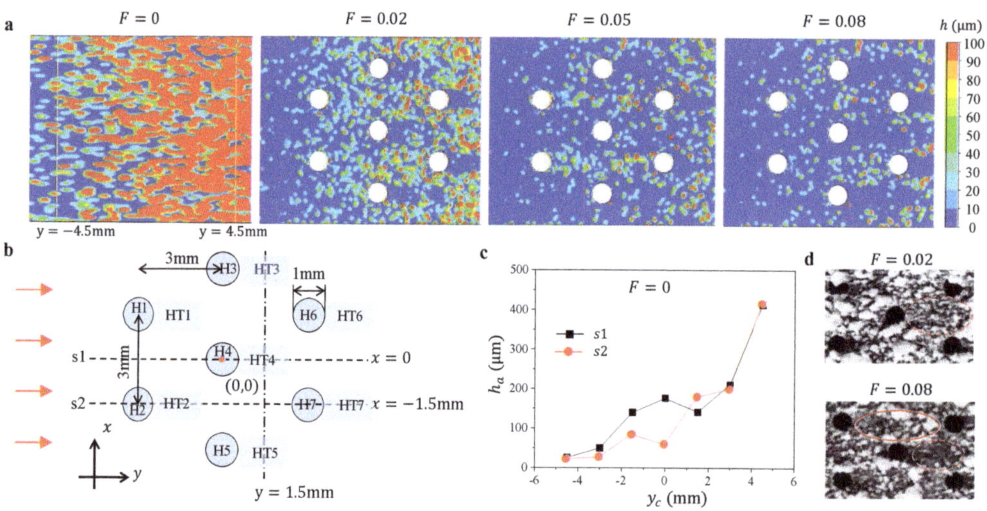

Figure 6. (**a**) Contours of particle deposition thickness h with different blowing ratio F. (**b**) Schematic of the film cooling holes arrangement and the labeled areas. (**c**) The variation of averaged deposition thickness h_a with the y coordinate y_c of the center point of the chosen areas along s1 and s2. (**d**) Experiment snapshots of the particle deposition in the middle of the flat surface at $F = 0.02$ and $F = 0.08$.

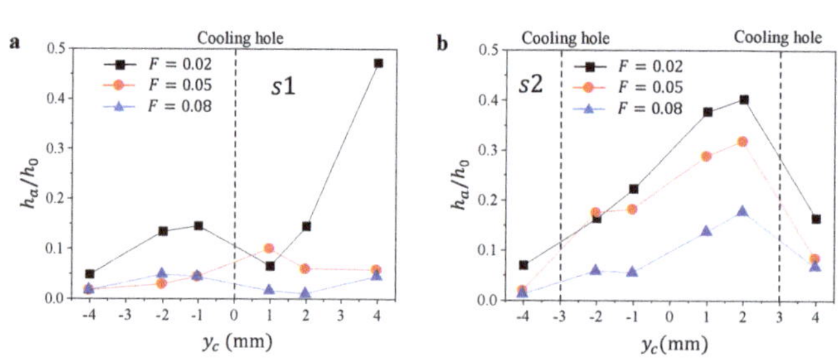

Figure 7. The variation of normalized thickness h_a/h_0 with y_c for different blowing ratio F along (**a**) s1 and (**b**) s2.

3.3. The Interaction between the Outflows of Multiple Film Cooling Holes

To obtain further insights into the particle deposition characteristics resulting from the interaction between outflows of multiple film cooling holes, we provide the detail of the flow fields from various view angles for further analysis.

In Figure 8, we show the flow fields accompanied with the contours of the particle concentration c in the $x-z$ cross sections located in different positions along the s1 and s2 (Figure 8e,f) with $F = 0.02$ (Figure 8a,c) and $F = 0.08$ (Figure 8b,d). Here, we denote the cross section at position 1 along s1 when $F = 0.02$ as "$F = 0.02 - s1 - 1$" for the convenience of description. We highlight the sections in which the cooling outflow occurs, and the generated vortices by cooling air.

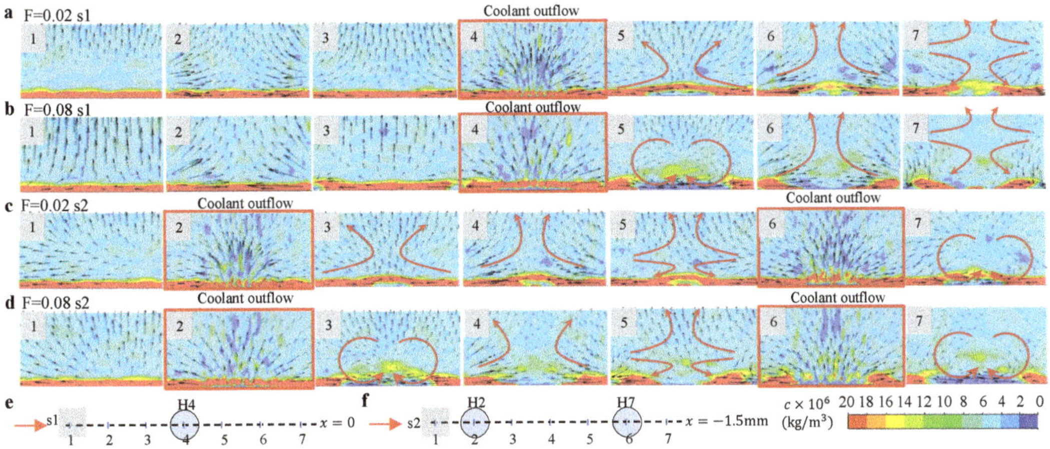

Figure 8. The contours of the particle mass concentration c with accompanied velocity vector field at different $x-z$ cross sections along s1 or s2 at different blowing ratios: (**a**) $F = 0.02$-s1, (**b**) $F = 0.08$-s1, (**c**) $F = 0.02$-s2, and (**d**) $F = 0.08$-s2. The cross sections where cooling outflow happens are highlighted. The schematics of the positions of the cross section along (**e**) s1 and (**f**) s2.

In "$F = 0.02 - s1 - 1$", we observe a high concentration of particles near the wall and upward flow of the fluid. The former one may be caused by the small velocity in the boundary layer, which induces the accumulation of the particles, and the latter one can be attributed to the coolant outflow in the downstream. In "$F = 0.02 - s1 - 2$" and "$F = 0.02 - s1 - 3$", we observe the flow in transverse direction that is caused by the cooling outflow from the neighboring cooling holes, namely, H1 and H2. In "$F = 0.02 - s1 - 4$", the outflow of the cooling gas significantly disturbs the distribution of the particle concentration in the boundary layer, and generates a pair of "kidney" shape vortices (clearly seen in "$F = 0.08 - s1 - 4$") in the downstream. This pair of vortices moves upward and induces another pair of vortices underneath in the downstream (as shown in "$F = 0.02 - s1 - 5$" and "$F = 0.02 - s1 - 6$"). Although these flow field variations have been discovered in many studies of film cooling, they illustrate the strong mixing near the wall caused by the cooling outflow. In Figure 8c,d, we observe a similar variation of flow field along s2 after two film cooling holes, but the concentration contours change by the interaction between the two cooling outflows, particularly shown in "$F = 0.08 - s2 - 2$" and "$F = 0.08 - s2 - 6$". The mixing caused by the outflow from H2 induces lower particle concentration in its tail region, which is further diluted by the cooling outflow from H7, resulting in a region with nearly zero particle concentration in the tail region (shown in "$F = 0.08 - s2 - 7$").

In Figure 9, we present the velocity field and concentration contour in $y-z$ cross sections along s1 and s2. Along s1, the incoming boundary layer flow with high particle concentration is lifted and diluted by the cooling air (Figure 9a,b). Along s2, we observe that the diluted flow by the cooling air from the first hole is further mixed with the cooling air from the second hole. In particular, a region without any particle appears behind the second hole as shown in Figure 9d. In Figure 9e,f, the streamlines of the cooling air originating from two cooling holes are respectively colored with grey and black to illustrate the interaction between the cooling flows from two holes along s2. The outflow from the first hole, overwhelmed by the flow in the mainstream, enters the boundary layer, and is further lifted by the outflow from the second hole.

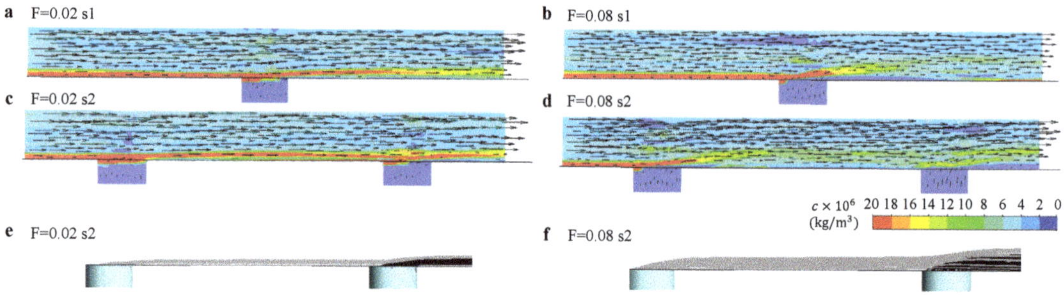

Figure 9. The velocity field and concentration contour in $y - z$ cross section along s1 or s2 at different blowing ratios: (**a**) $F = 0.02$-s1, (**b**) $F = 0.08$-s1, (**c**) $F = 0.02$-s2, and (**d**) $F = 0.08$-s2. The streamline of the cooling air originated from two cooling holes along s2 for (**e**) $F = 0.02$ and (**f**) $F = 0.08$. The streamline from the first hole is colored in grey, while that from second one is colored in black.

Based on these flow fields and concentration contours, we are able to understand the particle deposition characteristics under the interactions between the cooling outflows from the neighboring cooling holes. For the film cooling outflow, due to its low velocity, it enters into the boundary layer of the mainstream and induces the mixing in the mainstream, which not only dilutes the particle concentration in the boundary layer but also gives higher possibility of the particle deposition by the induced disturbance. Therefore, the film cooling outflow from one hole may not necessarily eliminate the particle deposition in its tail region (Figure 6a). The cooling outflow from the second hole lifts the diluted flow after the first one and supplies the boundary layer with a flow of zero particle concentration, leading to a tail region without particle deposition (Figure 6a). For the particle deposition in between these tail regions, it can be explained by the redistribution of the particles in the mainstream under the disturbance of the cooling flow, as shown in Figure 10. The particle concentration in the $x - z$ cross section is uniform along the x-axis before the film cooling holes ($y = -15$ mm) and it becomes uneven after the film cooling holes ($y = 1.5$ mm). The generated vortices induces the tail regions with lower particle concentration and those in between with higher particle concentration (Figure 10c,d), resulting in more particle deposition in areas between the tail regions (Figure 4a).

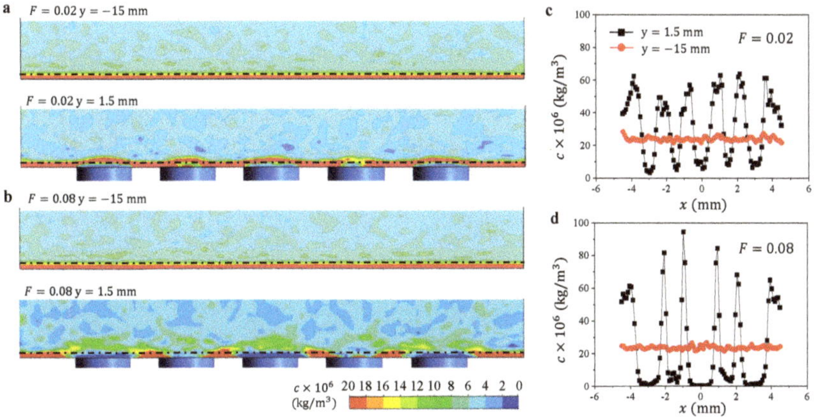

Figure 10. The contours of the particle concentration c in the $x - z$ cross section before the hole ($y = -15\,\text{mm}$, upper panel) and after the hole ($y = 1.5\,\text{mm}$, lower panel) for (**a**) $F = 0.02$ and (**b**) $F = 0.08$. The variation of particle concentration c with x along the dash-dot lines ($z = 0.5\,\text{mm}$) in (**a**) and (**b**) for (**c**) $F = 0.02$ and (**d**) $F = 0.08$.

4. Conclusions

In conclusion, the particle deposition in the vicinity of multiple film cooling holes are presented by simulation and validated by experiment. In general, the supply of cooling air through the hole reduces the particle deposition, and this effect increases with the blowing ratio. The interactions between the outflows of the film cooling holes reveals different deposition characteristics in different regions of the surface. The particle deposition occurs after the hole in the first row, but it can be eliminated in the tail regions of the holes starting from the second row. The flow field analysis from simulation indicates that the interaction between the outflows of different holes generates such characteristics of particle deposition. Due to the high particle concentration in the boundary layer near the wall, the cooling air from the hole in the first row dilutes the concentration, but induces the disturbance that facilitates particle deposition. The cooling air from the hole in the second row lifts the diluted flow and separates the particles from the wall, generating a tail region without deposition. These revealed particle deposition characteristics could be helpful in understanding the deposition mechanism in practical applications and benefit the design of the film cooling and transpiration cooling.

Author Contributions: Conceptualization, G.X., X.L. and D.L.; methodology, G.X., X.L., Y.P. and D.L.; investigation, Y.P. and J.H.; writing—review and editing, Y.P. and D.L.; supervision, G.X., X.L. and D.L. All authors have read and agreed to the published version of the manuscript.

Funding: This research received no external funding.

Data Availability Statement: All processed data in this study are included in this published article. Raw data will be provided on request from the corresponding author.

Acknowledgments: This work is supported by the National Science and Technology Major Project (2017-III-0003-0027).

Conflicts of Interest: The authors declare no conflict of interest.

Nomenclature

c	Particle mass concentration
D	Diameter of the hole
f	Blowing ratio
H	Hole
HT	Hole tail region
h	Particle deposition thickness
h_a	Averaged deposition thickness
h_0	Averaged deposition thickness for $F = 0$
P_c	Pressure of coolant chamber
T_c	Temperature of cooling air
T_{aw}	Adiabatic wall temperature
T_w	Surface temperature
u_c	Velocity of the coolant
u_g	Velocity of the mainstream flow
V_p	Initial velocity of particle
x	The transverse direction of main flow
y	The direction of mainsteam flow
y_c	The y coordinate of center point of the chosen areas
z	Vertical direction
ε	Particle capture efficiency
η	Cooling efficiency
η_a	Averaged cooling efficiency in transverse direction
ρ_c	Density of the coolant
ρ_g	Density of the mainstream flow

References

1. Bunker, R.S. A review of shaped hole turbine film-cooling technology. *J. Heat Transf.* **2005**, *127*, 441–453. [CrossRef]
2. Goldstein, R.; Eckert, E.; Burggraf, F. Effects of hole geometry and density on three-dimensional film cooling. *Int. J. Heat Mass Transf.* **1974**, *17*, 595–607. [CrossRef]
3. Kim, Y.S.; Jeong, J.Y.; Kwak, J.S.; Chung, H. The effects of hole arrangement and density ratio on the heat transfer coefficient augmentation of fan-shaped film cooling holes. *Energies* **2021**, *14*, 186. [CrossRef]
4. Ravelli, S.; Abdeh, H.; Barigozzi, G. Numerical Assessment of Density Ratio and Mainstream Turbulence Effects on Leading-Edge Film Cooling: Heat and Mass Transfer Methods. *J. Turbomach.* **2021**, *143*, 041002. [CrossRef]
5. Dupuy, D.; Perrot, A.; Odier, N.; Gicquel, L.Y.; Duchaine, F. Boundary-condition models of film-cooling holes for large-eddy simulation of turbine vanes. *Int. J. Heat Mass Transf.* **2021**, *166*, 120763. [CrossRef]
6. Zhang, J.; Zhang, S.; Chunhua, W.; Xiaoming, T. Recent advances in film cooling enhancement: A review. *Chin. J. Aeronaut.* **2020**, *33*, 1119–1136. [CrossRef]
7. Hamed, A.; Tabakoff, W.; Wenglarz, R. Erosion and deposition in turbomachinery. *J. Propuls. Power* **2006**, *22*, 350–360. [CrossRef]
8. Walsh, W.; Thole, K.; Joe, C. Effects of sand ingestion on the blockage of film-cooling holes. *Turbo Expo Power Land Sea Air.* **2006**, *4238*, 81–90.
9. Wolff, T.; Bowen, C.; Bons, J.P. The effect of particle size on deposition in an effusion cooling geometry. In Proceedings of the 2018 AIAA Aerospace Sciences Meeting, Kissimmee, FL, USA, 8–12 January 2018; p. 0391.
10. Demling, P.; Bogard, D.G. The effects of obstructions on film cooling effectiveness on the suction side of a gas turbine vane. *Turbo Expo Power Land Sea Air.* **2006**, *4238*, 547–556.
11. Ai, W.; Murray, N.; Fletcher, T.H.; Harding, S.; Lewis, S.; Bons, J.P. Deposition Near Film Cooling Holes on a High Pressure Turbine Vane. *ASME J. Turbomach.*, **2012**, *134*, 041013. [CrossRef]
12. Somawardhana, R.P.; Bogard, D.G. Effects of surface roughness and near hole obstructions on film cooling effectiveness. *Turbo Expo: Power for Land, Sea, Air.* **2007**, *47934*, 799–809.
13. Cheng-xiong, P.; Jing-zhou, Z.; Ke-nan, H. Numerical investigation of partial blockage effect on film cooling effectiveness. *Math. Probl. Eng.* **2014**, *2014*, 167193. [CrossRef]
14. Cardwell, N.; Thole, K.A.; Burd, S. Investigation of Sand Blocking within Impingement and Film-Cooling Holes. *ASME J. Turbomach.*, **2010**, *132*, 021020. [CrossRef]
15. Varney, B.; Barker, B.; Bons, J.; Wolff, T.; Gnanaselvam, P. Fine Particulate Deposition in an Effusion Plate Geometry. Turbo Expo: Power for Land, Sea, and Air. *Am. Soc. Mech. Eng.* **2019**, *58585*, V02DT47A017.
16. Wang, X.Y.; Pu, J.; Yuan, R.M.; Wang, J.H. Combined Influence of Surface Deposition and Hole-Blockage on Film-Cooling Performances. Turbo Expo: Power for Land, Sea, and Air. *Am. Soc. Mech. Eng.* **2016**, *49804*, V05CT19A019.

17. Rozati, A.; Tafti, D.K.; Sreedharan, S.S. Effects of Syngas Ash Particle Size on Deposition and Erosion of a Film Cooled Leading Edge. *ASME J. Turbomach.*, **2011**, *133*, 011010. [CrossRef]
18. Ai, W.; Laycock, R.G.; Rappleye, D.S.; Fletcher, T.H.; Bons, J.P. Effect of particle size and trench configuration on deposition from fine coal flyash near film cooling holes. *Turbo Expo Power Land Sea Air.* **2009**, *48845*, 561–571.
19. Lawson, S.; Thole, K. Effects of simulated particle deposition on film cooling. *J. Turbomach.* **2011**, *133*, 021009. [CrossRef]
20. Mensch, A.; Thole, K. Simulations of multiphase particle deposition on a gas turbine endwall with impingement and film cooling. *J. Turbomach.* **2015**, *137*, 111002. [CrossRef]
21. Wang, J.; Zhao, Z.; Tian, L.; Ren, X.; Sunden, B. Effects of Hole Configuration on Film Cooling Effectiveness and Particle Deposition on Curved Surfaces in Gas Turbines. *Appl. Therm. Eng.* **2021**, *190*, 116861. [CrossRef]
22. Wang, C.; Zhang, J.; Zhou, J. Investigation of Particle Deposition Characteristics in Vicinity of Laidback Fan-shaped Film Cooling Holes. *Flow Turbul. Combust.* **2016**, *97*, 591–607. [CrossRef]
23. Yang, L.; Chen, W.; Chyu, M.K. A convolution modeling method for pore plugging impact on transpiration cooling configurations perforated by straight holes. *Int. J. Heat Mass Transf.* **2018**, *126*, 1057–1066. [CrossRef]
24. Min, Z.; Huang, G.; Parbat, S.N.; Yang, L.; Chyu, M.K. Experimental investigation on additively manufactured transpiration and film cooling structures. *J. Turbomach.* **2019**, *141*, 031009. [CrossRef]
25. Liu, D.; Tao, Z.; Luo, X. Experimental investigation of heat transfer characteristics on turbine endwall with full coverage film cooling. *Appl. Therm. Eng.* **2018**, *140*, 295–303. [CrossRef]
26. Na, S.; Cunha, F.; Chyu, M.; Shih, T. Effects of Coating Blockage and Deposit on Film Cooling Effectiveness and Heat Transfer. In Proceedings of the 44th AIAA Aerospace Sciences Meeting and Exhibit, Reno, NV, USA, 9–12 January 2006; p. 24.

micromachines

MDPI

Article

Experimental Investigation on the Transpiration Cooling Characteristics of Sintered Wire Mesh in Plain Weave

Yubo Peng, Guoqiang Xu, Xiang Luo *, Jian He and Dongdong Liu *

National Key Laboratory of Science and Technology on Aero Engines Aero-Thermodynamics, The Collaborative Innovation Center for Advanced Aero-Engine of China, Beihang University, Beijing 100191, China; dream070141@buaa.edu.cn (Y.P.); 04822@buaa.edu.cn (G.X.); clhejian@126.com (J.H.)
* Correspondence: xiang.luo@buaa.edu.cn (X.L.); ddliu@buaa.edu.cn (D.L.)

Abstract: We experimentally investigate the transpiration cooling characteristics of a porous material, sintered wire mesh. Three samples with different porosities in a plain weave structure are tested with various blowing ratios in an open-loop wind tunnel with a heated mainstream flow. The temperature on the surface of the porous material is measured by an infrared camera to obtain the cooling efficiency. The measurements reveal nonuniform distributions of the surface temperature and the cooling efficiency in both the flow direction and the transverse direction. The averaged cooling efficiency on the surface first decreases and then increases with the blowing ratio, but increases and then decreases with the porosity of the material. The internal cooling by forced convection and its combination with the external film cooling from the transpiration cooling are considered to be attributed to those two cooling characteristics, respectively. Finally, we propose a modified blowing ratio to collapse the minima of the blowing ratio for all tested samples, providing an universal transition for the decreasing and increasing branches for all tested samples in the relation between averaged cooling efficiency and blowing ratio.

Keywords: sintered metal wire mesh; transpiration cooling; heat transfer; porous material

1. Introduction

Conventional cooling technologies such as film cooling and impingement cooling, although providing thermal protection of high-temperature components of aeroengine, have almost reached their performance limits. This not only makes improvement of their cooling performance much more difficult, but also becomes an obstacle in the ever-increasing demand for aeroengine efficiency.

Transpiration cooling uses porous material as a cooling wall which has a small pore size and extremely dense pore distribution with a typical size scale of 100 μm, inducing internal forced convection inside the material and external film cooling on the surface. The former reduces the overall average temperature of the solid framework of the porous material, while the latter generates a continuous and relatively uniform coolant film with low temperature on the surface of the porous material. This so-called "air membrane" is capable of separating the surface from the high temperature mainstream flow, inducing a further decrease in the surface temperature and, correspondingly, the improvement of cooling efficiency. Therefore, transpiration cooling has the highest theoretical cooling efficiency [1] and has shown its importance in various applications [2–5].

For transpiration cooling, several parameters—including the flow condition, the porous material and its geometric parameter—are deemed vital in determining the cooling characteristics. For the flow condition, one commonly investigated parameter is the blowing ratio, which measures the ratio of the mass flow rate between the coolant and the heated mainstream flow. The research of Jiang et al. [6] showed, that under the condition of constant thermal-physical properties, the increase in transpiration cooling blowing ratio greatly reduces the wall temperature and the convective heat transfer coefficient between the wall and the main stream. For a blowing ratio of 1%, the convection heat transfer coefficient was reduced by about 50%. Langener et al. [7] explored the transpiration cooling efficiency of C/C materials under the condition that the mainstream is subsonic or supersonic. The experiment revealed that the mainstream condition has little effect on the transpiration cooling efficiency in the current conditions. Subsequently, they selected air, argon and helium as the cooling medium for transpiration cooling. The experiments indicated that the specific heat capacity and blowing ratio of the cooling medium were the most important factors affecting transpiration cooling efficiency [3].

For the matrix porous material used for transpiration cooling, a variety of materials can be used as the matrix for transpiration cooling. Transpiration cooling based on sintered metal powder has been designed and applied in the nose cone of some hypersonic vehicles or scramjet combustors [3,8]. The porous materials synthesized using a ceramic matrix such as carbon/carbon has been applied for transpiration cooling in rocket engine thrust chamber cooling [9–12]. In recent years, sintered wire mesh has also been utilized as the liquid fuel injection panel of rocket engine thrust chamber [13] for transpiration cooling research. Some researchers also applied sintered wire mesh on the turbine blade wall of an aeroengine and preliminary explored the application possibility of turbine blade transpiration cooling [14–16].

To design and apply the transpiration cooling in practical applications, a full understanding of the structure parameters of the porous material for the transpiration cooling is required. Sintered metal powder, sintered wire mesh and C/C composites are several popular transpiration cooling matrix materials. The transpiration cooling structure of sintered metal powder has been widely studied. Liu et al. [17] investigated the effect of metal powder particle size on transpiration cooling efficiency. The cooling performance of the porous material with the smaller particle diameters was found to have better performance, mainly due to the enhanced convection heat transfer inside the material. Gan Huang et al. [18] studied the effect of liquid phase transition on the transpiration cooling of sintered metal powder. The results illustrated that the average cooling efficiency and the maximum wall temperature increased with the decrease in metal powder size from 600 μm to 90 μm. Langener et al. [19] explored the effect of material thickness and porosity on the cooling efficiency of composite carbon/carbon materials (C/C). It was found that material thickness and porosity do not affect cooling efficiency. Min et al. [20] experimentally studied a variety of transpiration cooling materials fabricated by 3D printing. The experiments demonstrated that—among the three geometric parameters of transpiration cooling structure, such as porosity, material internal fluid solid contact area and cold air outlet area—the material internal fluid solid contact area has the greatest impact on transpiration cooling efficiency.

Sintered wire mesh uses metal wires as the raw material. The metal wires are woven into a screen in a specific way, usually in the form of a Dutch or plain weave, and then the multi-layer wire mesh is reasonably arranged and manufactured via the sintering process. Sintered metal wire mesh porous material has good controllability: firstly, when weaving the wire mesh, the weaving method of different layers of wire mesh can be changed to adjust its mechanical properties to form a fiber-reinforced anisotropic material that can be adapted to loads in different directions; secondly, during the rolling process, the wire mesh can be deformed by varying the applied load, thereby changing the pore size, and then the porosity and permeability of the wire mesh [21].

Recently, Xu et al. [22] explored the transpiration cooling efficiency of sintered wire mesh in Dutch weave form with three kinds of porosity. The results showed that the cooling efficiency did not increase with the porosity. Ma et al. [23] studied the multi-layer structure of sintered wire mesh with different porosity combinations in the thickness direction. The experiment results indicated that the divergent cooling efficiency increased with the average porosity in the thickness direction of sintered wire mesh, and adjusting the order of wire mesh layers with different porosities will affect the transpiration cooling efficiency. However, for the sintered wire mesh in plain weave, which is another important weaving form for sintered wire mesh, its transpiration cooling characteristics have not been fully elucidated.

In this study, the sintered wire mech in plain weave is used as the matrix material for transpiration cooling. An open-loop wind tunnel with a heated mainstream flow is applied for experimental measurement. The surface temperature on three test samples with different porosity are captured by infrared camera to obtain higher cooling efficiency at various blowing ratios. The presented relation between the transpiration cooling efficiency and blowing ratio or porosity is capable of providing further insights in the transpiration cooling of sintered wire mesh for practical applications.

2. Materials and Methods

2.1. Sintered Wire Meshes

Sintered metal wire mesh is manufactured using stainless steel wires with the diameter of 140 μm. The manufacture process consists of two steps: Firstly, the wires are woven to produce a single-layer structure with uniform pattern, as shown in Figure 1. The weaving method of the sintered wire mesh used in this study is plain weave, one of the most commonly used weaving methods. Subsequently, several single-layer woven wire meshes are exposed sequentially to compression, rolling and vacuum sintering, to produce the ultimate sintered wire mesh with multiple-layer structure. For all the tested sample, the dimensions of the porous material is kept at 30 mm in width, 90 mm in length and 6 mm in thickness.

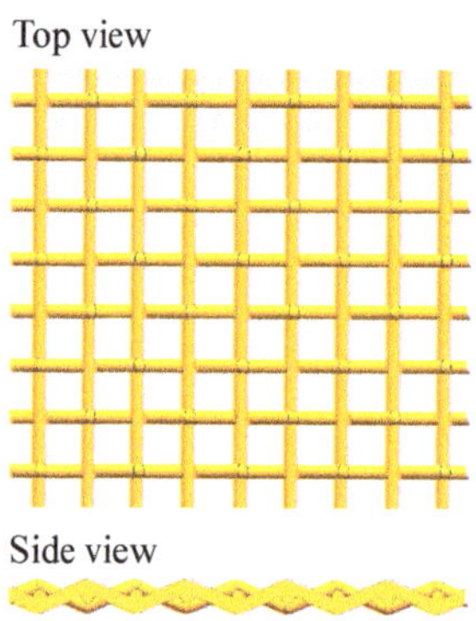

Figure 1. Schematic of the wire mesh structure. The diameter of the wire is 140 μm.

To obtain test samples with different porosity ε, i.e., the void fraction in the volume of the matrix, the quantity of the layers was changed from 40 to 54 in the manufacturing process. Typically, by increasing the layer quantity of the sintered wire meshed and compressing the matrix into a same size mode, the porosity of the manufactured sample decreases. The geometric detail of the sample is shown in Table 1.

Table 1. Geometric parameters of the tested sintered wire meshes.

No.	Diameter of Stainless Steel Wire (μm)	Quantity of Layers	Average Pore Diameter (μm)	Porosity ε
P1	140	40	165.5	56.5%
P2	140	42	107.8	46.6%
P3	140	54	71.8	38.9%

By compressing and sintering several layers of wire meshes into a porous material, the geometry deforms dramatically. For the unsintered wire mesh, the gap between the stainless steel wire is around 500 μm. For the sintered wire mesh, the average pore diameter changes from 165 μm to 71.8 μm by increasing the quantity of sintered layer from 40 to 54. Correspondingly, the porosity of the test sample varies from 56.5% to 38.9%. Therefore, as shown in Table 1, although diameter of the stainless steel wire does not determine the size of the pore, it preliminary sets the possible scale of the pore size in the sintered wire mesh.

2.2. *Experimental Apparatus*

The transpiration cooling experiment is conducted in a low-speed open-loop wind tunnel, as shown in Figure 2. The air flow of the experiment is provided by a compressor connected to an air reservoir. Two air flows extracted from the reservoir are controlled by two gas control valves separately. At each air flow channel, one FCI-ST98 thermal flowmeter accurately measures the flow rate with the uncertainty of ±0.5%. After the flowmeter, the air in the mainstream is heated by an electric heater to set temperature T_g ranging from 90 °C to 120 °C before it purges into the rectangular channel of test section, which is 30 mm wide and 45 mm high. For the cooling air, it is led into a coolant chamber for pressure stabilization, in which its temperature T_c and pressure P_c are measured. The transpiration cooling test piece, the sintered wire mesh, is assembled between the coolant chamber and mainstream channel, allowing the cooling air flow through this porous material and generating the transpiration cooling.

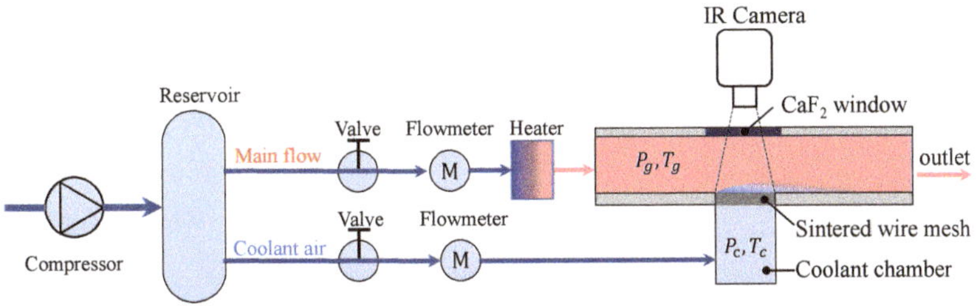

Figure 2. Schematic of the experiment apparatus for the measurement of the transpiration cooling characteristics of the sintered wire mesh.

Two calibrated K type thermocouples are fixed in the flow channels to measure the temperature of the coolant and hot air in the mainstream. To capture the two dimensional temperature distribution on the sintered wire mesh, an infrared camera (model: FLIR sc7700m) with focus on the surface of the test sample is set opposite the sample with a CaF_2 window as the observation window. It is a short wave infrared system with a receiving wavelength range between 2.5 μm and 5 μm and a measurement range between −20 °C and 1500 °C. Before the experiment, the camera was calibrated by a K type thermocouple to reach an uncertainty of ±0.5 °C. The resolution of the infrared measurement is 640 pixels × 512 pixels.

2.3. *Data Processing*

In the research on transpiration cooling, the blowing ratio F is commonly used to measure the mass flow ratio between the cooling air and hot air in the mainstream. The expression of F can be defined as following

$$F = \frac{\rho_c u_c}{\rho_g u_g}, \tag{1}$$

where ρ_c and ρ_g are the densities of the coolant and main flow, respectively, and u_c and u_g, respectively, the velocities of the coolant and main flow. However, it is difficult to accurately measure the density and flow velocity of the flow, especially for the coolant. Therefore, we use the blowing ratio according to the following formula

$$F = \frac{\dot{m}_c / A_c}{\dot{m}_g / A_g} = \frac{\dot{m}_c A_g}{\dot{m}_g A_c}, \tag{2}$$

where $\dot{m}_c$ and $\dot{m}_g$ are, respectively, the mass flow rates of the coolant and the main flow, and A_c and A_g, respectively, the flow area of the coolant channel and mainstream channel. Since the mass flow rate is measured by the thermal flowmeter, the uncertainty of the calculated blowing ratio F is very low. It has to be noted that the blowing ratio F defined in this study does not contain the geometric parameter of the test sample.

The transpiration cooling efficiency, which measures the cooling effectiveness, is defined as following

$$\eta = \frac{T_{wa} - T_w}{T_{wa} - T_c}, \tag{3}$$

where T_{wa} is the surface temperature of the test sample in the case of $F = 0$, and T_w the surface temperature of the test piece when $F \neq 0$.

3. Results and Discussion

3.1. The Effect of Blowing Ratio and Porosity on the Surface Temperature Distribution

When cooling air flows through the sintered wire mesh, it cools the surface by forced convection inside the medium and forming a air film between the hot air and the surface. Therefore, the temperature on the surface of the sintered wire mesh is much lower than that when $F = 0$. It has to be noted that the surface temperature T_{wa} on the surface of the sintered wire mesh is mostly uniform when $F = 0$. We measure the T_{wa} to calculate the cooling efficiency when $F > 0$: for $\varepsilon = 38.9\%$, $T_{wa} = 98.2\,°\mathrm{C}$; for $\varepsilon = 46.6\%$, $T_{wa} = 92.4\,°\mathrm{C}$; for $\varepsilon = 56.5\%$, $T_{wa} = 99.1\,°\mathrm{C}$. When $F > 0$, the surface temperature T_w significantly reduces under the cooling effect of the cooling air (as shown in Figure 3a).

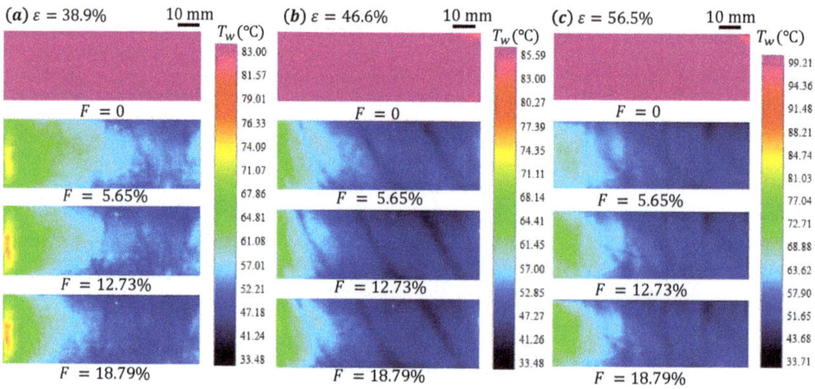

Figure 3. Temperature contour on the surface of the sintered wire mesh for different porosity and blowing ratios. The heated mainstream air flows from left to right, which is denoted as the x-axis. The legends are adjusted to show the distribution of the temperature on the surface when $F \neq 0$.

We observe that the temperature distribution on the surface of the sintered wire mesh becomes non-uniform. Typically, the surface temperature decreases along the flow direction, denoted as x-axis, in the mainstream. We also note the variation of surface temperature in the transverse direction, denoted as y-axis. The variation of temperature in x direction can be caused by either the decreasing of pressure along the mainstream or the entrance effect of the heat transfer. The former essentially provides a decreasing outlet pressure for the cooling flow through the porous material, inducing redistribution of the mass flow through the porous medium. Since the downstream has a lower pressure than the upstream, larger pressure drop in the downstream results in more cooling air flowing out in the downstream of the porous medium, leading to better cooling in that area. The latter is related to the increase in boundary layer thickness with the accumulating of the cooling air along the flowing direction, resulting in a relatively higher heat transfer efficiency in the upper part and a lower one in the lower part. For the uneven temperature distribution in the y direction, it can be generated by the boundary lay flow in the corners of the mainstream channel. Generally speaking, low velocity flow in the corner area reduces the heating effect of the hot air in the mainstream and generates low surface temperature near the corner areas.

With the increase in the blowing ratio F from 5.65% to 18.79%, the temperature on the surface of the sintered wire mesh decreases, suggesting the enhancing of the cooling effect by the cooling air. This phenomena are all observed on the surface of sintered wire mesh with different porosity ε, as shown in Figure 3.

To better understand the distribution of the surface temperature T_w on the test sample, we extract the T_w on the central line along x direction for each cases shown in Figure 3, and present the results in Figures 4 and 5. The data are arranged differently in these two figures to better illustrate the effects of blowing ratio F and porosity ε on the surface temperature T_w distribution of the sintered wire mesh.

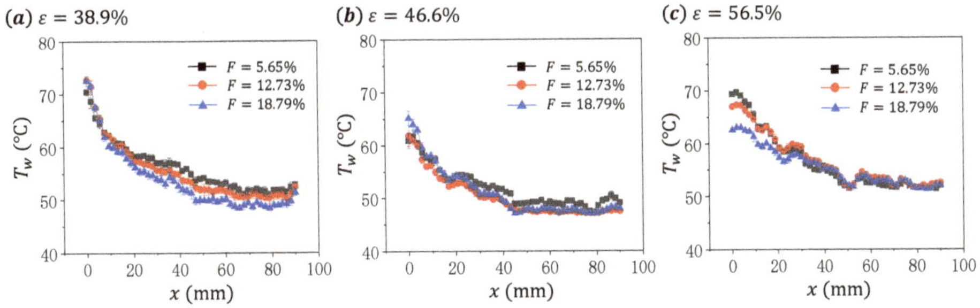

Figure 4. The distribution of surface temperature T_w along the central line of the test sample for different blowing ratio F at same porosity ε: (**a**) $\varepsilon = 38.9\%$, (**b**) $\varepsilon = 46.6\%$, and (**c**) $\varepsilon = 56.5\%$.

As shown in Figure 4, the surface temperature T_w generally decreases with x, and some fluctuations of T_w along the distribution. This may be caused by the rough surface of the sintered wire mesh generated during the manufacture. With the increase in blowing ratio F, we find different changing patterns for the three test samples. For the material with $\varepsilon = 38.9\%$, the increase in F causes insignificant variation of T_w in the upper part of test sample (around $x < 10$ mm), but significant decreasing of T_w in the lower part of test sample (around $x > 10$ mm). For the sample with $\varepsilon = 46.6\%$, the increase in F induces the temperature decrease in all the surface of the test sample. For the sample with $\varepsilon = 56.5\%$, we observe the decrease in T_w in the upper part of the sample (around $x < 20$ mm) with the increase in F. Therefore, the transpiration cooling characteristics of the sintered wire mesh change with the porosity.

It must be noted that for sintered wire mesh, its internal heat transfer characteristic generally increases with the mass flow rate of the cooling air [24]. When applying sintered wire mesh for transpiration cooling, the cooling characteristics are co-determined by the internal cooling and the external film cooling which have yet be fully understood. However, we can infer from Figure 4 that the cooling characteristics of the sintered wire mesh significantly depends on the porosity of the material.

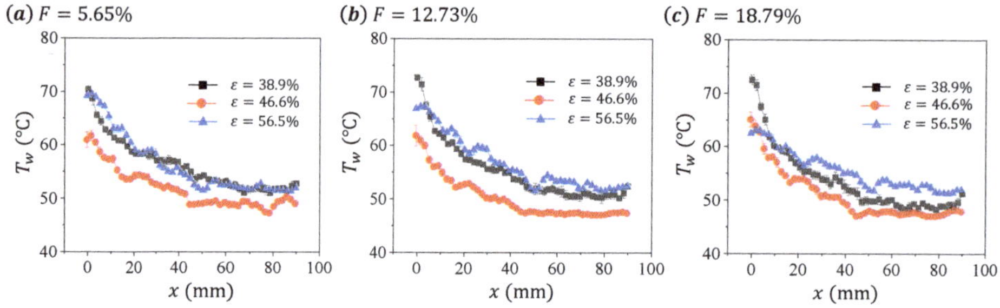

Figure 5. The distribution of surface temperature T_w along the central line of the test sample for different porosity ε at same blowing ratio F: (**a**) $F = 5.65\%$, (**b**) $F = 12.73\%$, and (**c**) $F = 18.79\%$.

In Figure 5, we compare the surface temperature T_w distribution for different test samples at same F. Similar distribution trends of T_w, decreasing with x, can be found among different samples in all presented F. Although we observe the difference in T_w among the samples with same ε, it has to be noted that the value of T_{wa} differs between samples. Therefore, simply comparing the surface temperature among samples with the same ε may be meaningless.

3.2. The Effects of Blowing Ratio and Porosity on the Cooling Efficiency

To demonstrate the effects of blowing ratio F and porosity ε on the cooling efficiency η of the test samples, we calculate the cooling efficiency η from surface temperature based on Equation (3) and present the result in Figures 6 and 7.

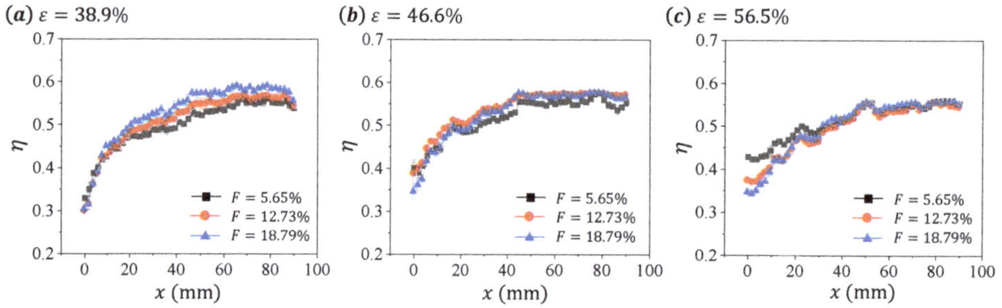

Figure 6. The distribution of cooling efficiency η on the central line of the test sample for different blowing ratio F at same porosity ε: (**a**) $\varepsilon = 38.9\%$, (**b**) $\varepsilon = 46.6\%$, and (**c**) $\varepsilon = 56.5\%$.

Since high surface temperature results in low cooling efficiency, the changing characteristics of η is inverse to that of T_w. For η, this increases with x first and then becomes saturated for all presented data in Figures 6 and 7. We note that the transition position between these two trends are all around $x \approx 50$ mm for all presented data. For the non-uniform distribution of η, this can be caused by either the decrease pressure in the mainstream channel or the entrance effect of the heat transfer. Since the pressure drop for the air flowing through the porous media varies between samples [24], the effect of pressure decrease on η may be different among samples. Therefore, we speculate that the entrance effect may play an more important role in the determining of the distribution of η, since the mass flow rate of the cooling air is generally low (less than 1%).

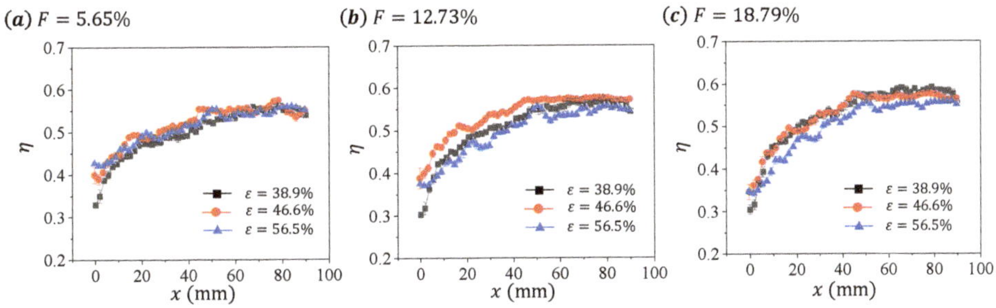

Figure 7. The distribution of cooling efficiency η on the central line of the test sample for different porosity ε at same blowing ratio F: (**a**) $F = 5.65\%$, (**b**) $F = 12.73\%$, and (**c**) $F = 18.79\%$.

For the transpiration cooling at same F among test samples, we observe insignificant variation at low F ($F = 5.65\%$), and significant differences at large F. In addition, the sample with $\varepsilon = 46.6\%$ has a higher η than other samples in all F (Figure 7), suggesting a non-monotonic changing characteristic between the cooling efficiency of the sample and its porosity.

3.3. The Effect of Blowing Ratio and Porosity on the Averaged Cooling Efficiency

To further compare the cooling efficiency among the samples at all tested blowing ratios, we calculate the averaged cooling efficiency η_a by averaging the cooling efficiency on the surface of the sample for one case. We plot the dependence of η_a with F for the three tested samples and present the result in Figure 8.

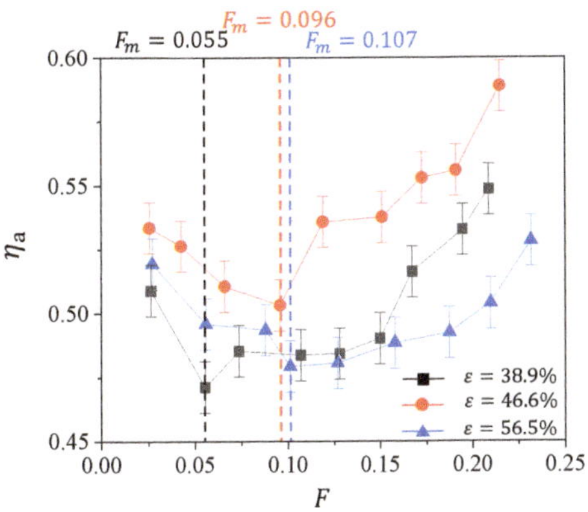

Figure 8. Variation of average cooling efficiency η_a with blowing ratio F for different porosity ε.

Generally, we observe that, for all the samples, η_a first decreases and then increases with F, giving minima of η_a in the curves of transpiration cooling characteristics. We highlight the values of F_m at the minima of η_a for all the samples, and find the monotonic relation between F_m and ε. For this transpiration cooling characteristics of sintered wire mesh, it resembles the characteristics of the film cooling on the endwall of gas turbine vane with full coverage of film cooling holes [25]. The film cooling efficiency on the endwall first decreases and then increases with the blowing ratio. This can be explained by the following: when blowing ratio is small, i.e., the velocity of the coolant is low, the cooling air easily enters into the low velocity region of the boundary layer in the mainstream channel, giving full coverage of the cooling air film on the endwall; with the increase in blowing ratio, the cooling air becomes difficulty to enter the boundary layer, giving rise to the decrease in cooling efficiency; with the further increasing of the blowing ratio, the cooling air with high velocity enhances the heat transfer on the surface, resulting in the increase in cooling efficiency. Therefore, the cooling efficiency decreases and then increases with the blowing ratio.

We speculate that this reasoning can also be applied in the transpiration cooling of the sintered wire mesh. Although the transpiration cooling consists of the internal cooling and external cooling, the internal cooling generally increases with the mass flow rate of the cooling air. Therefore, these transpiration cooling characteristics are mostly determined from the external cooling of the sintered wire mesh, which is essentially the film cooling. However, we still has to admit that both the internal cooling and external cooling determine the transpiration cooling of the sintered wire mesh.

Among the three samples, η_a of the sample with $\varepsilon = 46.6\%$ overwhelms that of the other samples in all tested F. Since the internal cooling monotonically increases with the decrease in porosity, this phenomenon can be explained from the side of external cooling. Smaller porosity causes enhanced heat transfer in the internal cooling, but the higher velocity and larger velocity difference in the cooling air, caused by the smaller size of the pore, may induce more intense mixing between the cooling air and mainstream flow and correspondingly reducing of the cooling efficiency. In addition, a large pore diameter may cause the decrease in the pore density on the surface, reducing the effect of external cooling. Therefore, the sample with $\varepsilon = 46.6\%$ has the highest cooling efficiency among the three tested samples. For the crossing of the heat transfer characteristic curves for the samples with $\varepsilon = 38.9\%$ and $\varepsilon = 56.5\%$, it can be the enhanced heat transfer of the internal cooling at high F.

To obtain further insights into the effect of geometry parameter on the transpiration cooling efficiency of the sintered wire mesh, we try to provide a modified blowing ratio F' that includes the factor of material's geometric parameter. For the blowing ratio used in this study, it does not consider the geometry factors of the porous medium, e.g., the porosity and the pore size. We can use the porosity to roughly estimate the flow area, which may give a better estimation of the cooling air velocity, or we can simply use the pore diameter, which essentially is the diameter of the hole in film cooling and determines the density of the hole on the surface. For the sake of convenience, we apply the normalized average pore diameter d/d_1 (as shown in Table 1) for the modified blowing ratio $F' = Fd/d_1$. Here, d is the average pore diameter, and d_1 the average pore diameter of sample P_1. In Figure 9, we show the variation of η_a with F'. We observe that the F_m for the three sample collapse around $F' = F^* \approx 0.055$. For $F' < F^*$, η_a decreases with F', while it increases with F' when $F' > F^*$, suggesting the proposed blowing ratio F' is practically applicable in unifying the transpiration cooling characteristics of the sintered wire mesh in this study. We also note that the differences in η_a among the three samples are insignificant when $F' < F^*$, while the differences becomes significant when $F' > F^*$. This can be interpreted from the combining effect of the internal cooling and external cooling. For the internal cooling, its heat transfer monotonically increases with the blowing ratio; for external cooling, its heat transfer first decreases and then increases with the blowing ratio. Therefore, on one hand, at low blowing ratio ($F' < F^*$), these two cooling parts compensate each other, making the effect of geometry parameter insignificant. On the other hand, for large blowing ratio ($F' > F^*$), these two cooling parts superimpose on each other, magnifying the effect of geometric parameter.

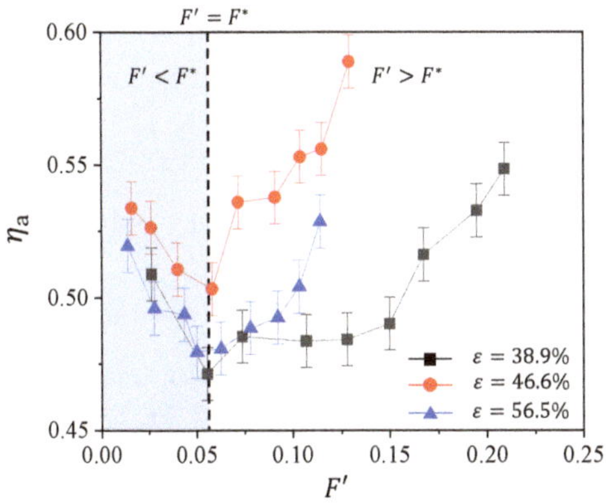

Figure 9. Variation of average cooling efficiency η_a with the modified blowing ratio F' for different porosity ε.

4. Conclusions

In summary, the transpiration cooling characteristics of the sintered wire mesh are systematically investigated and presented in this study. By measuring the surface temperature of the three test samples with different porosities at various blowing ratios, we observe the nonuniform distribution of surface temperature both along the flow direction and the transverse direction caused by the entrance effect and boundary layer in the corner, respectively. The transpiration cooling efficiency, which is determined by the internal forced convection and external film cooling, first decreases and then increases with the blowing ratio; however, it increases and then decreases with the porosity of the material. The former characteristic is caused by the external film cooling, while the latter is a combination effect from the internal and external cooling. We further propose a modified blowing ratio by the normalized pore diameter of the material to collapse the minima of F for all tested samples, which separates the decreasing and increase parts in the curve of transpiration cooling efficiency.

Author Contributions: Conceptualization, G.X., X.L. and D.L.; methodology, G.X., X.L., Y.P. and D.L.; investigation, Y.P. and J.H.; writing—review and editing, Y.P. and D.L.; supervision, G.X., X.L. and D.L. All authors have read and agreed to the published version of the manuscript.

Funding: This research received no external funding.

Institutional Review Board Statement: Not applicable.

Informed Consent Statement: Not applicable.

Data Availability Statement: All processed data in this study are included in this published article. Raw data will be provided on request from the corresponding author.

Acknowledgments: This work is supported by the National Science and Technology Major Project (2017-lll-0003-0027).

Conflicts of Interest: The authors declare no conflict of interest.

References

1. Eckert, E.R.G.; Livingood, J.N. *Comparison of Effectiveness of Convection-, Transpiration-, and Film-Cooling Methods with Air as Coolant*; National Advisory Committee for Aeronautics: Washington, DC, USA, 1954; Volume 1182.
2. Haidn, O.; Greuel, D.; Herbertz, A.; Ortelt, M.; Hald, H. Transpiration cooling appllied to C/C liners of cryogenic liquid rocket engines. In Proceedings of the 40th AIAA/ASME/SAE/ASEE Joint Propulsion Conference and Exhibit, Fort Lauderdale, FL, USA, 11–14 July 2004; p. 3682.
3. Liu, Y.Q.; Jiang, P.X.; Jin, S.S.; Sun, J.G. Transpiration cooling of a nose cone by various foreign gases. *Int. J. Heat Mass Transf.* **2010**, *53*, 5364–5372. [CrossRef]

4. Song, K.D.; Sang, H.C.; Scotti, S.J. Transpiration cooling experiment for scramjet engine combustion chamber by high heat fluxes. *J. Propuls. Power* **2006**, *22*, 96–102. [CrossRef]
5. Langener, T.; Wolfersdorf, J.V.; Steelant, J. Experimental Investigations on Transpiration Cooling for Scramjet Applications Using Different Coolants. *Aiaa J.* **2011**, *49*, 1409–1419. [CrossRef]
6. Jiang, P.X.; Yu, L.; Sun, J.G.; Wang, J. Experimental and numerical investigation of convection heat transfer in transpiration cooling. *Appl. Therm. Eng.* **2004**, *24*, 1271–1289. [CrossRef]
7. Langener, T.; Wolfersdorf, J.V.; Laux, T.; Steelant, J. Experimental Investigation of Transpiration Cooling with Subsonic and Supersonic Flows at Moderate Temperature Levels. In Proceedings of the AIAA/ASME/SAE/ASEE Joint Propulsion Conference & Exhibit, Hartford, CT, USA, 21–23 July 2008.
8. Huang, Z.; Zhu, Y.H.; Jiang, P.X.; Xiong, Y.B. Investigation of a Porous Transpiration-Cooled Strut Injector. *J. Propuls. Power* **2015**, *31*, 278–285. [CrossRef]
9. Ortelt, M.; Hald, H.; Herbertz, A.; Müller, I. Advanced Design Concepts for Ceramic Thrust Chamber Components of Rocket Engines. In Proceedings of the 5th European Conference for Aeronautics and Space Sciences (EUCASS), Munich, Germany, 1–5 July 2013.
10. Dahmen, W.; Gotzen, T.; Müller, S.; Rom, M. Numerical simulation of transpiration cooling through porous material. *Int. J. Numer. Methods Fluids* **2015**, *76*, 331–365. [CrossRef]
11. Prokein, D.; Dittert, C.; Böhrk, H.; Wolfersdorf, J.V. Transpiration Cooling Experiments on a CMC Wall Segment in a Supersonic Hot Gas Channel. In Proceedings of the American Institute of Aeronautics and Astronautics (AIAA) in 2018 International Energy Conversion Engineering Conference, Cincinnati, OH, USA, 9–11 July 2018.
12. Koenig, V.; Rom, M.; Mueller, S.; Selzer, M.; Schweikert, S.; Wolfersdorf, J.V. Numerical and Experimental Investigation of Transpiration Cooling with Carbon/Carbon Characteristic Outflow Distributions. *J. Thermophys. Heat Transf.* **2019**, *33*, 1–13. [CrossRef]
13. Zhu, Y.; Jiang, P.; Sun, J.; Xiong, Y. Injector Head Transpiration Cooling Coupled with Combustion in H_2/O_2 Subscale Thrust Chamber. *J. Thermophys. Heat Transf.* **2015**, *27*, 42–51. [CrossRef]
14. Moskowitz, S.L.; Lombardo, S. 2750 Deg F Engine Test of a Transpiration Air-Cooled Turbine. *J. Eng. Gas Turbines Power* **1971**, *93*, 238–248. [CrossRef]
15. Raj, R. Deposition Results of a Transpiration Air-Cooled Turbine Vane Cascade in a Contaminated Gas Stream. *J. Eng. Power Trans. Asme* **1983**, *105*, 826. [CrossRef]
16. Wolf, J.; Moskowitz, S. Development of the Transpiration Air-Cooled Turbine for High-Temperature Dirty Gas Streams. *J. Eng. Gas Turbines Power* **1983**, *105*, 821–825. [CrossRef]
17. Liu, Y.Q.; Jiang, P.X.; Xiong, Y.B.; Wang, Y.P. Experimental and numerical investigation of transpiration cooling for sintered porous flat plates. *Appl. Therm. Eng.* **2013**, *50*, 997–1007. [CrossRef]
18. Huang, G.; Zhu, Y.; Liao, Z.; Ouyang, X.; Jiang, P. Experimental investigation of transpiration cooling with phase change for sintered porous plates. *Int. J. Heat Mass Transf.* **2017**, *114*, 1201–1213. [CrossRef]
19. Langener, T.; Wolfersdorf, J.V.; Selzer, M.; Hald, H. Experimental investigations of transpiration cooling applied to C/C material. *Int. J. Therm. Sci.* **2012**, *54*, 70–81. [CrossRef]
20. Min, Z.; Huang, G.; Parbat, S.N.; Yang, L.; Chyu, M.K. Experimental Investigation on Additively Manufactured Transpiration and Film Cooling Structures. *J. Turbomach.* **2018**, *141*, 031009. [CrossRef]
21. Liu, P.S.; Chen, G.F. *Porous Materials: Processing and Applications*; Butterworth-Heinemann: Oxford, UK; Elsevier: Amsterdam, The Netherlands, 2014.
22. Xu, G.; Liu, Y.; Xiang, L.; Ma, J.; Li, H. Experimental investigation of transpiration cooling for sintered woven wire mesh structures. *Int. J. Heat Mass Transf.* **2015**, *91*, 898–907. [CrossRef]
23. Ma, J.; Luo, X.; Li, H.; Liu, Y. An Experimental Investigation on Transpiration Cooling Based on the Multilaminated Sintered Woven Wire Mesh Structures. *J. Therm. Sci. Eng. Appl.* **2016**, *8*, 031005. [CrossRef]
24. Liu, Y.; Xu, G.; Luo, X.; Li, H.; Ma, J. An experimental investigation on fluid flow and heat transfer characteristics of sintered woven wire mesh structures. *Appl. Therm. Eng.* **2015**, *80*, 118–126. [CrossRef]
25. Liu, D.; Tao, Z.; Luo, X. Experimental Investigation of Heat Transfer Characteristics on Turbine Endwall with Full Coverage Film Cooling. *Appl. Therm. Eng.* **2018**, *140*, 295–303. [CrossRef]

micromachines

MDPI

Article

Entropy Generation Analysis of Peristaltic Flow of Nanomaterial in a Rotating Medium through Generalized Complaint Walls of Micro-Channel with Radiation and Heat Flux Effects

Aamir Ali [1,*], Mehak Sajid [1], Hafiz Junaid Anjum [2], Muhammad Awais [1], Kottakkaran Sooppy Nisar [3,*] and C. Ahamed Saleel [4]

1 Department of Mathematics, Attock Campus, COMSATS University Islamabad, Kamra Road, Attock 43600, Pakistan; fa19-rmt-012@cuiatk.edu.pk (M.S.); awais@ciit-attock.edu.pk (M.A.)
2 Department of Mathematics, COMSATS University Islamabad, Park Road, Chak Shehzad, Islamabad 44000, Pakistan; mjunaidanjum@hotmail.com
3 Department of Mathematics, College of Arts and Sciences, Prince Sattam bin Abdulaziz University, Wadi Aldawaser 11991, Saudi Arabia
4 Department of Mechanical Engineering, College of Engineering, King Khalid University, Asir-Abha 61421, Saudi Arabia; ahamedsaleel@gmail.com
* Correspondence: aamir_ali@cuiatk.edu.pk or aamir_ali@ciit-attock.edu.pk (A.A.); n.sooppy@psau.edu.sa (K.S.N.)

Abstract: This study discusses entropy generation analysis for a peristaltic flow in a rotating medium with generalized complaint walls. The goal of the current analysis is to understand the fluid flow phenomena particular to micro devices. Nano materials with a size less than 100 nm have applications in micro heat exchangers to cool electronic circuits, blood analyzers, biological cell separations, etc. For this study, we considered the effects of radiation, viscous dissipation and heat flux on the flow of nanomaterial inside a cylindrical micro-channel. To investigate the slip effects on the flow, the second order slip condition for axial velocity, the first order slip condition for secondary velocity and the thermal slip conditions were used. The flow was governed by partial differential equations (PDE's), which were turned into a system of coupled ordinary differential equations (ODE's) that were highly non-linear and numerically solved using the NDSolve command in Mathematica. The impacts of different involved parameters on the flow field were investigated with the aid of graphical illustrations. Entropy generation and the Bejan number were given special attention, and it was found that they decreased as the Hartman number, rotation, and radiation parameters increased.

Keywords: entropy; peristaltic flow; heat flux; radiation; generalized complaint walls

1. Introduction

Peristalsis mainly refers to fluid flows driven by pressure gradients resulting from the movement of a wave along channel boundaries. The compression and expansion of the domain due to the propagation of the wave generates the flow. In the human body, urine flow, food transport through the digestive system, blood circulation (and others), all are due to peristalsis. Its application is also found in targeted drug delivery by the use of nano magnetic particles. The application of peristalsis can also be found in engineering, where it enables the construction of heart lung machines that operate on the principles of peristalsis. The study of thermal effects in peristaltic flows is motivated by its application to tumor growth, transportation of medical substances, such as a pill, or the transportation of nutrients to brain cells.

The research in this field can be traced back to pioneer works of Latham [1] and Shapiro et al. [2]. These studies provided a basic theoretical framework for investigating peristaltic flows with long wavelengths and low Reynolds numbers. Following their work, a great deal of research has been carried out by considering different flow configurations and a variety of Newtonian and non-Newtonian fluids. Yin et al. [3] studied the peristaltic flow of a viscous fluid in a circular tube. The mean flow and mean pressure gradients were reported to be proportionate with the square of the amplitude ratio in the absence of any wall motion. According to the numerical results, the mean axial velocity was dominated by the mean pressure gradient and the no-slip boundary condition. Asghar et al. [4] investigated the peristaltic flow of a reactive viscous fluid in a 2D setting, assuming viscosity to be temperature dependent. The existence of the Hartman layer, due to the presence of a magnetic field, and its consequent effects on the flow dynamics, were reported in [5]. Some other recent studies considering deforming walls and heat transfer effects in a peristaltic flow include [6–9] and references therein.

The issue of heat transfer arises in a wide range of engineering processes that use fluids such as refrigerants, water, engine oil, ethylene glycol, etc., as heat exchangers to maintain a desired working temperature. The poor thermal conductivity of the coolant fluids limits the effectiveness of these heat exchangers. It has been shown [10] that increasing the thermal conductivity of a fluid by introducing nanometer size particles into the base fluid used as a coolant can improve its thermal conductivity. The study of heat transfer in a peristaltic flow is motivated by its application in oxygenation and hemodialysis. Buongiorno [11] report their observations on convective transports in nanofluids, investigating various slip mechanisms attributed to the generation of relative velocity amongst the nanoparticles and the base fluid. Turkyilmazoglu [12] and Khan et al. [13] studied heat transfer effects in a nanofluid flow along stretching surfaces. Awais et al. [14] investigated the dynamical influence of nanoparticles on the flow of Oldroyd-B model polymeric liquids in the presence of nanoparticles. Other studies [15–17] consider flows in different geometrical configurations, investigating Brownian motions and thermophoresis effects on the flow dynamics and characteristics. In the existence of hydrodynamic slip and radiation impacts, Shashikumar et al. [18] conducted a thermodynamic study of Casson nanofluid in a porous microchannel. The Brownian motion and thermophoresis effects for Casson fluid on a stretched surface with non-Fourier heat flow were recently explored by Sreelakshmi et al. [19]. Fluid flow in small channels involving micro/nano materials holds importance in the study of micro ducts, micro pumps and valves, etc. Many scholars have studied these challenges in recent years as a result of these applications [20–24].

Complaint wall is defined as a wall that is deformable but is also stretchable, flexible and elastic in nature, with the ability to contain liquid in it. When a deforming force acts on this wall, it comes back to its original position due to its flexible and elastic nature. Elasticity is a characteristic of a body to restore itself once external deforming forces are removed, e.g., sponge, spring, rubber, etc. In contrast, plasticity is a material characteristic wherein material does not revert back to its original position when deforming force is removed, e.g., wood and glass. Deforming force is a force which brings change in shape, length or volume of the wall when applied. The amount of change is directly proportional to the force applied to the body. Every wall/medium has a limit to being elastic, which is termed as elastic limit; it means that a wall can bear a limited deformation force that is different for each medium to remain elastic, otherwise the deformation will be permanent. In complaint wall, all of these aspects are monitored. Movement of fluid is greatly dependent on complaint wall, as it manages the geometry of sinusoidal waves during peristalsis flow. The peristaltic flow of non-Newtonian fluid in a complaint walls channel was described by Ali et al. [25]. The analysis for MHD Jeffery fluid in a complaint walls channel was presented by Hayat et al. [26]. Heat transfer analysis of peristaltic flow in a complaint walls channel for viscous fluid was reported by Hayat et al. [27]. Nadeem et al. [28] discussed the trapping phenomenon in a rectangular channel with complaint walls for viscous fluid. The hall current investigations on peristaltic flow in a rotating medium were presented by Hayat et al. [29]. They considered the nanoparticles inside the channel of complaint walls. Recently, Awais et al. [30] presented the rheology of copper water nanomaterial in a channel by considering the generalized complaint walls and variable viscosity effects. In the branches of science, physiology and biology, magnetohydrodynamic (MHD) effect, which is critical for physiological liquids of peristaltic motion, has been discussed. Electric current causes a magnetic field due to the movement of conducting fluid particles changes in the fluid flow because of mechanical forces [31]. It has many applications in engineering, e.g., in aerodynamics, control of boundary layer, studies of plasma, energy removing from geothermal processes, MHD generator and inspection of oil; because of these applications, many inspectors focused on electrically conducting fluid passed through a porous medium with the magnetic field effects in peristalsis along transfer of mass and heat [32]. Kumar et al. [33] used the KKL model to investigate the effect of a magnetic dipole on the flow of a radiative nanofluid across a stretched surface. Alhumade et al. [34] examined the effects of nonlinear radiation on the flow over a stretched cylinder with the Cattaneo–Christove heat flux.

Heat transmission effects with the generation of entropy have been studied extensively since the pioneering work of Bejan [35], in which he presented a way to maximize the system's destruction, heat-transmission effects with the formation of entropy. Entropy is a measure of arbitrariness inside a system or the degree of molecular disorder. In a thermodynamic system, entropy generation could result in a loss of energy attributed to various processes, e.g., friction, viscosity and chemical reactions. In diverse processes, counting chemical vapor testimony instruments, combustion, turbo machinery, electronic cooling devices, heat exchangers, and solar collectors, entropy plays a vital role. Entropy generation minimization is essential to increase the system performance in terms of heat conductivity. Hayat et al. [36] explored the entropy generation phenomenon for peristaltic flow in a rotating medium. The entropy generation investigation on magnetohydrodynamics peristaltic flow of copper–water nanofluid under slip effects was described by Ali et al. [37]. They claim that when slip parameters are used, entropy production diminishes. Shashikumar et al. [38] studied slip effects and performed an entropy generation study for the flow of a nanofluid in a microchannel employing aluminium and titanium alloy nanoparticles. Entropy-generating processes are investigated in a number of energy-related applications, including geothermal energy and modern refrigeration equipment [39–44].

As the entropy generation and compliant walls consideration can affect the flow dynamics and thermal characteristics for a peristaltic flow, we conducted this investigation to incorporate these effects, which were missing in the previously reported theoretical investigations. Motivated by this, the objective of this research was to analyze the entropy generation on the peristaltic flow in a rotating medium with generalized complaint walls. The entropy analysis was used to study the thermodynamic irreversibility, which destroys the available energy. Further, we also used nanoparticles in this analysis due to their vast engineering applications, as discussed above. The effects of radiation, viscous dissipation, and thermal flux on the flow of nanofluid inside a micro-channel are investigated. The second order slip condition for axial velocity, first order slip condition for secondary velocity, and thermal slip conditions are used to explore the effects of slip parameter on the flow. The flow was governed in non-dimensional form by highly non-linear PDE's, which were then translated into a coupled system of ODE's. Mathematica's NDSolve tool was used to numerically solve the transformed system of coupled ODE's. Graphs are used to explore the effects of several key parameters on the flow field.

2. Mathematical Modelling

Consider an unsteady peristaltic nanofluid flow in a porous, rotating medium with angular speed along the $z-$axis. $2d$ is the width of the channel with generalized complaint walls having temperatures T_1 and T_0. The peristaltic flow arises due to the expansion of waves with wavelength λ, speed c, amplitude a, time t, and half channel width d aligned to the wall at $z = \pm\eta$ expressed through the relation.

$$z = \pm\eta(x,t) = \pm\left[d + a\sin\left(\frac{2\pi}{\lambda}\right)(x - ct)\right].$$

Physically the problem is presented in Figure 1.

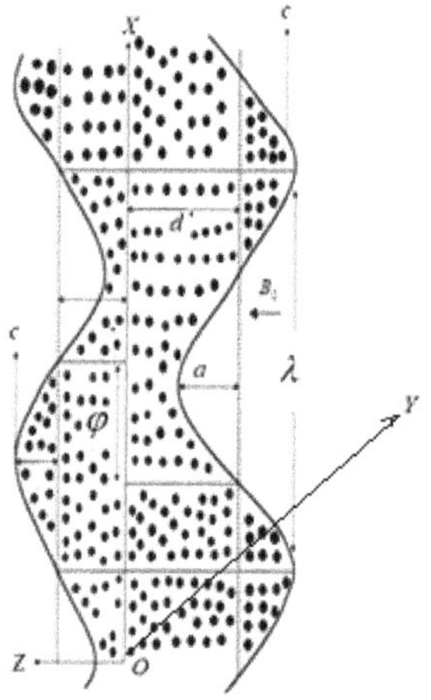

Figure 1. Schematic diagram of the problem.

For incompressible flow, the mass, momentum and energy equations are [29,30]:

$$\nabla\cdot\mathbf{V} = 0, \tag{1}$$

$$\rho_{nf}\left(\tfrac{d\mathbf{V}}{dt}\right) + \rho_{nf}[\boldsymbol{\Omega}\times(\boldsymbol{\Omega}\times \boldsymbol{r}) + 2(\boldsymbol{\Omega}\times\mathbf{V})] = \nabla\cdot\boldsymbol{\tau} + \boldsymbol{J}\times\boldsymbol{B} - \tfrac{\mu_{nf}}{K}\mathbf{V} \\ + g(\rho\beta)_{nf}(T - T_m), \tag{2}$$

$$(\rho c_p)_{nf}\left(\frac{dT}{dt}\right) = \kappa_{nf}\nabla^2 T + \boldsymbol{\tau}\cdot\boldsymbol{L} - \nabla q_r - \frac{1}{\sigma_{nf}}\boldsymbol{J}\cdot\boldsymbol{J} + Q_0, \tag{3}$$

where, $\frac{d}{dt}$ is material time derivative, $\mathbf{V} = \mathbf{V}[u(x,z,t), v(x,z,t), w(x,z,t)]$ is velocity field, ρ_{nf} is nanofluid density, $\boldsymbol{\Omega} = \Omega\hat{k}$ is angular velocity, $\boldsymbol{\tau}$ is Cauchy stress tensor, $\boldsymbol{B}$ is magnetic field, $\boldsymbol{J}$ is current density, $\boldsymbol{J}\times\boldsymbol{B}$ is Lorentz force, μ_{nf} is nanofluid viscosity, $(\rho\beta)_{nf}$ is coefficient of thermal expansion, $(\rho c_p)_{nf}$ is heat capacity of nanofluid, κ_{nf} is thermal conductivity of nanofluid, T is temperature of fluid, $\boldsymbol{\tau}\cdot\boldsymbol{L}$ is viscous dissipation term, q_r is thermal radiation term, $(\sigma)_{nf}$ is electric conductivity of nanofluid, $\boldsymbol{J}\cdot\boldsymbol{J}$ is joule heating term, and Q_0 is heat generation/absorption parameter. The Cauchy stress tensor is expressed as:

$$\boldsymbol{\tau} = -P\mathbf{I} + \mu_{nf}\mathbf{A}, \tag{4}$$

where P is pressure. $\mathbf{A}$ is Rivilin Ericksen tensor and is defined by the relation:

$$\mathbf{A} = (\nabla\cdot\mathbf{V}) + (\nabla\cdot\mathbf{V})^t.$$

The modified Ohm's law is given as:

$$\boldsymbol{J} = \sigma_{nf}\left[\boldsymbol{E} + (\mathbf{V}\times\boldsymbol{B}) - \frac{1}{en_e}(\boldsymbol{J}\times\boldsymbol{B})\right], \tag{5}$$

In above equations, $\frac{1}{en_e}$ is the hall factor where e the electron charge and n_e is the density of free electron. When there is electric field $\boldsymbol{E}$, the Lorentz force becomes:

$$\boldsymbol{J}\times\boldsymbol{B} = \left[\frac{\sigma_{nf}B_0^2}{1+\left(\frac{\sigma_{nf}B_0}{en_e}\right)^2}\left(-u + \left(\frac{\sigma_{nf}B_0}{en_e}\right)v\right), -\frac{\sigma_{nf}B_0^2}{1+\left(\frac{\sigma_{nf}B_0}{en_e}\right)^2}\left(v + \left(\frac{\sigma_{nf}B_0}{en_e}\right)u\right), 0\right]. \tag{6}$$

The expression for rotational force is:

$$\boldsymbol{\Omega} \times (\boldsymbol{\Omega} \times \boldsymbol{r}) + 2(\boldsymbol{\Omega} \times \mathbf{V}) = -\Omega(\Omega x + 2v)\hat{i} - \Omega(\Omega y - 2u)\hat{j} + 0\hat{k}. \tag{7}$$

The radiative diffusion in the Rosselend approximation is defined as:

$$q_r = -\frac{4\sigma^*}{3k^*}\frac{\partial T^4}{\partial x}, \tag{8}$$

where σ^* is the Stefan–Boltzmann constant, whereas k^* is the mean absorption coefficient. When the Taylor series expansion is applied to the T^4 about mean temperature of nanomaterial T_w, we get:

$$\nabla q_r = \frac{\partial q_r}{\partial x} + \frac{\partial q_r}{\partial z} = -\frac{16\sigma^* T_m^3}{3k^*}\left[\frac{\partial^2 T}{\partial x^2} + \frac{\partial^2 T}{\partial z^2}\right]. \tag{9}$$

The relation of viscous dissipation is:

$$\boldsymbol{\tau}\cdot\boldsymbol{L} = \mu_{nf}\left[2\left\{\left(\frac{\partial u}{\partial x}\right)^2 + \left(\frac{\partial w}{\partial z}\right)^2\right\} + \left(\frac{\partial u}{\partial z} + \frac{\partial w}{\partial x}\right)^2\right]. \tag{10}$$

We obtain the following system of equations in components form by substituting Equations (4), (6), (7), (9) and (10) into fundamental Equations (1)–(3):

$$\frac{\partial u}{\partial x} + \frac{\partial w}{\partial z} = 0, \tag{11}$$

$$\begin{gathered}\rho_{nf}\left(\frac{du}{dt} - 2\Omega v\right) = -\frac{\partial \hat{P}}{\partial x} + \mu_{nf}\left(\frac{\partial^2 u}{\partial x^2} + \frac{\partial^2 u}{\partial z^2}\right) + \frac{\sigma_{nf}B_0^2}{1+\left(\frac{\sigma_{nf}B_0}{en_e}\right)^2}\left(-u + \left(\frac{\sigma_{nf}B_0}{en_e}\right)v\right) \\ -\frac{\mu_{nf}}{K}u + g(\rho\beta)_{nf}(T - T_m),\end{gathered} \tag{12}$$

$$\begin{gathered}\rho_{nf}\left(\frac{dv}{dt} + 2\Omega u\right) = -\frac{\partial \hat{P}}{\partial y} + \mu_{nf}\left(\frac{\partial^2 v}{\partial x^2} + \frac{\partial^2 v}{\partial z^2}\right) - \frac{\sigma_{nf}B_0^2}{1+\left(\frac{\sigma_{nf}B_0}{en_e}\right)^2}\left(v + \left(\frac{\sigma_{nf}B_0}{en_e}\right)u\right) \\ -\frac{\mu_{nf}}{K}v,\end{gathered} \tag{13}$$

$$\rho_{nf}\frac{dw}{dt} = -\frac{\partial \hat{P}}{\partial z} + \mu_{nf}\left(\frac{\partial^2 w}{\partial x^2} + \frac{\partial^2 w}{\partial z^2}\right), \tag{14}$$

$$\begin{gathered}(\rho c_p)_{nf}\frac{dT}{dt} = \kappa_{nf}\left(\frac{\partial^2 T}{\partial x^2} + \frac{\partial^2 T}{\partial z^2}\right) + \mu_{nf}\left[2\left\{\left(\frac{\partial u}{\partial x}\right)^2 + \left(\frac{\partial w}{\partial z}\right)^2\right\} + \left(\frac{\partial u}{\partial z} + \frac{\partial w}{\partial x}\right)^2\right] \\ +\frac{16\sigma^* T_m^3}{3k^*}\left[\frac{\partial^2 T}{\partial x^2} + \frac{\partial^2 T}{\partial z^2}\right] + \frac{\mu_{nf}}{K}u^2 + \Phi.\end{gathered} \tag{15}$$

The centrifugal effect with modified pressure $\hat{P}$ can be expressed as:

$$\hat{P} = p - \frac{1}{2}\rho\Omega^2\left(x^2 + y^2\right). \tag{16}$$

The values for $\rho_{nf}, \mu_{nf}, (\rho c_p)_{nf}, \kappa_{nf}, (\rho\beta)_{nf}$ and σ_{nf} are presented in Table 1 where subscript p denotes particle of copper and f for base fluid. The numerical values of these quantities for nanoparticles has been provided in Table 2.

Table 1. Expressions for thermal features of nanofluid Ali et al. [38].

Properties	Nanofluid
Density	$\rho_{nf} = (1-\phi)\rho_f + \phi\rho_p,$
Viscosity	$\mu_{nf} = \frac{\mu_f}{(1-\phi)^{2.5}},$
Heat capacity	$(\rho c_p)_{nf} = (1-\phi)(\rho c_p)_f + \phi(\rho c_p)_p,$
Thermal conductivity	$\frac{\kappa_{nf}}{\kappa_f} = \frac{\kappa_p + 2\kappa_f - 2\phi(\kappa_f - \kappa_p)}{\kappa_p + 2\kappa_f + \phi(\kappa_f - \kappa_p)},$
Thermal expansion	$(\rho\beta)_{nf} = (1-\phi)(\rho\beta)_f + \phi(\rho\beta)_p,$
Electric conductivity	$\frac{\sigma_{nf}}{\sigma_f} = 1 + \frac{3\phi\left(\frac{\sigma_p}{\sigma_f} - 1\right)}{\left(\frac{\sigma_p}{\sigma_f} + 2\right) - \phi\left(\frac{\sigma_p}{\sigma_f} - 1\right)},$

Table 2. Numerical values of thermal properties of nanomaterial Ali et al. [37].

Physical Properties	Water (H_2O)	Copper (Cu)
$\rho(\text{kgm}^{-3})$	997.1	8933
$c_p\left(\text{Jkg}^{-1}\text{K}^{-1}\right)$	4180	385
$\kappa\left(\text{Wm}^{-1}\text{K}^{-1}\right)$	0.613	401
$\beta\left(\text{K}^{-1}\right) \times 10^{-5}$	21	1.67
$\sigma(\Omega\text{m})^{-1}$	0.05	5.96×10^7

2.1. *Thermo-Physical Properties*

The important relations for nano-material fluid are given in Table 1.

The expression for generalized complaint wall is:

$$L(\eta) = P - P_0 \tag{17}$$

and

$$L(\eta) = \left[-\tau\frac{\partial^2}{\partial x^2} + m'\frac{\partial^2}{\partial t^2} + d'\frac{\partial}{\partial t} + \beta'\frac{\partial^4}{\partial x^4} + k\right]\eta, \tag{18}$$

where, τ is the wall velocity, m' is the plate mass, d' is the wall-damping coefficient, β' is the stiffness in flexure, k is the stiffness effects.

$$\frac{\partial L}{\partial x} = \frac{\partial p}{\partial x}$$

$$\begin{aligned} \frac{\partial L}{\partial x} &= \mu_{nf}\left(\frac{\partial^2 u}{\partial x^2} + \frac{\partial^2 u}{\partial z^2}\right) + \frac{\sigma_{nf}B_0^2}{1+\left(\frac{\sigma_{nf}B_0}{en_e}\right)^2}\left(-u + \left(\frac{\sigma_{nf}B_0}{en_e}\right)v\right) \\ &-\frac{\mu_{nf}}{K}u + g(\rho\beta)_{nf}(T - T_m) - \rho_{nf}\left(\frac{du}{dt} - 2\Omega v\right). \end{aligned} \tag{19}$$

To convert Equations (11)–(15) into non-dimensional form, we utilize the following set of non-dimensional variables:

$$\begin{aligned} &x^* = \frac{x}{\lambda},\ y^* = \frac{y}{\lambda},\ z^* = \frac{z}{d},\ p^* = \frac{d^2\hat{p}}{c\mu\lambda},\ t^* = \frac{ct}{\lambda},\ u^* = \frac{u}{c},\ v^* = \frac{v}{c}, \\ &w^* = \frac{w}{c},\ \eta^* = \frac{\eta}{d},\ \delta = \frac{d}{\lambda},\ T_m = \frac{T_1 - T_0}{2},\ \theta = \frac{T - T_m}{T_1 - T_0},\ K_1 = \frac{K}{d^2}, \\ &T^* = \frac{\text{Re}\Omega d}{c},\ u = \frac{\partial\psi}{\partial z},\ w = -\delta\frac{\partial\psi}{\partial x}. \end{aligned} \tag{20}$$

where ψ is the stream function, T_m is the mean temperature, T_1 is the upper wall temperature, T_0 is the lower wall temperature. Utilizing these non-dimensional variables and stream function defined in (20) into Equations (11)–(15), the continuity Equation (11) is satisfied identically, while momentum and heat equations becomes:

$$\begin{gathered} Re\delta\left(\frac{\rho_{nf}}{\rho_f}\right)\frac{d}{dt}\left(\frac{\partial\psi}{\partial z}\right) - 2T'\left(\frac{\rho_{nf}}{\rho_f}\right)v = -\frac{\partial p}{\partial x} + Gr\left(\frac{(\rho c_p)_{nf}}{(\rho c_p)_f}\right)\theta \\ +\left(\frac{\mu_{nf}}{\mu_f}\right)\left[\delta^2\frac{\partial^3\psi}{\partial x^2\partial z} + \frac{\partial^3\psi}{\partial z^3} - \frac{1}{K_1}\frac{\partial\psi}{\partial z}\right] - \frac{\left(\frac{\sigma_{nf}}{\sigma_f}\right)M^2}{1+\left(\left(\frac{\sigma_{nf}}{\sigma_f}\right)m\right)^2}\left(\frac{\partial\psi}{\partial z} - \left(\frac{\sigma_{nf}}{\sigma_f}\right)mv\right), \end{gathered} \tag{21}$$

$$\begin{gathered} Re\delta\left(\frac{\rho_{nf}}{\rho_f}\right)\frac{dv}{dt} + 2T'\left(\frac{\rho_{nf}}{\rho_f}\right)\frac{\partial\psi}{\partial z} = -\frac{\partial p}{\partial y} \\ +\left(\frac{\mu_{nf}}{\mu_f}\right)\left[\delta^2\frac{\partial^2\psi}{\partial x^2} + \frac{\partial^2 v}{\partial z^2} - \frac{v}{K_1}\right] - \frac{\left(\frac{\sigma_{nf}}{\sigma_f}\right)M^2}{1+\left(\left(\frac{\sigma_{nf}}{\sigma_f}\right)m\right)^2}\left(v + A_1 m\frac{\partial\psi}{\partial z}\right), \end{gathered} \tag{22}$$

$$-Re\delta^2\left(\frac{\rho_{nf}}{\rho_f}\right)\frac{d}{dt}\left(\frac{\partial\psi}{\partial x}\right) = -\frac{\partial p}{\partial z} + \left(\frac{\mu_{nf}}{\mu_f}\right)\left[\delta^3\frac{\partial^3\psi}{\partial x^3} - \delta\frac{\partial^3\psi}{\partial z^2\partial x}\right], \tag{23}$$

$$\begin{gathered} \delta \mathrm{RePr}\left(\frac{\rho_{nf}}{\rho_f}\right)\left[\frac{\partial\theta}{\partial t} + \frac{\partial\psi}{\partial z}\frac{\partial\theta}{\partial x} + v\frac{\partial\theta}{\partial y} - \frac{\partial\psi}{\partial x}\frac{\partial\theta}{\partial z}\right] = \\ Gr\frac{(\rho\beta)_{nf}}{(\rho\beta)_f}\left[\delta^2\frac{\partial^2\theta}{\partial x^2} + \frac{\partial^2\theta}{\partial z^2}\right] + \left(\frac{\mu_{nf}}{\mu_f}\right)\frac{Br}{K_1}\left(\frac{\partial\psi}{\partial z}\right)^2 + \frac{4Rd}{3}\left[\delta^2\frac{\partial^2\theta}{\partial x^2} + \frac{\partial^2\theta}{\partial z^2}\right] \\ +\varepsilon_1 + \left(\frac{\mu_{nf}}{\mu_f}\right)Br\left[2\delta^2\left\{\left(\frac{\partial^2\psi}{\partial x\partial z}\right)^2 + \left(\frac{\partial^2\psi}{\partial z^2}\right)^2\right\} + \left(\frac{\partial^2\psi}{\partial z^2} - \delta^2\frac{\partial^2\psi}{\partial x^2}\right)^2\right]. \end{gathered} \tag{24}$$

In above non-dimensional Equations (21)–(24), M, Re, Ec, Pr, Rd, ε_1, m, Gr, and Br are the Hartman number, the Reynolds number, the Eckert number, the Prandtl number, the radiation parameter, the heat generation/absorption parameter, the Hall parameter, the Grashof number, and the Brinkman number, respectively, which are non-dimensional parameters and are mathematically defined as:

$$\begin{gathered} M = B_0 d\sqrt{\frac{\sigma_f}{\mu_f}}, \mathrm{Re} = \frac{\rho_f c d}{\mu_f}, Ec = \frac{d^2\mu_f}{(c_p)_f(T_1 - T_0)}, Pr = \frac{\mu_f (c_p)_f}{\kappa_f}, Rd = \frac{4\sigma^* T_m^3}{k^*\kappa_f}, \\ \varepsilon_1 = \frac{d^2\Phi}{\kappa_f(T_1 - T_0)}, m = \frac{\sigma_f B_0}{e n_e}, Gr = \frac{g(\rho\beta)_f d^2}{c\mu_f}(T_1 - T_0), Br = Ec\mathrm{Pr}. \end{gathered} \tag{25}$$

We get the following simplified form of Equations (21)–(24), when we apply the assumption of long wavelength and modest inertial forces to momentum and energy equations:

$$\begin{gathered} -2T'\left(\frac{\rho_{nf}}{\rho_f}\right)v = -\frac{\partial p}{\partial x} + \left(\frac{\mu_{nf}}{\mu_f}\right)\left[\frac{\partial^3\psi}{\partial z^3} - \frac{1}{K_1}\frac{\partial\psi}{\partial z}\right] \\ -\frac{\left(\frac{\sigma_{nf}}{\sigma_f}\right)M^2}{1+\left(\left(\frac{\sigma_{nf}}{\sigma_f}\right)m\right)^2}\left(\frac{\partial\psi}{\partial z} - \left(\frac{\sigma_{nf}}{\sigma_f}\right)mv\right) + Gr\left(\frac{(\rho c_p)_{nf}}{(\rho c_p)_f}\right)\theta, \end{gathered} \tag{26}$$

$$2T'\left(\frac{\rho_{nf}}{\rho_f}\right)\frac{\partial\psi}{\partial z} = -\frac{\partial p}{\partial y} + \left(\frac{\mu_{nf}}{\mu_f}\right)\left[\frac{\partial^2 v}{\partial z^2} - \frac{v}{K_1}\right] - \frac{\left(\frac{\sigma_{nf}}{\sigma_f}\right)M^2}{1+\left(\left(\frac{\sigma_{nf}}{\sigma_f}\right)m\right)^2}\left(v + A_1 m\frac{\partial\psi}{\partial z}\right), \tag{27}$$

$$\frac{\partial p}{\partial z} = 0, \tag{28}$$

$$\begin{gathered} Gr\frac{(\rho\beta)_{nf}}{(\rho\beta)f}\left[\frac{\partial^2\theta}{\partial z^2}\right] + \frac{4Rd}{3}\left[\frac{\partial^2\theta}{\partial z^2}\right] + \left(\frac{\mu_{nf}}{\mu_f}\right)\frac{Br}{K_1}\left(\frac{\partial\psi}{\partial z}\right)^2 \\ +\varepsilon_1 + \left(\frac{\mu_{nf}}{\mu_f}\right)Br\left(\frac{\partial^2\psi}{\partial z^2}\right)^2 = 0. \end{gathered} \tag{29}$$

Now, η amplitude ratio parameter with wall properties gives non-dimensional boundary conditions:

$$\begin{array}{ll}
\frac{\partial \psi}{\partial z} \pm \alpha_1\left(\frac{\mu_{nf}}{\mu_f}\right)\frac{\partial^2 \psi}{\partial z^2} \pm \alpha_2\left(\frac{\mu_{nf}}{\mu_f}\right)\frac{\partial^3 \psi}{\partial z^3} = 0 & \text{at } z = \pm\eta \\
\left[E_1\frac{\partial^3}{\partial x^3} + E_2\frac{\partial^3}{\partial x\partial t^2} + E_3\frac{\partial^2}{\partial x\partial t} + E_4\frac{\partial^5}{\partial x^5} + E_5\frac{\partial}{\partial x}\right]\eta = \frac{\partial p}{\partial x} & \text{at } z = \pm\eta \\
v \pm \beta_1\left(\frac{\mu_{nf}}{\mu_f}\right)\left[\frac{\partial v}{\partial z}\right] = 0 & \text{at } z = \pm\eta \\
\theta \pm \beta_2\left(\frac{\mu_{nf}}{\mu_f}\right)\left[\frac{\partial \theta}{\partial z}\right] = \pm\frac{1}{2} & \text{at } z = \pm\eta
\end{array} \tag{30}$$

Here, $\alpha_1, \alpha_2, \beta_1$, and β_2 are first order, second order, secondary velocity slip, and thermal slip parameters. The main aim was to eliminate the pressure from x and y components of the momentum equation; thus, the author used Equation (28), which already represented that pressure does not depends on z. Secondary flow is due to rotation, so pressure can be ignored from Equation (27). Thus, we have

$$\begin{array}{l}
\left(\frac{\mu_{nf}}{\mu_f}\right)\left[\frac{\partial^4 \psi}{\partial z^4} - \frac{1}{K_1}\frac{\partial^2 \psi}{\partial z^2}\right] - \frac{\left(\frac{\sigma_{nf}}{\sigma_f}\right)M^2}{1+\left(\left(\frac{\sigma_{nf}}{\sigma_f}\right)m\right)^2}\left(\frac{\partial^2 \psi}{\partial z^2} - \left(\frac{\sigma_{nf}}{\sigma_f}\right)m\frac{\partial v}{\partial z}\right) \\
+2T'\left(\frac{\rho_{nf}}{\rho_f}\right)\frac{\partial v}{\partial z} + Gr\left(\frac{(\rho c_p)_{nf}}{(\rho c_p)_f}\right)\frac{\partial \theta}{\partial z} = 0,
\end{array} \tag{31}$$

$$\left(\frac{\mu_{nf}}{\mu_f}\right)\left[\frac{\partial^2 v}{\partial z^2} - \frac{v}{K_1}\right] - 2T'\left(\frac{\rho_{nf}}{\rho_f}\right)\frac{\partial \psi}{\partial z} - \frac{\left(\frac{\sigma_{nf}}{\sigma_f}\right)M^2}{1+\left(\left(\frac{\sigma_{nf}}{\sigma_f}\right)m\right)^2}\left(v + \left(\frac{\sigma_{nf}}{\sigma_f}\right)m\frac{\partial \psi}{\partial z}\right) = 0, \tag{32}$$

At the end the simplified form of non-dimensional equations with boundary conditions is:

$$A_0\left(\frac{\partial^4 \psi}{\partial z^4} - \frac{1}{K_1}\frac{\partial^2 \psi}{\partial z^2}\right) - \frac{A_1 M^2}{1+(A_1 m)^2}\left(\frac{\partial^2 \psi}{\partial z^2} - A_1 m\frac{\partial v}{\partial z}\right) + 2T'A_2\frac{\partial v}{\partial z} + A_3\frac{\partial \theta}{\partial z} = 0, \tag{33}$$

$$A_0\left(\frac{\partial^2 v}{\partial z^2} - \frac{v}{K_1}\right) - \frac{A_1 M^2}{1+(A_1 m)^2}\left(v + A_1 m\frac{\partial \psi}{\partial z}\right) - 2T'A_2\frac{\partial \psi}{\partial z} = 0, \tag{34}$$

$$\left(A_4 + \frac{4}{3}Rd\right)\frac{\partial^2 \theta}{\partial z^2} + A_0\frac{Br}{K_1}\left(\frac{\partial \psi}{\partial z}\right)^2 + A_0 Br\left(\frac{\partial^2 \psi}{\partial z^2}\right)^2 + \varepsilon_1 = 0. \tag{35}$$

With boundary conditions:

$$\begin{array}{ll}
\frac{\partial \psi}{\partial z} \pm \alpha_1 A_0\frac{\partial^2 \psi}{\partial z^2} \pm \alpha_2 A_0\frac{\partial^3 \psi}{\partial z^3} = 0 & \text{at } z = \pm\eta \\
\left[E_1\frac{\partial^3}{\partial x^3} + E_2\frac{\partial^3}{\partial x\partial t^2} + E_3\frac{\partial^2}{\partial x\partial t} + E_4\frac{\partial^5}{\partial x^5} + E_5\frac{\partial}{\partial x}\right]\eta = \frac{\partial p}{\partial x} & \text{at } z = \pm\eta \\
v \pm \beta_1 A_0\frac{\partial v}{\partial z} = 0 & \text{at } z = \pm\eta \\
\theta \pm \beta_2 A_0\frac{\partial \theta}{\partial z} = \pm\frac{1}{2} & \text{at } z = \pm\eta
\end{array} \tag{36}$$

In the above equations, the constants that are used are given below:

$$\begin{array}{l}
A_0 = \frac{\mu_{nf}}{\mu_f},\ A_1 = \frac{\sigma_{nf}}{\sigma_f},\ A_2 = \frac{\rho_{nf}}{\rho_f},\ A_3 = Gr\cdot\frac{(\rho c_p)_{nf}}{(\rho c_p)_f},\ A_4 = \frac{(\rho\beta)_{nf}}{(\rho\beta)_f}, \\
E_1 = -\frac{\tau d^3}{\lambda^3 \mu_f c},\ E_2 = \frac{m_1 c d^3}{\lambda^3 \mu_f},\ E3 = \frac{d' d^3}{\lambda^2 \mu_f},\ E_4 = \frac{\beta' d^3}{\lambda^3 \mu_f c},\ E_5 = \frac{k d^3}{\lambda \mu_f c}.
\end{array} \tag{37}$$

2.2. Entropy Generation Analysis

An irreversible process in which two phenomena occur, i.e., thermal diffusion and fluid friction, results in entropy generation, which is basically loss of ability to do work. The volume fraction entropy generation for two-phase nanomaterial is:

$$Ngen = \frac{\kappa_{nf}}{\theta_0^2}\left[\left(\frac{\partial T}{\partial x}\right)^2 + \left(\frac{\partial T}{\partial z}\right)^2\right] + \frac{\mu_{nf}}{\theta_0}\left[2\left(\frac{\partial u}{\partial x}\right)^2 + \left(\frac{\partial w}{\partial z}\right)^2 + \left(\frac{\partial u}{\partial z} + \frac{\partial w}{\partial x}\right)^2\right] + \frac{16\sigma^* T_m^3}{3k^*\theta_0}\left[\frac{\partial^2 T}{\partial x^2} + \frac{\partial^2 T}{\partial z^2}\right] + \frac{\mu_{nf}u^2}{K\theta_0} + \frac{\Phi}{\theta_0}. \tag{38}$$

The first term on the right side is irreversible heat transfer, the second term is irreversible viscous dissipation, the third term is radiation effects, the fourth term is heat transfer analysis for convection, and the fifth term is heat generation/absorption effects. The entropy generation number in non-dimensional form is:

$$N_s = \frac{N_{gen}}{N_g} = \left(A_4 + \frac{4}{3}Rd\right)\left(\frac{\partial \theta}{\partial z}\right)^2 + A_0 Br\Lambda\left[\frac{1}{K_1}\left(\frac{\partial \psi}{\partial z}\right)^2 + \left(\frac{\partial^2 \psi}{\partial z^2}\right)^2\right] + \varepsilon_1 \tag{39}$$

where Ng is the rate of entropy generation and Λ is the temperature ratio, defined as:

$$N_g = \frac{\kappa_f (T_1 - T_0)^2}{\theta_0^2 d^2}, \Lambda = \frac{\theta_0}{(T_1 - T_0)} \tag{40}$$

2.3. Bejan Number Analysis

The Bejan number was introduced by professor Adrian Bejan from Duke University. The Bejan number is the proportion of heat transfer irreversibility to total entropy generation. Mathematically,

$$Be = \frac{\left(A4 + \frac{4}{3}Rd\right)\left(\frac{\partial \theta}{\partial z}\right)^2}{\left(A_4 + \frac{4}{3}Rd\right)\left(\frac{\partial \theta}{\partial z}\right)^2 + A_0 Br\Lambda\left[\frac{1}{K_1}\left(\frac{\partial \psi}{\partial z}\right)^2 + \left(\frac{\partial^2 \psi}{\partial z^2}\right)^2\right] + \varepsilon_1} \tag{41}$$

The variation of the Bejan number is $0 < Be < 1$, which implies that total entropy generation dominates heat transfer irreversibility in one case and that total entropy generation equals heat transfer irreversibility in the other.

3. Graphical Discussion

The focus of this work is on the analysis of entropy generation of dual-stage nanomaterial in a peristaltic motion considering thermal fluxes and radiation, along with the boundary condition of the generalized complaint wall in a rotating channel. The governing equations, derived in the preceding section, were solved in Mathematica using built-in solver NDSolve. In this section, we discuss the physical impacts of various parameters (Hall parameter (m), radiation parameter (Rd), permeability parameter (K_1), heat generation/absorption parameter (ε_1), Hartman number (M), rotation parameter (T'), first and second order velocity slip parameters (α_1, α_2), secondary velocity slip (β_1), and thermal slip parameters (β_2) on axial and secondary velocities (u and v), temperature distribution (θ), entropy production (Ns), and Bejan number (Be) for the given values of $t = 0.1$, $x = 0.2$, $\phi = 0.01$, $\varepsilon = 0.3$, $E_1 = 0.03$, $E_2 = 0.02$, $E_3 = 0.01$, $E_4 = 0.03$, $E_5 = 0.02$).

3.1. Axial Velocity Analysis

The effects of various flow parameters, such as Hartman number (M), nanoparticles volume fraction (ϕ), permeability parameter (K_1), first order velocity slip (α_1), second order velocity slip (α_2) and rotation parameter (T') on the axial velocity u are presented in Figures 2–7. The results are calculated for $\varepsilon = 0.3$, $m = 1$, $Gr = 3$, $Rd = 0.1$, $Br = 0.01$,

$K_1 = 0.5$, $M = 2$, $\beta_1 = 0.02$, $\beta_2 = 0.02$, $\alpha_1 = 0.01$, and $\beta_1 = -0.01$. Figure 2 shows curves representing spatial distribution of axial velocities u computed for several Hartman numbers M, ranging from 0 to 3. It is noted that, as the Hartmann number increased, the axial velocity decreased. Furthermore, the velocity gradients in the interior domain also decreased as M increased. These observations are consistent with the physical characteristics of magnetic force, i.e., Lorentz force is a decelerating force, and also with the previously reported experimental and theoretical observations. Figure 3 depicts that the axial velocity decreased with the increase in the values of volume fraction ϕ, due to strong resistive forces. The small change was observed at the starting and end points; however, at the middle of the graph, axial velocity distribution had great loss.

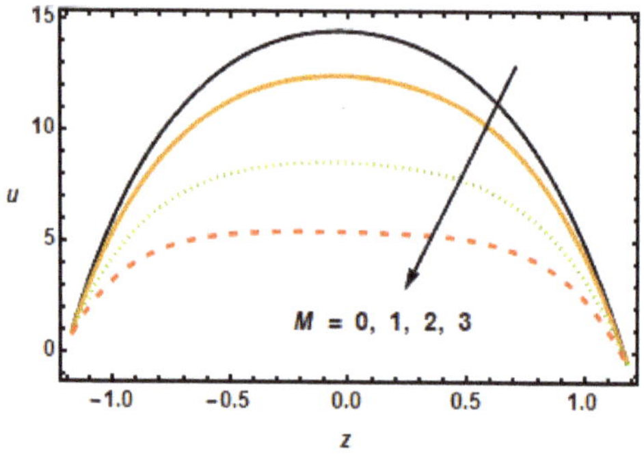

Figure 2. Effects of Hartmann number M on u.

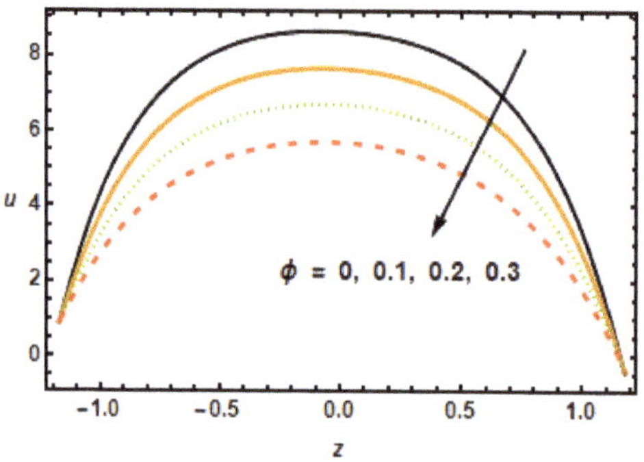

Figure 3. Effects of Volume fraction ϕ on u.

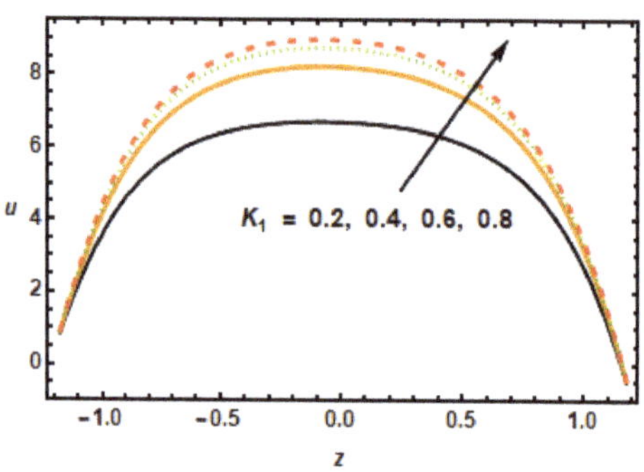

Figure 4. Effects of permeability K_1 on u.

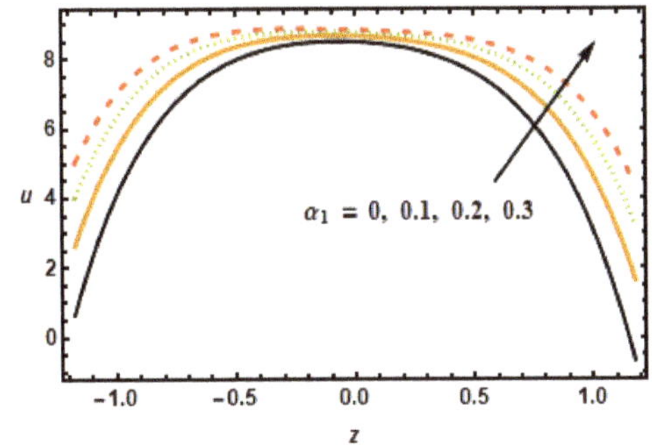

Figure 5. Effects of first order slip α_1 on u.

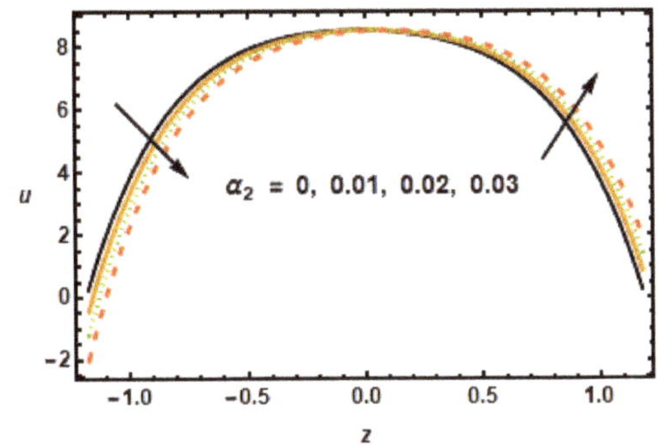

Figure 6. Effects of second order slip α_2 on u.

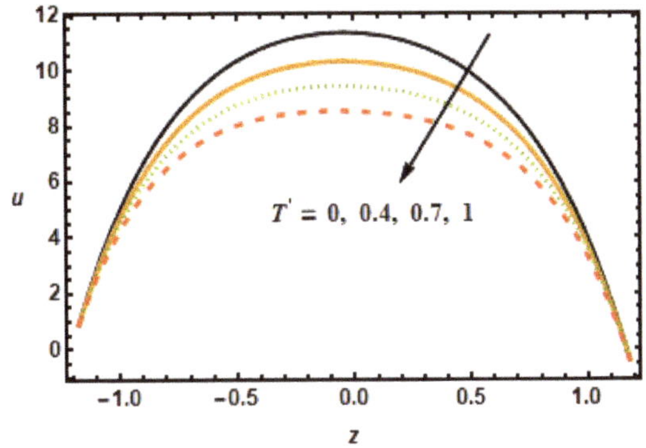

Figure 7. Effects of rotation parameter T' on u.

Figure 4 shows axial velocity plots for numerous values of permeability parameter K_1, representing the permeability of the porous medium. The plotted results show that the axial velocity rose with increasing values of the permeability parameter since, with increased permeability of porous medium K_1, frictional forces decreased, hence, the fluid accelerated because of higher energy budgets. The effects of slip parameter α_1 on the axial velocity distributions are shown in Figure 5. The results plotted in the figure, for different values of α_1, shows that the slip parameter affects flow dynamics in the locality of the boundary as expected. It is noted that increase in slip parameter α_1 had accelerating effects on the axial velocity; therefore, axial velocity increased with increasing α_1. When the slip effects were incorporated in the applied boundary condition, the energy losses at the boundaries decreased, which is why the slip parameter α_1 exhibited accelerating effects on the flow dynamics.

Similar observations are made in Figure 6, showing axial velocity for different values of second slip parameter α_2. In contrast to α_1, the second slip parameter α_2 is linked with the spatial change in the velocity gradients. It is, therefore, seen in plotted results that the axial velocities decreased in the lower half, since velocity gradients were decreasing in z, and increased in the upper half due to increasing velocity gradients. In Figure 7, we show axial velocities for different values of the rotation parameter T'. The results show that the increase in the values of rotational parameter T' resulted in a decrease in axial velocity, exhibiting an inverse relation between the velocity and rotation. Notice that the maximum velocity in the axial direction was achieved when rotation was zero, as, in that case, the retarding effects of the magnetic field were minimal.

3.2. Secondary Velocity Analysis

The dynamical effects of different parameters such as the Hartman number, the Hall parameter, the nanoparticle volume fraction, the permeability parameter, and the secondary slip parameter, and rotation parameter on the secondary velocity are given in Figures 8–13. The decelerating effects of the Hartman number M, quantifying magnetic field strength, on the secondary velocity v are given in Figure 8. Consistent with the observations made in Figure 2, the secondary velocity also decreased with an increase in Hartman number M because of increased Lorentz force, which impeded the flow, as discussed above. Figure 9 illustrates velocity profiles generated for various values of the Hall parameter m to investigate the influence of the Hall parameter on secondary velocity. The results revealed that the Hall parameter had an accelerating influence on the flow, as seen by the increasing trend in velocities as the Hall parameter increased. By the decay of magnetic damping forces, the thermal conductivity of two-stage nanomaterial decreased, which resulted in higher secondary velocities.

Hall effect is necessary for the manufacturing of secondary velocity because $m = 0$ means having no secondary velocity. To analyze the effects of nanoparticle volume fraction ϕ on the secondary velocity, we plotted velocity curves corresponding to different values of ϕ, shown in Figure 10. The results showed that the secondary velocity decreased for increasing values of ϕ. This is consistent with the physical effects associated with volume fraction ϕ, i.e., for higher values of volume fractions ϕ, the inter-particle interactions increased, resulting in a loss of energy. Figure 11 shows velocity curves plotted, corresponding to different values of the permeability parameter K_1. The velocity curves may be observed in the figure to have a rising tendency when the permeability parameter K_1 increased. The increment in fluid velocities resulted from decreased hindrance experience by the fluid at large values of permeability parameter K_1.

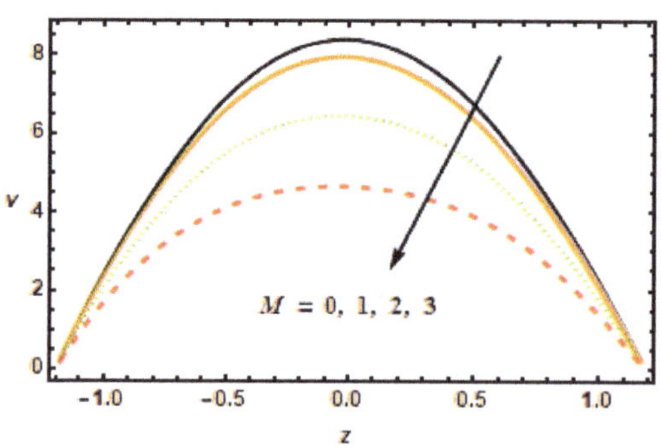

Figure 8. Dependence of v on Hartman number M.

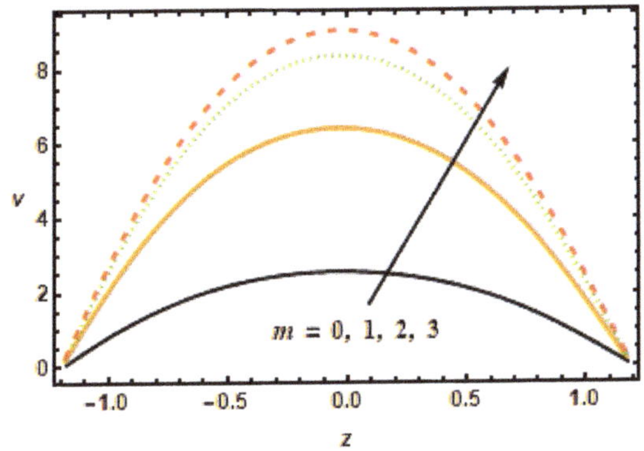

Figure 9. Dependence of v on Hartman number m.

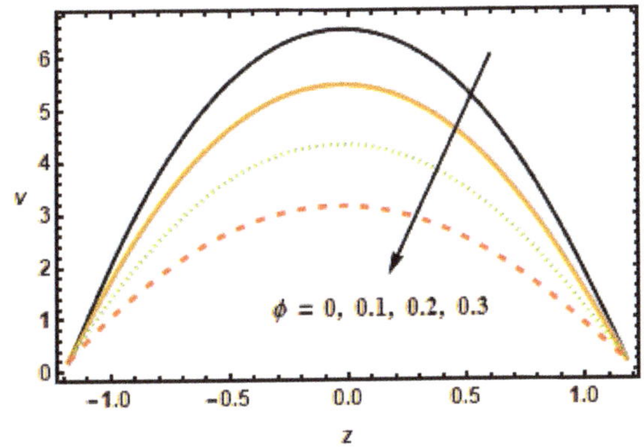

Figure 10. Dependence of v on volume fraction ϕ.

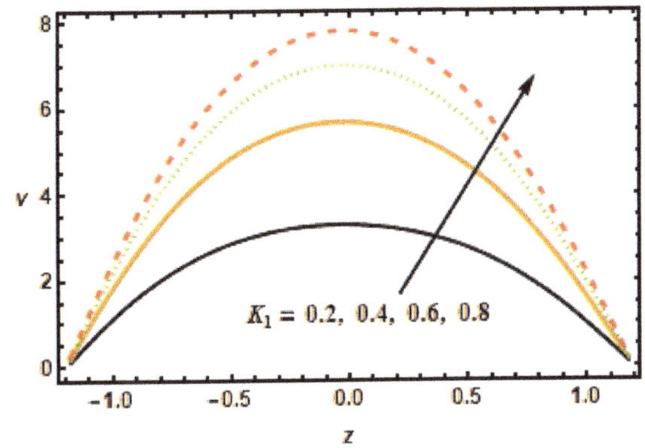

Figure 11. Dependence of v on permeability K_1.

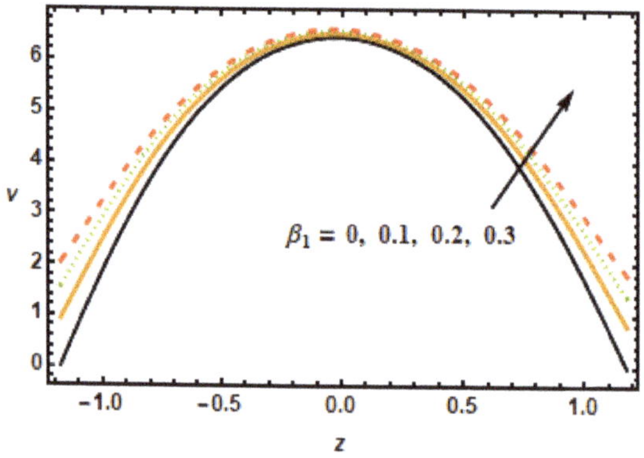

Figure 12. Dependence of v on secondary slip β_1.

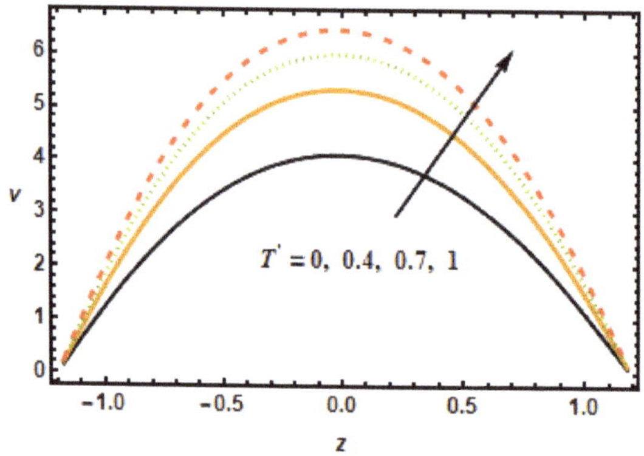

Figure 13. Dependence of v on rotation parameter T'.

In Figure 12, we show results in terms of secondary velocity for several values of the secondary slip parameter. In agreement with the observations made in Figure 5, secondary velocity increased with increasing values of the slip parameter. As shown in Figure 5, the effect of the slip parameter was observed to be localized for secondary velocities as well. This sketch represents a very different trend compared to the other parameters. Figure 13 depicts the influence of the rotation parameter T' on secondary velocity. The secondary velocity rose as the rotation parameter grew larger, due to greater inertial forces, as shown in the displayed figure.

3.3. *Temperature Analysis*

As discussed in the introduction, thermal properties of the base fluids could be considerably enhanced with the presence of nanoparticles. In this section, we explore the effects of the Hartman number M, the nanoparticle volume fraction ϕ, the permeability parameter K_1, the thermal slip parameter β_2 and the radiation parameter Rd on thermal characteristics of the flow. Figure 14 shows temperature profiles for diverse values of the Hartman number. As the values of the Hartman number increased, fluid velocities decreased, as shown in Figures 2 and 8. This decrement in velocities resulted in less inter-particle interaction, due to which temperature profiles showed a decreasing trend with increasing values of the Hartman number. The effects of the volume fraction ϕ on the thermal profiles were investigated, as shown in Figure 15. The results showed a drop in temperature profiles at higher values of the volume fraction ϕ. For higher volume fractions, the effective viscosity of the mixture increased, due to which the thermal conductivity was reduced; hence, the temperatures dropped at large values of the volume fraction ϕ. The main point is that the absence of copper and the addition of nanoparticle volume fraction in base fluid resulted in increased thermal conductivity of the fluid; thus, by enhancing nanoparticles, the fluid capacity was increased to accumulate the consequent temperature fall.

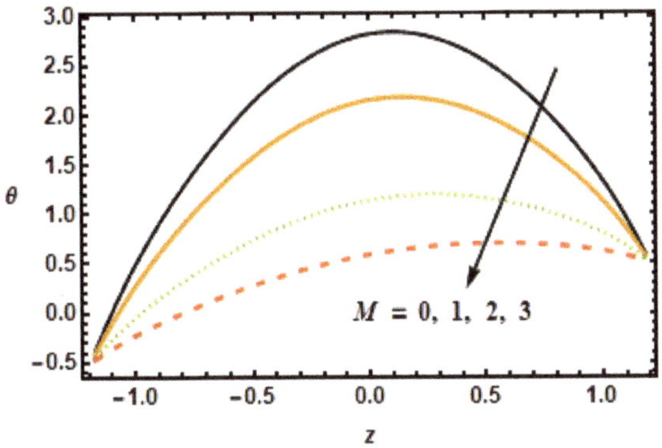

Figure 14. Effects of Hartman number M on θ.

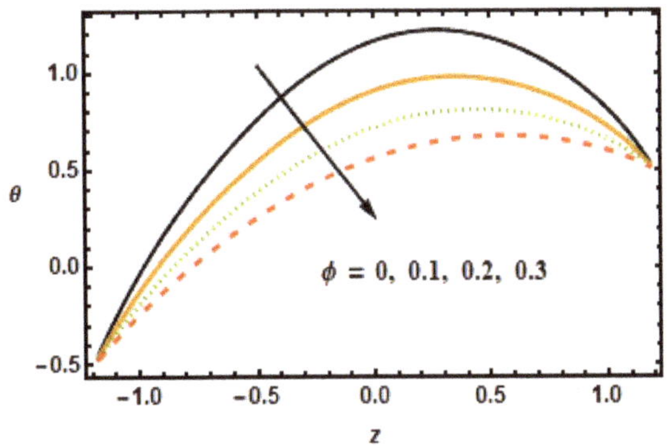

Figure 15. Effects of volume fraction ϕ on θ.

The effect of the permeability parameter on temperature was observed. It can be observed in Figure 16 that temperature differences enlarged at the center of the graph. By increasing the porosity, the system leaves the temperature. Figure 16 shows temperature profiles at different porosity levels. Notice the decreasing trend in temperature curves corresponding to increased values of the permeability parameter. This is due to the fact that fluid experiences less resistance for large porosity. Notice that the two parameters, the Hartman number M and the volume fraction ϕ, have similar trends for velocities and temperature distribution, whereas permeability of porous medium shows different behavior for temperature compared to that of axial and secondary velocities. The effects of the slip parameter β_2 on the thermal characteristics of the fluid is shown in Figure 17. The plotted results show that the slippage condition enhanced the heat transfer rate. This is because, for large values of the slip parameter, the velocity gradients rose, which, in turn, enhanced inter-particle interactions, generating more heat. For the radiation parameter Rd, the temperature profiles showed a decreasing trend, as is depicted in the results plotted in Figure 18. The temperature in the inner domain lowered when the radiation parameter Rd increased because energy absorption reduced as the radiation parameter increased.

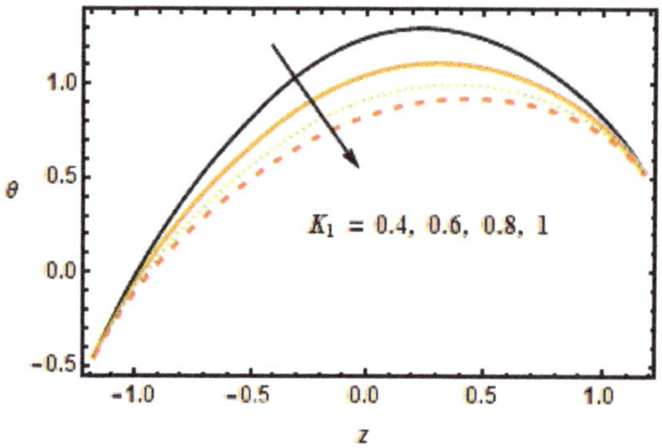

Figure 16. Dependence of θ on K_1.

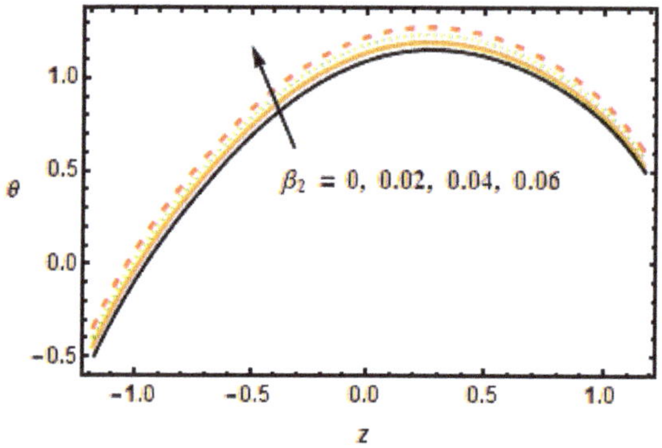

Figure 17. Dependence of θ on thermal slip.

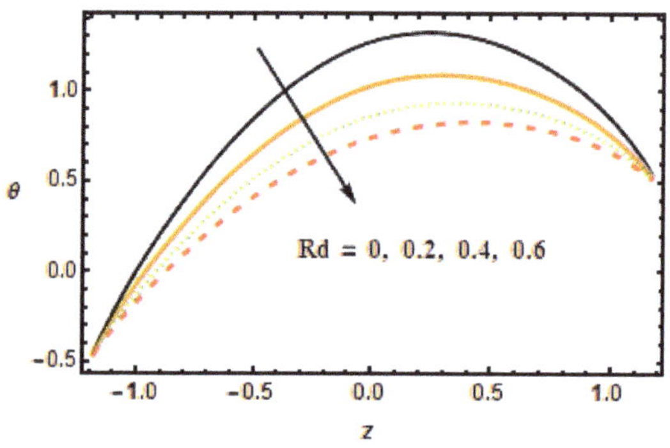

Figure 18. Effects of radiation Rd on θ.

3.4. Entropy Generation Analysis

The impact of entropy generation on different parameters, i.e., the nanoparticle volume fraction parameter, the radiation parameter, the Hall parameter, the heat generation/absorption parameter, the rotation parameter, and the Hartmann number (ϕ, Rd, m, ε_1, T', and M) is discussed in Figures 19–24. The results plotted in Figure 19 show the entropy production number Ns for diverse values of the Hartmann number M. In the plotted results, the entropy generation number exhibited an inversely proportional association with the Hartmann number, i.e., Ns dropped as the value of M increased. Notice that, for diverse values of the Hartmann number, the difference between the entropy was maximum at the boundaries, whereas, in the center of the domain, this difference was the lowest. Figure 20 shows spatial variation of the entropy parameter Ns for dissimilar values of the Hall current parameter m. As shown above, velocity gradients increased at higher values of m; it is, therefore, seen in the figure that the entropy generation number Ns increased with m.

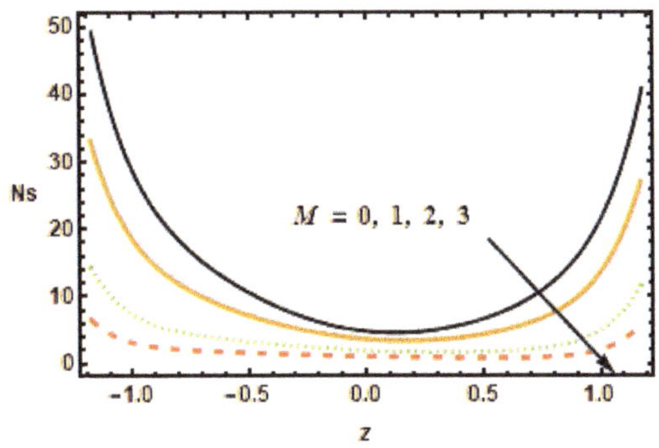

Figure 19. Effects of Hartman number M on Ns.

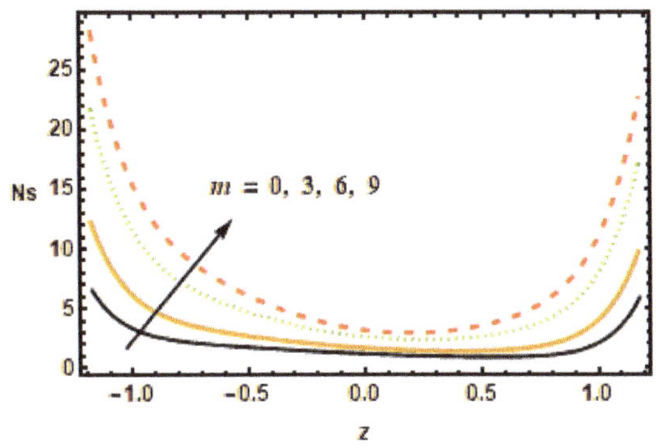

Figure 20. Effects of Hall parameter m on Ns.

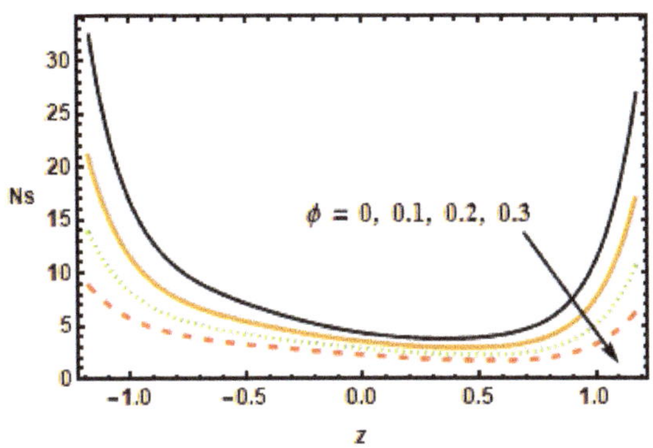

Figure 21. Effects of volume fraction ϕ on Ns.

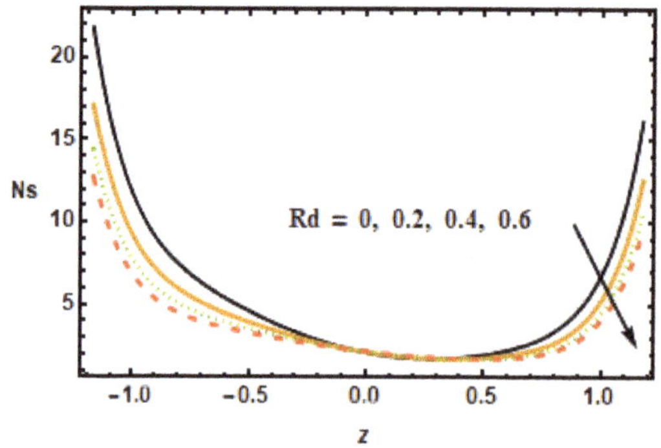

Figure 22. Effects of radiation parameter *Rd* on *Ns*.

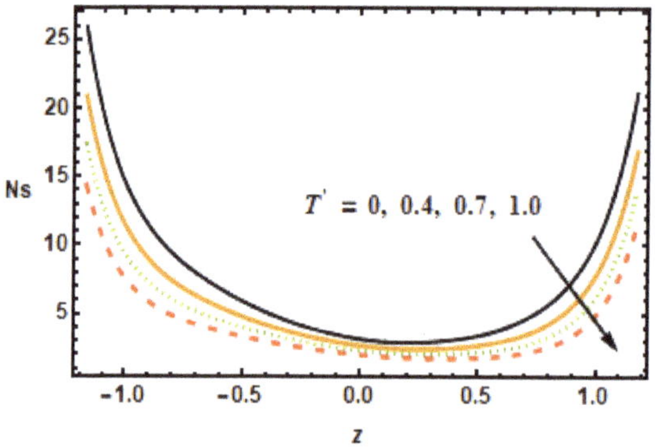

Figure 23. Effects of rotation parameter T' on *Ns*.

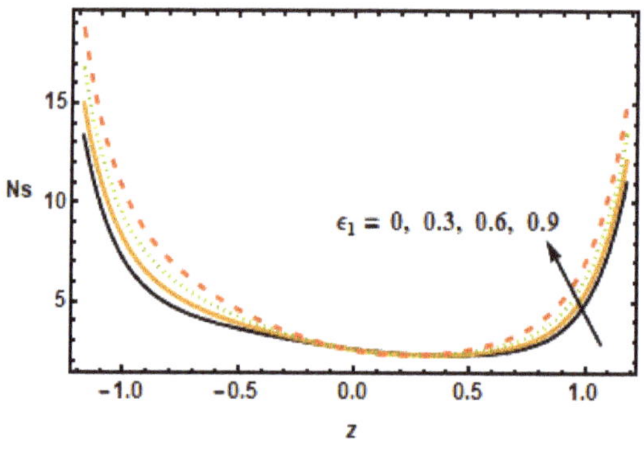

Figure 24. Effects of heat generation ε_1 on *Ns*.

To understand the effect of the nanoparticle volume fraction ϕ on the entropy production number *Ns*, results were plotted for different volume fractions ϕ. It is noted that, by increasing the value of the nanoparticle volume fraction ϕ, the entropy generation number *Ns* decreased. Similarly for the radiation parameter *Rd*, the entropy generation number decreased due to small thermal gradients resulting from a loss of thermal energy due to radiation.

It is depicted in Figure 23 that *Ns* reduced for rising values of the rotation parameter T', which means that the entropy generation number increased in the absence of rotation. The results in Figure 24 show that, for large values of heat generation or the absorption parameter ε_1, Ns increased at the boundaries but remained constant at the center of the domain. Notice that the difference was minimal for smaller values of ε_1, evident from the observation that the black and yellow curves resided very close to each other. It is also worth noting that the entropy generation trend for all of the parameters showed a decreasing trend, with the exception of the heat generation/absorption parameter, which showed an increasing trend. Furthermore, the Hartman number, the volume fraction and the rotation influence axial velocity and entropy generation changed in a similar manner.

3.5. Bejan Number Analysis

Figures 25–30 are portrayed to depict the influence of the Bejan number Be on various parameters, including the heat generation/absorption parameter, the volume fraction, the Hall parameter, the rotation, the Hartman number and radiation. Figure 25 shows the inverse relation of the Bejan number against the Hartman number M, i.e., by increasing values of M across the channel, the Be values dropped. It is clearly seen that the Bejan number decreased at the starting and ending point, but it remained the same at the region from -0.5 to 0. Figure 26 shows that Be magnified across the boundaries with increasing values of the Hall parameter m but Be was reduced in the middle of the channel due to reduced strengths of the magnetic field, which reduced the fluid acceleration.

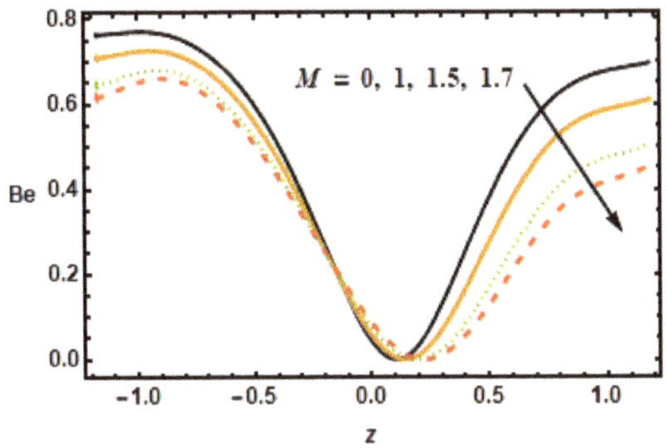

Figure 25. Effects of Hartman number M on Be.

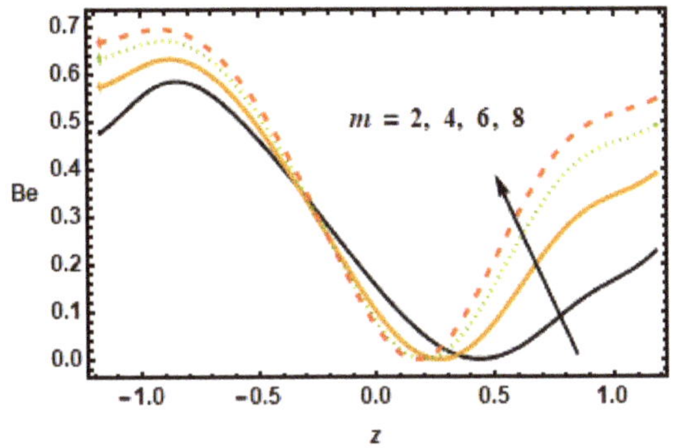

Figure 26. Effects of Hall parameter m on Be.

The plot in Figure 27 revealed that the Bejan number Be decelerated for greater values of nanoparticle volume fraction ϕ. It is so because of the irreversibility of heat transfer, total entropy generation is greater than entropy generation. At the center of the channel, the trend changes whereas at the right end of channel Bejan number is almost constant for increasing values of ϕ. In Figure 28, it is noted that the increasing value of the radiation parameter *Rd* resulted in a reduced Bejan number *Be* due to the fact that temperature gradients drop as the radiation parameter *Rd* increased.

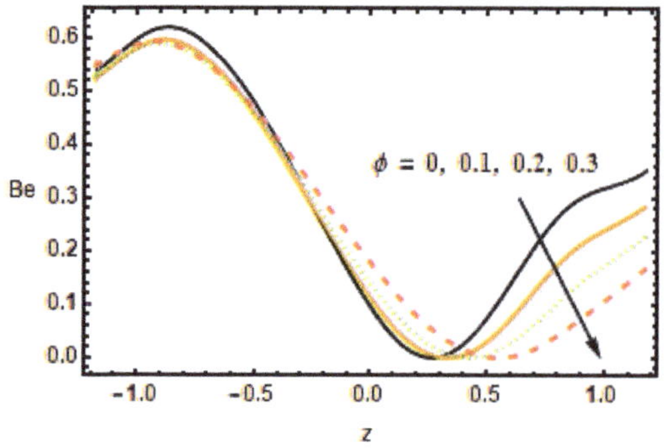

Figure 27. Effects of volume fraction ϕ on *Be*.

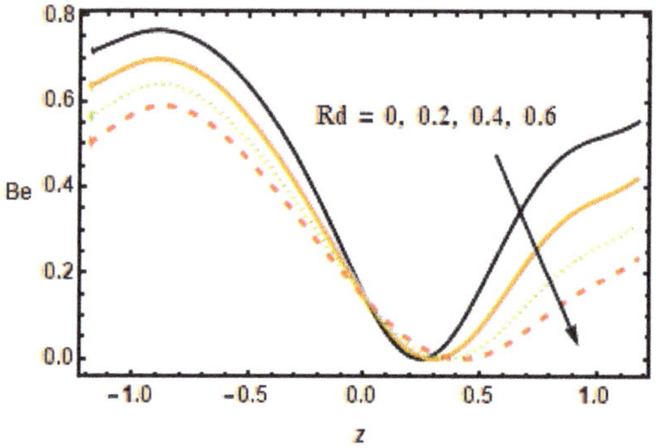

Figure 28. Effects of radiation parameter *Rd* on *Be*.

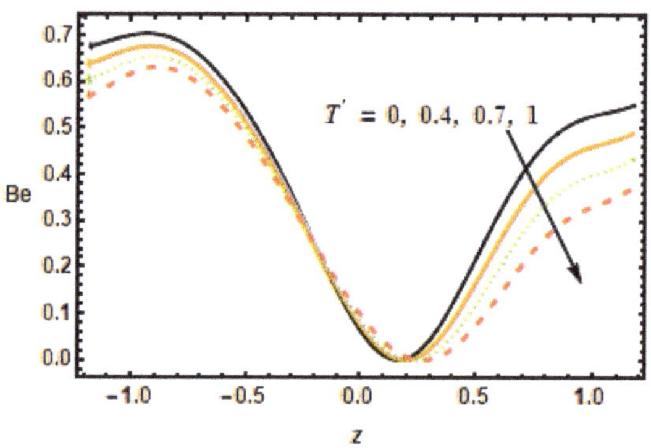

Figure 29. Effects of rotation parameter T' on *Be*.

Rotation parameter T' has similar behavior, such as the Hartman number M having an inverse relation with the Bejan number due to angular velocity across the boundaries of the channel. It is seen in Figure 30 that the heat flux parameter ε_1 increased the Bejan number *Be* along the boundaries, whilst the dependence remained constant in the center of the channel. The Bejan number had an increasing and decreasing trend for the Hall Effect and the volume fraction, respectively. Rotation, radiation and the Hartman number affected the Bejan number and entropy generation number in a similar manner. Furthermore, changes in the temperature distribution, entropy production and the Bejan number due to the Hartman number was similar.

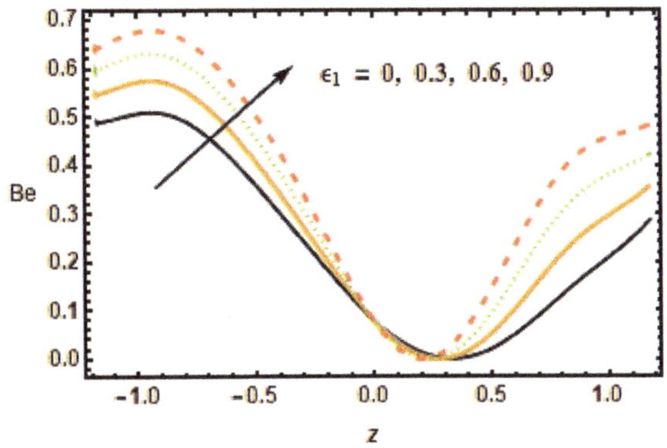

Figure 30. Effects of heat generation ε_1 on *Be*.

4. Conclusions

The peristaltically moving flow of nanomaterial in a porous rotating channel with generalized complaint walls was investigated in this article. The impact of the magnetic field, the Hall parameter, the heat source/sink, thermal radiation, Joule heating, and boundary slip on the velocity and temperature profiles was discussed in detail. The major goal of this research was to look at how entropy generation and the Bejan number are affected by varying physical conditions. The following are the most important consequences:

- The axial velocity was inversely proportional to the Hartman number, the volume fraction of nanoparticles, and the rotation parameters. It decreased when these parameters were increased.
- The axial velocity increased when the porosity and first order slip parameters were increased.
- Secondary velocity decreased with the increase of the Hartmann number and the nanoparticle volume fraction.
- Increasing the Hall parameter, porosity parameter, secondary velocity slip parameter, and rotation parameter improved secondary velocity.
- The temperature profile was enhanced for only the thermal slip parameter, demonstrating temperature rises due to slip effects.
- The temperature dropped for increasing values of the Hartman number, the Hall parameter, porosity, and the radiation parameters.
- Increasing the Hall parameter and the heat generation/absorption parameter enhanced the amount of entropy generation.
- Entropy was reduced for large values of the Hartman number, the nanoparticles volume fraction, the radiation parameter, and the rotation parameter.
- The alternative behavior for the Hall parameter and the nanoparticles volume fraction was represented by the Bejan number.
- Due to the pressure drop throughout the length of the channel, the Bejan number rose in relation to the heat generation/absorption parameter.
- The Bejan number and entropy generation had the same behavior when it came to physical parameters such as the rotation parameter, the radiation parameter, and the Hartman number.

Author Contributions: Conceptualization, A.A. and K.S.N.; methodology, M.S. and H.J.A.; software, M.S. and M.A.; validation, M.A. and C.A.S.; formal analysis, A.A. and M.S.; investigation, M.A. and K.S.N.; resources, K.S.N. and C.A.S.; writing—original draft preparation, M.S. and H.J.A.; writing—review and editing, A.A., K.S.N. and C.A.S.; supervision, A.A.; funding acquisition, K.S.N. and C.A.S. All authors have read and agreed to the published version of the manuscript.

Funding: The authors extend their appreciation to the Deanship of Scientific Research at King Khalid University, Saudi Arabia for funding this work through the Research Group Program under Grant No: RGP 2/26/43.

Data Availability Statement: Not applicable.

Conflicts of Interest: The authors declare no conflict of interest.

Nomenclature

List of Symbols		**Greek Symbols**	
T	Fluid temperature (K)	ϕ	Nanoparticles volume fraction
T_0	Lower wall temperature (K)	ρ	Fluid density (kg m^{-3})
T_1	Upper wall temperature (K)	ψ	Stream function (m^2 s^{-1})
T_m	Mean Temperature of nanofluid	μ	Dynamic Viscosity (kg m^{-1} s^{-1})
c_p	Specific heat (J kg k^{-1})	v	Kinematic Viscosity (m^2 s^{-1})
Ω	Angular speed	σ	Electric conductivity (Sm^{-1})
Ns	Entropy generation	κ	Thermal conductivity (W m^{-1} k^{-1})
Be	Bejan number	R	Dimensionless heat flux parameter
Pr	Prandtl number	ε_1	Heat generation/absorption parameter
Ec	Eckert number	δ	Wave number
Br	Brinkman number	**Subscript**	
M	Hartmann number	nf	Nanofluid
u, v, w	x, y, z velocity component (m s^{-1})	F	Base fluid
Re	Reynolds number	P	Particle of nanofluid
B_o	Strength of magnetic field (T)	eff	Effective

References

1. Latham, T.W. Fluid Motion in a Peristaltic Pump. Master's Thesis, Massachusetts Institute of Technology, Cambridge, MA, USA, 1966.
2. Shapiro, A.H.; Jaffrin, M.Y.; Weinberg, S.L. Peristaltic pumping with long wavelengths at low Reynolds number. *J. Fluid Mech.* **1969**, *37*, 799–825. [CrossRef]
3. Yin, F.; Fung, Y.C. Peristaltic waves in circular cylindrical tubes. *J. Appl. Mech.* **1969**, *36*, 57987. [CrossRef]
4. Asghar, S.; Hussain, Q.; Hayat, T. Peristaltic flow of reactive viscous fluid with temperature dependent viscosity. *Math. Comput. Appl.* **2013**, *18*, 198–220. [CrossRef]
5. Asghar, S.; Minhas, T.; Ali, A. Existence of a Hartmann layer in the peristalsis of Sisko fluid. *Chin. Phys. B* **2014**, *23*, 054702. [CrossRef]
6. Ali, A.; Asghar, S.; Awais, M. Thermophoresis and concentration effects in a fourth grade peristaltic flow with convective walls. *J. Cent. South Univ.* **2017**, *24*, 1654–1662. [CrossRef]
7. Laila, R.; Khan Marwat, D.N.; Khan, K.; Ali, A.; Shah, Z. Flow in a two dimensional channel with deforming and peristaltically moving walls. *SN Appl. Sci.* **2019**, *1*, 1534. [CrossRef]
8. Ali, A.; Ali, Y.; Khan Marwat, D.N.; Awais, M.; Shah, Z. Peristaltic flow of nanofluid in a deformable porous channel with double diffusion. *SN Appl. Sci.* **2020**, *2*, 100. [CrossRef]
9. Ali, A.; Awais, M.; Al-Zubaidi, A.; Saleem, S.; Khan Marwat, D.N. Hartmann boundary layer in peristaltic flow for viscoelastic fluid: Existence. *Ain Shams Eng. J.* **2022**, *13*, 101555. [CrossRef]
10. Choi, S.U.S.; Eastman, J.A. *Enhancing Thermal Conductivity of Fluids with Nanoparticles*; Argonne National Lab: Lemont, IL, USA, 1995.
11. Buongiorno, J. Convective transport in nanofluids. *J. Heat Transf.* **2006**, *128*, 240–250. [CrossRef]
12. Turkyilmazoglu, M. The analytical solution of mixed convection heat transfer and fluid flow of a MHD viscoelastic fluid over a permeable stretching surface. *Int. J. Mech. Sci.* **2013**, *77*, 263–268. [CrossRef]
13. Khan, J.A.; Mustafa, M.; Hayat, T.; Alsaedi, A. Three-dimensional flow of nanofluid over a nonlinearly stretching sheet: An application to solar energy. *Int. J. Heat Mass Transf.* **2015**, *86*, 158–164. [CrossRef]

14. Awais, M.; Hayat, T.; Muqaddass, N.; Ali, A.; Aqsa; Awan, S.E. Nanoparticles and nonlinear thermal radiation properties in the rheology of polymeric material. *Results Phys.* **2018**, *8*, 1038–1045. [CrossRef]
15. Ali, A.; Ali, Y.; Kumam, P.; Babar, K.; Ahmad, A.; Shah, Z. Flow of a Nanofluid and Heat Transfer in Channel with Contracting/Expanding Walls. *IEEE Access* **2019**, *7*, 102427–102436. [CrossRef]
16. Ali, A.; Akhtar, J.; Anjum, H.J.; Awais, M.; Shah, Z.; Kumam, P. 3D nanofluid flow over exponentially expanding surface of Oldroyd-B fluid. *Ain Shams Eng. J.* **2021**, *12*, 3939–3946. [CrossRef]
17. Ali, A.; Maqsood, M.; Anjum, H.J.; Awais, M.; Sulaiman, M. Analysis of heat transfer on MHD Jeffrey nanofluid flow over non-linear elongating surface of variable thickness. *Z. Angew. Math. Mech.* **2022**, *102*, e202100250. [CrossRef]
18. Shashikumar, N.S.; Prasannakumara, B.C.; Archana, M.; Gireesha, B.J. Thermodynamics analysis of a Casson nanofluid flow through a porous microchannel in the presence of hydrodynamic slip: A model of solar radiation. *J. Nanofluids* **2019**, *8*, 63–72. [CrossRef]
19. Sreelakshmi, T.K.; Abraham, A.; Raju, C.S.K.; Prasannakumara, B.C.; Chethan, A.S. Brownian motion and thermophoresis significance of Casson fluid flow in an elastic-radiated surface with a suspension of transpiration and non-Fourier heat flux. *Waves Random Complex Media* **2021**, 1–20. [CrossRef]
20. Gad-el-Hak, M. The fluid mechanics of microdevices—The Freeman scholar lecture. *J. Fluids Eng.* **1999**, *121*, 5–33. [CrossRef]
21. Tanaka, Y. A peristaltic pump integrated on a 100% glass microchip using computer controlled piezoelectric actuators. *Micromachines* **2014**, *5*, 289–299. [CrossRef]
22. Guo, X.; Qi, H. Analytical solution of Electro-Osmotic peristalsis of fractional Jeffreys fluid in a micro-channel. *Micromachines* **2017**, *8*, 341. [CrossRef]
23. Yasmin, H.; Iqbal, N.; Tanveer, A. Engineering applications of peristaltic fluid flow with hall current, thermal deposition and convective conditions. *Mathematics* **2020**, *8*, 1710. [CrossRef]
24. Forouzandeh, F.; Arevalo, A.; Alfadhel, A.; Brokholder, D.A. A review of peristaltic micropumps. *Sens. Actuator A* **2021**, *326*, 112602. [CrossRef]
25. Hayat, T.; Javed, M.; Ali, N. MHD peristaltic transport of a Jeffery fluid in a channel with complaint walls and porous space. *Transp. Porous Media* **2008**, *74*, 259–274. [CrossRef]
26. Ali, N.; Hayat, T.; Asghar, S. Peristaltic flow of a Maxwell fluid in a channel with compliant walls. *Chaos Soliton Fract.* **2009**, *39*, 407–416. [CrossRef]
27. Hayat, T.; Javed, M.; Hendi, A.A. Peristaltic transport of viscous fluid in a curved channel with compliant walls. *Int. J. Heat Mass Transf.* **2011**, *54*, 1615–1621. [CrossRef]
28. Nadeem, S.; Raiz, A.; Ellahi, R. Peristaltic flow of viscous fluid in a rectangular duct with complaint walls. *Comput. Math. Model.* **2014**, *25*, 404–415. [CrossRef]
29. Hayat, T.; Rafiq, M.; Alsaedi, A. Investigation of hall current and slip conditions on peristaltic transport of Cu-water nanofluid in a rotating medium. *Int. J. Ther. Sci.* **2017**, *112*, 129–141. [CrossRef]
30. Awais, M.; Kumam, P.; Parveen, N.; Ali, A.; Shah, Z.; Thounthong, P. Slip and hall effects on peristaltic rheology of Copper-water nanomaterial through generalized complaint walls with variable viscosity. *Front. Phys.* **2020**, *7*, 249. [CrossRef]
31. Ferraro, V.C.A.; Plumpton, C. *An Introduction to Magneto Fluid Mechanics*; Clarendon Press: Oxford, UK, 1966.
32. Soundalgekar, V.M.; Takhar, H.S. MHD forced and free convection flow past a semi-infinite plate. *AIAAJ* **1977**, *15*, 457458. [CrossRef]
33. Kumar, R.N.; Suresha, S.; Gowda, R.J.P.; Megalamani, S.B.; Prasannakumara, B.C. Exploring the impact of magnetic dipole on the radiative nanofluid flow over a stretching sheet by means of KKL model. *Pramana* **2021**, *95*, 180. [CrossRef]
34. Alhumade, H.; Radhika, M.; Sowmya, G.; Manjunatha, P.T.; Alam, M.M.; Ahmad, I.; Gorji, M.R.; Prasannakumara, B.C. Effect of nonlinear radiation on flow and heat transfer of dusty fluid over a stretching cylinder with Cattaneo Christov heat flux. *Int. J. Mod. Phys. C* **2021**, *32*, 2150156. [CrossRef]
35. Bejan, A. Entropy generation minimization: The new thermodynamics of finite-size devices and finite-time processes. *J. Appl. Phys.* **1996**, *79*, 1191–1218. [CrossRef]
36. Hayat, T.; Rafiq, M.; Ahmad, B.; Asghar, S. Entropy generation analysis for peristaltic flow of nanoparticles in a rotating frame. *Int. J. Heat Mass Transf.* **2017**, *108*, 1775–1786. [CrossRef]
37. Ali, A.; Shah, Z.; Mumraiz, S.; Kumam, P.; Awais, M. Entropy generation on MHD peristaltic flow of Cu-water nanofluid with slip conditions. *Heat Transf.-Asian Res.* **2019**, *48*, 4301–4319. [CrossRef]
38. Shashikumar, N.S.; Gireesha, B.J.; Mahanthesh, B.; Prasannakumara, B.C.; Chamkha, A.J. Entropy generation analysis of magneto-nanoliquids embedded with aluminium and titanium alloy nanoparticles in microchannel with partial slips and convective conditions. *Int. J. Numer. Methods Heat Fluid Flow* **2019**, *29*, 3638–3658.
39. Bibi, A.; Xu, H. Entropy generation analysis of peristaltic flow and heat transfer of a Jeffery nanofluid in a horizontal channel under magnetic environment. *Math. Probl. Eng.* **2019**, *2019*, 2405986. [CrossRef]
40. Abbasi, F.M.; Shanakhat, I.; Shehzad, S.A. Analysis of entropy generation in peristaltic nanofluid flow with Ohmic heating and Hall current. *Phys. Scr.* **2019**, *94*, 025001. [CrossRef]

41. Hayat, T.; Nawaz, S.; Alsaedi, A. Entropy generation analysis of peristaltic flow of magneto-nanoparticles suspended in water under second-order slip conditions. *Sci. Iran.* **2020**, *27*, 3434–3446. [CrossRef]
42. Khan, A.; Shah, Z.; Alzahrani, E.; Islam, S. Entropy generation and thermal analysis for rotary motion of hydromagnetic Casson nanofluid past a rotating cylinder with joule heating effect. *Int. Commun. Heat Mass Transf.* **2020**, *119*, 104979. [CrossRef]
43. Bibi, F.; Hayat, T.; Farooq, S.; Khan, A.A.; Alsaedi, A. Entropy generation analysis in peristaltic motion of Sisko material with variable viscosity and thermal conductivity. *J. Therm. Anal. Calorim.* **2021**, *143*, 363–375. [CrossRef]
44. Awais, M.; Kumam, P.; Memoona; Ali, A.; Shah, Z.; Alrabaiah, H. Impact of activation energy on hyperbolic tangent nanofluid with mixed convection rheology and entropy optimization. *Alex. Eng. J.* **2021**, *60*, 1123–1135. [CrossRef]

micromachines

MDPI

Article

Investigation of the Dynamics of Cavitation Bubbles in a Microfluidic Channel with Actuations

Xiaopeng Shang * and Xiaoyang Huang

School of Mechanical and Aerospace Engineering, Nanyang Technological University, Singapore 639798, Singapore; mxhuang@ntu.edu.sg
* Correspondence: xpshang@ntu.edu.sg

Abstract: This work presents experimental and numerical studies on the dynamics of cavitation bubbles in a nozzle-shaped microfluidic channel with PZT (lead-zirconate-titanate) actuations. It is found that a cloud of bubbles can be generated near the center of the microfluidic channel when the actuation voltage is larger than a threshold at 1 kHz. After being generated, the bubbles under actuations oscillate radially with violent expansion and compression, and simultaneously translate upstream towards the opening of the nozzle. Along with radial oscillation and translation, the bubbles undergo frequent and drastic coalescence and breakup, leading to vigorous churning of surrounding liquids. The pressure variation and distribution in the microchannel are calculated by numerical simulation in Ansys Fluent, and results show that there is a low-pressure zone inside the microfluidic channel within each cycle of the actuation period, which is responsible for bubble generation observed in the experiments. The method of bubble generation in this study is novel and can be applied for the enhancement of heat and mass transfer in microfluidic operations.

Keywords: bubble dynamics; coalescence; breakup; cavitation bubble; microfluidics

1. Introduction

Microfluidics, which is involved in the science and technology of manipulating small amounts of fluids in networks of micro channels, has been broadly applied in different fields, such as chemistry, medicine and biology, owing to its simplicity, high sensitivity, low cost, and low sample consumption, etc. [1]. Heat and mass transfer exist widely in microfluidic operations. Due to small geometric dimensions, extremely low flow rates, and dominance of viscosity over inertia, the Reynolds number (Re) in microfluidics is usually very low, leading to laminar or even creeping flows. Thus, finding how to enhance the heat and mass transfer in microfluidic flows, so as to improve sample mixing, biochemical reaction and flow control, etc., remains an attractive topic.

A number of studies aiming at enhancing heat and mass transfer in microfluidics have been reported, and the developed methods are typically divided into two categories, passive and active methods. In the former, enhancement of heat and mass transfer relies on the advection flow caused by complex configurations or geometries of microchannels, such as herringbone structures, zigzag turns, serpentine channels and flow focusing systems and so on [2,3]. In spite of simple operation and low costs, a passive method requires a complex geometry design and very high flow rates to produce flow disturbances in a microchannel. If the requirements for the flow rate or working fluids are changed, a new design might be needed. In the latter category, an external actuation, such as magnetic forcing, electrokinetic forcing, an acoustic field, and even the thermal effect, is introduced to perturb the microfluidic flows to enhance the heat and mass transfer [4–6]. Park and Wereley [7] developed an opto-electric vortex generator with two coplanar electrodes. When a laser spot is focused on any region of exposed electrodes, a complex, three-dimensional vortex flow is formed rapidly. This technique is promising to produce desirable fluid flows and improve chemical and biological reactions through enhancing mass transfer in microchannels. Ahmed et al. [8] designed an acoustic micro-mixer by introducing an external air bubble within a "horse-shoe" structure of a microfluidic channel. Acoustic waves induce vibration of the trapped air bubble at its resonance frequency and cause a bulk flow around the air–liquid interface. Thermal disturbance can also act as the driving force to enhance the heat and mass transfer in microfluidic channels. Kim et al. [9] created multi-vortices by using the natural convection in conjunction with alternating heating, and the resultant vortical flow is successfully used for the mixing of samples for polymerase chain reaction (PCR) tests. Active methods have fewer limits on flow conditions and higher efficiency compared to passive methods. However, special equipment and elaborate manipulation are usually indispensable, which causes inconvenience in real applications. Another downside is that some active methods are primarily for fluids or particles with specific properties. For example, the electrokinetic and magnetic methods have requirements regarding the conductivity and magnetism of working fluids, respectively. The thermal method may damage the reagents and samples which are sensitive to the temperature variation; acoustic methods need to introduce and trap external gas in advance and there is also a problem of temperature rise at high acoustic frequency. Therefore, despite so many types of passive and active methods, development of novel and simple methods for the enhancement of heat and mass transfer in microfluidic systems remains to be an attractive research topic.

Bubble dynamics has been broadly applied in lots of microfluidic processes, such as micropumping, micromixing, flow control and manipulation of micron-sized objects [10–12]. Air bubbles of micron sizes are introduced into a microfluidic channel integrated with PZT (lead-zirconate-titanate) transducers. The surface of an air bubble exposed to PZT actuations functions as a vibrating membrane, causing a bulk movement in liquid near the air–liquid interface. This effect is known as cavitation microstreaming [13]. Rasouli and Tabrizian [14] reported an ultra-fast micromixer for the synthesis of organic nanoparticles based on cavitation microstreaming of oscillatory bubbles. The straightforward fabrication of the device coupled with efficient energy consumption, high controllability, and rapid mixing renders this mixer a practical platform for a myriad of nano and biotechnological applications. Guan and Sun [15] demonstrated a microfluidic mixing platform for low- and high-viscosity fluids by integrating oscillating bubbles actuated by acoustic forces with a zigzag microstructure. Under proper operating conditions, they achieved a mixing efficiency as high as 0.95 for highly viscous glycerol within a time less than 1 s. Kaba et al. [16] for the first time employed the cavitation microstreaming effect to enhance chemical lysis and DNA extraction in a microfluidic channel. This method overcomes the limit of ineffective mass transport due to a low Reynolds number, and exhibits excellent performance compared to the traditional methods. The aforementioned methods using acoustically actuated oscillatory bubbles, although effective, usually suffer from the inconvenient bubble-trapping process and unstable bubble structure during operation [17]. In addition, it is a complex and far from easy task to generate necessary bubbles directly inside a microfluidic channel due to the challenges in energy concentration, viscous surface force and low number of cavitation nuclei [13,17].

Recently, we reported a novel design of a microfluidic channel with PZT actuations for the direct generation of bubbles [18]. The dynamics of a single bubble under actuations was investigated and the mechanism of bubble translation was discussed by studying the interaction force between an oscillating bubble and a boundary wall. In the present study, we investigate the dynamics of a group of cavitation bubbles, which are more frequently observed in the microfluidic channel with actuations. The dynamic behaviors of bubbles under actuations, including oscillation, coalescence and breakup, etc., are recorded and analyzed using high-speed photography. Numerical simulation is also carried out using the commercial CFD software Ansys Fluent, and the pressure field in the microfluidic channel with and without actuations is investigated.

2. Materials and Methods

2.1. Experimental Setup

As shown in Figure 1, the microfluidic channel contained two parts—a circular chamber of 16 mm in diameter and a nozzle with an acoustic resonator profile to achieve high-amplitude pressure fluctuations. The microfluidic channel was constructed from two PMMA (polymethylmethacrylate) plates sandwiching a dry adhesive layer 300 μm in thickness (Arclad 8102 transfer adhesive, Adhesives Research, Inc., Glen Rock, PA, USA). Two inlets were connected at a 60-degree angle to the circular chamber by a straight channel 10 mm in length and 1 mm in width, and another straight channel was connected to the other end of the chamber to form the outlet. A PZT disk was attached to the bottom of the circular chamber to provide actuation. The PZT disk consisted of a piezoelectric ceramic layer 15 mm in diameter and a brass sheet 22 mm in diameter. The height of the channel was controlled by the thickness of the adhesive layer. The geometry of the entire configuration was precisely manufactured by a laser cutting machine (Universal M-300 Laser Platform, Universal Laser Systems Inc., Arizona, AZ, USA). Finally, the inlets and outlet were attached to the plate using epoxy glue (Araldite, Huntsman Advanced Materials, Los Angeles, CA, USA).

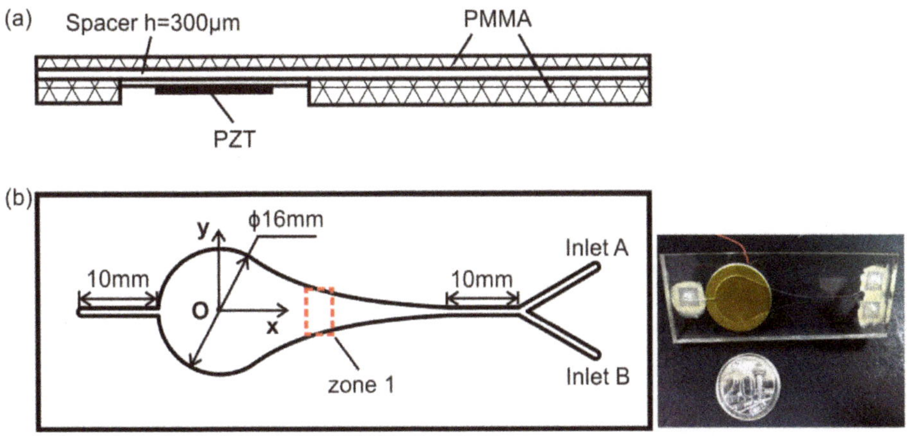

Figure 1. Schematic illustration of the microfluidic channel. (**a**) Side view. (**b**) Top view. The rectangular region zone 1 is selected to record the dynamic behavior of bubbles.

The schematic of the experimental setup is illustrated in Figure 2. The working fluid, deionized (DI) water, was supplied to the microfluidic channel through two inlets. The flow rate was controlled by two syringe pumps (KD Scientific Inc., Holliston, MA, USA). The piezoelectric disk was driven by an external signal generator (33120A, Hewlett Packard, Palo Alto, CA, USA) and an amplifier (790, PCB Piezotronics, Depew, NY, USA). A high-speed CCD camera (Phantom V711, Vision Research, Wayne, NJ, USA) was mounted on top of the chamber to record the flow field and bubble motion. In the experiments, we focused on the motion of the generated bubbles moving through zone 1 (indicated in Figure 2). The actuation frequency was fixed at 1 kHz and the applied voltage was varied within the range of 120 to 150 V. The flow rate was 10 mL/h in total.

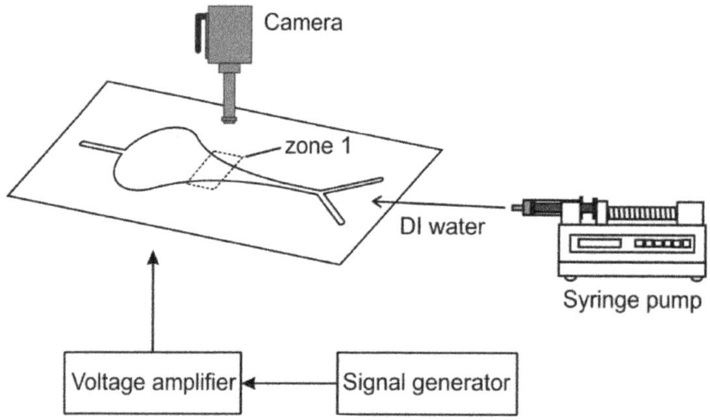

Figure 2. Schematic illustration of the experimental setup.

2.2. Numerical Simulation

In order to explore the mechanism of bubble generation and oscillation in the microfluidic channel under actuations, three-dimensional numerical simulation was conducted by using the commercial software Ansys Fluent.

The bubbles generated inside the channel are associated with cavitation of DI water under low-frequency actuations. Thus, the pressure variation and distribution are key issues to understand the mechanisms of bubble generation in the microfluidic chamber. Due to small dimensions of the microfluidic channel, it is difficult to measure the pressure experimentally. Here, numerical simulations using Ansys Fluent were performed to study the pressure variation and distribution in the chamber. The configuration and dimensions of the microfluidic channel in the numerical simulation were identical to the experimental device. The computational domain and meshing scheme are shown in Figure 3. There were tens of layers of meshes along the channel height direction and the meshes inside the chamber and at the Y junction were specifically refined.

Under the assumptions of an incompressible Newtonian fluid, the governing equations of continuity and momentum are expressed as,

$$\nabla \cdot \vec{V} = 0, \tag{1}$$

$$\frac{\partial\left(\rho\vec{V}\right)}{\partial t} + \nabla \cdot \left(\rho\vec{V}\vec{V}\right) = -\nabla p + \mu\nabla^2\vec{V} \tag{2}$$

where $\vec{V}$ is the velocity vector, p the pressure, ρ the density and μ the dynamic viscosity.

In accordance with the experiments, the top wall was stationary and the bottom wall with a PZT disk was set as a moving boundary of the PZT disk. The vibration of the PZT disk is described by the displacement $\zeta(x, y, t)$ [19],

$$\zeta(x,y,t) = \begin{cases} A_0 \sin(2\pi f t)\left[1 - \left(\sqrt{x^2+y^2}/R_c\right)^2\right]^2 & \text{within PZT disk} \\ 0 & \text{outside PZT disk} \end{cases} \tag{3}$$

where the amplitude of vibration A_0 is 0.5 μm, the frequency imposed to the PZT disk f is 1.0 kHz, and R_c is radius of the circular chamber, which is 8 mm in our experiment.

The boundary conditions for the inlet and outlet were set as the velocity inlet and pressure outlet, respectively. All the boundary conditions are listed in Table 1. To examine the mesh independence, we tested three levels of meshes, based on the same meshing scheme, with a total of 438,648 (coarse), 877,296 (medium), and 1,754,592 (fine) cells. The mesh independence tests showed that the simulation with medium grids provided a good trade-off between accuracy and computational cost [18], and it was thus used in the rest of the present study.

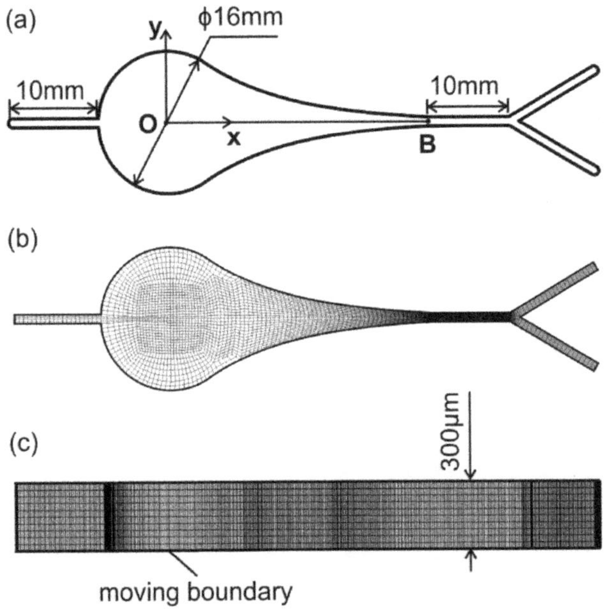

Figure 3. (**a**) Geometry and dimension of the numerical simulation domain. (**b**) Meshing scheme in x–y plane (not to scale). (**c**) Meshing scheme in the height direction (not to scale).

Table 1. Boundary conditions used in the numerical simulation.

Component	Boundary Condition	Description
Inlet	Velocity inlet	0.00926 m/s
Outlet	Pressure outlet	Reference pressure: 101,325 Pa
Fluid		Density ρ = 998.2 kg/m^3 Viscosity μ = 0.001003 Pa·s
Wall (Bottom)	Moving	
Wall (Others)	Stationary	

The governing equations Equations (1) and (2) were solved by the finite volume method (FVM) in Ansys Fluent and the displacement of the moving boundary described by Equation (3) was fulfilled by the user-defined-function (UDF). The following solution methods and numerical schemes were used in the calculations: PRESTO! for pressure discretion and second order upwind for momentum discretion; PISO scheme for pressure-velocity coupling; and the first order implicit time-dependent solution for unsteady formulations.

3. Results and Discussion

3.1. Visualization of Bubble Generation by Fluorescent Dyes

In order to visualize the generation and evolution of bubbles, the degassed DI water solution was supplied to the microfluidic channel from inlet A (indicated in Figure 1b), and the same DI solution with fluorescent dye was supplied to inlet B.

The images of the bubbles in the flow at different times are presented in Figure 4. Without the actuation, the upper fluid (without fluorescent dye) and lower fluid (with fluorescent dye) flow separately through the chamber, and no bubbles are observed. As shown in Figure 4a, after the actuation is switched on, a group of small bubbles are found to be generated in the area near the center of the chamber. After being generated, the bubbles travel upstream against the hydrodynamic flow from the inlets (Figure 4b), which results from the bubble–wall interaction. The bubbles finally arrive at some position close to the nozzle and stay in relative equilibrium (Figure 4c). In this process, the violent oscillation, coalescence and breakup of the bubbles cause strong churning of the bulk flow, enhancing the mixing of fluids inside the microfluidic chamber. It was also observed in our experiment that the bubbles can be generated to form as a single bubble, or as a group of bubbles, depending on the applied voltage and the degree of the fluid degassing. Since the dynamics of a single bubble were reported in Ref. [18], we present the study of multiple cavitation bubbles in the next section.

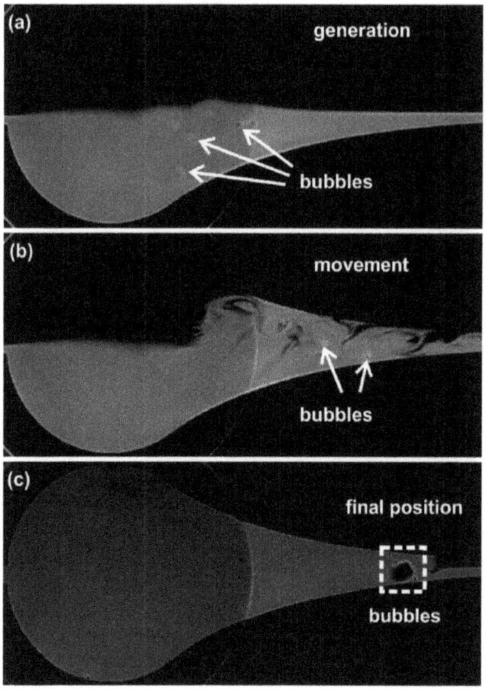

Figure 4. Visualization of the bubbles generated in the channel with actuations at frequency 1.0 kHz and voltage 150 V. (**a**) Bubbles are generated when the actuation is switched on. (**b**) The bubbles travel upstream against the hydrodynamic flow. (**c**) The bubbles arrive at some position close to the nozzle. The flow rate is 10 mL/h in total and the flow direction is from right to left.

3.2. Oscillation and Translation of Bubbles

The motions of a group of bubbles moving through zone 1 were recorded by the camera at 12,000 fps, with 12 images captured in per cycle based on the actuation frequency of 1 kHz. The time interval between two consecutive images is one sixth of one period in Figures 5 and 6. It shows that under strong actuations, the bubbles undergo explosive growth and violent collapse alternatively, and during each cycle, the volumes of bubbles are varied by several orders of magnitude: in Figure 5a, the diameter of the largest bubble is 0.45 mm, while in Figure 5c–e, the bubbles are fragmented or even dissolved owing to the compression of actuation. It can also be found that the bubbles undergo drastic coalescence and breakup under the action of actuations. Such a response with the fragmentation or dissolution of bubbles during the collapse phase is termed as "transient cavitation", which is distinguished from the "stable cavitation" [20]. After their diameters exceed the channel height, the generated bubbles will lose the spherical shape and demonstrate flat surfaces lubricated by thin liquid films near the top and bottom channel walls.

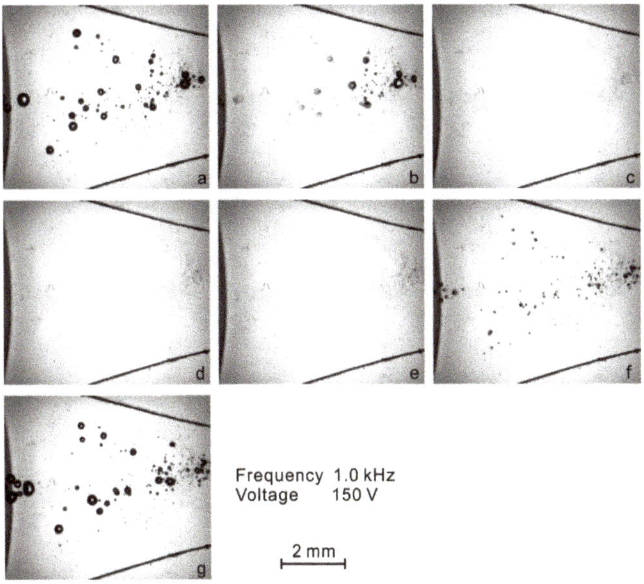

Figure 5. (**a–g**) Photographic series of the dynamic behavior of bubbles under actuations in one cycle. The time interval between two successive images is 1/6 T, or 0.167 ms. The actuation voltage and frequency are 150 V and 1 kHz, respectively.

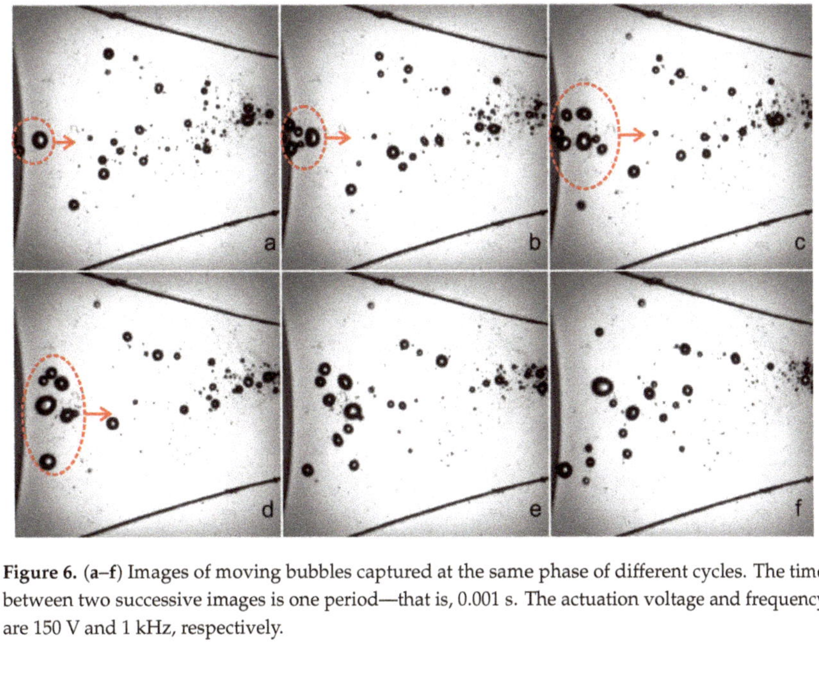

Figure 6. (**a–f**) Images of moving bubbles captured at the same phase of different cycles. The time between two successive images is one period—that is, 0.001 s. The actuation voltage and frequency are 150 V and 1 kHz, respectively.

The time interval between two consecutive images is one period—that is, 0.001 s. At high frequencies, especially in standing waves, bubbles can self-organize into a variety of structures such as streamers, clusters or bubble layers, and evolve on a timescale much larger than the acoustic period. However, in the present study, the bubble clouds are distributed randomly in space without certain structures due to low frequency. For the case of 1 kHz actuation in water, the actuated pressure is almost uniformly distributed in the microfluidic channels, and no acoustic standing waves or travelling waves exist in the microfluidic chamber and thus, the bubbles here cannot be organized into a certain pattern under actuation.

3.3. *Bubble Coalescence and Breakup*

It is known that at low frequencies, the primary Bjerknes force, which results from radiation of the oscillatory pressure field, is very weak, while the secondary Bjerknes force, which is the mutual interaction between oscillating bubbles, is dominant at a short distance. The secondary Bjerknes force can cause the attraction or repulsion of bubbles depending on the phase difference of two bubbles [18]. Therefore, for the bubbles subjected to an oscillatory pressure field, coalescence is further complicated due to the secondary Bjerknes force.

The phenomena of coalescence involving two bubbles were photographed and the typical process was exemplified in Figure 7. It is observed that, during the compression phase, a small bubble approaches a large one quickly under the action of the secondary Bjerknes force, and subsequently, at the expansion phase, the two bubbles grow in size until they collide with each other. When the surfaces of two bubbles touch each other, a saddle-shaped "bridge" is built at first and the large bubble immediately "swallows" the small one with the growth of its volume. After the surface area of the resultant cusp decreases, the bubble coalescence is finished. Finally, the new bubble contracts and expands regularly with periodic actuations. The whole process of bubble coalescence occurs during an extremely short time, which is about one third of a period (0.33 ms).

Traditional bubble models such as the well-known Rayleigh–Plesset equation usually assume that the oscillating bubble maintains spherical [21]. However, when a bubble is collapsing steeply, the interface between the dense liquid and less dense gas is violently accelerated in the inward direction and thus, the bubble's spherical shape is distorted due to the Rayleigh–Taylor instability. Under strong actuations, the shape instability can even cause the breakup of an oscillating bubble.

An example of bubble breakup is demonstrated in Figure 8. Contrary to the process of coalescence, as the bubble collapses sharply, a cusp is first expelled from the base bubble and then a "neck" connecting the two parts of the bubble is thinned quickly. When the interfacial tension is no longer able to keep the bubble intact, a small bubble is shed from the main one. The whole process of bubble breakup is even faster than coalescence, and takes only one fourth of a period, namely 0.25 ms. Two mechanisms for bubble breakup in fluids have been reported—the inertial mechanism, when the shear stress at the interface overcomes the surface tension, and the resonance mechanism, when the exciting frequency matches the natural frequency. The natural frequency of a micron-sized bubble is higher than 10 kHz, far beyond the applied actuation frequency (1 kHz), and thus, here the bubble breakup is ascribed to the great shear stress caused by large velocity gradients on the two sides of the interface.

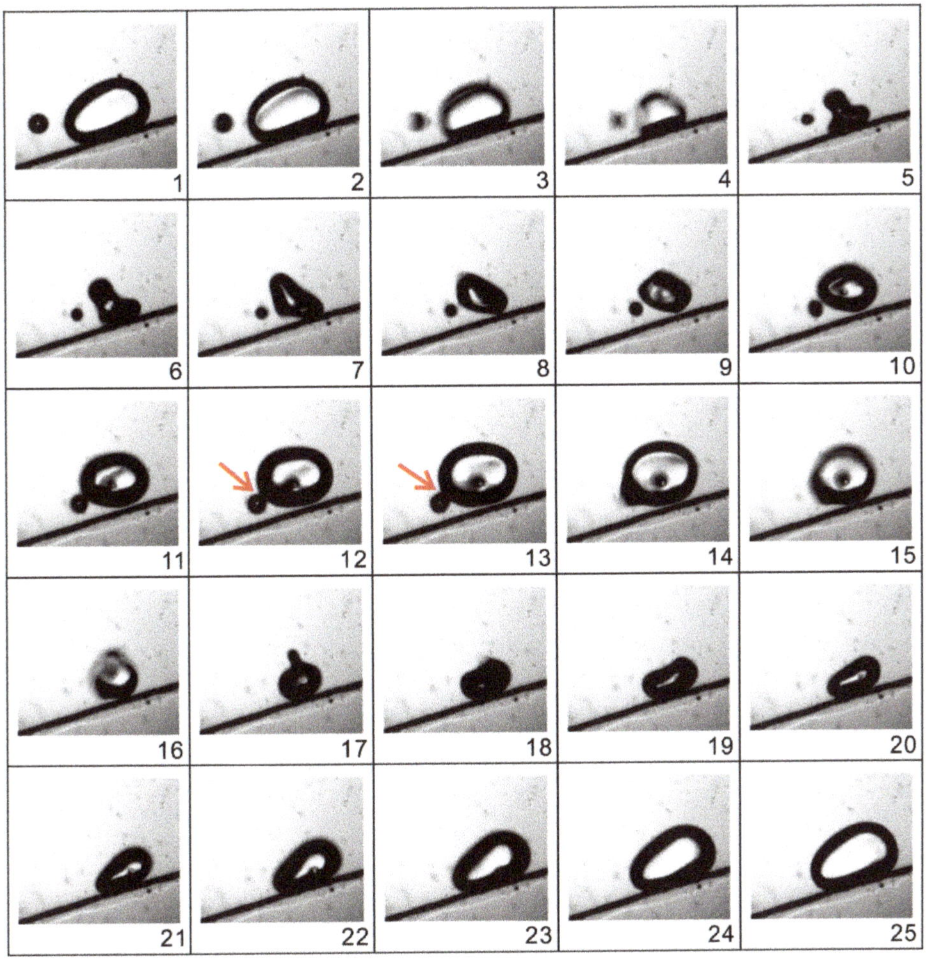

Figure 7. Coalescence of two oscillating bubbles. One small bubble is attracted by a large one and then the two bubbles merge to form a single one. The time between two successive images is 1/12 T, or 0.083 ms. The actuation voltage and frequency are 150 V and 1 kHz, respectively.

When bubbles are translated forward, coalescences and breakups happen frequently. Figure 9 shows an example of two bubbles undergoing a series of breakup and coalescence processes in a very short timeframe (three cycles). At first, one bubble breaks up into two smaller ones during the collapse phase and immediately the two resultant bubbles merge into another large bubble, and during the collapse phase of the next cycle, the new large bubble ruptures again into several satellite ones. As a result, after alternating breakup and coalescence over several cycles, the bubble clouds merge into a single bubble. The whole process of successive breakup and coalescence of the bubbles causes vigorous churning of the surrounding liquid, which plays a critical role in mixing the dyed fluids, as shown in Figure 4.

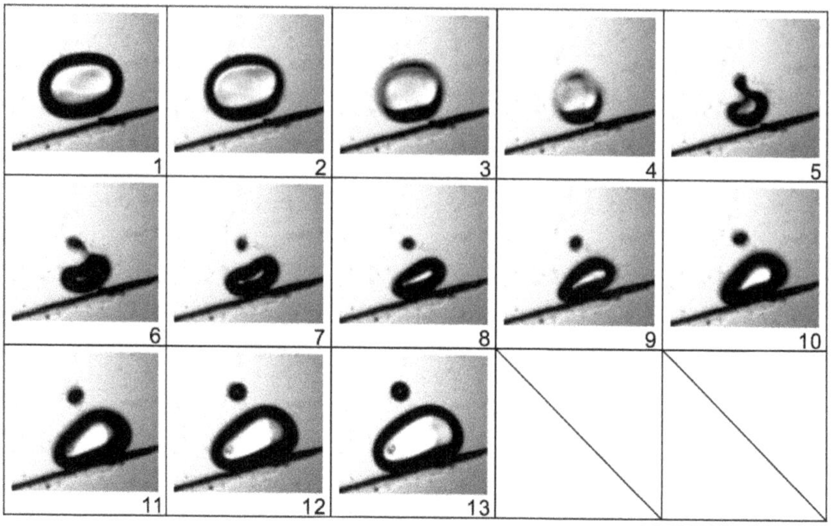

Figure 8. Breakup of one oscillating bubble at the phase of violent collapse. The time between two successive images is 1/12 T, or 0.083 ms. The actuation voltage and frequency are 150 V and 1 kHz, respectively.

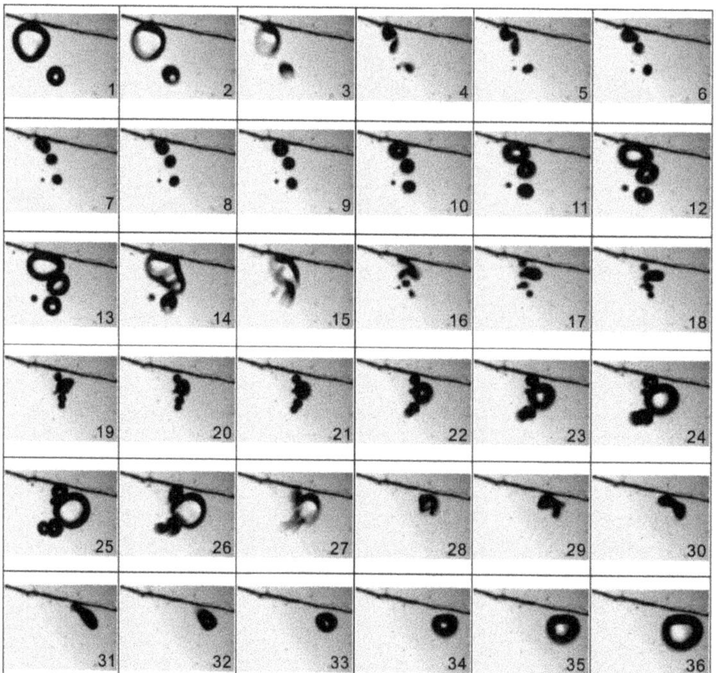

Figure 9. An example of bubbles undergoing alternative coalescence and breakup in the process of translation. The time between two successive images is 1/12 T, or 0.083 ms. The actuation voltage and frequency are 150 V and 1 kHz, respectively.

3.4. Bubble Generation by Low Pressure

The simulation results are plotted in Figure 10, which shows the pressure distributions with and without actuations. It can be seen that without actuations (Figure 10a), the DI water flows through the fluidic chamber from right to left with a constant flow rate (10 mL/h). The pressure exhibits a monotonic decrease from the inlet to outlet, and finally is equal to ambient pressure (101,325 Pa) at the outlet, and the pressure drop is slight in the chamber. When the actuation is switched on, the pressure in the channel fluctuates periodically following the actuation. The minimum pressure, which is our focus, in the channel occurs at 0.75 T (T = $1/f$ = 0.001 s) of each period. A typical pressure distribution at 0.75 T is presented in Figure 10b. It is found that, instead of decreasing monotonously, a low-pressure zone occurs in the region of the circular chamber where we observed the bubble generation in the experiments. The minimum pressure is as low as 469 Pa, which is critical to the bubble generation.

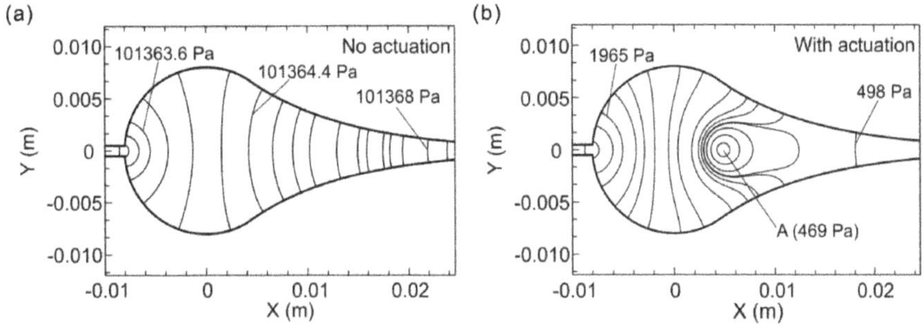

Figure 10. The pressure distributions inside the microfluidic channel by numerical simulation. (**a**) Pressure distribution without actuation. (**b**) Pressure distribution under actuation with frequency f = 1.0 kHz at 0.75 T (T = $1/f$ = 0.001 s). The pressure at Point A is 469 Pa.

Cavitation refers to a process of nucleation from the liquid to the gas phase when pressure falls below some critical value at almost a constant temperature. In real liquid, lots of nuclei or microbubbles exist inevitably, freely suspended or attached to solid boundaries or motes. Those nuclei act at the starting point of cavitation and significantly limit the tensile strength of liquids. It is known that excess pressure is imposed on a microbubble's surface due to the existence of surface tension, under which a nucleus tends to remain inactive and stable. Thus, cavitation cannot occur unless the applied actuation counteracts the surface tension effect. Considering an oscillating pressure field with the amplitude of P_a, the pressure imposed on the bubble is equal to $P_0 + P_a \sin(\omega t)$, where P_0 is the static pressure. A microbubble will undergo very weak oscillation and tiny volume pulsation unless the amplitude of the applied pressure fluctuation, P_a, is above a threshold, which is called the Blake threshold [22],

$$P_a^B = P_0 - P_v + P_0\sqrt{\frac{4}{27}\frac{\alpha_S^3}{1+\alpha_S}} \tag{4}$$

where $\alpha_S = 2\sigma/(P_0 R_0)$ is the Laplace tension of the bubble in ambient conditions, P_0 = 101,325 Pa and P_v = 3169 Pa are the ambient pressure and vapor pressure, respectively, σ = 0.072 N/m is the surface tension and R_0 is the initial radius of the bubble.

Once the amplitude of the pressure fluctuation is higher than the Blake threshold, the minimum pressure is lowered to the critical value, so that the surface tension pressure cannot maintain the bubble's equilibrium. As such, microbubbles under actuations may expand to a multiple of the initial size and collapse steeply afterwards, undergoing a considerable volume pulsation—bubbles are observed to be "generated" in liquid. The microbubbles in real liquid have characteristic sizes ranging from several microns to tens of microns [22]. Taking a typical microbubble 10 μm in radius (R_0 = 10 μm), the Blake threshold P_a^B can be calculated to be 100,111 Pa, corresponding to the threshold pressure on the bubble $P_L = P_0 - P_a^B$, equal to 1214 Pa. The minimum pressure (469 Pa) from the simulation in the low-pressure zone is significantly below this threshold pressure, and thus the nuclei with radii larger than 10 μm can be "generated" in our experiments.

3.5. Volume Pulsation by Pressure Fluctuation

The pressure fluctuation over one period at the point P of zone 1 is plotted in Figure 11. It shows that the pressure fluctuates periodically with its amplitude being about 1 bar following a sinusoidal curve. Under this large amplitude pressure fluctuation, the bubble is expanded corresponding to the minimum pressure, while the bubble is squeezed to a collapse phase at the maximum pressure. These observations can be seen in the inset bubble photos in Figure 11. It should be pointed out that the simulation results are obtained under the conditions of a single phase and incompressible fluid. Once the bubbles are generated in the fluid, these conditions are not applicable and the pressure fluctuation amplitude will be different. However, as the bubble photos show, the actuated pressure fluctuations in the fluid with bubbles have sufficient amplitude to induce a large volume oscillation of the bubbles.

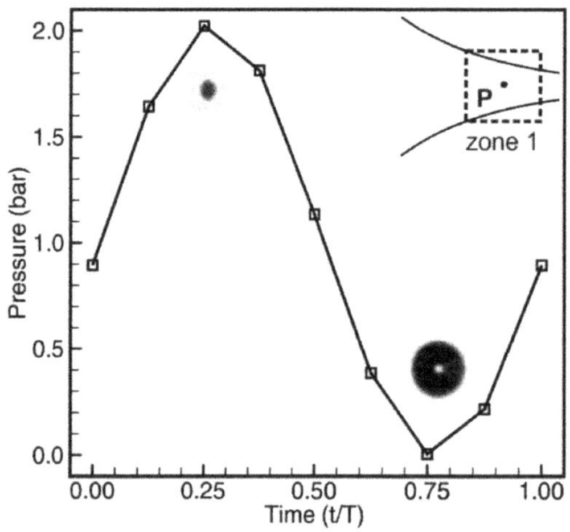

Figure 11. Pressure fluctuation in one cycle at point P (0.013, 0, −0.00015) in zone 1. T = $1/f$ = 0.001 s.

4. Conclusions

In this study, the dynamics of cavitation bubbles in a microfluidic channel under actuations are investigated experimentally and numerically. The bubble generation is visualized by tracing fluorescent dyes, and the motions, coalescence and breakup of bubbles are recorded by a high-speed camera. The mechanisms of the bubble dynamics, including generation and oscillation, are studied using numerical simulation.

Bubbles are observed to occur near the center of a chamber under actuations. After being generated, the bubbles translate upstream to a position in a nozzle and simultaneously undergo explosive growth and violent collapse. In this process, the frequent occurrence of the coalescence and breakup of bubbles results in the mixing enhancement of surrounding fluids. Representative examples of bubble coalescence and breakup are provided in detail. The coalescence of oscillating bubbles may be ascribed to the secondary Bjerknes force, and the breakup of an oscillating bubble is caused by the shape instability at the collapse phase. Numerical simulation has been conducted to investigate the pressure variation and distribution inside the microfluidic channel. Simulation results show that during each actuation cycle, there is a low-pressure zone which is exactly the area in which bubble generation is observed in the experiment. The minimum pressure in this zone is found to exceed the Blake threshold corresponding to the hydrodynamic cavitation of pre-existing nuclei, leading to the phenomena of bubble generation in the experiment.

This novel method of bubble generation has the potential to be widely applied for the enhancement of heat and mass transfer in microfluidic channels.

Author Contributions: Conceptualization, X.H. and X.S.; methodology, X.S.; investigation, X.S.; writing—original draft preparation, X.S.; writing—review and editing, X.H.; supervision, X.H. All authors have read and agreed to the published version of the manuscript.

Funding: This research received no external funding.

Institutional Review Board Statement: Not applicable.

Informed Consent Statement: Not applicable.

Data Availability Statement: All processed data in this study are included in this published article. Raw data will be provided on request from the corresponding author.

Conflicts of Interest: The authors declare no conflict of interest.

References

1. Whitesides, G.M. The origins and the future of microfluidics. *Nature* **2006**, *442*, 368–373. [CrossRef] [PubMed]
2. Mach, A.J.; Kim, J.H.; Arshi, A.; Hur, S.C.; Carlo, D.D. Automated cellular sample preparation using a Centrifuge-on-a-Chip. *Lab Chip* **2011**, *11*, 2827–2834. [CrossRef] [PubMed]
3. Shang, X.; Huang, X.; Yang, C. Mixing enhancement by the vortex in a microfluidic mixer with actuation. *Exp. Therm. Fluid Sci.* **2015**, *67*, 57–61. [CrossRef]
4. Glasgow, I.; Lieber, S.; Aubry, N. Parameters influencing pulsed flow mixing in microchannels. *Anal. Chem.* **2004**, *76*, 4825–4832. [CrossRef] [PubMed]
5. Wu, Z.; Li, D. Micromixing using induced-charge electrokinetic flow. *Electrochim. Acta* **2008**, *53*, 5827–5835. [CrossRef]
6. Tatlisoz, M.M.; Canpolat, C. Pulsatile flow micromixing coupled with ICEO for non-Newtonian fluids. *Chem. Eng. Process. Process Intensif.* **2018**, *131*, 12–19. [CrossRef]
7. Park, C.; Wereley, S.T. Rapid generation and manipulation of microfluidic vortex flows induced by AC electrokinetics with optical illumination. *Lab Chip* **2013**, *13*, 1289–1294. [CrossRef] [PubMed]
8. Ahmed, D.; Mao, X.; Shi, J.; Juluri, B.K.; Huang, T.J. A millisecond micromixer via single-bubble-based acoustic streaming. *Lab Chip* **2009**, *9*, 2738–2741. [CrossRef] [PubMed]
9. Kim, S.J.; Wang, F.; Burns, M.A.; Kurabayashi, K. Temperature-Programmed Natural Convection for Micromixing and Biochemical Reaction in a Single Microfluidic Chamber. *Anal. Chem.* **2009**, *81*, 4510–4516. [CrossRef] [PubMed]
10. Wang, C.; Rallabandi, B.; Hilgenfeldt, S. Frequency dependence and frequency control of microbubble streaming flows. *Phys. Fluids* **2013**, *25*, 022002. [CrossRef]
11. Segers, T.; Versluis, M. Acoustic bubble sorting for ultrasound contrast agent enrichment. *Lab Chip* **2014**, *14*, 1705–1714. [CrossRef] [PubMed]
12. Su, P.A.Q.; Lim, K.Y.; Ohl, C.D. Cavitation bubble dynamics in microfluidic gaps of variable height. *Phys. Rev. E* **2009**, *80*, 047301.
13. Shang, X. Vortex Generation and Bubble Dynamics in a Microfluidic Chamber with Actuations. Ph.D. Thesis, Nanyang Technological University, Singapore, 2017.
14. Rasouli, M.R.; Tabrizian, M. An ultra-rapid acoustic micromixer for synthesis of organic nanoparticles. *Lab Chip* **2019**, *19*, 3316–3325. [CrossRef] [PubMed]
15. Guan, Y.; Sun, B. Versatile microfluidic mixing platform for high- and low-viscosity liquids via acoustic and chemical microbubbles. *Micromachines* **2019**, *10*, 854. [CrossRef] [PubMed]
16. Kaba, A.M.; Jeon, H.; Park, A.; Yi, K.; Baek, S.; Park, A.; Kim, D. Cavitation-microstreaming-based lysis and DNA extraction using a laser-machined polycarbonate microfluidic chip. *Sens. Actuators B Chem.* **2021**, *346*, 130511. [CrossRef]
17. Peng, T.; Zhou, M.; Yuan, S.; Jiang, B. Trapping stable bubbles in hydrophobic microchannel for continuous ultrasonic microparticle manipulation. *Sens. Actuators A Phys.* **2021**, *331*, 113045. [CrossRef]
18. Shang, X.; Huang, X.; Yang, C. Bubble dynamics in a microfluidic chamber under low-frequency actuation. *Microfluid. Nanofluid.* **2016**, *20*, 14. [CrossRef]
19. Timoshenko, W.K.S.; Woinowsky-Krieger, S. *Theory of Plates and Shells*; McGraw-Hill: New York, NY, USA, 1959; pp. 51–72.
20. Brennen, C.E. *Cavitation and Bubble Dynamics*; Oxford University Press: London, UK, 1995; p. 89.
21. Plesset, M.S.; Calif, P. The dynamics of cavitation bubbles. *J. Appl. Mech.* **1949**, *16*, 277–282. [CrossRef]
22. Leighton, T.G. *The Acoustic Bubble*; Academic Press: San Diego, CA, USA, 1994; pp. 72–128.

micromachines

MDPI

Article

Flow and Heat Transfer Performances of Liquid Metal Based Microchannel Heat Sinks under High Temperature Conditions

Tao Wu *, Lizhi Wang, Yicun Tang, Chao Yin and Xiankai Li

Yangzhou Collaborative Innovation Research Institute Co., Ltd., Yangzhou 225006, China; lychee@mail.ustc.edu.cn (L.W.); tangyicun13@163.com (Y.T.); fortunatetomato@163.com (C.Y.); wootall62442@outlook.com (X.L.)

* Correspondence: wutao_921006@163.com

Abstract: Developments in applications such as rocket nozzles, miniature nuclear reactors and solar thermal generation pose high-density heat dissipation challenges. In these applications, a large amount heat must be removed in a limited space under high temperature. In order to handle this kind of cooling problem, this paper proposes liquid metal-based microchannel heat sinks. Using a numerical method, the flow and heat transfer performances of liquid metal-based heat sinks with different working fluid types, diverse microchannel cross-section shapes and various inlet velocities were studied. By solving the 3-D steady and conjugate heat transfer model, we found that among all the investigated cases, lithium and circle were the most appropriate choices for the working fluid and microchannel cross-section shape, respectively. Moreover, inlet velocity had a great influence on the flow and heat transfer performances. From 1 m/s to 9 m/s, the pressure drop increased as much as 65 times, and the heat transfer coefficient was enhanced by about 74.35%.

Keywords: microchannel heat sink; liquid metal; working fluid; cross-section shape; inlet velocity

1. Introduction

In many industrial applications, such as rocket nozzles, miniature nuclear reactors and solar thermal generation, one of the major concerns is heat dissipation [1]. If the large amount heat cannot be removed quickly, the devices or systems cannot function or, worse, can be damaged. Hence, high-efficient heat transfer technology is important. In the aforementioned applications, the challenges of heat removal are attributed to high heat flux density, limited space for heat transfer equipment and accompanying high temperature. In this case, liquid metal-based microchannel heat sinks are well suited.

Microchannel heat sink has attracted worldwide attention from researchers because of its advantages, including high heat transfer coefficient, direct integration on the substrate and compactness. Since the pioneering work by Tuckerman and Pease [2], a lot of research has been conducted to study the thermal performance and hydraulic characteristics of various microchannel heat sinks. Wu and Little [3] tested silicon and glass microchannels with nitrogen, hydrogen and argon as the working fluids. The experimental data was fitted to give Nusselt number and friction factor correlations. Xu et al. [4] investigated the pressure drop characteristics in aluminum and silicon microchannels. It was found that the Poiseuille number remains constant in laminar flow region, as in conventional channels, but its value is different for different materials. The Poiseuille number was determined to be constant in laminar flow by Judy et al. [5] for fused silica and stainless steel microchannels using water, isopropanol and methanol as working fluids. Qu and Mudawar [6] manufactured a copper microchannel and tested it at different heat fluxes. The obtained pressure drop and temperature distribution were validated with a mathematical model developed by the authors, using the classical theory of conventional channels. Lee et al. [7] tested silicon microchannels with water and conclude that classical correlations can be applied to determine both the heat transfer and fluid flow characteristics of microchannel heat sinks. Liu and Garimella [8,9] conducted water flow tests on plexiglass microchannel samples, and found that the friction factor data are in agreement with conventional correlations for both laminar and turbulent flow. According to the research, it can be concluded that the single-phase fluid flow and heat transfer behaviors in microchannels are similar to those under normal scale and convectional theory is still appropriate to the flow characteristics [10,11].

In order to improve the cooling performance of microchannel heat sinks, the effect of geometrical factors including cross-sectional shape, microchannel pattern, manifolds and input/output ports have been widely studied [12]. The studies show that changing the geometrical factors could result in the optimization of average temperature, pressure drop, heat transfer coefficient, Nusselt number, as well as flow and temperature uniformity.

Jing and He [13] numerically studied the hydraulic and thermal performances of microchannel heat sinks with three different channel cross-sectional shapes: rectangle, ellipse and triangle. The calculating results show that the triangular-shaped microchannel has the smallest hydraulic resistance and the smallest convective heat transfer coefficient. Gunnasegaran et al. [14] studied the flow and convective heat transfer characteristics of water in rectangular, trapezoid and triangular shaped microchannels and found that for a given cross-sectional shape, the channel with smaller hydraulic diameter had larger convective heat transfer coefficient and pressure drop. Xia et al. [15] researched fluid flow in microchannels with different header shapes and inlet/outlet arrangements. They concluded that I-type had better flow uniformity than other cases due to the symmetrical flow distribution. Moreover, the effect of walls friction decreased the static pressure, which ultimately led to uniform velocity distribution. Kumar et al. [16,17] summarized the works focusing on the influence of inlet and outlet arrangements on the thermal and hydraulic performance. It was observed that vertical supply and collection system showed lower temperature non-uniformity and better liquid flow distribution.

Ahmed et al. [18] investigated the effect of geometrical parameters on laminar water flow and forced convection heat transfer characteristics in grooved microchannel heat sink. By means of numerical calculations, they found that with the optimum groove tip length ratio, groove depth ratio, groove pitch ratio and grooves orientation ratio, Nusselt number and friction factor could be improved by 51.59% and 2.35%, respectively. Zhu et al. [19] compared the thermal and hydraulic performance of wave microchannel heat sinks with wavy bottom ribs and side ribs design. The results showed that the up–down wavy design exhibited a better performance when small wavelengths were adopted. Wang et al. [20] proposed a new double layer wave microchannel heat sink and porous-ribs, they found that the porous-ribs exhibited obvious superiority at a low pumping power, and the wavy microchannel became domain at a high pumping power. Ermagan et al. [21] replaced the conventional walls with superhydrophobic walls in wave microchannel heat sinks to reduce the pressure loss. It was found that the beneficial effect of superhydrophobic walls on the comprehensive performance weakened as the waviness of the channel increased.

Gong et al. [22] designed a porous/solid compound fins microchannel heat sink to enhance the cooling performance. They found that the viscous shear stress decreased at the interfaces between fluid and porous fins, which resulted in the reduction of pressure drop. Ghahremannezhad et al. [23] proposed a new porous double-layered microchannel heat sink and found that the new microchannel heat sink showed not only good heat transfer performance, but also exhibited lower pressure drop. Li et al. [24] compared the thermal performance and flow characteristics of five heat sink designs, including porous-ribs single-layered, solid-ribs single-layered, solid-ribs double-layered, porous-ribs double-layered and mixed double-layered. It was found that the mixed double-layered microchannel heat sink processed a combination of low pressure drop and high thermal performance.

Besides geometric parameters, the proper selection of working fluid type is also a crucial factor to the performances of microchannel heat sinks [25,26]. Conventional cooling liquids, like water, could not meet the rapidly increasing demand of high-density heat dissipation because of its low thermal conductivity. In order to enhance the effective thermal conductivity, the method of adding conductive nano metal particles such as copper or aluminum to the solution and suspension is proposed by researchers. Jang and Choi [27] and Farsad et al. [28] showed numerically that water-based nanofluids enable microchannel heat sinks to dissipate heat fluxes as high as 1350–2000 W/cm^2. Sohel et al. [29,30] showed analytically that 0.5–4.0 vol% CuO nanofluid flow in a copper microchannel heat sink having circular channels provides far better heat transfer performance and lower friction factor than Al_2O_3 and TiO_2 nanofluids. Sivakumar et al. [31] also showed that CuO nanofluid provides better heat transfer coefficient enhancement than Al_2O_3 nanofluid. Salman et al. [32,33] showed numerically that dispersing SiO_2 nanoparticles in ethylene glycol (base liquid) provides the highest Nu, followed by ZnO, CuO and Al_2O_3, and Nu increases with decreasing nanoparticles size. Kumar et al. [34] conducted thermofluidic analysis of Al_2O_3-water nanofluid cooled, branched wavy heat sink microchannels using a numerical method. The results showed that apart from disruption of the boundary layer and its reinitialization, vortices were formed near the secondary channel, which improved thermal performance. The heat transfer coefficient increased with increasing nanofluids concentrations for any investigated Reynolds number. Wang et al. [35] numerically investigated the forced convection in microchannel heat sinks using multi-wall carbon nanotube-Fe_3O_4 hybrid nanofluid as coolant working fluid. According to the results, the heat sinks, which consist of metallic foam, have better cooling performance and are able to decrease the surface temperature. However, the performance improvements brought by nanofluids are still limited due to the base fluids used. Particularly, the mixed solution may easily be subject to additional troubles such as susceptibility to fouling, particle deposition or conglomeration, degeneration of solution quality and flow jamming over the channels [36,37]. Besides, the low evaporation point of these liquids may imply potential dangers in preventing the device from burning out, since the liquids may easily escape to the ambient air.

Apart from nanofluids, liquid metal could also lead to an excellent cooling capacity because of its high thermal conductivity. In fact, liquid metals such as sodium, sodium-potassium alloy and lithium have been used as coolant in nuclear engineering for a long time. Nowadays, liquid metal cooling has been widely used in nuclear-powered plants with many different metals having been tried. Liquid metal cooling is also used in accelerators, solar power generation, LEDs and lasers [38–40]. Since Liu and Zhou [41] proposed for the first time to use low-melting-point liquid metals as an ideal coolant for the thermal management of computer chips, a lot of research has been performed to investigate the heat transfer capability. Miner and Goshal [36] carried out analytical and experimental work on liquid metal flow in a pipe. Their results indicated that the heat transfer is enhanced in both laminar and turbulent regimes using liquid metal coolant. Goshal et al. [42] used a GaIn alloy-based heat sink in a cooling loop and achieved a thermal resistance of 0.22 K/W for the entire system. Hodes et al. [43] studied the optimum geometry for water-based and Galinstan-based heat sinks in terms of minimum thermal resistance. It was shown that in the optimized configurations, Galinstan can reduce the overall thermal resistance by about 40% compared to water. Zhang et al. [44] carried out a follow-up work and according to their experimental data, liquid metal could enhance the convective heat transfer due to the superior thermophysical properties.

Currently, studies on liquid metal cooling or microchannel heat sink are widely conducted, but research on microchannel heat sinks using liquid metal as the working fluid is scarce. Some related investigations aim at dealing with the heat dissipation problem of computer chips or electronic devices at room temperature. However, in some applications, such as rocket nozzles, miniature nuclear reactors and solar thermal generation, high-density heat needs to be dissipated or transferred in a limited space under high temperature. In order to solve the problem of high-density heat dissipation under high temperature and limited space, this paper takes liquid metal-based microchannel heat sinks as object, establishing 3D physical and mathematical numerical models, so that flow and heat transfer performances with diverse working fluid types, various microchannel cross-section shapes and different inlet velocities could be obtained and analyzed.

2. Modeling and Numerical Methods

2.1. Physical Model

The structure of the investigated microchannel heat sink is shown in Figure 1. The cylindrical inlet and outlet passages are connected with 10 microchannels by inlet and outlet manifolds. In the cooling process, liquid metal enters the inlet passage and then flows into the microchannels through inlet manifold. After that, the liquid metal successively flows through the outlet manifold and outlet passages and finally flows out of the heat sink.

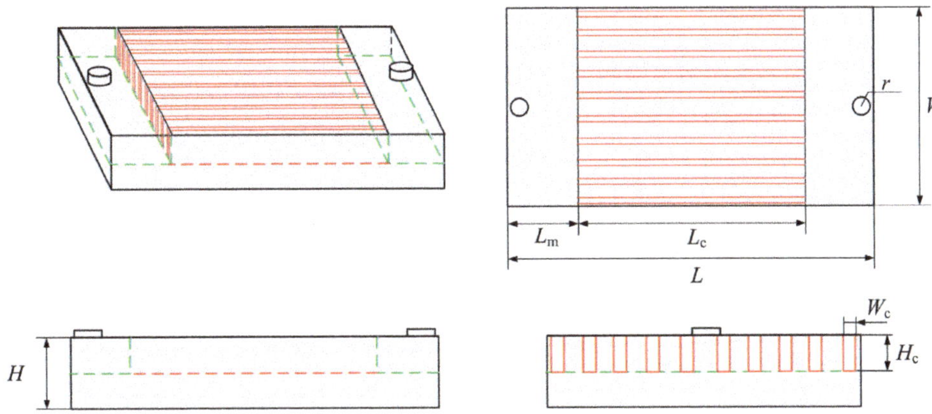

Figure 1. The structure of microchannel heat sink.

The overall size of the heat sink is $L \times W \times H$ = 19 mm × 10 mm × 2 mm. The radius and height of inlet and outlet passages are 0.5 mm and 0.5 mm, respectively. Both the inlet and outlet manifolds are 3 mm long. All the microchannels have the same size of $L_c \times W_c \times H_c$ = 13 mm × 0.4 mm × 1 mm. Besides rectangular microchannel cross-section shape, three cross-section shapes (circle, trapezoid and parallelogram) are also studied in this paper. Different microchannel cross-section shapes are shown in Figure 2. The four different cross-section shapes have the same hydraulic diameter (D_h = 0.57 mm).

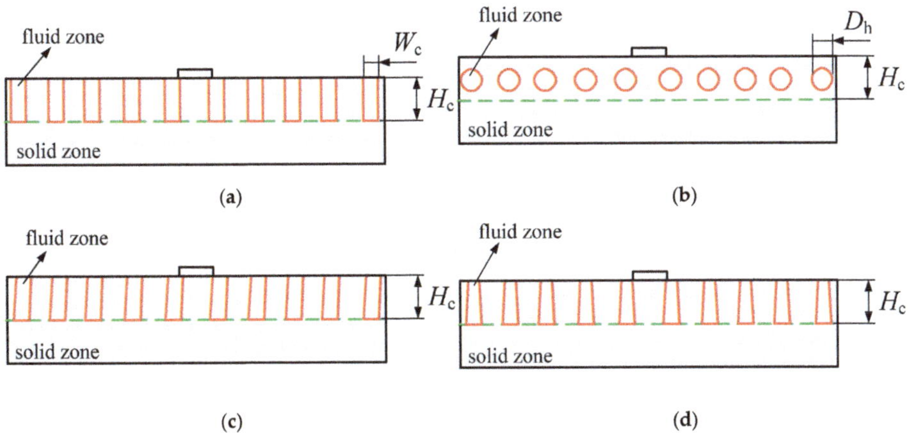

Figure 2. Diagram of different microchannel cross-section types. (**a**) Rectangle; (**b**) circle; (**c**) parallelogram; (**d**) trapezoid.

In order to investigate the flow and heat transfer performances of the microchannel heat sink under high temperature conditions, silicon carbide (SiC) was applied for solid wall materials and alkalis with high melting point were selected as the liquid metal coolant. The thermo-physical properties of SiC were assumed to be constant. The density, specific heat capacity and thermal conductivity of SiC were 3220 kg/m^3, 800 J/(kg·K) and 80 W/(m·K) respectively.

As for the alkalis, sodium (Na), potassium (K), sodium–potassium alloy (Na-K) and lithium (Li) were considered in this paper. For these alkalis, their thermo-physical property correlations as well as the range of validity of each correlation were given [45,46]. For all the correlations, temperature (T), density (ρ), specific heat capacity (Cp), thermal conductivity (k) and dynamic viscosity (η) were respectively given in K, kg/m^3, J/(kg·K), W/(m·K) and Pa·s.

(1) Sodium (Na)

Density (370 K $\leq T \leq$ 1100 K),

$$\rho = T\left[T\left(0.9667 \times 10^{-9} \cdot T - 0.46005 \times 10^{-5}\right) - 0.1273534\right] + 954.1579 \tag{1}$$

Specific heat capacity (370 K $\leq T \leq$ 1100 K),

$$Cp = 1630.14 - 0.4631 \cdot (1.8T) + 0.14284 \times 10^{-3} \cdot (1.8T)^2 \tag{2}$$

Thermal conductivity (370 K $\leq T \leq$ 1100 K),

$$k = 93.9892 - 0.032503 t_f + 3.6197 \times 10^{-6} \cdot t_f^2 \tag{3}$$

$$t_f = 1.8(T - 273.15) + 32 \tag{4}$$

Dynamic viscosity (370 K $\leq T \leq$ 1100 K),

$$\eta = \frac{0.11259 \times 10^{-3} \cdot e^{\frac{749.08\rho}{1000T}}}{(1000/\rho)^{0.3333}} \tag{5}$$

(2) Potassium (K)

Density (334 K $\leq T \leq$ 2270 K),

$$\begin{aligned} \rho = 841.5 - 0.2172 \cdot (T - 273.15) - 2.7 \times 10^{-5} \cdot (T - 273.15)^2 \\ +4.77 \times 10^{-9} \cdot (T - 273.15)^3 \end{aligned} \tag{6}$$

Specific heat capacity (373 K $\leq T \leq$ 1423 K),

$$Cp = 4186.8\left[0.2004 - 0.8777 \times 10^{-4} \cdot (T - 273.15) + 1.097 \times 10^{-7} \cdot (T - 273.15)^2\right] \tag{7}$$

Thermal conductivity (373 K $\leq T \leq$ 1300 K),

$$k = 81.021 - 8.1812 \times 10^{-2} \cdot T + 4.2521 \times 10^{-5} \cdot T^2 - 1.0589 \times 10^{-8} \cdot T^3 \tag{8}$$

Dynamic viscosity (334 K $\leq T \leq$ 2000 K),

$$\ln(\eta) = -6.4846 - 0.42903 \cdot \ln(T) + 485.32/T \tag{9}$$

(3) Sodium-Potassium (NaK)

The density and specific heat capacity were defined on the basis of additive laws by means of weight fraction (wt) weighting. The dynamic viscosity was based on the mole fraction (at) weighting.

Density (273 K $\leq T \leq$ 1573 K),

$$\rho = \frac{1}{\frac{wt_{Na}}{\rho_{Na}} + \frac{wt_K}{\rho_K}} \tag{10}$$

Specific heat capacity (273 K $\leq T \leq$ 1573 K),

$$Cp = wt_{Na} \cdot Cp_{Na} + wt_K \cdot Cp_K \tag{11}$$

Thermal conductivity (273 K $\leq T \leq$ 1573 K),

$$k = 15.0006 + 30.2877 \times 10^{-3} \cdot T - 2.08095 \times 10^{-5} \cdot T^2 \tag{12}$$

Dynamic viscosity (273 K $\leq T \leq$ 1573 K),

$$\eta = \frac{1}{\frac{at_{Na}}{\eta_{Na}} + \frac{at_K}{\eta_K}} \tag{13}$$

(4) Lithium (Li)
Density (455 K $\leq T \leq$ 1500 K),

$$\rho = 278.5 - 0.04657T + 274.6(1 - T/3500)^{0.467} \tag{14}$$

Specific heat capacity (455 K $\leq T \leq$ 1500 K),

$$Cp = 4754 - 0.925T + 2.91 \times 10^{-4} \cdot T^2 \tag{15}$$

Thermal conductivity (455 K $\leq T \leq$ 1500 K),

$$k = 22.28 + 0.05T - 1.243 \times 10^{-5} \cdot T^2 \tag{16}$$

Dynamic viscosity (455 K $\leq T \leq$ 1500 K),

$$\ln(\eta) = -4.16435 - 0.6374 \cdot \ln(T) + 292.1/T \tag{17}$$

2.2. Mathematical Model

2.2.1. Governing Equations and Boundary Conditions

In order to establish the mathematical model of flow and heat transfer processes in the microchannel heat sink, the following assumptions are made in this paper:

(1) Both the fluid flow and heat transfer are steady.
(2) The fluid flow is incompressible and single phase.
(3) There is no slip between fluid and wall.
(4) Radiation heat transfer and viscous dissipation effect are neglected.

According to previous research [47], when the size of the microchannel is over 200 μm, the continuous medium hypothesis of fluid is still applicable. Hence, the governing equations for continuity, momentum and energy could be written as follows [48].

Continuity equation,

$$\frac{\partial u}{\partial x} + \frac{\partial v}{\partial y} + \frac{\partial w}{\partial z} = 0 \tag{18}$$

where u, v, w in m/s are respectively the velocity in x, y, z direction

Momentum equations,

$$u\frac{\partial u}{\partial x} + v\frac{\partial u}{\partial y} + w\frac{\partial u}{\partial z} = -\frac{1}{\rho}\frac{\partial p}{\partial x} + \frac{\eta}{\rho}\left(\frac{\partial^2 u}{\partial x^2} + \frac{\partial^2 u}{\partial y^2} + \frac{\partial^2 u}{\partial z^2}\right) \tag{19}$$

$$u\frac{\partial v}{\partial x} + v\frac{\partial v}{\partial y} + w\frac{\partial v}{\partial z} = -\frac{1}{\rho}\frac{\partial p}{\partial y} + \frac{\eta}{\rho}\left(\frac{\partial^2 v}{\partial x^2} + \frac{\partial^2 v}{\partial y^2} + \frac{\partial^2 v}{\partial z^2}\right) \tag{20}$$

$$u\frac{\partial w}{\partial x} + v\frac{\partial w}{\partial y} + w\frac{\partial w}{\partial z} = -\frac{1}{\rho}\frac{\partial p}{\partial z} + \frac{\eta}{\rho}\left(\frac{\partial^2 w}{\partial x^2} + \frac{\partial^2 w}{\partial y^2} + \frac{\partial^2 w}{\partial z^2}\right) \tag{21}$$

where p in Pa is pressure, ρ in kg/m^3 is density and η in Pa·s is viscosity.

Energy equations, in the fluid region,

$$u\frac{\partial T}{\partial x} + v\frac{\partial T}{\partial y} + w\frac{\partial T}{\partial z} = \frac{\lambda_f}{\rho Cp}\left(\frac{\partial^2 T}{\partial x^2} + \frac{\partial^2 T}{\partial y^2} + \frac{\partial^2 T}{\partial z^2}\right) \tag{22}$$

in the solid region,

$$\frac{\partial^2 T}{\partial x^2} + \frac{\partial^2 T}{\partial y^2} + \frac{\partial^2 T}{\partial z^2} = 0 \tag{23}$$

where T in K represents temperature, λ_f in W/(m·K) is thermal conductivity and Cp in J/(kg·K) is specific heat capacity.

The calculating domain is divided into two parts: fluid domain and solid domain, as shown in Figure 3. The boundary conditions are set as follows. The velocity inlet and pressure outlet boundary conditions are applied for the inlet and outlet of the heat sink respectively. The constant heat flux boundary is set for the bottom surface. The coupled wall boundary conditions are applied for the interfaces between fluid domain and solid domain (microchannels' surrounding surfaces, inlet and outlet manifold's bottom surfaces). The thermal insulation wall boundary conditions are set for the other surfaces. To be clear, the 'inlet velocity' mentioned in this paper refers to the fluid velocity of the heat sink's inlet, which is the set velocity of the velocity inlet boundary.

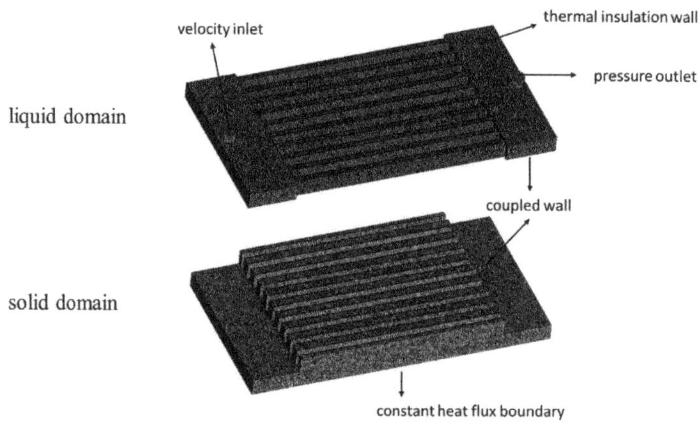

Figure 3. Computational domain and boundary conditions.

The commercial software ANSYS Fluent was used to perform the numerical calculations. When the calculations were complete, the velocity, pressure and temperature fields in solid and fluid domains of the heat sink could be obtained and the flow and heat transfer performances could be characterized by the following parameters.

Mean heat transfer coefficient h_m [49],

$$h_m = \frac{Q}{A_w\left(T_w - T_f\right)} \tag{24}$$

where A_w is the heat transfer area between the fluid and the walls; T_f and T_w are the average temperature of the fluid and the walls respectively; Q is the heat exchange capacity which could be calculated as,

$$Q = qA_b \tag{25}$$

where q and A_b are the heat flux and area of the heat sink bottom surface.

Mean flow resistance coefficient f [50],

$$f = \frac{\Delta P}{(\rho v^2/2)(L_c/D_h)} \tag{26}$$

where ΔP is the pressure drop, ρ and v are the density and velocity of the fluid.

2.2.2. Mesh Independence and Model Validation

In order to ensure that the solutions are independent of grid size, three different grid systems with cell number of 1,627,410, 3,480,055 and 5,741,015 are generated and used to perfume the mesh independence study. Under the same boundary conditions, the fluid pressure drop calculated by the latter two grid systems varies only by 0.9%, as shown in Table 1. In order to save calculating time, the grid system with cell number of 3,480,055 is finally adopted in this paper.

Table 1. Mesh independency analysis.

Cell Number	Calculated Pressure Drop	Relative Difference
1,627,410	9709.92 Pa	/
3,480,055	10,320.50 Pa	5.9%
5,741,015	10,414.30 Pa	0.9%

Besides the mesh independence test, the validity of aforementioned numerical method was verified by means of theoretical results obtained by Mortensen et al. [51]. Figure 4 shows the comparison of hydraulic resistances between the theoretical and numerical results. The definition equation of the hydraulic resistance is,

$$R_H = \frac{\Delta P}{V_{fr}} \tag{27}$$

where R_H is the hydraulic resistance, ΔP is the pressure drop, V_{fr} is the volume flow rate.

Compared to the theoretical values, the mean relative difference of the calculating values are 2.9%, 3.6% and 0.38% with liquid Na, liquid K and liquid Li, respectively; it can be concluded that the numerical results are agreeable with the existing theoretical results, which means the numerical method applied in this study has satisfactory accuracy.

The heat transfer calculating method is verified by comparing the result of Adeel Muhammad et al. [52]. Table 2 shows the comparison of total thermal resistance of the whole microchannel heat sink with the channel height varies from 3 mm to 9 mm. The maximum deviation is found to be less than 1%, which indicates the heat transfer calculating method applied in this paper is reliable.

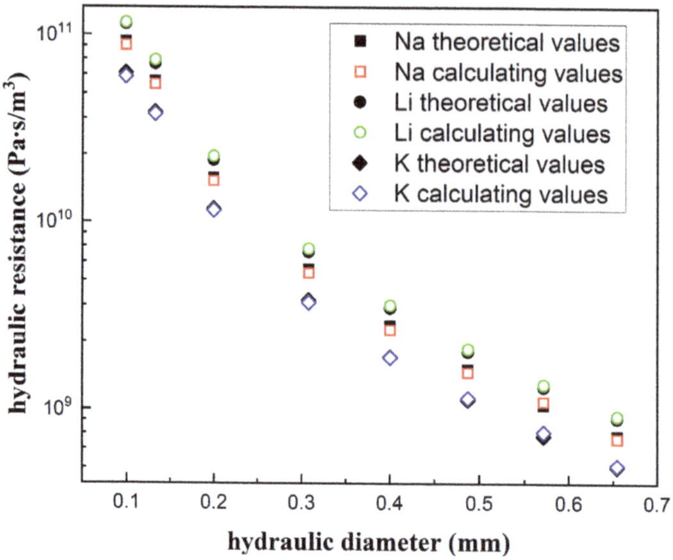

Figure 4. Comparison between theoretical and numerical results of dimensionless hydraulic resistance.

Table 2. Comparison of total thermal resistance.

Microchannel Height	R_{Thm} (K/W) Present	R_{Thm} (K/W) Ref. [52]	Relative Difference
3	0.1174	0.1184	0.84%
4	0.09944	0.1002	0.76%
5	0.09076	0.09156	0.87%
6	0.08648	0.08718	0.80%
7	0.08429	0.08478	0.58%
9	0.08282	0.08346	0.77%

The total thermal resistance of the heat sink is defined as,

$$R_{Thm} = \frac{T_{\max} - T_{in}}{Q} \tag{28}$$

where $T_{\max}$ is the maximum temperature of the whole heat sink, T_{in} is the inlet temperature of the liquid metal, Q is the heat exchange capacity.

3. Results and Discussions

3.1. Effects of Working Fluid

Seven alkalis are considered in this paper: sodium (Na), potassium (K), Na22-K78 alloy (0.22 weight fraction of Na, 0.78 weight fraction of K), Na44-K56, Na52-K48, Na56-K44 and lithium (Li).

Taking a rectangular microchannel cross-section, heat flux 200 W/cm^2, inlet temperature 600 K and inlet velocity 3 m/s as typical condition, Figure 4 shows the temperature distributions of the heat sink with different working fluids (temperature contours have the same scale). It could be seen that taking Li as the working fluid has the best cooling effect while taking K as the working fluid has the worst. For the heat sinks with Na-K alloy as the working fluid, their cooling performances are intermediate between the Na-based and K-based heat sinks. Moreover, the bigger the weight fraction of Na, the better the cooling effect.

For all the heat sinks shown in Figure 5, the temperature gradient is mainly in the fluid flow direction. In the direction perpendicular to the fluid flow, the temperature distribution shows good uniformity, the working fluid and the heat sink base basically have the same temperature. Take the heat sink with Li as an example, Figure 6 shows the temperature and velocity distributions at the center section of microchannels (1.5 mm high from the bottom surface). It can be seen that the difference of temperature between fluid and solid regions is very small, which indicates the liquid alkali-based microchannel heat sink has excellent heat exchange ability.

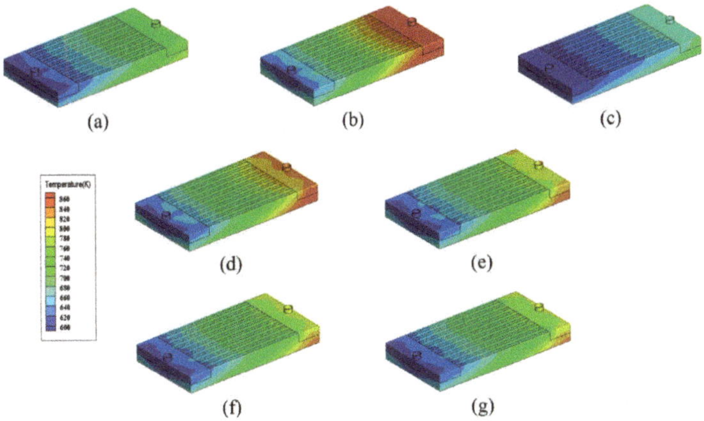

Figure 5. Temperature distributions of the heat sink with different working fluids. (**a**) Na; (**b**) K; (**c**) Li; (**d**) Na22-K78; (**e**) Na44-K56; (**f**) Na52-K48; (**g**) Na56-K44.

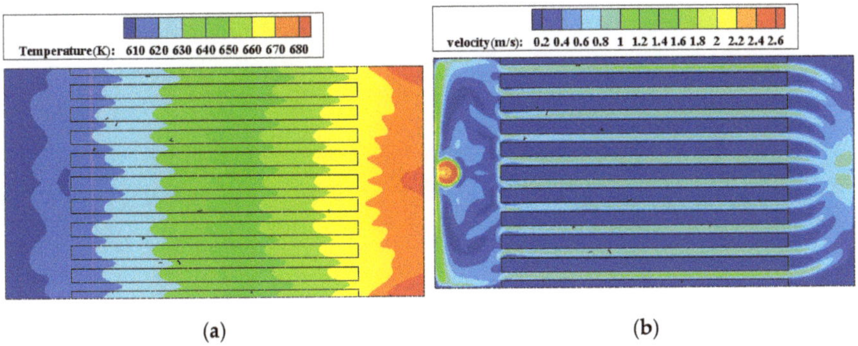

Figure 6. Temperature and velocity distributions at the center section of microchannels. (**a**) Temperature distribution; (**b**) velocity distribution.

Figure 7 shows the mean heat transfer coefficients and Nusselt numbers under various conditions with rectangular microchannel cross-section, inlet temperature 600 K and inlet velocity 3 m/s. For all the investigated alkalis, taking Li as the working fluid obtains the highest mean heat transfer coefficient while taking K as the working fluid gets the lowest. When the working fluid is Na, the mean heat transfer coefficient is the second highest, slightly lower than that when the working fluid is Li. Affected by the thermal conductivity, the changing tendency of Nusselt number is different with that of mean heat transfer coefficient. Liquid Na-K alloy-based heat sinks have bigger Nusselt numbers because of Na-K alloys' smaller thermal conductivities. On the contrary, the relatively big thermal conductivity of Na leads to a small Nusselt number. Moreover, with the same working fluid, both Nusselt number and mean heat transfer coefficient vary slightly with the change of heat flux. This is because under different heat fluxes, the working fluid has different temperatures, which leads to the variation of thermo-physical properties as well as the heat transfer performances.

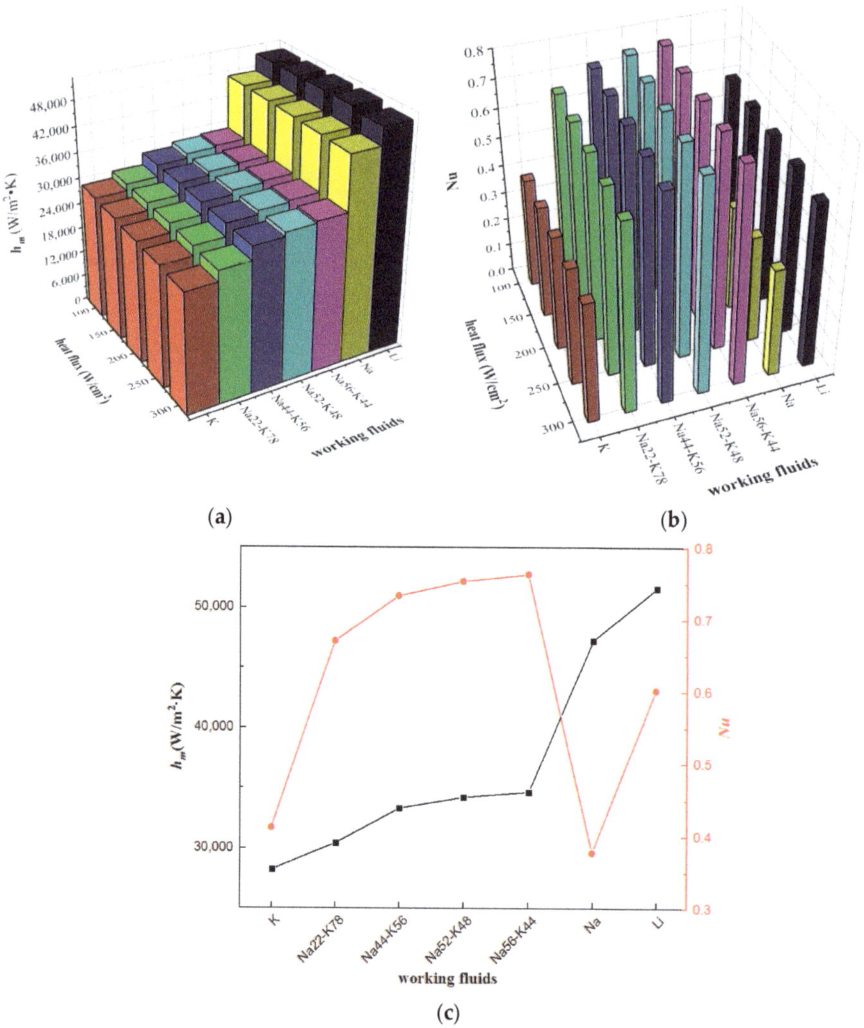

Figure 7. Mean heat transfer coefficients and Nusselt numbers with different working fluids. (**a**) Mean heat transfer coefficient; (**b**) Nusselt number; (**c**) Mean heat transfer coefficient and Nusselt number under heat flux 200 W/cm^2.

The convective heat transfer of liquid metals in microchannels depends on the liquid metals' thermal properties. According to the convective heat transfer theory, Nusselt number, Nu, is associated with Reynold number, Re, and Prandtl number, Pr. The considered alkalis' different thermo properties (thermal conductivity, specific heat capacity and viscosity) results in different Re and Pr, which leads various Nu as shown in Figure 7b. Moreover, based on the definition equation of Nu, the mean heat transfer coefficient can be calculated as,

$$h_m = \frac{Nu \cdot k}{D_h} \quad (29)$$

Influenced by thermal conductivity, the relative magnitude of mean heat transfer coefficient among the investigated alkalis is different with that of *Nu*. Because of bigger thermal conductivities, the mean heat transfer coefficients of liquid Li and Na are obviously higher than liquid K and liquid Na-K alloys.

Figure 8 shows the pressure drops and flow resistance coefficients under various conditions with rectangular microchannel cross-section, inlet temperature 600 K and inlet velocity 3 m/s. It could be seen that the pressure drop and the flow resistance coefficient changes in different degrees with the change of heat flux, among which the biggest change happens when the working fluid is K. This is because the changes of K's temperature and thermo-physical properties are the greatest. Furthermore, the flow resistance coefficient among all the investigated alkalis has no great difference, while the pressure drop is significantly smaller when the working fluid is Li. This is because the pressure drop is positively associated with the flow resistance as well as the working liquid's density, as shown in Equation (26). Li's density is the smallest among the considered liquid metals, which leads to the smallest pressure drop. Considering flow and heat transfer performances synthetically, Li is the optimum working fluid among all the investigated alkalis.

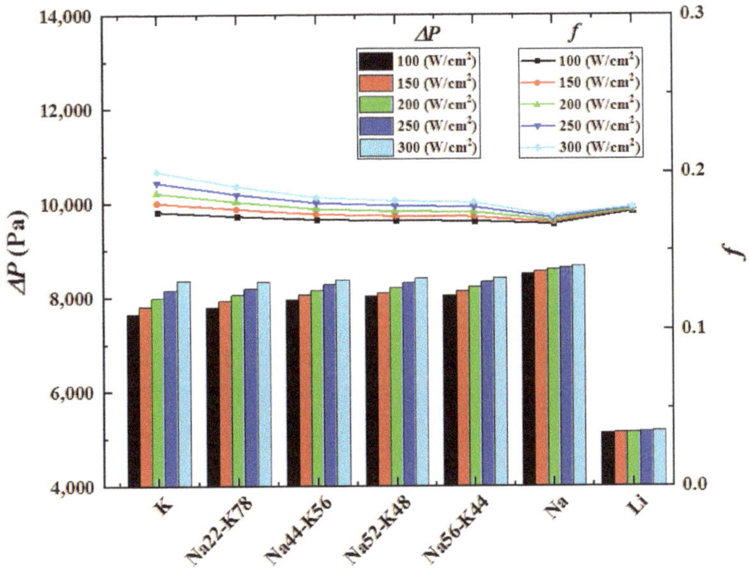

Figure 8. Pressure drops and flow resistance coefficients with different working fluids.

For comparative analysis, pumping power (pressure drop) should be kept consistent [53–56]. Therefore, in order to obtain the optimum working fluid, the mean heat transfer coefficient, h_m, is compared under the same pressure drop. Figure 9 shows the changing trend of h_m with different pressure drop at rectangular microchannel cross-section, inlet temperature 600 K, heat flux 200 W/cm^2. Obviously, since using liquid Li could get the biggest heat transfer coefficient, it is the optimum working fluid among all the investigated alkalis.

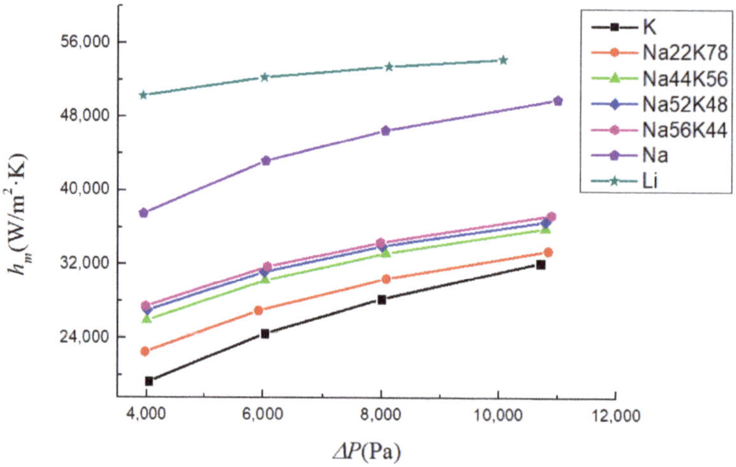

Figure 9. Variation of mean heat transfer coefficient (h_m) as a function of pressure drop (ΔP).

3.2. Effects of Microchannel Cross-Section

In order to study the effects of microchannel cross-section on the flow and heat transfer performances, four microchannel cross-section shapes (rectangle, circle, trapezoid and parallelogram, as shown in Figure 2) are considered in this paper. Figure 10 shows temperature distributions of the heat sink with different microchannel cross-section shapes when working fluid is Li, heat flux is 200 W/cm^2, inlet temperature is 600 K and inlet velocity is 3 m/s. Obviously, the temperature distribution is hardly influenced by the change of microchannel cross-section shape.

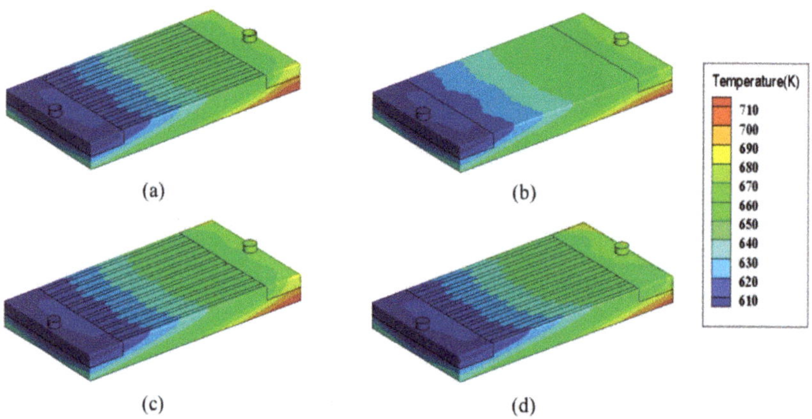

Figure 10. Temperature distributions of the heat sink with different microchannel cross-section types. (**a**) Rectangle; (**b**) circle; (**c**) trapezoid; (**d**) parallelogram.

Figure 11 shows the mean heat transfer coefficients and Nusselt numbers under various conditions with working fluid Li, inlet temperature 600 K and inlet velocity 3 m/s. For the Nusselt number, its value depends on the channel geometry [13,50]. Within the study scope shown in Figure 11, using rectangular, trapezoidal or parallelogram cross-section shaped microchannel has basically the same Nusselt number, while using circular microchannel could obtain significantly higher Nusselt number. This is because under the same hydraulic diameter, the area of circular cross-section is smaller than that of rectangular, trapezoidal and parallelogram cross-section, hence, the working fluid has bigger Reynold number inside the circular microchannel, which leads to the higher Nusselt number. According to the equation, $h_m = \mathrm{Nu} \cdot k/D_h$, the variation of the mean heat transfer coefficient is accordance with Nusselt number. So, as shown in Figure 11a, using a circular cross-section obtains the highest heat transfer coefficient. Compared to the other three cross-section shapes, using circular cross-section could increase the mean heat transfer coefficient by about 14,000 W/(m^2·K), which indicates the heat sink with circle microchannel cross-section has the best flow performance.

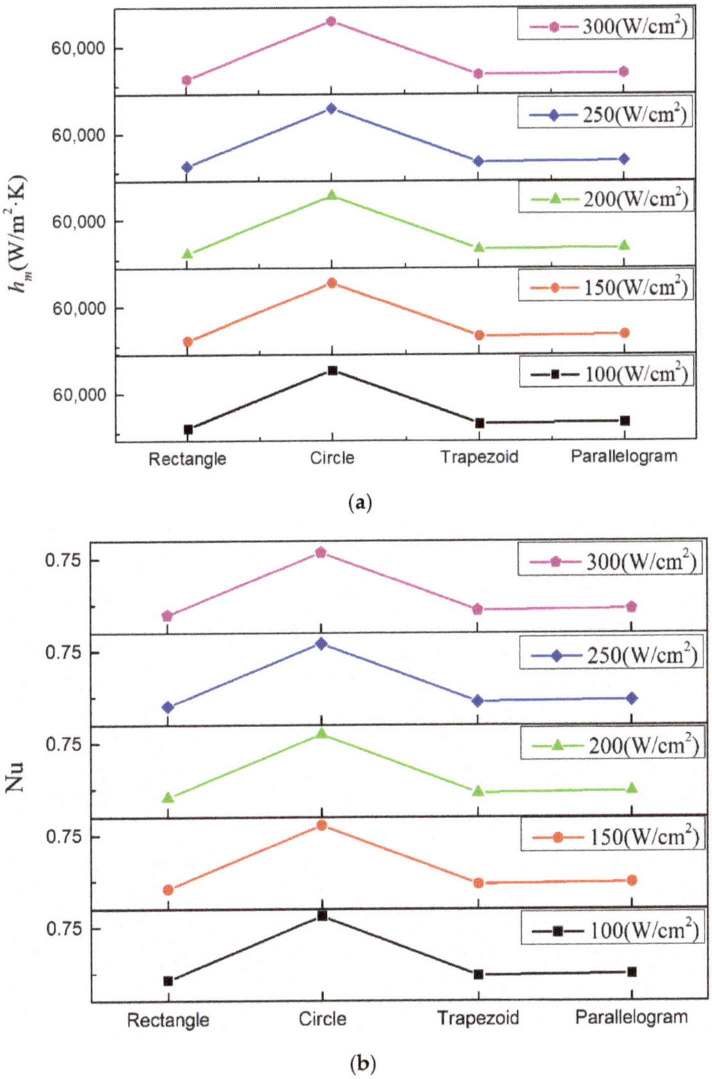

Figure 11. Mean heat transfer coefficients and Nusselt numbers with different micro channel cross-section types. (**a**) Mean heat transfer coefficient; (**b**) Nusselt number.

Figure 12 shows the pressure drops and flow resistance coefficients under various conditions with working fluid Li, inlet temperature 600 K and inlet velocity 3 m/s. Obviously, the heat sink with parallelogram cross-section has the lowest pressure drop and flow resistance coefficient while the heat sink with circular cross-section has the highest. This is because with the same hydraulic diameter, the investigated four microchannel cross-section shapes have different cross-sectional area, which leads to different velocity magnitudes as well as the pressure drop. Among the four different kinds of microchannels, the circular microchannel's cross-sectional area is the smallest. Hence, under the same heat sink's inlet velocity, the working fluid within the circular microchannel has the highest speed (as shown in Figure 13), which leads to the highest pressure drop.

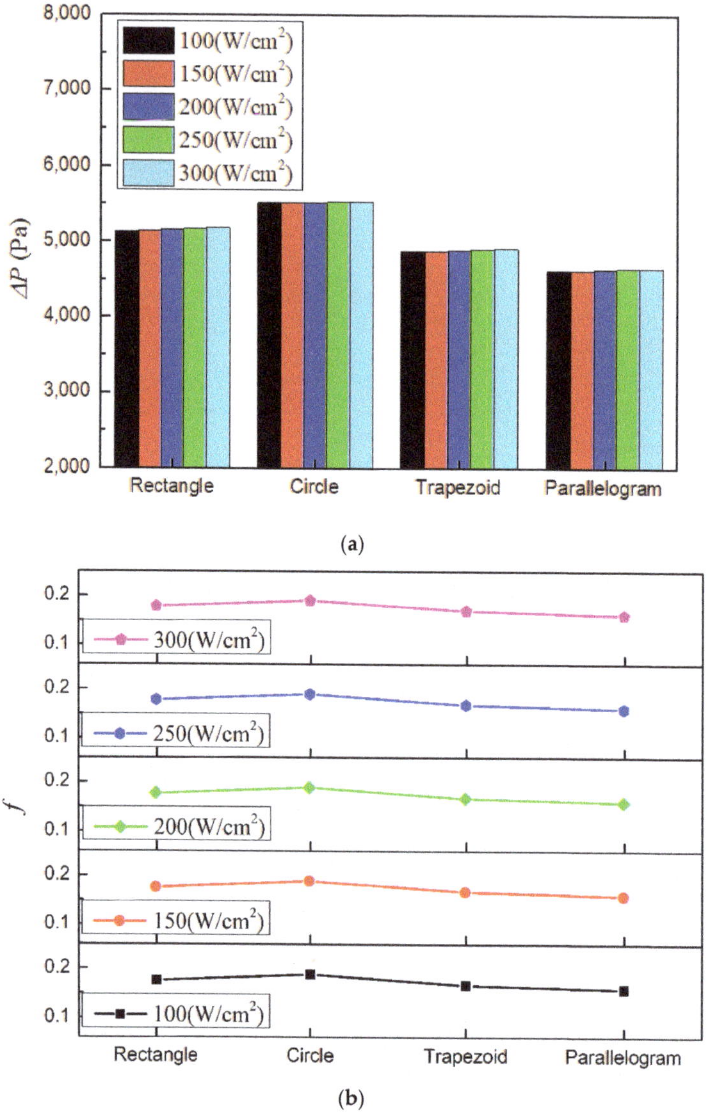

Figure 12. Pressure drops and flow resistance coefficients with different microchannel cross-section types. (**a**) Pressure drop; (**b**) flow resistance coefficient.

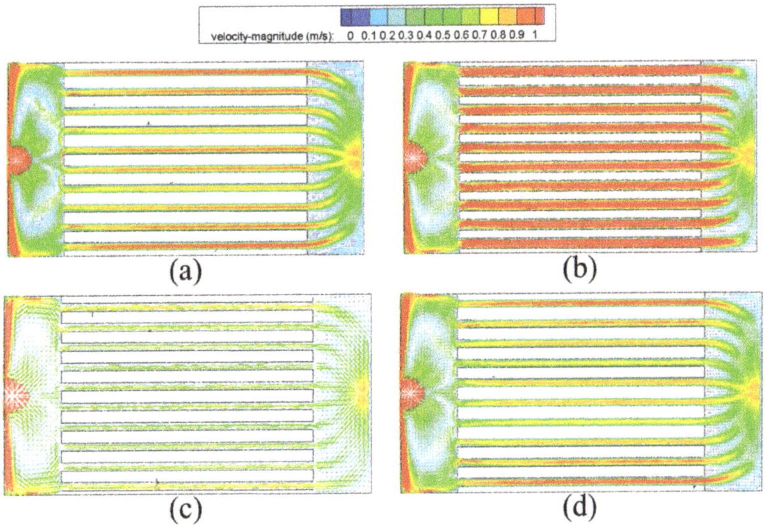

Figure 13. Velocity vectors at the center section of the heat sink with different microchannel cross-section types. (**a**) Rectangle; (**b**) circle; (**c**) trapezoid; (**d**) parallelogram.

In order to find the best cross-section shape, the performance evaluation criteria *PEC*, which represents the comprehensive performance of flow and heat transfer, is calculated. The *PEC* is defined as [57],

$$PEC = \frac{Nu/Nu_0}{(f/f_0)^{1/3}} \tag{30}$$

where Nu_0 and f_0 are the benchmark Nusselt number and mean flow resistance coefficient. In this paper, *Nu* and *f* of the heat sink with rectangular microchannel cross-section under heat flux 100 W/cm^2 are appointed as Nu_0 and f_0.

Figure 14 shows the *PEC* values of heat sinks with different microchannel cross-section shapes. Obviously, using circular microchannel could obtained the highest *PEC* values, indicating circle is the best choice for microchannel cross-section shape.

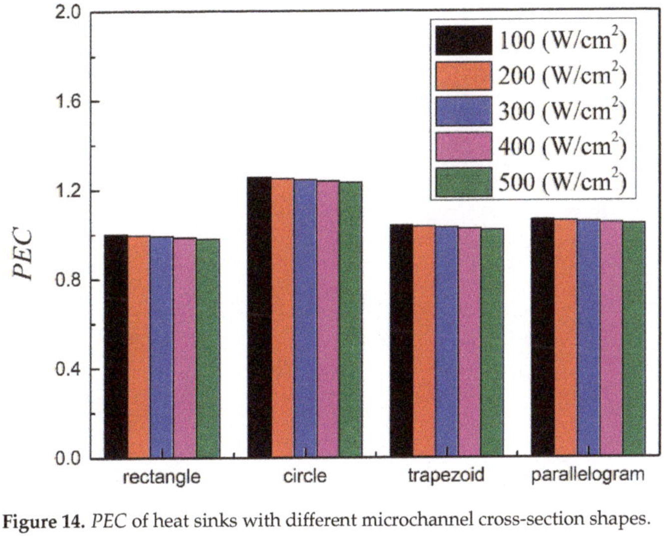

Figure 14. *PEC* of heat sinks with different microchannel cross-section shapes.

3.3. Effects of Inlet Velocity

Five inlet velocities (1 m/s, 3 m/s, 5 m/s, 7 m/s and 9 m/s) are considered in this paper so that the effects of inlet velocity on the flow and heat transfer performances could be investigated. Figure 15 shows temperature distributions of the heat sink with different inlet velocities under rectangular microchannel cross-section, heat flux 200 W/cm^2, inlet temperature 600 K and Li as the working fluid. It could be seen that the inlet velocity has a significant effect on the temperature distribution. With the increase of inlet velocity, the temperature of the heat sink becomes obviously lower. Under the inlet velocity of 1 m/s, the high temperature region covers most of the heat sink, indicating the heat sink's poor cooling effect. When the inlet velocity increases to 3 and 5 m/s, the high temperature region becomes much smaller. When the inlet velocity reaches 7 and 9 m/s, the entire heat sink is basically at a relatively low temperature level.

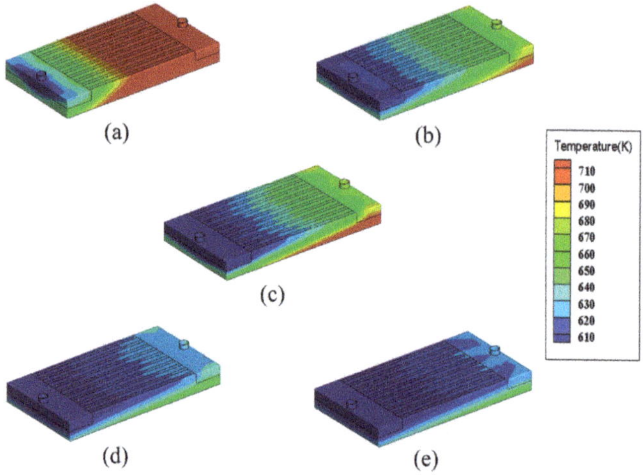

Figure 15. Temperature distributions of the heat sink with different inlet velocities. (**a**) 1 m/s; (**b**) 3 m/s; (**c**) 5 m/s; (**d**) 7 m/s; (**e**) 9 m/s.

Figure 16 shows the mean heat transfer coefficients and Nusselt numbers under various conditions with working fluid Li, rectangular microchannel cross-section and inlet temperature 600 K (using pressure drop under corresponding inlet velocity as x-axis label so that the heat transfer performance could be compared properly [53–56]). For all the investigated heat fluxes, both the heat transfer coefficient and Nusselt number increase with the increase of inlet velocity. From 1 m/s to 9 m/s, the heat transfer coefficient and Nusselt number, respectively, enhance 74.35% and 80.5% at most. Moreover, the increasing rates of the heat transfer coefficient and Nusselt number are considerably high at low inlet velocity, but quite low at high inlet velocity. This implies that under small inlet velocity, the cooling ability of the heat sink could be improved by increasing inlet velocity. However, when the inlet velocity is high enough, continuing to increase the inlet velocity brings little benefit.

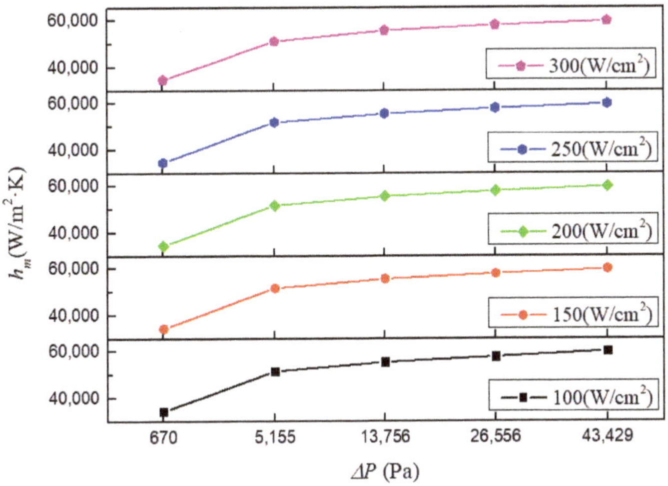

(a)

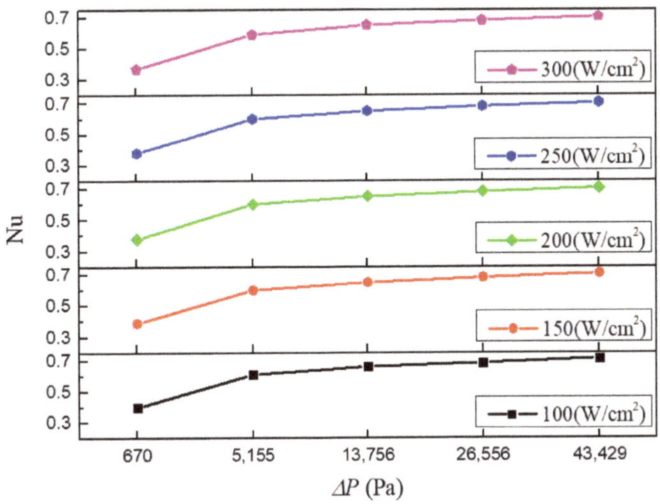

(b)

Figure 16. Mean heat transfer coefficients and Nusselt numbers with different inlet velocities. (**a**) Mean heat transfer coefficient; (**b**) Nusselt number.

Figure 17 shows the pressure drops and flow resistance coefficients under various conditions with working fluid Li, rectangular microchannel cross-section and inlet temperature 600 K. It could be seen that the pressure drop rises rapidly with the increase of inlet velocity. From 1 m/s to 9 m/s, the pressure drop increases as much as 65 times. As for the flow resistance coefficient, it decreases gradually with the increase of the inlet velocity. This is because although the flow resistance coefficient is proportional to the pressure drop (ΔP), it is inversely related to the square of velocity (v^2) (according to Equation (26)).

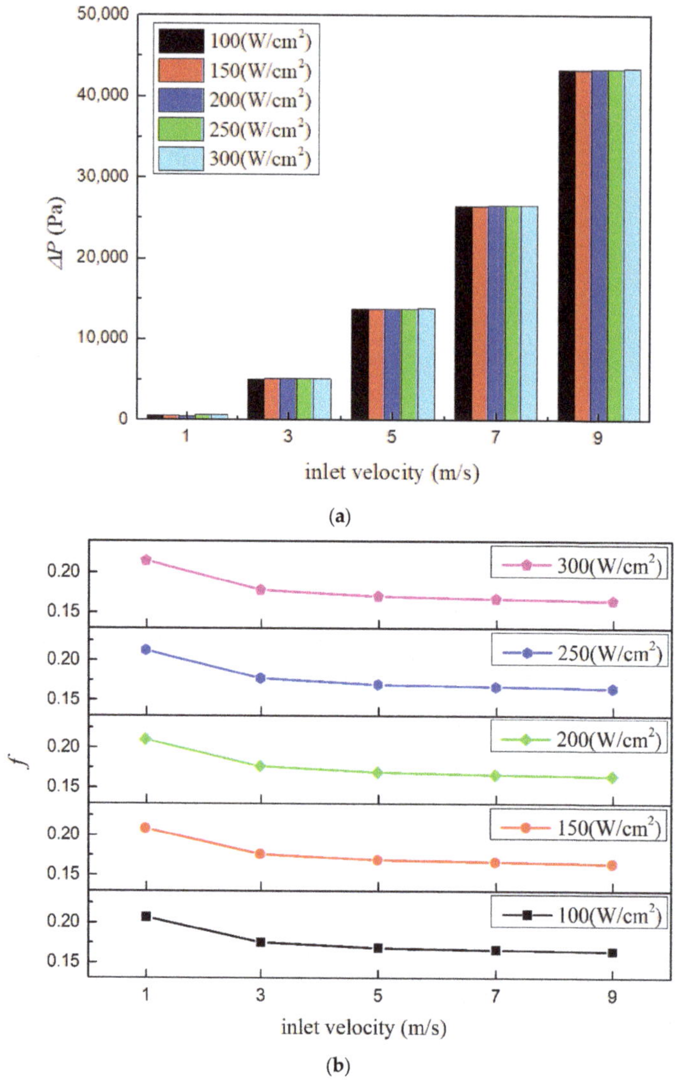

Figure 17. Pressure drops and flow resistance coefficients with different inlet velocities. (**a**) Pressure drop; (**b**) flow resistance coefficient.

4. Conclusions

Numerical investigations of flow and heat transfer performances in liquid metal-based microchannel heat sinks are presented in this paper. During the simulation processes, different working fluid types, microchannel's cross-sectional geometries and inlet velocities are considered so that their influences on the heat sink's characteristics can be analyzed. Based on the calculating results, some conclusions were drawn as follow:

(1) Among all the seven investigated alkalis, lithium is the best option for working fluid because the lithium-based microchannel heat sink has the best cooling ability and the lowest pressure drop.

(2) For the four considered microchannel cross-section types (rectangle, circle, trapezoid and parallelogram), utilizing a circular microchannel cross-section obtains a higher

mean heat transfer coefficient, while using a parallelogram obtains the lowest pressure drop. Considering flow and heat transfer performances comprehensively, the circle is the optimum choice for microchannel cross-section shape because using a circular microchannel has the highest PEC value.

(3) Inlet velocity has a significant influence on the heat sink's flow and heat transfer performances. When the inlet velocity rises from 1 m/s to 9 m/s, the heat transfer coefficient enhances 74.35% at most, while the pressure drop increases up to 65 times. In order to obtain a favorable overall performance, the inlet velocity should be selected carefully.

Author Contributions: Conceptualization, T.W. and L.W.; methodology, Y.T. and T.W.; software, T.W.; validation, T.W. and Y.T.; formal analysis, T.W.; investigation, T.W.; resources, L.W. and C.Y.; data curation, T.W.; writing—original draft preparation, T.W.; writing—review and editing, C.Y. and X.L.; visualization, T.W.; supervision, C.Y. and X.L.; project administration, X.L. All authors have read and agreed to the published version of the manuscript.

Funding: This research received no external funding.

Data Availability Statement: The data presented in this study are available on request from the corresponding author. The data are not publicly available due to privacy.

Conflicts of Interest: The authors declare no conflict of interest.

Nomenclature

A	heat transfer area (m^2)
at	mole fraction
Cp	specific heat capacity (J/(kg·K))
D_h	hydraulic diameter (mm)
f	flow resistance coefficient
H	height (mm)
h_m	mean heat transfer coefficient (W/(m^2·K))
k	thermal conductivity (W/(m·K))
L	length (mm)
Nu	Nusselt number
PEC	performance evaluation criteria
Pr	Prandtl number
p	pressure (Pa)
Q	heat exchange capacity (W)
q	heat flux (W/m^2)
Re	Reynold number
R_H	hydraulic resistance (Pa·s/m^3)
R_{Thm}	thermal resistance (K/W)
r	radius of inlet and outlet passages (mm)
T	Kelvin temperature (K)
t_f	Fahrenheit temperature (°F)
u	velocity in x direction (m/s)
V_{fr}	volume flow rate (m^3/s)
v	velocity in y direction (m/s)
W	width (mm)
w	velocity in z direction (m/s)
wt	weight fraction
x	Cartesian coordinate (m)
y	Cartesian coordinate (m)
z	Cartesian coordinate (m)
Greek symbols	
ΔP	pressure drop (Pa)
η	dynamic viscosity (Pa·s)

λ_f thermal conductivity (W/(m·K))
ρ density (kg/m^3)
Subscripts
b heat sink bottom
c microchannel
f fluid
K potassium
m manifold
Na sodium
w wall

References

1. Gangtao, L.; Issam, M. Review of single-phase and two-phase nanofluid heat transfer in macro-channels and micro-channels. *Int. J. Heat Mass Transf.* **2019**, *136*, 324–354.
2. Tuckerman, D.B.; Pease, R. High-performance heat sinking for VLSI. *IEEE Electr. Device Lett.* **1981**, *1*, 126–129. [CrossRef]
3. Wu, P.Y.; Little, W.A. Measurement of friction factor for the flow of gases in very fine channels used for micro miniature Joule-Thomson refrigerators. *Cryogrnics* **1983**, *23*, 273–277.
4. Xu, B.; Ooi, K.T.; Wong, N.T.; Choi, W.K. Experimental investigation of flow friction for liquid flow in micro channels. *Int. Comm. Mass Transf.* **2000**, *27*, 1165–1176. [CrossRef]
5. Judy, J.; Maynes, D.; Webb, B.W. Characterization of frictional pressure drop for liquid flows through micro channels. *Int. J. Heat Mass Transf.* **2002**, *45*, 3477–3489. [CrossRef]
6. Qu, W.; Mudawar, I. Experimental and numerical study of pressure drop and heat transfer in a single-phase micro-channel heat sink. *Int. J. Heat Mass Transf.* **2002**, *45*, 2549–2565. [CrossRef]
7. Lee, P.S.; Garimella, S.V.; Liu, D. Investigation of heat transfer in rectangular micro channels. *Int. J. Heat Mass Transf.* **2005**, *48*, 1688–1704. [CrossRef]
8. Liu, D.; Garimella, S.V. Investigation of liquid flow in micro-channels. *J. Thermophys. Heat Transf.* **2004**, *18*, 65–72. [CrossRef]
9. Kandlikar, S.G.; Garimella, S.; Liu, D. *Heat Transfer and Fluid Flow in Mini Channels and Micro Channels*, 2nd ed.; Elsevier Science: Kidlington, UK, 2013.
10. TAdams, M.; Abdel-Khalik, S.I.; Jeter, S.M.; Qureshi, Z.H. An experimental investigation of single-phase forced convection in micro channels. *Int. J. Heat Mass Transf.* **1998**, *41*, 851–857. [CrossRef]
11. Adams, T.M.; Dowling, M.F.; Abdel-Khalik, S.I.; Jeter, S.M. Applicability of traditional turbulent single-phase forced convection correlations to non-circular micro-channels. *Int. J. Heat Mass Transf.* **1999**, *42*, 4411–4415. [CrossRef]
12. Yousef, A.; Mohammad, Z.T.; Mohammad, M.H.; Nima, G. Effect of a micro heat sink geometric design on thermo-hydraulic performance: A review. *Appl. Therm. Eng.* **2020**, *170*, 114974.
13. Dalei, J.; Lei, H. Numerical studies on the hydraulic and thermal performances of microchannels with different cross-sectional shapes. *Int. J. Heat Mass Transf.* **2019**, *143*, 118604.
14. Gunnasegaran, P.; Mohammed, H.A.; Shuaib, N.H.; Saidur, R. The effect of geometrical parameters on heat transfer characteristics of microchannels heat sink with different shapes. *Int. Commun. Heat Mass Transf.* **2010**, *37*, 1078–1086. [CrossRef]
15. Xia, G.D.; Jiang, J.; Wang, J.; Zhai, Y.L.; Ma, D.D. Effects of different geometric structures on fluid flow and heat transfer performance in microchannel heat sinks. *Int. J. Heat Mass Transf.* **2015**, *80*, 439–447. [CrossRef]
16. Kumar, R.; Singh, G.; Mikielewicz, D. A new approach for the mitigating of flow maldistribution in parallel microchannel heat sink. *J. Heat Transf.* **2018**, *140*, 072401. [CrossRef]
17. Kumar, R.; Singh, G.; Mikielewicz, D. Numerical study on mitigation of flow maldistribution on parallel microchannel heat sink: Channels variable width versus variable height approach. *J. Electron. Packag.* **2019**, *141*, 021009. [CrossRef]
18. Ahmed, H.E.; Ahmed, M.I. Optimum thermal design of triangular, trapezoidal and rectangular grooved microchannel heat sinks. *Int. Commun. Heat Mass Transf.* **2015**, *66*, 47–57. [CrossRef]
19. Zhu, J.-F.; Li, X.-Y.; Wang, S.-L.; Yang, Y.-R.; Wang, X.-D. Performance comparison of wavy microchannel heat sinks with wavy bottom rib and side rib designs. *Int. J. Therm. Sci.* **2019**, *146*, 106068. [CrossRef]
20. Wang, S.-L.; Chen, L.-Y.; Zhang, B.-X.; Yang, Y.-R.; Wang, X.-D. A new design of double-layered microchannel heat sinks with wavy microchannels and porous-ribs. *J. Therm. Anal. Calorim.* **2020**, *141*, 547–558. [CrossRef]
21. Ermagan, H.; Rafee, R. Numerical investigation into the thermo-fluid performance of wavy microchannels with superhydrophobic walls. *Int. J. Therm. Sci.* **2018**, *132*, 578–588. [CrossRef]
22. Gong, L.; Li, Y.; Bai, Z.; Xu, M. Thermal performance of micro-channel heat sink with metallic porous/solid compound fin design. *Appl. Therm. Eng.* **2018**, *137*, 288–295. [CrossRef]
23. Ghahremannezhad, A.; Xu, H.; Nazari, M.A.; Ahmadi, M.H.; Vafai, K. Effect of porous substrates on thermohydraulic performance enhancement of double layer microchannel heat sinks. *Int. J. Heat Mass Transf.* **2019**, *131*, 52–63. [CrossRef]
24. Li, X.Y.; Wang, S.L.; Wang, X.D.; Wang, T.-H. Selected porous-ribs design for performance improvement in double-layered microchannel heat sinks. *Int. J. Therm. Sci.* **2019**, *137*, 616–626. [CrossRef]

25. Hussien, A.A.; Abdullah, M.Z.; Mohda, A.N. Single-phased heat transfer enhancement in micro/minichannels using nanofluids: Theory and applications. *Appl. Energy* **2012**, *89*, 150–155.
26. Chen, Z.; Qian, P.; Huang, Z.; Luo, C.; Liu, M. Study on flow and heat transfer of liquid metal in a new top-slotted microchannel heat sink. *IOP Conf. Ser. Earth Environ. Sci.* **2021**, *624*, 012054. [CrossRef]
27. Jang, S.P.; Choi, S.U.S. Cooling performance of a microchannel heat sink with nanofluids. *Appl. Therm. Eng.* **2006**, *26*, 2457–2463. [CrossRef]
28. Farsad, E.; Abbasi, S.P.; Zabihi, M.S.; Sabbaghzadeh, J. Numerical simulation of heat transfer in a micro channel heat sink using nanofluids. *Heat Mass Transf.* **2011**, *47*, 479–490. [CrossRef]
29. Sohel, M.R.; Saidur, R.; Sabri, M.F.M.; Kamalisarvestani, M.; Elias, M.M.; Ijam, A. Investigating the heat transfer performance and thermophysical properties of nanofluids in a circular micro-channel. *Int. Commun. Heat Mass Transf.* **2013**, *42*, 75–81. [CrossRef]
30. Sohel, M.R.; Khaleduzzaman, S.S.; Saidur, R.; Hepbasli, A.; Sabri, M.F.M.; Mahhubul, I.M. An experimental investigation of heat transfer enhancement of a minichannel heat sink using Al_2O_3-H_2O nanofluid. *Int. J. Heat Mass Transf.* **2014**, *74*, 164–172. [CrossRef]
31. Sivakumar, A.; Alagumurthi, N.; Senthilvelan, T. Experimental investigation of forced convective heat transfer performance in nanofluids of Al_2O_3/water and CuO/water in a serpentine shaped micro channel heat sink. *Heat Mass Transf.* **2016**, *52*, 1265–1274. [CrossRef]
32. Salman, B.H.; Mohammed, H.A.; Kherbeet, A.S. Heat transfer enhancement of nanofluids flow in microtube with constant heat flux. *Int. Commun. Heat Mass Transf.* **2012**, *39*, 1195–1204. [CrossRef]
33. Salman, B.H.; Mohammed, H.A.; Munisamy, K.M.; Kherbeet, A.S. Three-dimensional numerical investigation of nanofluids flow in microtube with different values of heat flux. *Heat Transf. Asian Res.* **2015**, *44*, 599–619. [CrossRef]
34. Kumar, R.; Tiwary, B.; Singh, P.K. Thermofluidic analysis of Al_2O_3-water nanofluid cooled branched wavy heat sink. *Appl. Therm. Eng.* **2022**, *201*, 117787. [CrossRef]
35. Wang, J.; Xu, Y.-P.; Qahiti, R.; Jafaryar, M.; Alazwari, M.A.; Abu-Hamdeh, N.H.; Issakhov, A.; Selim, M.M. Simulation of hybrid nanofluid flow within a microchannel heat sink considering porous media analyzing CPU stability. *J. Petro. Sci. Eng.* **2022**, *208*, 109734. [CrossRef]
36. Miner, A.; Ghoshal, U. Cooling of high-power-density microdevices using liquid metal coolants. *Appl. Phys. Lett.* **2004**, *85*, 506–508. [CrossRef]
37. Buongiorno, J. Convective transport in nanofluids. *ASME J. Heat Transf.* **2006**, *128*, 240–250. [CrossRef]
38. Deng, Y.; Jiang, Y.; Liu, J. Liquid metal technology in solar power generation—Basics and applications. *Sol. Energy Mater. Sol. Cells* **2021**, *222*, 110925. [CrossRef]
39. Deng, Y.; Liu, J. A liquid metal cooling system for the thermal management of high power LEDs. *Int. Commun. Heat Mass Transf.* **2010**, *37*, 788–791. [CrossRef]
40. Ma, K.; Liu, J. Liquid metal cooling in thermal management of computer chips. *Front. Energy Power Eng. China* **2007**, *1*, 384–402. [CrossRef]
41. Liu, J.; Zhou, Y.X. A Computer Chip Cooling Method Which Uses Low Melting Point Metal and Its Alloys as the Cooling Fluid. China Patent 02131419.5, 10 October 2002.
42. Ghoshal, U.; Grimm, D.; Ibani, S.; Johnston, C.; Miner, A. High performance liquid metal cooling loops. In Proceedings of the IEEE 21st Annual Semiconductor Thermal Measurement and Management Symposium, San Jose, CA, USA, 15–17 March 2005; pp. 16–19.
43. Hodes, M.; Zhang, R.; Lam, L.S.; Wilcoxon, R.; Lower, N. On the potential of Galinstan-based minichannel and minigap cooling. *IEEE Trans. Compon. Packag. Technol.* **2014**, *4*, 46–56. [CrossRef]
44. Zhang, R.; Hodes, M.; Lower, N.; Wilcoxon, R. Water-based microchannel and Galinstan-based minichannel cooling beyond 1 kW/cm^2 heat flux. *IEEE Trans. Compon. Packag. Technol.* **2015**, *5*, 762–770.
45. Jarger, W. Heat transfer to liquid metals with empirical models for turbulent forced convection in various geometries. *Nucl. Eng. Des.* **2017**, *319*, 12–27.
46. IAEA. *Thermophysical Properties of Materials for Nuclear Engineering: A Tutorial and Collection of Data*; International Atomic Energy Agency: Vienna, Austria, 2002.
47. He, Z.; Yan, Y.; Zhang, Z. Thermal management and temperature uniformity enhancement of electronic devices by micro heat sinks: A review. *Energy* **2021**, *216*, 119223. [CrossRef]
48. Wu, T. *Numerical Heat Transfer*; Xi'an Jiaotong University Press: Xi'an, China, 2001.
49. Holman, J.P. *Heat Transfer*, 9th ed.; McGraw-Hill: New York, NY, USA, 2002.
50. Kandikar, S.G.; Garimella, S.; Li, D.Q.; Colin, S.; King, M.R. *Heat Transfer and Fluid Flow in Minichannels and Microchannels*; Elsevier: Oxford, UK, 2006.
51. Mortensen, N.A.; Okkels, F.; Bruus, H. Reexamination of Hagen-Poiseuille flow: Shape dependence of the hydraulic resistance in microchannels. *Phys. Rev.* **2005**, *71*, 057301. [CrossRef]
52. Muhammad, A.; Selvakumar, D.; Wu, J. Numerical investigation of laminar flow and heat transfer in a liquid metal cooled mini-channel heat sink. *Int. J. Heat Mass* **2019**, *150*, 119265. [CrossRef]
53. Ambreen, T.; Saleem, A.; Park, C.W. Numerical analysis of the heat transfer and fluid flow characteristics of a nanofluid-cooled micropin-fin heat sink using the Eulerian-Lagrangian approach. *Powder Technol.* **2019**, *345*, 509–520. [CrossRef]

54. Ambreen, T.; Saleem, A.; Park, C.W. Pin-fin shape-dependent heat transfer and fluid flow characteristics of water- and nanofluid-cooled micropin-fin heat sinks: Square, circular and triangular fin cross-sections. *Appl. Therm. Eng.* **2019**, *158*, 113781. [CrossRef]
55. Ambreen, T.; Saleem, A.; Ali, H.M.; Shehzad, S.A.; Park, C.W. Performance analysis of hybrid nanofluid in a heat sink equipped with sharp and streamlined micro pin-fins. *Powder Technol.* **2019**, *355*, 552–563. [CrossRef]
56. Ambreen, T.; Saleem, A.; Park, C.W. Analysis of hydro-thermal and entropy generation characteristics of nanofluid in an aluminium foam heat sink by employing Darcy-Forchheimer-Brinkman model coupled with multiphases Eulerian model. *Appl. Therm. Eng.* **2020**, *173*, 115231. [CrossRef]
57. Chai, L.; Xia, G.D.; Wang, H.S. Numerical study of laminar flow and heat transfer in microchannel heat sink with offset ribs on sidewalls. *Appl. Therm. Eng.* **2016**, *92*, 32–41. [CrossRef]

MDPI AG
Grosspeteranlage 5
4052 Basel
Switzerland
Tel.: +41 61 683 77 34

Micromachines Editorial Office
E-mail: micromachines@mdpi.com
www.mdpi.com/journal/micromachines

www.ingramcontent.com/pod-product-compliance
Lightning Source LLC
Chambersburg PA
CBHW061758210926
51850CB00006B/588

9783725876693